Heterochrony

The Evolution of Ontogeny

Heterochrony

The Evolution of Ontogeny

Michael L. McKinney

The University of Tennessee
Knoxville, Tennessee

and

Kenneth J. McNamara

Western Australian Museum
Perth, Western Australia
Australia

Plenum Press • *New York and London*

Library of Congress Cataloging in Publication Data

McKinney, Michael L.
   Heterochrony: the evolution of ontogeny / Michael L. McKinney and Kenneth J.
McNamara.
     p.   cm.
   Includes bibliographical references (p.) and index.
   ISBN 0-306-43638-8
   1. Heterochrony (Biology). I. McNamara, Kenneth. II. Title.
QH395.M34   1991   91-6371
575 – dc20   CIP

QH
395
M734
1991

ISBN 0-306-43638-8

© 1991 Plenum Press, New York
A Division of Plenum Publishing Corporation
233 Spring Street, New York, N.Y. 10013

Printed in the United States of America

We dedicate this book to our wives, Victoria M. McKinney and Susan Radford, for sharing so much of their ontogenies with ours.

Preface and Preview

Many readers will have noticed the recent trend toward quotations at the beginnings of chapters in scientific books. Often these quotes are sappy, dippy little things as if the authors of the book were struggling for profundity. Perhaps they want to give their self-indulgent esoteric musings some hint of relevance to the rapidly deteriorating human condition, to cast a shadow of importance over the gray wall of clinical facts to follow. Most pathetic of all is the hopeless clod who quotes himself. He is, in effect, baldly stating that he has, with his own mind, assembled something profound, clever, or at least amusing to his fellow primates. This kind of conceit will detract from his scientific standing. Nor should humor be attempted, at any cost; it represents a distraction to the clear, analytical thought necessary for Scientific Progress.

M. L. McKinney, from ''Jawless Fish and Hairless Apes: Ruminations from Edge of the Photic Zone'' (unpublished)

PREFACE

In the past, the study of heterochrony has been more conducive to lexicography (or even cryptography) than to improving our understanding of evolution and development. This book is an attempt to correct that. We have tried to say, as directly as possible, what we think heterochrony is, how it works, and what role it plays in evolution. It is written for anyone (students, colleagues in any field) who has an interest in learning more about it. Thus, we have tried to organize the book in a logical, straightforward progression, used boldfaced (for primary emphasis) and italicized (for secondary emphasis) terms, and included a glossary of terms. We have tried to keep the style informal. Parts are bound to seem elementary to specialists in those areas, but our purpose is to integrate a broad range of information, not excavate deeply in just one area.

In part, then, the book is a primer. However, a book of this nature should not only educate but also stimulate, so that, after presenting the basics, we have tried to delve into the knotty, problematic areas. Inevitably, this grades into speculation at some points, but this is a necessary part of scientific progress; it helps point out areas of future research, if only by irritating people. The problem of course is just where speculation stops being creative inference and becomes irresponsible nonsense. We have tried to take a moderate course, but it is particularly difficult in this area of research because so little is known about major aspects of it. Yet, if science is to tie all the loose ends together, it is going to take some creative thought. In any case, even the developmentally informed reader is likely to find a few heresies, and surprising twists and turns in the plot that follows. Just remember our goal is to inform and stimulate, not assert.

The number of "loose ends" extends far beyond the "black box" of development and genes when we consider the role of development in evolution, how development can interact with ecology (affecting life history tactics and behavior, not only size and morphology), and its ultimate role in macroevolutionary patterns. How much does development "constrain" evolution? (While acknowledging that "constraint" is an overused and misleading term; e.g., why not say that selection "constrains" development?) The role of development is one of the (perhaps "the") fastest growing and most exciting areas of inquiry in evolutionary theory. As Futuyama (1988) has discussed in his presidential address to the Society for the Study of Evolution, we greatly need an understanding of the origin of variation to have a complete theory of evolution. Since Darwin, the emphasis has been on selection, yet selection can obviously only act upon variation created by genes and developmental processes. With the growth of knowledge in genetics and developmental biology, it is time to examine the other side of the dialectic: how much is directionality and rate of evolution "controlled" (or the milder word, "constrained") by the production of variation. That is, how much is "internally" controlled? While this book (and others like it) addresses this, the jury is still out. We need much more empirical information, especially about the dynamics of development. In a sense, this requires a shift of focus from the genetic level to the cellular (and its collective unit: the tissue) level (e.g., Buss, 1987). Cells are the basic unit of the organism, yet clearly metazoans (and metaphytes) do not consist of or develop by freely moving groups of cells. To understand why (and in what way) they do not, is to understand the dynamics of ontogeny. In this book, such a cellular approach (especially Chapter 3) is promoted.

Aside from the growth of new information on development, another stimulus to the rising interest in its role in evolution is the rise of interest among evolutionists in hierarchies. It is easy to talk about hierarchies, but people have long known (at least implicitly) that complex phenomena have levels, each with emergent properties. It is only by learning the specifics of those emergent properties that we can create a theory of evolution that will do what truly complete theory must do: understand the connections (or, interactions) among the levels in the hierarchy. The attractive thing to us (and, perhaps, many others) is that the "properties" of development are at an intermediate level which allows such linkage: above the genes but below the individual organism interacting with its ecosystem.

PREVIEW

We begin (Chapter 1) with an overview of the history of ideas about heterochrony. This is followed (Chapter 2) by a discussion of what heterochrony is: how does one classify and analyze it? We hope that many who have been "turned off" by the past profusion of terms will be helped by our approach. In particular, we use bivariate graphs (both size–shape and size–age) as heuristic tools; these are also useful in distinguishing between age-based and size-based heterochrony. This encompasses the key relationship between allometry and heterochrony. We also consider a number of problems, such as determining ontogenetic age and dealing with noncomparable ontogenetic stages. After introducing such basic terms and ideas in Chapter 2, Chapter 3 attempts to answer the question of what causes heterochrony. This is a much more difficult question than has often been implied. Obviously, genetic changes are the ultimate cause but concepts such as "rate" and "timing" genes have been greatly oversimplified. There are many kinds of "regulatory genes," at many levels in a regulatory hierarchy, in which not only developmental, but also many other kinds (e.g., physiological) of genes and processes are regulated. Further, many heterochronic genes are not "programmed" to regulate the "rate" of developmental processes but can affect them indirectly, by changing the flow of information via (e.g.) change in structural parameters.

When we turn to the cellular (tissue) levels of heterochronic causation (also Chapter 3), we find that here too there has been much oversimplification. In particular, "heterochrony" in the past has been limited to late-stage allometric (size–shape) changes in ontogeny. Because of this, heterochrony has been little more than a "taxonomy" of patterns relating simple shape changes between ancestral and descendant ontogenies. Yet, we show [extending the earlier work of Hall (1983, 1984)] that the same kinds of rate/timing changes, when occurring earlier in ontogeny, can cause radical changes in morphology, by changing the time of tissue induction, to create new tissues, eliminate old ones, or just change tissue configuration. This kind of heterochrony we call "differentiative heterochrony" to denote its occurrence before differentiation. Later, more traditional kinds are called "growth heterochronies." Again, we use heuristic tools to illustrate the concepts, this time in the form of morphogenetic trees to represent development as cellular assembly with shared, then diverging, histories. In keeping with the cellular view, heterochrony in this light is redefined as change in rate or timing of cell dialogues. Thus, we spend some time on biochemical mediators and other aspects of how cells communicate, and how changes in this communication can occur (often, not by change in the usual "rate/timing" genes, as noted).

In Chapter 4, we leave the classification and mechanisms of ontogenetic change (i.e., the production of variation) and begin to examine how that variation is acted upon. We say "begin" because this is a transitional chapter in that environment is seen not only as a "sieve" selecting intrinsically produced variation, but also as acting directly upon ontogeny to cause heterochronic variation itself. This is an often neglected aspect of heterochrony (indeed of ontogenetic variation per se): that variation (e.g., ecophenotypic plasticity) is directly "shaped" by the environment, and not totally programmed by the genes. Thus, we discuss heterochronic variation that is produced by a variety of environmental stimuli, and some different ways that development can be labile. (Indeed, selection can act to favor lability itself.) Thus, there are at least three somewhat often-neglected points underlying this chapter: (1) that developmental variation can be directly created by the environment, (2) that such nonprogrammed variation can be selected for, and (3) heterochronic variation is incremental. This last point is emphasized because heterochrony has had a long (and not very rewarding) association with saltations (because of Goldschmidt and others) and many are not used to thinking in terms of intrapopulational polymorphisms being heterochronic variants (we call them "heterochronic morphotypes"). Yet, such variants are often produced by minor rate and timing changes (e.g., sexual dimorphism as an extreme intrapopulational example that results from heterochronic differences).

In Chapter 5, we turn to the broad view, that of evolutionary patterns produced by the interaction between the environment and ontogeny. Like any change, evolution has two basic parameters, direction (space) and time (rate), and we spend much of this chapter on them. After an initial section on the methods of evolutionary reconstruction (especially determining phylogenetic polarity), we show how directionality has been ("strongly"?, a relative term for now) influenced by the "intrinsic" forces of ontogeny. At the core of this documentation is McNamara's key concept of the "heterochronocline," which provides some of the best direct empirical evidence on how ontogenetic change is reflected in evolutionary patterns. We show, in a number of lineages, that ontogenetic traits (morphologically, both "local" and "global") are often modified in systematic, clinal patterns across distinct environmental (spatial) and temporal vectors. We discuss a number of different types of such clines and their utility in characterizing and explaining evolutionary patterns. We then turn to the role of heterochrony in affecting rates, suggesting that, for the most part, heterochronic variations are minor, as already noted. But "saltations" are still a real possibility. Finally, we focus on the agents of selection on heterochronic variations of all kinds. In particular, predation seems to have been a major force driving the formation of heterochronoclines.

The often problematic issue of targets of selection is the general subject of Chapter 6. Specific targets we discuss are size, shape, and life history (in Chapter 7, we discuss a fourth target: behavior). Heterochrony affects all of these, and because rate and timing changes often change coadapted suites in the ontogeny, it is often unclear whether some traits change because they are actually under selection, or just part of a coadapted suite where selection is acting only on some traits in the suite. For instance, how often is body size change an incidental by-product of selection for life history tactics (e.g., early maturation)? Or change in shape a result of allometric extrapolation for body size change? Such questions are really at the heart of the constraint debate because they directly ask how "easily" traits can be decoupled from one another. If they are easily decoupled, then developmental constraint is much less than if trait suites are "locked in" and traits are often "dragged along" in spite of neutral or even negative selection. Rhetoric about "constraints" aside, such questions are best answered by the direct methods of quantitative genetics as applied to selection response experiments in living animals [see especially Cheverud (1984) and Riska (1989)]. Also in Chapter 6, we note that targets often change through time. Heterochronically speaking, this broaches the topic of whether groups show more peramorphosis ("overdevelopment") or paedomorphosis ("juvenilization") through time. Again, there are insufficient data to draw complete conclusions, but there is strong evidence that the relative frequency of pera- to paedomorphosis varies among groups (perhaps because of differences in developmental programs) and through time within the same group.

Having a separate chapter, Chapter 7, on behavioral and human heterochrony may seem a little anthropocentric, this being a book written by hominids. However, there are at least two good reasons for it. One, the study of the evolution of behavior has often neglected the ontogenetic view even though it is now clear that such a view can be invaluable. Behavior, like morphology and life history, not only undergoes predictable ontogenetic change, but these changes occur as coadapted suites (of component behaviors). Furthermore, such suites are amenable to heterochronic changes among and within the suites, so that a descendant adult may have behaviors present only in the ancestral juvenile (paedomorphosis). Yet, this approach is only just now being systematically taken [Irwin's (1988) study is something of a groundbreaking work].

A second reason for this chapter is that human heterochrony has been the locus of a great number of misconceptions. Indeed, it is almost a proxy for all the gibberish generated in the study of heterochrony writ large. There is far too much to go into here but the major point is that humans are *not* neotenic apes. That humans are neotenic is one of the most widely circulated bits of misinformation in both the scientific and popular literature. Neoteny is the process of growing *slower*. Yet humans do not grow particularly slow (relative to either the chimp or our ancestors, as is now being discovered with growth lines on fossil hominid teeth). What we do is *delay* the offset of virtually all developmental events (growth phases) so that we are in each phase longer. This is hypermorphosis (we call it "sequential" hypermorphosis because a number of sequential phases are affected). The key point is that the time spent in a growth phase is distinct from the rate of growth in it. The lack of this distinction is one of the main reasons for the confusion, such that people speak of humans having a "slowed" maturation; yet such events (or any single event) is not "slowed," it is delayed (or early, depending on the change). Another reason for this "neotenous" misconception is our superficial resemblance (as adults) to a juvenile chimp (especially the skull). The problem here is that such a superficial, subjective shape criterion is often not a good basis of ontogenetic comparison. In this case, our "chimplike" shape results from a greatly enlarged brain (via hypermorphosis, or prolonged brain growth), combined with a number of probably complex growth field changes in the jaw and facial area. Thus, to get the "complete" heterochronic story, investigators should look at developmental events of component growth fields. As Atchley (1987) has said, each organ may often be best looked at as having its own ontogeny.

In short, hypermorphosis seems to best explain most of those traits that make us human: large body size, large brain, long learning stage and life span, although there are also a number of morphologically "local" nonhypermorphic growth field "adjustments" (as expected, given the novel pressures of cultural adaptations). We also attempt to relate heterochrony to various aspects of race, sex, and human behavior. This last is obviously a difficult prospect, contrary to Montagu's (1981) attempt to "explain" dance, song, trustworthiness, and virtually everything else in terms of "neoteny."

In the penultimate chapter, Chapter 8, we take on the broad view of ontogeny in evolution. This is more than just a synthesis of the major ideas and themes of the other chapters, because still other ideas and themes emerge when these are combined (synergistically). After some theoretical soul-searching (on why development is becoming a focus of interest), we discuss the evolution of constraint. The message is that, despite all the talk about constraint in evolution, few workers explicitly point out that constraints themselves evolve. We try to show that, because evolution often "builds upon" preexisting ontogenies, developmental contingencies accumulate, increasing constraint through evolutionary time. [Levinton (1988) calls this the "evolutionary ratchet."] This "hardening" has a number of implications that we discuss at length. The major ones are that, as "hardening" increases, the role of intrinsic processes in evolutionary directionality and rates diminishes. However, this is somewhat misleading in that, as "hardening" evolves, there is a simultaneous decrease in the number of directions that evolution can go. In a sense, in accumulating the (ontogenetically and phylogenetically) early contingencies, intrinsic processes have already determined the basic trajectory of the biosphere's evolution. Indeed, one might say that the biosphere itself has become "hardened" in that, as contingencies within its components ("ontogenies") accumulated, so have the interactions (contingencies) among those components. More recent, mainly environmentally selected ("extrinsic") ontogenetic contingencies are mostly minor permutations on the basic architecture of the earth's biota. Superficially paradoxical is that, at the same time that directionality is diminished via "hardening," evolutionary rates are increasing. That is, there is more change (in ontogenies) going on, but the change is more minor because the accumulation of ontogenetic contingencies makes viable "jumps" less feasible. This is due to what might be called "the first law of evolving complex systems": the more parts there are, the more there are to change, but the (relatively) less important each local change is (being smaller relative to a larger whole). This kind of positive feedback has been elegantly described in DeAngelis et al. (1986).

The bottom line is that, while the debate rages on over the "intrinsic versus extrinsic" control of evolutionary directions and rates, the relative role of each has changed. Intrinsic control is diminished, but only after having "set" much of the trajectory, often via rapid rates. Incidentally, none of this requires the now-discarded process of "recapitulation via evolution by terminal addition." What we are pointing out is that the evolution of complexity (and life) has occurred because the system has a memory, implying that earlier contingencies are conserved and built upon by later-occurring ones. This does not require that all change occur at the end of an ontogenetic process ("terminal addition"), only that earlier-occurring ones are progressively less likely to be successful [and because ontogeny is a multiplicative, branching process (in many ways, not just mitosis), the odds against successful earlier changes may often go up exponentially].

Obviously, all this is not to say that ontogeny has little role in directing evolution today. Even though changes in evolutionary direction and rates are not as profound as in the past, development is still more than an "on-demand parts supplier" for environmental selection to operate upon. This is seen most clearly in the heterochronoclines of McNamara where heterochronic variants grade into evolutionary lineages. Perhaps the most dramatic role for development is in the origin of such lineages (and clades) where paedomorphosis and peramorphosis play a role in "innovation." (We give a number of reasons why, contrary to de Beer's assertions and "common widsom," paedomorphs do not generally have more evolutionary potential than peramorphs.) We try to work such innovation into a general theory of ontogeny and ecology by combining it with paleontological observations, such as nearshore origination of most major clades, and ecological observations, such as that r-selection (e.g., nearshore, unstable regimes) may select for certain kinds of heterochrony [e.g., progenesis (early offset of growth) producing small size and early reproduction]. The subsequent diversification into more stable, offshore environments is often accompanied by larger body size, increased longevity, and other K-selected traits produced by the heterochronic mechanism of hypermorphosis (delayed offset of growth). The attraction of such a general theory is that it not only links genes, cells, development, and ecology into an evolutionary framework, but also includes aspects of development often overlooked. Thus, aside from selection on the ontogeny of morphology (size and shape), it is able to take in the ontogeny of life history events (e.g., maturation, death), and behavior. Whether our proposed, specific scenarios are correct is less important than that they are testable, especially by modern ecologists (e.g., Ebenman and Persson, 1988). One intriguing (and perhaps partly fanciful) notion to follow from such a theory is that heterochrony is a way of scaling the timing of intrinsic (ontogenetic) events to the tempo of extrinsic ones. This leads to a discussion of intrinsic time and, perhaps surprisingly, body size, which may often turn out to be a good estimator of intrinsic time.

READER COMMENTS AND ACKNOWLEDGMENTS

Comments

We consider this book as something of a "status report" and would greatly appreciate comments (positive or negative) by any readers, students, and colleagues alike. Should there be enough interest to justify a revised edition, this will help us to improve it. Chapters 1, 4, 5, and 6 (except body size and some life history text) were written mainly by McNamara and comments on those might best be sent to him. McKinney wrote Chapters 2, 3, 7, 8, and 9, (body size, life history in Chapter 6), the Glossary, and the (probably too-long) Preface and Preview.

Acknowledgments

We would like to thank the following people (in no particular order) for kindly reading drafts of chapters (in one version or another) of this book and offering their comments: Brian Hall (Dalhousie University), Brian Shea (Northwestern University), John Gittleman (University of Tennessee), Simon Conway Morris (Cambridge University), Jim Hanken (University of Colorado), Brian Tissot (Oregon State University), Neil Blackstone (Yale University), Michael Green (University of Tennessee), and Bill Calder (University of Arizona). As always, this does not mean they agree with the end product. In addition, K.J.M. wishes to thank the following: Alex Baynes for help with literature and provision of computer facilities. The Department of Geology, University of Western Australia, is thanked for computer support, especially Drs. Nick Rock and David Haig.

Drs. William Peters (mayflies) and Alan Rayner (fungi) supplied information in their fields of expertise. Kris Brimmel of the Western Australian Museum assisted with some of the diagrams. Special thanks to Susan Radford for her encouragement and support. M.L.M. would like to thank Dan Frederick and Tony Tingle, University of Tennessee, for assistance with some figures and photography.

Ken and I also thank the nice people at Plenum Publishing who helped us turn amorphous thoughts and back-of-the-envelope sketches into something fit for public consumption: Amelia McNamara, Mariclaire Cloutier, and the highly professional production staff. Finally, Ken and I would both like to thank our small children for permitting us to work on the book when they were at school, out playing, or otherwise had no need for us.

Michael L. McKinney

Knoxville, Tennessee

Contents

Chapter 5
Heterochrony in Evolution: Direction, Rates, and Agents

Chapter 6
Heterochrony and Targets of Selection

Chapter 7
Behavioral and Human Heterochrony

Chapter 8
Epilogue and Synthesis: Ontogeny in Evolution

Chapter 9
Latest Developments in Heterochrony

Heterochrony

The Evolution of Ontogeny

Chapter 1

Heterochrony
A Historical Overview

*Only three men have ever understood it. One was
Prince Albert, who is dead. The second was a
German professor who has gone mad. I am the
third, and I have forgotten all about it.*
Lord Palmerston (on timing of governmental
change in European politics)

1. INTRODUCTION

In 1859 Charles Darwin published *The Origin of Species by Means of Natural Selection*. No single book before or since has had such a profound effect on evolutionary theory. To most present-day biologists it is seen as the bright light that burst upon the Dark Ages of biology that preceded it. While no biologist in his right mind could do otherwise than regard this work as being of pivotal importance in modern evolutionary thought, from the perspective of those who have been exploring the role of changes in developmental regulation in evolution, this work could also be seen in another, less flattering, light. Even though Darwin drew heavily on the detailed embryological studies made in the early 19th century, this *intrinsic* aspect of evolution has been largely neglected in the neo-Darwinian view of more recent times. Rather, focus has been on the *extrinsic* factor of natural selection. Thus, Darwin's work has, albeit indirectly, been instrumental in closing the door on a facet of biological theory that had dominated scientific thought in the late 18th and early 19th centuries—the developmental relationships among organisms.

With the increasing emphasis on the external forces of natural selection to explain the diversity of life, pre-Darwinian ideas that arose from detailed embryological studies soon fell into disrepute. Their demise was assured by the diversity of views about development. While we do not wish to subscribe overtly to the views of the 19th century biologists, who were as fascinated by similarities and differences that arise during ontogeny as we are, many useful concepts on the subject of heterochrony have been formed as

1

a result of their work. We shall not dwell at length on these early workers, as others, most notably S. J. Gould in his masterly tome *Ontogeny and Phylogeny,* have covered much of this ground in great detail. We shall merely provide a brief overview of their basic ideas and show how these relegated the study of heterochrony, until the last decade, to the backwaters of biology.

2. VON BAER AND THE NATURALPHILOSOPHIE SCHOOL

The relationship between ontogeny and phylogeny was dominantly viewed from a single perspective throughout the 19th century—an increase in complexity from "primitive" life forms through the animal kingdom to mankind, the pinnacle of Creation. During their development, "higher" life forms passed through the adult stages of life forms lower on the scale of organismic complexity. Thus, for example, during its development the human embryo was considered to pass through a fish stage, then a reptilian stage, before finally attaining our mammalian state. This was the concept of **recapitulation** which, in various guises, was the cornerstone of developmental biology throughout the 19th century. As far back as the time of Aristotle, this concept of increasing complexity in the biological world was an important aspect of philosophy. As Gould (1977) has discussed, to Aristotle, it was the sequence of increasingly more "complex" souls—nutritive, sensitive, then rational—that entered the human embryo during successive stages of development. The nutritive soul Aristotle compared with plants, the sensitive with animals, and the rational with humans. Although Needham (1959) considered Aristotle to be the great-great-grandfather of recapitulation, Gould (1977) has quite rightly rejected this pedigree. Little can be gained from trying to compare such broad philosophical concepts with specific ideas that arose from embryological studies in the latter part of the 18th and the early part of the 19th centuries.

During the early 19th century the concept of recapitulation arose as "an inescapable consequence of a particular biological philosophy" (Gould, 1977, p. 35), that of *Naturalphilosophie* which existed in Germany at this time. The uncompromising belief that its adherents had in "developmentalism"—a unidirectional flux, from lower to higher, from initial chaos to man—formed an integral part of their philosophy. The first notable proponent of recapitulation of this school was Lorenz Oken (1809–1881). An anatomist and embryologist who is best known for his work on the embryology of the pig and the dog, Oken proposed a classification of animals based on a linear addition of organs during development. He believed that, starting from nothing, there was a general, progressive increase in complexity and number of organs leading to the most complex structure. This occurred by the successive addition of organs in a predetermined sequence. Man was seen as a microcosm of the whole world, the summit of Nature's achievement. Oken (1847, pp. 491–492) wrote:

> During its development the animal passes through all stages of the animal kingdom. The foetus is a representation of all animal classes in time. At first it is simple vesicle, stomach, or vitellus, as in the Infusoria. Then the vesicle is doubled through the albumen and shell, and obtains an intestine, as in the Corals.

It obtains a vascular system in the vitelline vessels, or absorbents, as in the Acalephae.

With the blood-system, liver, and ovarium, the embryo enters the class of bivalved Mollusca.

With the muscular heart, the testicle, and the penis, into the class of Snails.

With the venous and arteriose hearts, and the urinary apparatus, into the class of Cephalopods or Cuttle-fish.

With the absorption of the integument, into the class of Worms.

With the formation of branchial fissures, into the class of Crustacea.

With the germination or budding forth of limbs, into the class of Insects.

With the appearance of the osseus system, into the class of Fishes.

With the evolution of muscles, into the class of Reptiles.

With the ingress of respiration through the lungs, into the class of Birds. The foetus, when born, is actually like them, edentulous.

Another whose thoughts paralleled those of Oken in many ways was J. F. Meckel, perhaps the most influential Naturphilosoph in spreading the doctrine of recapitulation (Russell, 1916). Like Oken, he considered that a single developmental pattern governed nature. However, he differed from Oken in not relating it purely to addition of new organs. He viewed development more as an increase in specialization and coordination of the organism. Meckel (1821, p. 514) wrote:

> The development of the individual organism obeys the same laws as the development of the whole animal series; that is to say, the higher animal, in its gradual development, essentially passes through the permanent organic stages that lie below it.

A strong belief in recapitulation was also evident in the school of French transcendental morphologists, led by Etienne Geoffroy Saint-Hilaire. One of this group was the medical anatomist Etienne Serres, who studied the comparative anatomy of the vertebrate brain. He based his views on recapitulation on the study of the nervous systems of vertebrates and invertebrates. "Lower" animals he considered to be the permanent embryos of "higher" animals. Like the German Naturphilosophs, Serres saw in development a "step by step march to perfection." Serres extended the concept of recapitulation (Gould, 1977) to the consideration of developmental arrests. For instance, he compared the state of undescended testicles of some men with the permanent state in fishes. Likewise, malformations of the heart he saw as comparable with the stages achieved by "lower" animals.

With its extension to include developmental arrests and the formation of teratologies, the theory of recapitulation increased in respectability. However, among embryologists of the day there were opponents. Perhaps the most eminent of these was Karl Ernst von Baer. In his monumental *Entwickelungsgeschichte der Thiere* published in 1828, von Baer set out to demolish many of the basic tenets of recapitulation. von Baer's principal argument was that the theory of recapitulation carried a fundamental error. He argued that arrested embryonic stages are not comparable with the adult state of "lower" organisms. Many embryonic features are specializations. Furthermore, while there may be one or two similarities between an embryonic structure of one organism and the adult of another, many other structures will not be comparable. Even more significant is that many features that occur in adults of "higher" organisms can occur in embryos of "lower" organisms. von Baer argued that the embryonic vertebrate is always an undeveloped vertebrate. It can never be considered as representing any "lower" adult form. Perhaps most significant was von Baer's assertion that embryological development is not simply, as Gould (1977) puts it, "a climb up the ladder of perfection"; rather, it is a differentiation from more generalized characters to more complex ones. Consequently, general features of a large group of animals appear earlier in development. The more specialized the character, the later in ontogeny it occurs. Thus, the embryo of a "higher" animal never resembles the adult of a "lower" animal, only its embryo.

The first 19th century biologist to attempt to apply the theory of recapitulation to the fossil record was Louis Agassiz. By doing so he established what Gould (1977, p. 66) has termed the "threefold parallelism," comprising embryonic growth, structural gradation, and geologic succession. Thus, an embryo repeats not only a series of *living* "lower" forms, but also the history of its type as shown by the fossil record. In his notable work *Recherches sur les poissons fossiles,* published between 1833 and 1843, Agassiz noted the parallelism between the ontogenetic development of characters such as the tail in "higher" teleost fish (from diphycercal through heterocercal to homocercal) (Fig. 1-1) and the development of the same feature through the fossil record. Likewise, in his studies of echinoids during the late 1830s and early 1840s (see Clark, 1946, for references) he noted how juveniles of some irregular echinoids, that as adults have a mass of fine, hairlike spines, have few, relatively large spines, like adults of cidaroid echinoids which long predate the irregulars (see McNamara, 1988a, for a more recent discussion of this phenomenon).

What separated the early 19th century recapitulationists from those who followed, such as Haeckel and Hyatt, was an inability to provide an effective mechanism for recapitulation. There was surprisingly little attempt to even search for a mechanism. But this may more be due to other, philosophical pressures of the day. It was the easy option to fall back on the great Creator to explain such phenomena:

> There exists throughout the animal kingdom the closest correspondence between the gradation of their types and the embryonic changes their respective representatives exhibit throughout. And yet what genetic relation can there exist between the Pentacrinus of the West Indies and the Comatulae, found in every sea; what between the embryos of Spatangoids and those of Echinoids . . . what between the Tadpole of a Toad and our Menobranchus; what between a young Dog and our Seals, unless it be the plan designed by an intelligent creator [Agassiz, 1857, 1962 ed., p. 119].

3. HAECKEL AND THE BIOGENETIC LAW

The significance of the publication of Darwin's *Origin of Species* to recapitulation was that for the first time there was an evolutionary framework against which the theory of recapitulation could be tested. Agassiz's threefold parallelism pointed the way toward

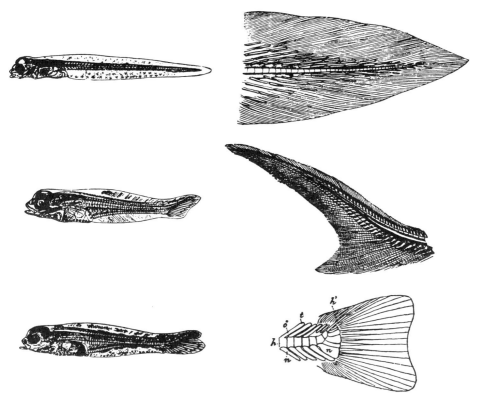

Figure 1-1. The ontogeny of a teleost, the flatfish *Pleuronectes,* showing the change in the tail from diphycercal through heterocercal to homocercal (top to bottom), compared with the tail shapes of adult fishes arranged from "primitive" (top) to "advanced" (bottom): *Protopterus* with a diphycercal tail; a sturgeon with a hetero-cercal tail; and a salmon with a homocercal tail. Such comparisons were used by Agassiz in his support for recapitulation. Reproduced from Schmidt (1909).

a more meaningful appraisal of recapitulation in terms of evolution by its promotion of the fossil record as a testimony of the ubiquity of recapitulation (even though Agassiz himself was no evolutionist). In terms of evolution, recapitulation could only, as Gould (1977) has argued, work on the basis of two fundamental assumptions. The first is that "evolutionary change occurs by the successive addition of stages to the end of an unaltered, ancestral ontogeny." The second is that "the length of an ancestral ontogeny must be continuously shortened during the subsequent evolution of its lineage." Gould calls the first of these "the principle of **terminal addition**" (Fig. 1-2), the second, "the principle of **condensation**" (Fig. 1-3). Much of the debate surrounding recapitulation in the latter half of the 19th century centered on attempts by a number of biologists to demonstrate these two principles.

Foremost among these were three biologists, each of whom published a major work on the subject in the same year, 1866—Ernst Haeckel, Alpheus Hyatt and Edward Drinker Cope. The most influential of these works and the one whose influence is still felt today was Haeckel's *Generelle Morphologie der Organismen.* It was in this work that

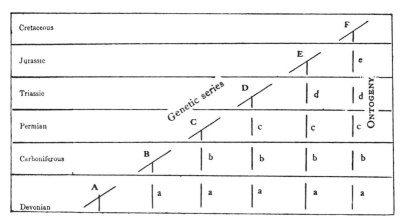

Figure 1-2. The "principle of terminal addition," as perceived by Smith (1914, Fig. 3).

Haeckel proposed his now-famous **biogenetic law**. While generally abbreviated to "ontogeny recapitulates phylogeny," he actually wrote (Haeckel, 1866, Vol. 2, p. 300): "Ontogeny is the short and rapid recapitulation of phylogeny . . . ." Haeckel's influence was far-reaching, influencing not only biological and paleontological thought, but also aspects of politics, sociology, and religion. In their most extreme form, his statements came to be associated with concepts of "racial purity" and other doctrines of national socialism that emerged in Germany in the early part of this century. To explain the principle of terminal addition, Haeckel drew on Lamarckian views of the heritability of acquired characters. Such characters, he believed, were more likely to be expressed in the adult stage, as it is far more "permanent" than the transient juvenile stages.

The standard explanation offered by Haeckel and his contemporaries for the principle of condensation was either that there was an acceleration of developmental rate later

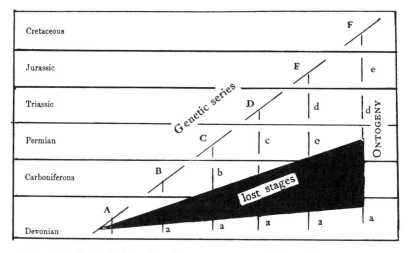

Figure 1-3. The "principle of condensation," as perceived by Smith (1914, Fig. 4).

in ontogeny, or that certain stages would be cut out of the ontogenetic sequence altogether, allowing the remaining stages to be passed through more quickly. This made room for the terminal additions. Haeckel called the repetition of past phylogenetic stages in ontogenetic stages of descendants *palingenesis*. Exceptions to this, i.e., the introduction of different features into the ontogenetic sequence, he termed *caenogenesis*. Haeckel thus saw ontogeny as little more than the mechanistic effect of phylogeny: ontogenetic stages that are under the control of phylogenetic processes. Of the many new terms that Haeckel introduced (which included "ecology," "ontogeny," and "phylogeny") one was **heterochrony**. Haeckel used this term to explain the displacement in time or change in order of succession of particular organs, such as reproductive organs.

More so than Haeckel, the second character in the triad, Edward Drinker Cope, concerned himself with the mechanisms of recapitulation. As Gould (1977) has argued, Cope and fellow paleontologist Alpheus Hyatt afforded to recapitulation a status that it had never enjoyed before, nor was ever to achieve again. To Cope, species represented the modification of structures already in existence. New genera, on the other hand, arose, he believed, by additions to *or* subtractions from the ancestral ontogenies. Cope thought that as recapitulation occurred there was an acceleration in individual development, with ancestral growth stages being repeated at successively shorter and shorter intervals. A combination of terminal addition and acceleration would result in recapitulation.

Cope, like others, recognized that there were cases whereby there was less development in an ancestral form. Here the later ontogenetic stages would have failed to realize their potential and were deleted. These two aspects of "more" and "less" growth were enunciated by Cope as his laws of "acceleration" and "retardation," respectively. Where Cope and Haeckel fundamentally differed in their approaches to the question of recapitulation, was that Haeckel saw it as primarily affecting the *whole* organism (a view that is still prevalent today), whereas Cope more perceptively saw it as operating on *individual* organs. Even though he accepted the validity of both "acceleration" and "retardation," Cope, like his contemporaries, saw acceleration, producing recapitulation, as the more important evolutionary force. Cope actually tried to offer an explanation for this disparity, invoking directional changes in atmospheric conditions, producing a rise in levels of oxygen, greater metabolic rates and so accelerated development.

One man, more than any other, contributed to the decline of recapitulation in the 20th century. He was the third in our triad, Alpheus Hyatt. His extreme views on the biogenetic law not only influenced a cohort of other paleontologists of the day, but were also to play a critical role in contributing to the downfall of the biogenetic law itself. A student of mollusks, his most notable works were on the Arietidae (Hyatt, 1889) and a paper on "the phylogeny of acquired characteristics" (1893). However, it was a short communication read to the Boston Society of Natural History on February 21, 1866, that was to influence so profoundly his own works and those of his contemporaries well into the 20th century:

> Mr. A. Hyatt made a communication upon the agreement between the different periods in the life of the individual shell, and the collective life of the Tetrabranchiate Cephalopods. He showed that the aberrant genera beginning the life of the Nautiloids in the Paleozoic age, and the aberrant genera terminating the existence of the Ammonoids in the Cretaceous Period, are morphologically similar to the youngest pe-

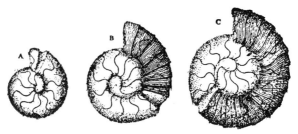

Figure 1-4. The Hyatt view of ammonoid evolution, by terminal addition. Reproduced from Smith (1914, Fig. 1).

riod and the period of decay of the individual; the intermediate normal forms agreeing in a similar manner with the adult period of the individual. He also pointed out the departure of the whorl among aberrant Ammonoids from its complete development among the normal forms, its final appearance as a straight tube in the Baculite, and the close connection between this morphological degradation of the whorl and the production of the degradational features in the declining period of the individual, demonstrating that both consisted in the return of embryonic or prototypical characteristics of the form, and partly of the structure.

Hyatt firmly believed that evolutionary series of species passed through a "life cycle" like that of an individual. He particularly stressed the importance of "senile" characters, these becoming the normal adults of tomorrow (Fig. 1-4). Independently of Cope, Hyatt developed a concept of acceleration to explain recapitulation. However, whereas Cope considered that *both* acceleration and retardation played a part in the theory of recapitulation, Hyatt saw only acceleration. This, ultimately, was to be the undoing of Haeckel's "biogenetic law." Hyatt underwent some extraordinary logical convolutions in his attempts to explain all changes as being caused by acceleration. Thus, if a seemingly juvenile feature was retained into the adult stage, this was really a deceptive gerontic condition that had appeared beyond the normal adult stage (which had been conveniently lost) and which mimicked the ancestral juvenile stage!

As we have mentioned, the effect of Hyatt's ideas on his contemporaries was quite profound (Donovan, 1973). Beecher's (1897) classification of the Trilobita, for instance, was based on the premise that the ontogeny of later trilobites encapsulated the early history of the entire group. He thus considered that ". . . the process of acceleration or earlier inheritance has pushed forward certain characters until they appear in the protaspis, thus making it more and more complex" (Beecher, 1897). He (Beecher, 1893) interpreted brachiopod evolution in a similar light. However, as we discuss in the following section, Beecher was fully cognizant of many examples of "retardation" in brachiopods, as he was in trilobites. While other groups, such as bivalves (Jackson, 1890) and echinoderms (Jackson, 1912), were also viewed in this same recapitulatory light, the turn of the century saw the first chinks beginning to appear in the armor of the extreme recapitulationists.

Unlike Hyatt, whose ideas on ammonite evolution were derived largely from studies of museum collections (Donovan, 1973), the Russian paleontologist A. P. Pavlov, by careful stratigraphic collecting of the ammonite *Kepplerites,* demonstrated that new characters often appear first in young stages, and then spread to the adult stages of later forms (Pavlov, 1901). He also documented similar phenomena in belemnites, gastropods, and vertebrates. But Hyatt's stranglehold on ammonite workers was tight. English workers in the first two decades of the 20th century, such as S. S. Buckman, A. E. Buckman, and L. F. Spath, based much of their ideas on ammonite phylogeny in terms of recapitulation.

There is a general belief that when it came to the biogenetic law, most paleontologists of this era, with the notable exception of the likes of Pavlov, were blinkered idealogues who saw nothing but recapitulation. Yet there were serious dissenters, even among those who have always been unjustly lumped with the extreme recapitulationists. James Perrin Smith, professor of paleontology, was one of these. Cited by Donovan (1973) as one of the proponents of recapitulation, he in fact became a cogent adversary of this school of thought. In a largely ignored paper entitled "Acceleration of development in fossil Cephalopoda" (Smith, 1914), Smith tried to point out, by citing many examples, that in ammonites retardation of development was a more common phenomenon than recapitulation. His concept of the biogenetic law was, like that of Cope, much broader than originally envisaged by Haeckel, and certainly far broader than conceived by Hyatt. In a scathing conclusion to his paper, he castigated his fellow paleontologists for their "overzealous" acceptance of Hyatt's ideas:

> It may be that, when this paper is read by ardent members of the "Hyatt school" of paleontologists and adherents of the biogenetic law, they will be inclined to call the writer a deserter from the camp, and to suggest that the paper ought to have been entitled, "Why recapitulation does not recapitulate." The writer is still a firm believer in the biogenetic law, but that law is not such a simple thing as it was once thought to be. In the youth of every theory everything is beautifully clear, and ideally simple. As time goes on we are compelled to drop one idea after another, until it almost seems that the whole will be lost. When sceptics concerning the recapitulation theory throw up to us that ontogeny does not *always* recapitulate, we are prepared to admit this, even to go further and admit that it does not *often* recapitulate. In fact, the writer would be prepared to go still further, and to state that, in the sense in which the term has been used by most adherents of the theory, it *never* recapitulates. Our over-zealous friends have claimed too much, and have done more to prevent general acceptance of the theory than a host of enemies." [Smith, 1914, p. 26]

4. GARSTANG, DE BEER, AND THE MODERN HETEROCHRONIC SYNTHESIS

For all the urging of people like Smith, the domination of the concept of recapitulation, particularly in ammonite studies, survived well into the 1930s, and its effects lingered well on into the 1970s. [One of us (K.J.M.) well recalls that "ontogeny recapitulates phylogeny" was taught to him in the early 1970s.] But the death knell for the biogenetic law came not from paleontological studies, but from neontology. Perhaps because paleontologists had carried the banner of recapitulation so high and for such a long time it took longer for them to let it go. Even though paleontologists like Otto Schindewolf were demonstrating countless examples of retardation as "proterogenesis" (e.g., Schindewolf, 1929), it took the dramatic statements of Walter Garstang, working on tunicate larvae, to effectively kill the biogenetic law.

What Garstang (1922, 1928, 1946) proposed was that rather than recapitulating phylogeny, ontogeny actually *created* it. From this arose his term **paedomorphosis**, meaning the retention of a "childlike" (*paedo* = child) morphology by adults. It is tempting to suggest that what was needed to elevate "retardation" to respectability was a suitable term to describe it, such as paedomorphosis. But already terms to describe such retention of juvenile characters by descendant adults had been coined in the 19th century. For instance, Julius Kollman introduced the term **neoteny** (Kollman, 1885) to describe the retention of juvenile features in the axolotl *Ambystoma*, a salamander that reproduces as an aquatic larval form. Garstang was later (Garstang, 1951) to describe this process in verse:

> The Axolotl and the Ammocoete
> Amblystoma's a giant newt who rears in swampy waters,
> As other newts are wont to do, a lot of fishy daughters:
> These Axolotls, having gills, pursue a life aquatic,
> But, when they should transform to newts, are naughty and erratic.
>
> They change upon compulsion, if the water grows too foul,
> For then they have to use their lungs, and go ashore to prowl:
> But when a lake's attractive, nicely aired, and full of food,
> They cling to youth perpetual, and rear a tadpole brood.
>
> And newts Perennibranchiate have gone from bad to worse:
> They think aquatic life is bliss, terrestrial a curse.
> They do not even contemplate a change to suit the weather,
> But live as tadpoles, breed as tadpoles, tadpoles altogether!

Two years after Kollman introduced the term "neoteny," Giard (1887) proposed the term **progenesis** to indicate the process whereby the onset of sexual maturation was early, relative to the ancestor. The reason Garstang's voice was heard, when others before him had been largely ignored, was because he spoke within a setting of evolutionary theory in Mendelian genetic terms. Moreover, Garstang ascribed to paedomorphosis a major role in macroevolution. He argued that vertebrates may well have evolved from a pelagic invertebrate larval form by paedomorphosis. This would have occurred by an early larval stage becoming precociously sexually mature. Free-swimming tunicate larvae, for instance, possess all the fundamental attributes of a vertebrate: a notochord; dorsal hollow nerve cord; gill slits; and postanal propulsive tail (see Raff and Kaufman, 1983, for discussion).

At the same time that Garstang was promoting paedomorphosis, Louis Bolk, professor of human anatomy at Amsterdam, was developing his ideas of *fetalization* in human evolution. In a series of papers, most notably Bolk (1926), he put forward the idea that many human features were the product of retention of juvenile primate features: for instance, a flat face, loss of pigmentation, absence of brow ridges, small teeth, late eruption of teeth and many more (see Gould, 1977, p. 357, for a full list). Bolk was not, as Gould has shown, a proponent of Darwinian views of evolution. Rather than extrinsic factors directing evolution, he saw intrinsic forces, arising from developmental changes, as being most important. As he is quoted in Gould (1977, p. 362): "Evolution is for organized nature what growth is for the individual: and (for the former) as for the latter, outer factors have only a secondary influence. They can never play a creative role, only one of modelling what is already there . . . ."

From the 1930s until the late 1970s the pendulum swung from an adherence to recapitulation to a promotion of paedomorphosis as the dominant factor in the relationship between ontogeny and phylogeny. The main driving force in this was Gavin de Beer. In his two books *Embryology and Evolution*, published in 1930, and *Embryos and Ancestors*, first published in 1940, Garstang set out to demonstrate the all-pervasive influence of paedomorphosis, particularly in the animal kingdom. He saw it as being responsible for the origin of many groups of animals, including numerous groups of invertebrates, including insects, and vertebrates, such as man, chordates, flightless birds, and flying fish. Just as importantly, perhaps, de Beer tried to establish embryology as a potent force in evolutionary theory. His books, however, failed to have as much influence on the place of the relationship between ontogeny and phylogeny in modern evolutionary theory as might have been expected. Moreover, there are a number of deficiencies with de Beer's approach to heterochrony.

In particular, de Beer produced a complicated system of eight categories: caenogenesis, adult variation, neoteny (and paedogenesis), hypermorphosis, deviation, retardation, reduction, and acceleration. Only four of these are actually heterochronic processes: **paedogenesis, neoteny, hypermorphosis,** and **acceleration.** de Beer was confused over the distinction between paedogenesis and neoteny, effectively combining them because they produce the same end result of paedomorphosis. Yet as Gould (1977) and McNamara (1978) have pointed out, they are quite different processes that produce a similar morphological result—paedomorphosis. Acceleration and hypermorphosis result in recapitulation.

Gould's (1977) more recent reappraisal of heterochrony has been important in promoting the view that there is no known reason why paedomorphosis should be any more common than recapitulation. de Beer had argued that paedomorphosis was the overwhelmingly dominant process. Gould looked at the timing of development of somatic features and reproductive organs. Thus, he saw de Beer's acceleration as accelerating somatic features, and his paedogenesis (which Gould preferred to call progenesis) acceleration of reproductive organs. In contrast, neoteny was associated with retarded somatic features, and hypermorphosis with retarded reproductive organ development. This approach has resulted in some confusion in that acceleration is associated with both recapitulation (when applied to somatic features) and paedomorphosis (when applied to reproductive organs). This was clarified in a quantitative approach to the question by Alberch *et al.* (1979). Their revised classification has formed the basis for research on heterochrony in the last decade.

Alberch *et al.* (1979) showed that between ancestor and descendant, development can either be reduced (resulting in **paedomorphosis**) or increased (resulting in what they termed **peramorphosis**). Each could be produced by three processes, involving: developmental rate change, change in onset time of development, or change in its offset time. In the case of paedomorphosis, reduced rate is **neoteny**; delayed onset time is **postdisplacement**; and earlier offset is **progenesis**. For the opposing case of peramorphosis, increased rate is **acceleration**; earlier onset **predisplacement**; and delayed offset **hypermorphosis**. These six processes could therefore describe all heterochronic processes (see Chapter 2 for a more complete discussion of classification and terms). Whereas Alberch *et al.* retained the use of the term *recapitulation* for the phylogenetic effect of peramorphosis, McNamara (1986a) advocated that the term should finally be put to rest, for not only is it an inaccurate word to describe what is really going on in peramorphosis, but there is no reason why peramorphosis cannot also be used in a phylogenetic sense. The introduction of the term *reverse recapitulation* by Alberch *et al.* (1979) may therefore be unnecessary.

The impetus created by Gould (1977) and Alberch *et al.* (1979) in the last decade has resulted in a great increase in the number of papers describing heterochrony, both in fossil (see McNamara, 1988b, for synthesis) and living organisms; attempts to develop Gould's (1977) suggestion that different forms of heterochrony may be associated with different life history strategies (McKinney, 1986, 1988b); changing frequencies of heterochrony through time (McNamara, 1986b; Anstey, 1987); and the role of heterochrony in evolutionary trends (McNamara, 1982b). The momentum that is being generated by papers written in the last decade has the potential to return the study of intrinsic factors in evolution back to their rightful place in evolutionary theory.

Chapter 2

Classifying and Analyzing Heterochrony

The brain is the medium by which time is
converted into space and space into time.
W. Gooddy, *Time and the Nervous System*

1. INTRODUCTION

The historical proliferation of terms discussed in Chapter 1 created a formidable obstacle to the study of heterochrony for many years. It was in Gould's (1977) *Ontogeny and Phylogeny* that the first successful simplification was achieved with the establishment of the "clock model." Shortly thereafter, this semiquantitative scheme was modified in a widely cited paper by Alberch *et al.* (1979). Since then, McNamara (1986a) and McKinney (1986) have presented schemes based on allometry alone, with only partial success. More recently, McKinney (1988a) has attempted to present a simplified version of the Alberch *et al.* (1979) scheme. Each of these will be reviewed briefly because all are in use. However, lest the reader despair of a tyranny of schemes replacing the tyranny of terms, be aware that all schemes are very similar, with concepts and terms that deviate little from the original presentation of Gould (1977). The process has been mainly one of theoretical refinement and methodological simplification.

What is striking is the apparent simplicity to which the former terminological morass can be resolved. In brief, the comparison of ontogenies (on which heterochronic classification depends) always reveals changes of only three kinds: beginning of growth, end of growth, and rate of growth of an organ, trait, or whatever it is that changes between the ontogenies compared (usually measured in the form of size or shape change). All but the allometric schemes of McNamara (1986a) and McKinney (1986) are based on changes in shape and/or size with ontogenetic age. The allometric schemes omit age and are based on shape and/or size change of a trait as a function of body size. This makes them less informative in some respects and misleading if care is not taken. However, they have some practical use and clarify the intuitively obvious yet usually unarticulated relationship between heterochrony and allometry. Further, the size-based approach complements the growing awareness that many physiological and ecological phenomena are more size-determined than age-determined, since size is a better metric of "intrinsic age" than external time. The age-based approach is reviewed first, followed by a section on finer points and problems of that approach. Last is a section on the allometric view, introducing both bivariate and multivariate methods. In the next chapter, we look at the genetic and developmental processes which underlie the phenotypic expressions classified here.

2. AGE-BASED VIEWS OF HETEROCHRONY

2.1. Clock Model

Gould (1977) made the distinction between ontogenetic changes in size and shape with time (age) by using "hands" on a clock to represent size and shape, and a horizontal bar to represent age (Fig. 2-1). Heterochrony is shown when the hands are moved relative to the ancestral state. For example, in progenesis a descendant species would stop growing at a younger age and therefore retain, as an adult, the "less developed" size and shape of the ancestral juvenile (Fig. 2-1). This retention of the ancestral juvenile shape in the descendant adult is **paedomorphosis** (literally, "child formation"). In neoteny, paedomorphosis occurs by slowing down change in shape only. Size change and age at growth cessation both follow the ancestral pattern (Fig. 2-1). Note that the paedomorphic adult descendant need not resemble the ancestral juvenile *per se;* it need only resemble the ancestor at an earlier, preadult stage. Hypermorphosis is the opposite of progenesis in that growth stops later and size and shape change go beyond that of the ancestor. Similarly, acceleration is the opposite of neoteny in that shape change (only) goes beyond that of the ancestor. In going beyond the shape change of the ancestor, the opposite of paedomorphosis occurs. Gould called this recapitulation, but as we shall see, it is more appropriately called **peramorphosis** (i.e., "overdevelopment"). Finally, Gould defined proportioned giants and dwarfs as descendants which changed in size only and not shape; that is, they grow isometrically so that they retain the same shape as the ancestors at a different size. This is not common because scaling laws usually require changes in shape (positive or negative allometry discussed shortly) to accompany size change.

The heterochronic changes just outlined do not have to affect the entire organism (in fact they usually do not). The length of an organ such as horn may grow at a faster rate in the descendant (i.e., be accelerated) while shoulder height might not. This is an example of **dissociated** heterochrony, whereby different organs (or growth fields) can undergo heterochronies (or remain unaffected) independent of what is going on elsewhere. This process has long been known as "mosaic evolution" (traits evolving at different rates) though not often conceived in heterochronic terms. (The mechanisms and evolutionary implications of this very important aspect are explored in later

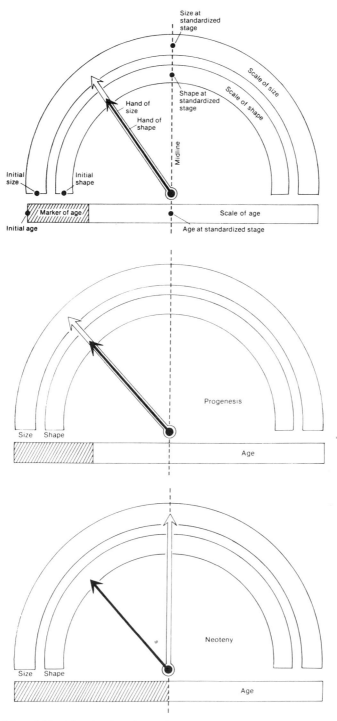

Figure 2-1. "Clock" model (top) for classifying heterochrony, showing hands for size and shape, and bar for age. The lower two examples show progenesis and neoteny. From Gould (1977).

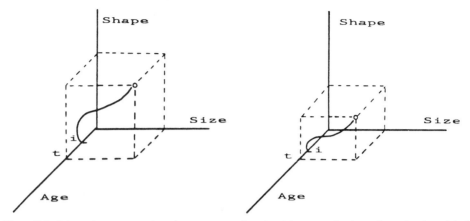

Figure 2-2. Schematic representation of two ontogenetic trajectories sensu the three-dimensional model of Alberch *et al.* (1979), where i is the time of initiation of a developmental process and t is the time of termination. The trajectory on the left might be considered as descendant from that on the right, or vice versa. From Atchley (1987).

chapters). Where the whole organism is affected by the same change, it is called **global** heterochrony.

2.2. Three-Dimensional Model

Alberch *et al.* (1979) supplemented Gould's (1977) ideas in several important ways. They greatly improved the quantitative rigor of the approach, providing a meaningful mathematical background for the concepts. They also disassembled Gould's clock and created three axes from the two hands and age bar, allowing a clearer representation of the size–shape–age changes involved. Heterochrony could be precisely plotted as a point in space (Fig. 2-2). While conceptually useful, this scheme is very difficult to apply operationally, as discussed shortly.

They also designated three "parameters" to describe the changes. The *offset parameter* designated the time at which growth terminated. This thus refers to progenesis and hypermorphosis, which undergo negative and positive perturbations, respectively, as described above. A hypermorphic organism or organ (where dissociated) for instance grows for a longer time, such that size and/or shape becomes greater. The *rate parameter* refers to rate of change in shape, i.e., neoteny (negative perturbation) and acceleration (positive perturbation). Thus, like Gould, they did not explicitly allow for changes in the rate of size change, except in the relatively trivial case of size change with no shape change (proportional dwarfism and giantism). This is an important omission and is perhaps related to the somewhat artificial necessity of distinguishing between size and shape; an attempt to clarify this is discussed below. Finally, they added two new processes, pre- and postdisplacement, which are negative and positive perturbations of the *onset parameter,* respectively. In predisplacement the process (shape and/or size change) begins earlier; in postdisplacement it begins later. Gould (1977) had acknowledged that such changes in onset timing could occur but had subsumed them under rate changes.

2.3. Review of Heterochronic Terms

Based on the groundwork above, it is possible to classify all heterochronic changes, the entire morass of "processes" debated in the past, into just three kinds of basic changes: changes in *rate,* changes in *onset* time, and changes in *offset* time. As each change can either be an increase or decrease, a total of only six kinds of "pure" heterochrony can be discerned (Fig. 2-3): **neoteny** (slower rate), **acceleration** (faster rate), **postdisplacement** (late onset), **predisplacement** (early onset), **progenesis** (early offset), and **hypermorphosis** (late offset). Paedomorphosis ("underdevelopment" in descendant adults) is caused by the first of each preceding pair: slower growth (neoteny), a later onset (postdisplacement) or earlier termination of growth (progenesis). Peramorphosis is the opposite pattern, with opposite processes. Again, all these processes may be seen globally or only in local growth fields. Also note that late offset in one stage may cause late onset of the succeeding stage. Whether this is hypermorphosis or postdisplacement (or both) depends on the stages referred to.

Notice that offset time is usually associated with sexual maturation since most growth essentially terminates then. However, this is not always true. In organisms with indeterminate growth, there will often be a decrease in growth rate after maturation. More important, growth fields and tissues grow at different rates and terminate growth at different times, often not related to sexual maturation. For example, neural cells finish dividing long before sexual maturation. These cases do not invalidate the terms. They simply point out the importance of specifying what we are comparing. Thus, humans have delayed offset (are hypermorphic) of brain growth relative to apes even though neuron growth offset is not directly linked to sexual maturation.

In retrospect, it is no surprise that all heterochrony can be distilled into changes in rate, onset, and offset timing of trait change between ontogenies. If we consider any trait

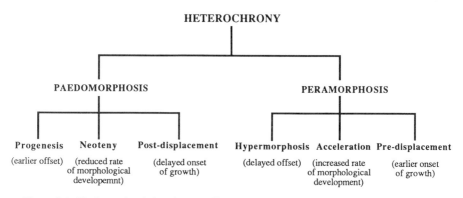

Figure 2-3. The hierarchical classification of heterochrony. Modified from McNamara (1986a).

change with age in an ancestral ontogeny, and assume that that trait continues to exist and change in the ontogeny of the descendant, what else can change but the rate and/or duration (onset and offset) of that change? From this perspective, we can see that heterochrony is similar to a derivative in calculus: it represents a change in a change, a (sometimes phylogenetic) change in the ontogenetic change of size and/or shape. Finally, we add here that the Alberch *et al.* (1979) paper made the key distinction between phylogenetic and morphological expression of heterochronic change. Thus, the morphological expression of peramorphosis involves the phylogenetic expression of recapitulation.

2.4. Two-Dimensional Model

To reduce operational difficulties, McKinney (1988a) presented graphical interpretations of the Alberch *et al.* (1979) scheme, which reduced the depiction of heterochrony to two axes, size or shape versus age, instead of size, shape, and age simultaneously. Thus, as shown in Fig. 2-4, onset time, offset time, and growth rate can be visualized as growth curves of traits, which can be measured in terms of size or shape, as a function of time (age). A main benefit of this approach is that it is much easier to visualize and analyze relationships on a simple x–y coordinate grid. Changes in growth rate are seen to be changes in the trait during a given period of time (i.e., changes in slope of curve), while changes in onset timing are shown as changes in the time at which the growth curves begin. Note that in this "pure" example the curves themselves are similar in shape but stop growth, i.e., level off, at the same time, so that the postdisplaced descendant is less developed, or paedomorphic, for that trait at the same ages. Finally, "pure" changes in offset timing alter only the time (and thus size or shape) at which the descendant curve levels off. Curves can be any shape, as discussed below, but will usually follow logistic patterns.

An example using mammalian data illustrates two major operational points (Fig. 2-5). The first is that body size (or shape) can be considered a "trait" just as well as any growth field within the body. Also, real curves often do not follow "pure" patterns, even when they are simple logistic forms: they combine changes in rate with changes in offset, for example. Changes in real growth are often more complex and do not follow simple trajectories. This is especially true in traits like body size and large growth fields wherein many smaller, changing growth gradients are subsumed.

What is lost in reducing the number of dimensions? The only loss of information is the axis which compared shape and size change together as a function of age. However, this is not much of a loss for at least three reasons. For one thing, changes in size and shape can still be compared simply by plotting the growth curves of both on the same time axis. This in fact is what the (also two-dimensional) clock model did in a less quantitative way. In addition, the whole subfield of allometry already exists for just this reason, the direct comparison of size and shape as discussed in the last major section of this chapter. We will see that much can be gained from comparing just these two in isolation from age (which is why the independently evolved subfield of allometry already exists in its advanced state). Third, the biological interpretability of the three-dimensional point in space is unclear to begin with. This is especially true given the difficulties

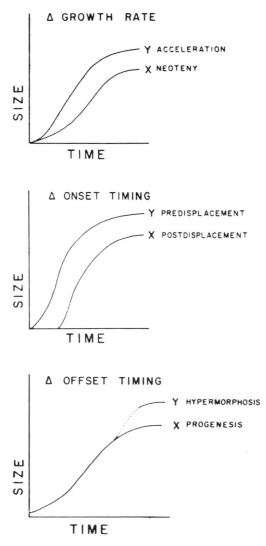

Figure 2-4. The six major types of heterochrony in terms of size versus age plots. From McKinney (1988a). The ancestral trajectory would be located between *y* and *x*.

of measuring or even defining shape, and even size in the comparison of ontogenies as discussed below ("the parameterization problem").

2.5. Summary

Figure 2-6 summarizes the different types of heterochrony using schematic drawings of hypothetical ontogenies. These are very useful in avoiding the problems of quan-

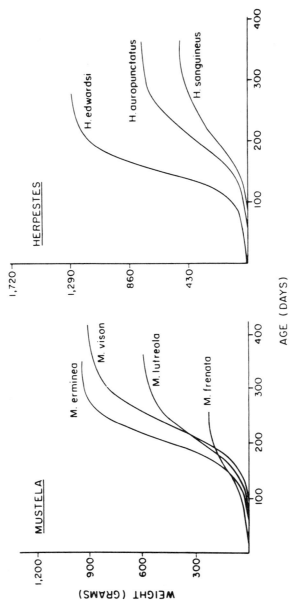

Figure 2-5. Carnivore body size growth patterns reflecting a variety of heterochronic processes. From McKinney (1988a).

tifying size and shape. Note that this figure is greatly modified from McNamara's (1986a) often-cited work which, unfortunately, is invalid for diagnosis of age-based heterochrony because it is based on size and not age. See Fig. 2-6 caption for discussion.

3. FINER POINTS AND PROBLEMS

On the surface, heterochronic classification is straightforward. One simply compares equal-age individuals of related species to isolate heritable alterations in their ontogenies, and using inferred ancestor–descendant relationships, assigns labels to the changes using the definitions above. However, in practice there are a large number of further considerations which can complicate analyses.

3.1. Problem of Shape and Size

The comparison of ontogenies is unavoidably tied to the concepts of size and shape. As long as subjective visual inspection of organisms is all one wants (e.g., left of Fig. 2-6), there is usually no problem. However, any attempts to quantify these useful but highly subjective notions can be difficult. Except in the simplest cases (such as sphericity for shape where axial ratios will do), size and shape require fairly complex multivariate methods to quantify them (discussion in Section 4.4). Even here some interpretability and information is often lost, although the methods can still serve as excellent ways to summarize the patterns. Atchley (1987) has referred to this difficulty of analyzing size and shape as the **parameterization problem**. For instance, in the classic paper by Alberch *et al.* (1979), they defined acceleration and neoteny as increases and decreases in the growth rate of shape, and proportional giantism and dwarfism as increases and decreases in the growth rate of size. Conceptually, this makes sense but in reality both body size and body shape are usually "accelerated" together and the distinction becomes confusing. We would simply say that size and/or shape are accelerated (further discussion below).

One solution to the parameterization problem is to simply analyze plots of age change in the basic components of size and shape. That is, changes in organ (or body) length, area, or weight can be plotted (e.g., Fig. 2-5 for body size but any local trait, such as limb length, could be plotted this way to show "acceleration" or other changes in that trait—see McKinney, 1988a, for discussion and examples). As Atchley (1987) has so well discussed, *organs are often best viewed as having their own, separate ontogenetic trajectories*. With such plots it is not only easier to see what is going on between the ontogenies, but because there is so much covariation in organic growth, much information is still summarized. Further, where dissociations do occur (covariation patterns change), they are more directly seen (e.g., skull length may be accelerated in one species while skull width is not).

A classic example of the hazards of qualitative shape inference, without such explicit dissection of change, is the so-called "neoteny" of human development relative to the apes (and by extension our ancestors). Literally hundreds of scientific articles mis-

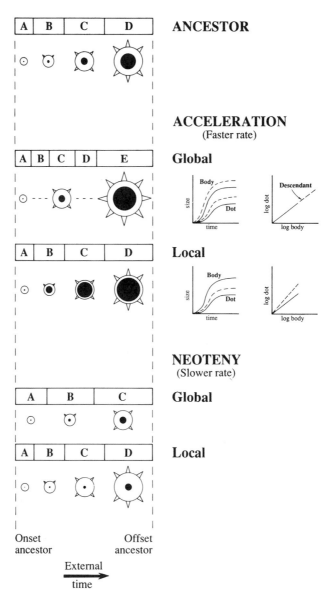

Figure 2-6. Summary of the six basic kinds of "true" (age-based) heterochronies. For comparison, allometric manifestations (size-based heterochronies) are also shown. Note how changes occur in morphology and "stages" (A, B, C, etc.). Note also how change can be either global or local. In local, only the "dot" organ is affected. In global, body size, spike, and dot organs are affected.

leadingly cite the notion that humans have a "neotenic" skull shape, illustrated by the similarity of the bulbous human skull with that of the juvenile chimpanzee (e.g., Gould, 1977). However, this superficial "shape" neoteny is brought about by different heterochronic processes acting on different growth fields. The bulbous cranium of the human is produced by prolonged brain growth (hypermorphosis). This is accompanied by a

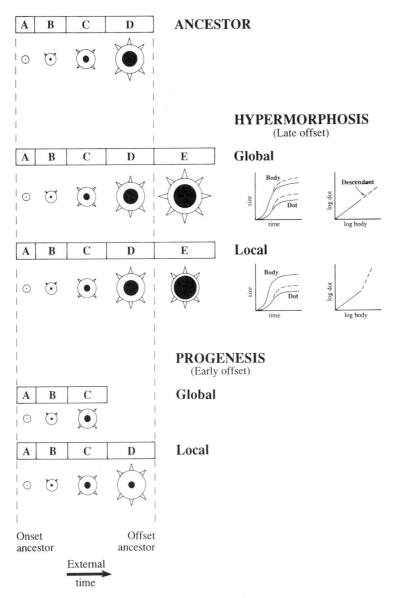

Figure 2-6. *(continued)*

complex of nonhypermorphic changes in the dental/mandibular growth field, reducing the prognathism (Chapter 7). To call the skull "shape" neotenous is to conflate pattern with process, and grossly mislead about what is really occurring. This problem is aggravated by imprecise thinking: neoteny (retarded rate) is often equated with hypermorphosis as "retarded" sexual maturation. Yet, as discussed later, delay in offset time is not a "retardation" or slowing of anything. It is a prolongation of *duration* of growth; *rate* of

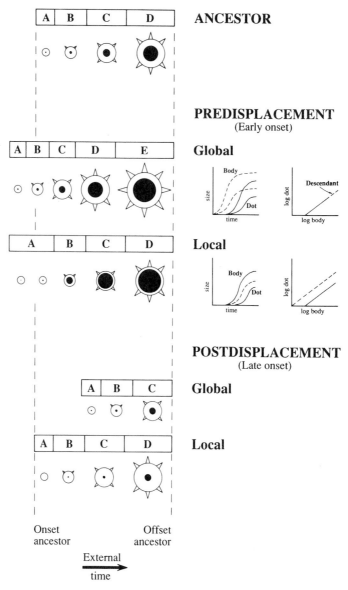

Figure 2-6. *(continued)*

growth during that delay is not specified (see also Shea, 1988). This point is closely related to the problem of complex growth curves discussed briefly in Section 3.3, but see especially Chapter 7. (Humans illustrate the problem nicely: we prolong the fetal phase, thus prolonging rapid fetal rates; delay is thus the ultimate cause. In so doing, just the opposite of "neotenic" rates occur.)

Even if we compare only simple traits among ontogenies, there are some important subtleties. While the most common metrics involve continuous quantitative trait

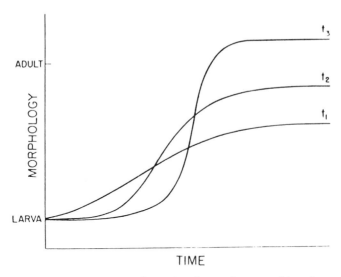

Figure 2-7. Suggested (by Wassersug and Hoff, 1982) evolution of metamorphic pathways in amphibians. Given that t_1 is the ancestral trajectory, such pathways illustrate: (1) the progressive divergence of adult and larval morphology ("overdevelopment" or peramorphosis) and (2) an increase in the rate of metamorphosis (acceleration). In this case the "trait" (y axis) is a suite of "qualitative" characters. Modified from Alberch (1987).

changes (e.g., Fig. 2-5), meristic traits are useful in some contexts. For instance, one may find a case of reduction in the number of trilobite segments, as discussed in later chapters. In addition to continuous and discontinuous quantitative examples, we must acknowledge the existence of continuous and discontinuous qualitative traits. For example, changes in coloration could be continuous yet difficult to quantify. Discontinuous qualitative traits are illustrated by ontogenetic changes in character states such as "level of development or differentiation." Thus, some amphibian evolution can be viewed in terms of primitive versus more derived metamorphic processes (Alberch, 1987). In such cases, "acceleration" (or hypermorphosis, etc.) of development can occur (Fig. 2-7). Again, these could be global or affect only local traits or trait suites; qualitative or quantitative; discrete or continuous.

In sum, any trait which can be either quantitatively or qualitatively characterized as a series of states having some vector of direction (e.g., size, level of differentiation) can be compared between ontogenies. However, the distinction between such dichotomies may be more apparent than real. For instance, quantitative and qualitative differences may reflect mainly the resolving power of the observer. Thus, the qualitative states of coloration could be translated into a very quantitative light wavelength scale if instrumentation is used. Similarly, degrees of "morphological differentiation" might be quantitatively characterized by cell spatial positions.

3.2. Trait Vectors and the Problem of Noncomparable Traits

Whether we use quantitative or qualitative "traits" to define the states in the ontogenetic comparison, there must be some meaningful vector or direction. As long as we can say that the descendant species has more or less of something than the ancestor, produced by faster, sooner, or later acting processes, we can make meaningful inferences. Statistically, we could say that quantitative states represent interval data such that exact values are involved, while qualitative states represent ordinal data wherein we can talk about more or less of something but cannot say how much more or less. In either case, the states can be continuous (e.g., width for quantitative, amount of pigmentation for qualitative) or discrete (e.g., number of segments for quantitative, larval stages for qualitative). In all such cases, we can meaningfully speak of acceleration, early onset, and so on of the various stages, e.g., acceleration of pigmentation, late offset of larval development. As noted, discreteness or qualitativeness is a matter of choice or technology. For example, in Fig. 2-6, we might choose to compare states A–D as discrete qualitative "degrees of development" (as in Fig. 2-7) or we could measure number of "spike" organs or width of "dot" organ (quantitative meristic and continuous, respectively). Related to this are "organizational" heterochronies. These are changes in rate or timing in the organization of parts into a whole, i.e., timing of connectedness. The human brain is a good example, and will be discussed in Chapter 7.

However, the statistician will also point out the existence of nominal data, where there is no vector, no sequence of more or less. This situation can occur in the comparison of ontogenies if the descendant species undergoes a change so deviant from the ancestor that it cannot be meaningfully related to ancestral states, i.e., it does not fit into the sequence. Alberch (1985a) discusses this problem of noncomparable states in some detail, making the crucial point that where ontogenetic sequences are not causal but strictly temporal, the order of ontogenetic states can be altered, making homologous comparisons impossible. For instance, the four-digited foot of a salamander species is not the same four-digited morphology as that which a closely related five-digited species goes through.

Nevertheless, we believe such noncomparable states are much rarer than comparable ones. The reason heterochrony is so abundant to begin with is that developmental events are so contingent upon one another that it is extremely difficult to significantly alter them. This conservatism means that ontogenetic trajectories will usually be altered in recognizable ways, leaving the vector of development intact. Furthermore, even where "new" structures appear, that are not directly comparable to ancestral states, the analyst can often assign a new meaningful vector of comparison. As noted above, such vectors as "degree of development" are highly arbitrary and qualitative in many cases to begin with.

Most importantly, even when incomparable novelties are generated, they are often still attributable to heterochronic processes. It is only that they are not translated in a simple linear (proportional) way to our scale of resolution (as extensively discussed in Chapter 3). For example, change in tissue induction timing can cause simple allometric organ changes classifiable in the traditional way but it can also result in incomparable novelties through major changes in cell interactions and resulting spatial juxtapositions. Yet the key causal process, change in timing of certain cell (and biochemical) level processes, can still be diagnosed in heterochronic terms (as change when compared to the ancestral events at that level). It is still a change relative to the ancestor whether it occurs at our convenient frame of reference or not. As another example, mammalian coat markings are generated in such a way that spotted and striped species show different (nonhomologous) developmental patterns in these markings. Yet, as shown by Murray (1981), these markings are created by a surprisingly simple set of "generating rules" of embryo size and chemical diffusion parameters. Simple timing changes in body size or rate of chemical diffusion (e.g., lowered, or neotenic, pigment production by cells) could lead to these novel states. If we cannot homologize macro-scale phenotypic states between species because of the fine scale of causation, it does not alter the fact that heterochrony may have been behind the changes. The phenotypic homologies are classificatory constructs to begin with, imposed by the observer. No doubt such usage as acceleration of chemical processes is outside the range of the original formulation of heterochronic terminology, but this reflects the past fixation with heterochronic expressions at our own scale. We need to focus on lower level scales as well as larger scales to understand the full process. A major point of Chapter 3 is that the cellular level serves as the nexus connecting lower and upper levels of the hierarchy. It is perhaps the best overall level to focus on for explanatory insight.

3.3. Problem of Complex Growth Curves and Sequential Heterochrony

The "pristine" logistic growth curves of Fig. 2-4 and 2-6 stand in contrast to the complexity often seen in close scrutiny of growth curves of real organs and individuals. Human growth for instance roughly resembles the logistic, but when we examine the **velocity curve** (gain per year) complexities emerge (Fig. 2-8). There have been many attempts to fit such growth curves with this or that equation. (An entire journal, *Growth*, is largely devoted to this endeavor.) In terms of heterochrony, this would provide a powerful descriptive tool as rate or timing parameters could provide exact comparative values. However, no single curve has proven fully adequate for the simple reason that cellular replication is not only variable through time but temporally varies in growth gradients across the body. For instance, organ allometries (from birth to adult) in the human body range from increases of 500% (pituitary gland) to 4000% (somatic muscles; Goss, 1972). It is true of course that a number of curves can be found to fit growth data. But this simply reflects the mathematical principle that virtually *any* set of points can be approximated by equations of *many* forms if three or more free parameters are allowed (von Bertalanffy, 1968), as most such equations do.

Most useful, therefore, are curves which have few parameters and are interpretable in a biologically meaningful way. The best hope in this direction seems to be multiphasic curves wherein the "phases" are based on biological phases of growth. Such curves show promise (Koops, 1986), especially versions that specifically incorporate growth field heterogeneity by the use of probability density functions (Piantadosi, 1987). Human body growth (weight, height) generally follows a triphasic pattern with these three phases: conception to about 2 years, 2 years to puberty, and puberty–growth termination. Among the differing growth field and somatic velocities characteristic of each phase, the first phase is the time of most neural growth and the last is the adolescent growth spurt of sex hormone-induced somatic growth (Fig. 2-8). As a result, the triple logistic is one of the most common curves applied to human growth, among many others (Harrison *et al.*, 1988). Again, such reification greatly oversimplifies the actual process even though the equation is quite complex.

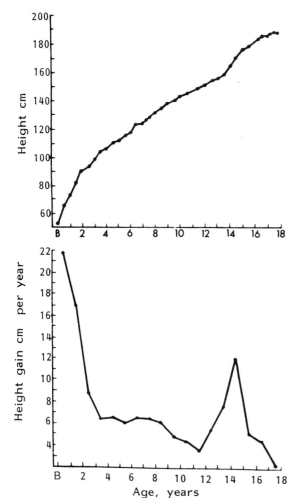

Figure 2-8. Growth in height (top) of a human male from birth to age 18 years. Bottom graph is "velocity" curve in height gain per year. From Harrison *et al.* (1988).

The point is that given such complexity, we must not expect even simple heterochronic changes, especially where only local growth fields are affected, to necessarily result in simple curve changes. This is not to say that analysis is impossible or even difficult. Local growth curves have already been discussed and serve to specify trajectories to compare. Another approach that to our knowledge has never been used, is the application of velocity curves such as the one in Fig. 2-8. These would allow direct specification of rate and timing changes. As an example, Shea (1983) elegantly showed that gorillas are generally produced by acceleration of late-state growth of the chimpanzee pattern ("body acceleration"—Fig. 2-9). There is little or no alteration of timing of growth. This is shown more clearly in the velocity curves of the two species: gorillas have a higher velocity (accelerated rate) at every part of the trajectory. Yet both follow the same timing pattern of growth to a surprising degree (same valleys and peaks on the velocity curve), indicating the timing of the various "spurt" and "lag" phases are the same (Fig. 2-9). Even more enlightening is the comparison of the velocity curves with the human velocity curve (Fig. 2-8). Humans have the same general curve shape as the apes (Fig. 2-8 is for height but the weight curve is very similar) except for a major shift in timing: the adolescent peak is at about 14 years versus 8–9 for the apes, diagnostic of hypermorphosis in us. We will discuss this further in Chapter 7, but for now note that each phase in our multiphasic curve shows late offset. This is, we show, "sequential hypermorphosis." Such **sequential heterochronies** may be common in highly integrated multiphasic development. See also the Glossary.

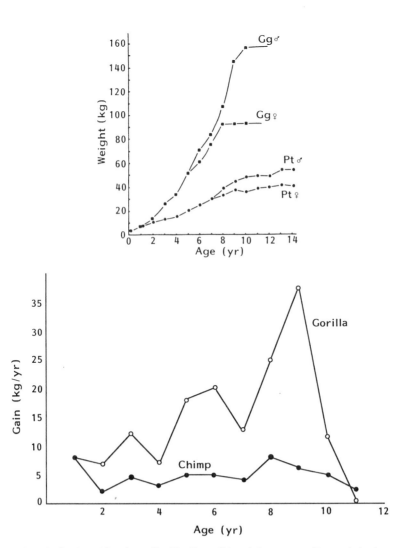

Figure 2-9. Growth of male and female gorillas (*Gorilla gorilla*) and chimpanzees (*Pan troglodytes*) as measured by cumulative weight gain and gain per year (males only). Cumulative curve from Shea (1983).

Finally, we note that precise curve shape may not be crucial for general comparisons. Where the purpose is to simply outline gross timing or rate changes, curve complexities may be overlooked, especially if curve shapes between ontogenies are similar.
One may thus speak nonrigorously of acceleration or delayed offset (for example) in a
descendant. As in many scientific analyses, the degree of precision depends on the goals
of the investigator.

3.4. Problem of Determining Ontogenetic Age

Changes in rate and timing of ontogenetic events can, by definition, only be determined where the age of compared individuals is known. More graphically, the trajectory
of a trait must be plotted as a function of age (external chronological time). In living
organisms, this should pose few problems, although surprisingly few studies so far have
done this. Instead they rely on allometry alone, looking at changes in a trait as a function
of body size. As will be discussed below, this is accurate as long as body size in compared
species grows at similar rates. However, this is demonstrably not always the case even in
many closely related species (e.g., Shea, 1983; Emerson, 1986; Jones, 1988).

For neontologists, this simply calls for incorporation of age data into their studies
(admittedly, such longitudinal, instead of cross-sectional, data are often very difficult to
obtain, even in neontology). However, paleontologists obviously face a much greater
problem. Jones (1988) has discussed this in some detail, outlining methods to obtain
ontogenetic age data from fossils. Growth increments in shells, tests, and bones seem to
offer the best evidence. Such study is especially well developed in molluskan taxa, using
isotopic signals. Very promising techniques for age determination in vertebrates using
bones and teeth have not been much explored. One notable exception is the emerging
use of tooth enamel increments in hominid fossils (Chapter 7).

Also, Benton (1988a) has discussed the growing recent interest in age in a number
of fossil vertebrate studies. Shape of certain skeletal elements and degree of ossification
have been used to infer developmental stages at different sizes. For instance, some fossil
pelycosaurs, once thought to be juveniles on the basis of small size alone, are, upon
closer study, adults (size paedomorphs). This was inferred from the change in shape of
the ends of limb bones from concave to convex when juvenile cartilage was replaced by
bone. Similarly, small dinosaurs were shown to be adults on the basis of knee width
allometry: large dinosaurs have large knees as juveniles in anticipation of larger adult
needs (as with large-footed puppies). Dwarf dinosaurs have relatively small knees at
small sizes. However, this whole approach, using shape and ossification, is still incomplete. While it is an improvement over the simple use of size to estimate age, it (so far)
has been limited to recognition of adults at a given size (size at sexual maturation,
generally speaking). This is still insufficient for diagnosis of the different kinds of
heterochrony because absolute age is still missing. For example, a dwarf adult dinosaur
may have attained its small (paedomorphic) size either through growing slower (neoteny) and maturing at the ancestral age or growing at the ancestral rate but maturing
earlier. Refined analysis of growth increments in bone might be a way to improve on this
(Jones, 1988). Similarly incomplete data are common in many groups, for example
echinoids, where the size at which genital pores appear, indicates size at maturation
(McKinney, 1984, 1986).

3.5. Problem of Determining Phylogeny

Ancestor–descendant relationships are obviously crucial in determining heterochronic changes among taxa. Fink (1988) has advocated the use of cladistic methods for phylogenetic reconstruction in a heterochronic context. Schoch (1986) has thoroughly reviewed the use of cladistics (and other methods) in a paleontological context. This is a problem that goes far beyond heterochronic inference so we defer to those sources and others, simply pointing out that the investigator use whatever lines of evidence are at hand to reconstruct phylogenetic relationships. Since virtually all evolutionary events are analyzed ex post facto, the evidence is inevitably circumstantial and therefore subject to dispute.

A major point is that significant heterochronic reconstruction can be done even where phylogeny is uncertain. If comparison of two species shows that one has relatively more rapid ontogenetic development of a certain trait, we may infer that there has been a change in rate of growth although we do not know if the relatively accelerated species is the descendant (in which case acceleration occurred) or the ancestor (in which case neoteny occurred). However, just knowing that a rate change has occurred, i.e., that one is relatively accelerated to another, can often be of more theoretical interest than historical reconstruction itself. For example, because we often know something about the habitat of each species, we can make useful observations concerning the cause of the rate differences between the species (Shea, 1983; Emerson, 1986). Thus, we may inspect why an organ or organism is *relatively* accelerated in an environment and another is not. Is it a matter of selection on form and function, or life history, or is it a matter of developmental constraint (e.g., McKinney, 1986)? This point, that the polarity of comparison itself yields very useful information, may be obvious but it is sometimes missed by workers whose focus is fixed on historical reconstruction alone. This is discussed further in Chapter 5.

3.6. Problem of Determining Heritability of Changes

As with age determination, this problem can be a nonproblem when dealing with living organisms. Again, however, the rich mine of heterochrony in the fossil record presents greater difficulties, although this is not limited to heterochronic investigations.

The first question is: are we sure that nonheritable ontogenetic (e.g., ecophenotypic) changes can mimic the genetic ones of heterochrony? The answer is a resounding yes, as shown by an elegant study by Meyer (1987). As seen in Fig. 2-10, cichlid fish underwent pronounced acceleration or neoteny of shape, depending on their diet. This important paper is a major attempt to unify heterochrony with such plasticity, and includes a good discussion of heterochrony as intraspecific variation as also discussed later in this book. As noted above, the use of "shape" to diagnose heterochrony can be misleading, but Meyer's use (Fig. 2-10) seems justified. Here "shape" includes a whole suite of clearly functional changes in the jaw and not just a superficial form similarity. In a more general sense, it is well known that organisms in stressed environments grow at a slower pace and may reach smaller sizes (i.e., can mimic body size neoteny).

There are two lines of thought relevant to this problem in heterochronic studies in the fossil record. One, while organisms may undergo environmentally induced morpho-

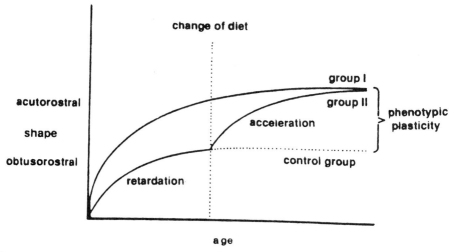

Figure 2-10. Heterochronic phenotypic effect of diet on intraspecific ontogenetic trajectories of cichlid fishes. From Meyer (1987).

logical changes for some time, they often undergo compensatory growth in the latter stages of ontogeny so that the genetically programmed morphology is realized (see Atchley, 1984, for discussion). Thus, even though environmental conditions may cause a temporary acceleration or slowing of growth rates (for example) relative to the genetically programmed rate, the adult form may ultimately converge on that programmed for anyway because as proportionately lower (or higher) rates act later on. However, this is often contingent on a change to more favorable environmental conditions. Further, such compensatory growth seems to be less common in invertebrates than vertebrates. Vermeij (1978), among others, has discussed how changed environmental conditions early in life can alter the entire ontogenetic trajectory even if they return to "normal." For example, shell coiling parameters in gastropods are affected by growth rate and these parameters, in turn, determine the trajectory of later shape change (slow growth leads to a larger dome).

Two, reports of stressed faunas and other types of recognizable faunas in the fossil record are rare (LaBarbera, 1986). This is not because they are impossible to observe, since reported cases are clear enough (e.g., Mancini, 1978). However, given the high abundance of ecophenotypic dwarfing in living invertebrates (Tissot, personal communication and in preparation), one wonders if paleontologists are often misdiagnosing it as genetic change; either way it is very difficult to prove. Sometimes there are clues to aid the investigator. For example, McKinney *et al.* (1990) observed living specimens of a related species to obtain information on the ontogenetic timing and growth rate of a peramorphic organ in an Eocene echinoid. In addition, they used four other lines of evidence, such as the fact that the heterochronically altered state was diagnostic of some living genera.

3.7. Nomenclatural Misconceptions

It is very often said that maturation (or even death) is "accelerated." This is not wrong but it is misleading because of its imprecision. It is much more accurate to say that a life history event is early (or late, rather than retarded). The reason is that acceleration and retardation are rate concepts and onset and offset of life history events are timing concepts.

It is also sometimes stated that progenetic organisms tend to be smaller than the ancestor while neotenic ones are larger. While it is true that globally progenetic forms will be smaller, since body growth terminates with the earlier maturation, it is also true that globally neotenic (and postdisplaced) forms will be smaller if they mature at the usual time (since they grow more slowly, or start later for postdisplaced). This misconception is widespread because slower growth (neoteny) is wrongly synonymized with delayed maturation (hypermorphosis). This is a good example of the imprecise terminology just noted since it seems to arise because "retarded" growth is assumed (wrongly) to generally correlate with "retarded" maturation, i.e., conflation of a rate with a time concept. In fact, there is no firm evidence that slow growth is even usually correlated with any change in duration of growth. This situation reaches its ultimate absurdity in human heterochrony where a truly alarming number of authors state the oxymoron (we paraphrase) that "neoteny results in *faster growth* by prolongation of the rapidly growing fetal growth stage." We hope to clarify (in Chapter 7) that the true situation is that humans are generally (and quite simply!) hypermorphic and this delay extends to all of the (roughly three as noted above) major human growth phases. Rapid and slow phases alike, the end of each phase is delayed. The point is that heterochronic jargon is useful and not misleading as long as we are precise in our statements.

4. SIZE-BASED HETEROCHRONY

Allometry ("alloios" is Greek for "different"; literally "different measure") is the study of size and shape. It has been very effective as a describer of change, but it has not been particularly helpful as an explainer. This is somewhat ironic in that Huxley (1932) originally developed the concept and original techniques for the purpose of exposing underlying rules of growth. However, the rules are not as simple as he had hoped, owing largely to the composite nature of growth fields discussed above (and especially Chapter 3). Indeed, he realized this early, having originally used the term "heterogony" in 1932 to describe the process of differential growth but changing to "allometry" in 1936 (Huxley and Tessier, 1936) to connote empirical measurement when discussing relative growth. Nevertheless, those rules are becoming more fully exposed, if more slowly than anticipated, and heterochrony plays a major role. Indeed, allometry and heterochrony can be conceptually linked quite easily since heterochrony addresses trait change relative to time and allometry views trait change relative to other traits. In short, allometry compares heterochronic results to one another; it is the pattern created by heterochronic processes. Hence, in a sense, allometry is size-based heterochrony, even though this is a nontraditional way of perceiving it. Since allometry is based on relative change, it is sometimes difficult to infer the underlying absolute (temporal) changes, but this view has its uses. It is especially important in directly observing the changes of morphologically local shape "components" discussed above.

In this section, we first sketch the mechanics and basic concepts of bivariate allometric techniques that are in use. Next, we relate these to the heterochronic concepts discussed above. We then propose the use of heterochronic terms to allometric patterns, with the express caveat that they may not reflect the "true" heterochronic processes involved. Finally, we outline some multivariate methods for allometric analysis and their role in these concepts.

4.1. Bivariate Allometry

Huxley's (1932) allometric formula, $\mathbf{y} = \mathbf{b}\mathbf{x}^{\mathbf{K}}$, relates quantitative changes in trait x (often body size) to those of trait y as a power function (review in Gould, 1966). $\mathbf{K}$ is the ratio of specific growth rates and $\mathbf{b}$ is a coefficient that is operationally defined, although it may have some biological meaning as the "geometric similarity" ratio (White and Gould, 1965). The reasons why this simple function is generally such a good fit to biological growth are related to heterochrony and are discussed below.

Computation of the parameters on bivariate trait data is usually done with log-transformation. Since the equation may be rewritten as:

$$\log y = k(\log x) + \log b$$

we may perform linear regression on bivariate plots of log-transformed data (log x versus log y) so that the slope will estimate k and the y intercept will estimate log b (so b is the antilog of that value). Although this may result in parameter values different from those obtained if the untransformed data points were fitted directly with a power function (Zar, 1968), the difference is almost always slight and linear regression is computationally much easier and reliable, even with computers.

Least-squares linear regression is usually employed by most curve-fitting programs but, unfortunately, this is not appropriate for morphological (and many other kinds of) data. The reason is that the least-squares algorithm fits a line which minimizes the vertical deviations to the points. This is appropriate where x is a truly "independent" variable (i.e., it assumes all error is due to variable y, and x is measured without error). However, since morphological traits are all part of the same growth system, none are really more "independent" or more fully determined. Thus, a curve-fitting technique is needed that acts more symmetrically in estimating the line, treating both variables equally. Reduced major axis (RMA) is frequently used since it minimizes the deviations of areas perpendicular to the line (Fig. 2-11). Even though most computer programs do not use RMA, the RMA parameters are readily derived from other output. The RMA slope is equal to the standard deviation of y over that of x, and the RMA y-intercept is equal to: mean of y − (mean of x) (slope). (For a fuller discussion see Imbrie, 1956; Davis, 1986; B. Jones, 1987).

A computational shortcut is that the RMA slope is equal to the least-squares slope divided by the correlation coefficient. This also shows that where correlation coefficients are high, the curve-fitting algorithm used is not that important, since the RMA slope (and intercept) will be nearly the same as the least-squares value(s), where the coefficient is near one. A high correlation coefficient is very common since we are correlating parts of the same organism, i.e., "growth system"; even more, both parts form time series data,

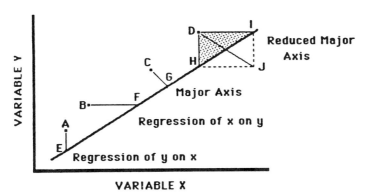

Figure 2-11. Schematic illustration of the different quantities minimized by various curve-fitting techniques. Unlike least-squares which minimizes the linear distance and does not act symmetrically, reduced major axis minimizes the area of triangles symmetrically about the line. Modified from B. Jones (1987).

which tend to correlate more often among themselves because succeeding states are not as free to vary as in nontime serial data (McKinney and Oyen, 1989). Where parameters must be calculated with extreme precision, e.g., medical research, an even more rigorous, though somewhat complicated method of curve fitting to allometric data has recently been developed by Mattfeldt and Mall (1987). A common problem is small sample size and nonnormal distributions. Plotnick (1989) has shown how "bootstrapping" methods can greatly improve RMA estimates in such cases.

Once the parameters are computed, the line may be drawn through the plotted points. We recommend plotting log-transformed values on an arithmetic scale rather than the mathematically equivalent plotting of raw values on a logarithmic scale since the latter may crowd points and obscure important visual relationships (McKinney and Schoch, 1985). In many cases, the investigator wishes to compare the slopes and intercepts of the shape changes (regressions) of different species or other grouping (especially where heterochrony is to be inferred, Fig. 2-12). There are a number of statistical tests of significance (Imbrie, 1956; Davis, 1986; B. Jones, 1987). Most are based on the Z-score.

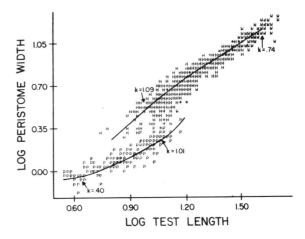

Figure 2-12. Allometric plot of ontogenetic shape change in three fossil echinoid species: *Oligopygus phelani* (P), *O. haldemani* (H), and *O. wetherbyi* (W), from the Eocene of Florida. k = specific growth ratio, which changes as seen in curvilinearity of trajectories. *Wetherbyi* is allometrically hypermorphic to *haldemani*; whether it is "truly" hypermorphic is not known. From McKinney (1984).

If x and y are measured in the same units (e.g., mm, kg) and the fitted curve has a slope (k) greater than one, then there is **positive** allometry between the traits, meaning that y becomes relatively larger than x with growth. The converse of course happens with **negative** allometry, k less than one, so that x becomes relatively larger. Where k equals one, **isometry** occurs and no shape change is seen with size increase. Sometimes, k is not constant, indicating that the specific growth rates being compared are not changing in concert (usually termed **complex** allometry). This will result in a curvilinear log–log plot which can be fitted by a polynomial equation. Taking the first derivative of this equation will allow an estimate of the slope (and hence k) at any point (i.e., size of x and y) to be made (Fig. 2-12); see McKinney (1984) for further discussion.

Where x and y growth is compared in the same individual at different ages, the ontogenetic curve is **longitudinal** (see Cock, 1966, for fuller discussion on types of allometry). **Cross-sectional** comparisons include trait measurements on different individuals (of either known or unknown age) spanning a significant part of the ontogenetic sequence. **Static** comparisons also observe different individuals but these are of similar age, such as adults, so that only a small part of the ontogenetic series is seen. Obviously, cross-sectional, and especially longitudinal, data provide the most complete estimates of a species' ontogenetic shape change. Longitudinal data are most desirable because they not only cover a broad part of the ontogeny but do not mix the trajectories of different individuals. However, owing to the practical difficulty of obtaining longitudinal data, cross-sectional studies are more common (e.g., Fig. 2-12), even in studies of living species. Species vary greatly in how much individuals follow similar trajectories. Hence, the "noise" introduced by such mixing varies greatly as well. Sometimes it is necessary to look at interspecific static adult comparisons, especially in fossil vertebrates, because fuller data are often not available. In such cases, the estimated k and b parameters almost certainly vary from those of more complete data since much of what we observe is size-variational "noise" among adults (Cheverud, 1982). However, if interpreted with caution, some meaningful inferences can be made (Gould, 1974, for the "Irish Elk"; McKinney and Schoch, 1985, for titanotheres).

4.2. Relating Allometry to Heterochrony

We concur with Blackstone (1987a and 1987b) that allometry and heterochrony can be best understood, both separately and in conjunction, by focusing on the cellular level. Rate changes in particular can be interrelated. Katz (1980; see also Laird, 1965) showed how a model of cell growth (binary fission) could produce the relationships of the allometric equation such that b represents the number of centers of cell division (germinal centers) and k is the relative frequency of cell division in parts x and y. However, application of these ideas must be made with care. For example, Kember (1978) argues that different rates of growth in long bones is more closely tied to number of cells in the proliferative zone than to rate of division itself (see Shea *et al.*, in press, for further discussion).

Nevertheless, it is clear that change in rate of mitosis and, therefore, rate of specific growth is an important way of altering shape and thus provides a good relationship between allometry and heterochrony. We begin with the allometric relationships (the manifestation of change) and work backwards to isolate the processes leading to that phenotypic pattern. Consider that the allometric equation is actually the solution to the differential equation:

$$dy/dty = (k)(dx/dtx)$$

(See von Bertalanffy, 1968, for the three-step derivation from this to the allometric solution.) This shows that Huxley's derivation was based on the notion of self-multiplication of living tissue (change in each trait standardized for both time and amount of tissue, called the specific growth rate). Importantly, the constant k asserts that there is an invariant, linear relationship (i.e., proportional, not necessarily equal) between mitotic activity in the two traits. This simplicity is perhaps surprising given the complexity of growth, with gradients and complex growth curves as discussed above (and see especially Chapter 3). No doubt this is because of the highly integrated systems (e.g., hormonal) of late stage growth (and the large majority of evolutionary allometric work is on just such "late," postnatal aspects). As noted, complex allometry (Fig. 2-12) does occur, showing that the relative mitosis is not always constant, but even here, change is generally smooth and clearly "controlled" in some fashion.

By rearranging the above equation, we can also see that k is the ratio of specific growth rates of x and y:

$$k = (dy/dty)/(dx/dtx)$$

To the allometrician interested in heterochrony, this has troublesome implications. Recall that heterochronic changes are expressed in trait versus time plots, so that dy/dt or dx/dt would be the ancestral ontogenetic rate of growth and an increase or decrease in that rate in the descendant would represent acceleration or neoteny, respectively. This leads to the problem that the allometric slope of the log–log plot, i.e., k, can be altered by changing the rate of growth in *either* y or x. Thus, if a descendant species has a higher allometric ontogenetic slope, it may have been obtained by: (1) increasing the growth rate of y and not altering x (acceleration of y, as might be initially intuited), (2) decreasing the growth rate of x (neoteny) without changing y, (3) accelerating y more than x, (4) accelerating y and retarding x, (5) retarding both x and y, but y is more neotenic, and so on (with the converse occurring for slope decrease). The point of course is that only relative rate changes are shown by the allometry but absolute changes are involved in the heterochrony. Therefore, classifying heterochrony on the basis of shape change, or relative change in sizes, alone can be misleading (Fig. 2-13).

A corollary to the above is that if growth rate of x remains at least roughly the same in the species being compared, then the heterochronic inferences will be correct. Any increase or decrease in slope (k) will of necessity result from an increase or decrease in the specific growth rate of y, respectively. Therefore, x can serve as a proxy for time against which change in y can be compared. Hence, allometric "acceleration" will mean growth rate acceleration. Further, timing change can also be inferred since larger values

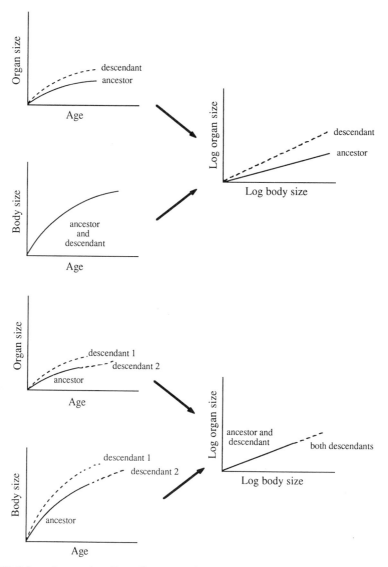

Figure 2-13. Schematic examples of how allometry can be related to heterochrony. Top: acceleration of organ growth with no change in body growth results in allometric acceleration. Bottom: acceleration of both organ and body growth (descendant 1) results in allometric hypermorphosis. The importance of knowing ontogenetic age is shown in both examples. In the top case, the same allometric acceleration can be produced by no change in organ growth if body size in the descendant slows (relative to ancestor) but organ size does not. In the bottom case, prolonged growth (hypermorphosis) of body and trait in descendant 2 produces the same allometric hypermorphosis as does body and trait acceleration of descendant 1.

of x must represent longer times of growth [x = rate(time of growth) = $(x/t)t$; since rate is the same, then t must be greater for larger x]. However, we cannot say whether the longer time was from earlier offset or later onset time (or both).

Body size is often used as the x trait in many allometric regressions. (More accurately, some partial metric of "body size"—a multivariate notion, such as body weight or body length—is used. This is generally acceptable because most of these multiple components of body size are highly covariant.) If both body size and trait size are accelerated by similar amounts in a descendant, the resulting allometric plot will show a larger organism which follows the same allometric trajectory, but extends beyond the smaller species (Fig. 2-12). This will mimic the plot which would occur if the descendant simply grew for a longer time at the same rate (hypermorphosis). Acceleration of body and trait or hypermorphosis of body and trait will show the same plot. Since the Alberch *et al.* (1979) scheme considered only rate change of size *or* shape growth and not simultaneous size *and* shape change, Shea (1983, 1988) distinguished "rate hypermorphosis" (and "rate hypomorphosis" = rate progenesis) resulting from faster body and trait growth from "time" hypermorphosis and progenesis resulting from longer time of body and trait growth. [The separation of size from shape rate changes by Alberch *et al.* is all the more curious since: (1) size rate changes usually cause attendant shape changes for reasons of both selection and genetic covariance and (2) they did allow for concurrent size and shape changes in timing.] In any case, we do not use Shea's terminology because his "rate" hyper- and hypomorphosis are not changes in offset time at all. Rather, they are changes in rate only, of body size with attendant shape change. His use of offset terms is due to the allometric plot: rate hypermorphs grow larger, i.e., greater offset *size,* but heterochronically this is still acceleration and the use of offset timing terms will likely confuse people. Instead we suggest that such cases be called body and trait acceleration, where the allometric trajectory is extended by faster growth, and body and trait hypermorphosis where the same rate of growth is extended for a longer time (Fig. 2-13).

So far we have shown that allometric depictions of heterochrony can mislead where *x* variables differ in rate of growth. Therefore, one must ask how common such variations are in these variables (especially body size) among closely related species (those most likely to be used in ontogenetic comparisons of heterochronic analysis). While surprisingly little work has explicitly addressed this question, the answer seems to be that variations of rate are quite common. The first indication of this was Shea's (1983) excellent study which showed that the gorilla is in many ways an "accelerated" chimpanzee. It grows faster, including body size, but not for longer time periods (Fig. 2-9). Emerson (1986) noted this in frogs. This is also seen in a compilation of mammalian age data (Gittleman, 1986; Fig. 2-5). Perhaps the most startling example is from Jones (1988) who shows that similar-sized bivalves can differ in age by 100 years or more! The most complete view of this is seen in Fig. 2-14 which shows that mammals of the same mass can mature at times differing by a huge factor, of *ten* or more.

We emphasize this point because a myriad of articles and books have stated that larger organisms grow longer. This stems from Bonner's (1988, most recently) plot, cited and republished by many authors, of length versus generation time showing a strong general correlation. This correlation is indeed real and is in fact also shown in Fig. 2-14. The problem, however, is one of scale. The general correlation of both curves occurs because animals of vastly different sizes are compared (Bonner's is most drastic of all, being a "bacterium to whale" scale). This indicates a real physical constraint on the time it takes to ontogenetically "build" progressively larger animals. The "noise" (**vertical allometry**: Calder, 1984) of these plots, however, shows that body size is a very poor judge (worthless really) of growth period for most related (and even unrelated) animals. Larger animals not only do not always grow longer, they often mature earlier (e.g., in Fig. 2-14 some 60,000-g mammals mature in about 400 days while some much smaller,

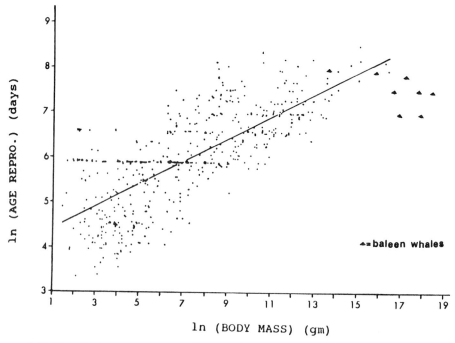

Figure 2-14. Plot of body mass versus age at first reproduction in 547 mammalian species. Modified from Wootton (1987).

1000-g mammals mature in as long as 1000 days). The gross physical constraints are thus modified by numerous secondary ontogenetic processes, often involving hetero- chronic mechanisms of rates of mitosis, hormonal control, and so on. These secondary mechanisms are of course ultimately the result of local environmental selection upon them. Note how major these "secondary" and "local" factors can be. In the example just cited, the earlier-maturing larger mammals are over 50 times larger yet mature in less than half the time. Such a massive deviation is far too large to be dismissed as "noise" and must represent very important developmental information. (Admittedly, such ex- treme cases usually do not represent closely related forms, but the well-documented chimp–gorilla example indicates that jumps of a factor of 2–3 are easily possible when comparing adult size of chimps versus gorillas. Further, rate and timing changes ulti- mately underlie differences among even distantly related forms, especially in such basic hormonally controlled events as maturation and somatic growth which have been very conservative in their essential mechanisms.)

4.3. Allometric Heterochrony

Given the incomplete picture offered by the allometric plots, it is clear that onto- genetic age is necessary to confidently discuss the types of heterochrony involved in changes among species. This means that the attempts of McKinney (1986) and McNamara (1986a) to infer heterochrony from allometric plots and size–shape relations alone are not valid. Nevertheless, allometric plots are of themselves very informative for many purposes. Description of shape change among traits is of course one use, but many have also argued that size itself (usually body size) is a useful and valid variable against which to plot trait changes (e.g., Calder, 1984; Peters, 1983). One major reason for this is that body size is a metric of an organism's "internal" age which is often more telling than "external" time. This is because many, perhaps most, ontogenetic events are *size*-specific rather than age-specific. That is, they are "programmed" to occur at a certain size rather than age (see Werner and Gilliam, 1984, for thorough review). This has led to ecological and evolutionary studies of size-structured rather than age-structured populations (Ebenman and Persson, 1988). Therefore, depending on the goals of an investigation, trait change as a function of body size may be as (or more) interesting or informative than trait change with time. There is also the matter of overwhelming abundance and simple practicality: bivariate morphological plots are ubiquitous, being used in all kinds of contexts, from taxonomic to evolutionary objectives. Blackstone and Yund (1989) provide an excellent comparison between size-based and age-based heterochronies.

Because of the sometime utility of the size-based view, McKinney (1988a) proposed the application of heterochronic terms to allometric plots with the strict understanding that they may not reflect underlying "true" heterochronies. These **allometric heterochronies**, as shown in Fig. 2-15, conform to the heterochronic classification discussed above with the difference that the x variable is body size instead of time (age). Thus, allometric progenetic species grow to a decreased offset size and not offset time, accelerated species show greater trait change per unit body size and not per unit time, and so on. (The trajectories are of course rectilinear because of the log–log nature of most allometric graphs but they need not be for complex allometry.) These terms are given in Table 2-1, with the diagnostic criteria, along with synonymous terms (from Shea, 1988). As noted, where rate change in body size is similar between the two species, these plots may also reflect the true heterochronies involved. But unless we can show this, we must use these as only "agnostic" labels of convenience and measures of size-determined ontogenetic events.

The same caveats apply to the larger categories of paedomorphosis and peramorphosis. We can observe size ("allometric") paedomorphosis where the descendant resembles the ancestor when the latter is at a smaller size. But unless we have an age marker (or perhaps a measure of sexual maturation but even here maturation may occur at a different age), we cannot be sure that "true" paedomorphosis occurred. For this, the descendant must resemble the ancestor when the latter is at a younger age. As always, the converse applies to peramorphosis. A practical way around this is to rely on a "loose" definition of paedomorphosis: the descendant adult resembles the ancestral preadult (in size or shape), regardless of age. Markers of sexual maturation (e.g., echinoid gonopore appearance) would then suffice.

4.4. Multivariate Allometry

While "size" and "shape" are conceptually simple concepts, they are very difficult to precisely define and analyze (Bookstein *et al.*, 1985). Nevertheless, there are a number

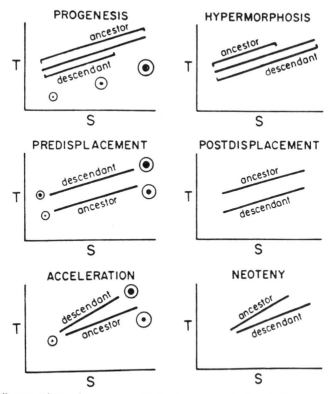

Figure 2-15. "Allometric" heterochrony, as classified by the ontogenetic plots of related species. S = body size (weight, length, and so on) and T = trait measurement. S and T may or may not be log-transformed. Light circle schematically illustrates body size, dark circle is a "trait" (e.g., organ size). From McKinney (1988a).

of multivariate methods which can assist as describers of size and shape change by providing summary variables that resolve the many traits into a few. However, we quickly point out that such approaches by their nature often lose in resolution of detail what they gain in synthesis, so that care must be taken. Often the choice of bivariate versus multivariate analysis depends on the goals of the analyst. Indeed it is often

Table 2-1. Terms of Allometric "Heterochrony" as Defined by Ontogenetic Trajectories of Descendant Species Relative to Ancestral[a]

	Trajectory of descendant species relative to ancestor		
Allometric term	Slope (or curve shape)	y-intercept	Adult body size
Allometric progenesis (= ontogenetic scaling)	Same	Same	Smaller
Allometric hypermorphosis (= ontogenetic scaling)	Same	Same	Larger
Allometric neoteny (= shape dissociation)	Lower	Same	May vary
Allometric acceleration (= shape dissociation)	Higher	Same	May vary
Allometric predisplacement (= shape dissociation)	Same	Greater	May vary
Allometric postdisplacement (= shape dissociation)	Same	Less	May vary

[a] These are illustrated in Fig. 2-15. Synonyms (parentheses) are from Shea (1988).

(perhaps usually) best to carry out both to obtain a complete view. Importantly, as long as the original data consist only of morphological variables (as they usually do), they may tell us little (if anything) more in multivariate than bivariate form when it comes to inferring the underlying age-based processes of heterochrony.

By far the most commonly applied multivariate methods, in heterochronic contexts, are principal components analysis (PCA) and the related technique, factor analysis (FA). These have been much discussed, in many places, so our goal here is to provide only a general outline of multivariate analysis and, especially, relate it to bivariate methods of allometry, and heterochrony. For more in-depth treatment in this context, the reader is referred to Shea's (1985a) excellent review and especially Tissot (1988). (Recently, shape analysis by Fourier and eigenshape analysis has also been used; for discussion in a heterochronic context, see Foster and Kaesler, 1988).

Figure 2-16 illustrates the basic operation of PCA and FA, which is to fit lines (variously called "axes," "factors," or "components") through the cloud of points representing ontogenetic trait measurements. In the example, only three traits are shown for reasons of clarity and to show the relation to bivariate comparisons, but *n*-dimensional comparisons are possible, for *n* traits. The technique acts to resolve the variation distributed along all the *n* axes into one or a few axes. Much as a bivariate regression line "summarizes" change between two traits, the multivariate axes summarize change among many traits, often in biologically meaningful ways. Mathematically, these axes are fitted by factoring a matrix of correlation coefficients or covariances of the traits. High correlations and covariances among all the traits (e.g., Fig. 2-16) will result in a first axis which accounts for ("fits") most of the variation among the traits. Graphically, this amounts to an axis which traverses the longitudinal part of the cloud, where most of the variation resides. The second axis is perpendicular to the first axis (in PCA and some forms of FA) and it accounts for the second highest amount of variation. Third, fourth,

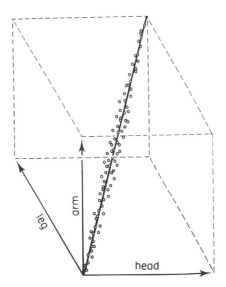

Figure 2-16. Three-dimensional graph showing correlations for three measurements. Line through longitudinal point cloud axis is the "size-determined" axis. From Gould (1981).

Table 2-2. Results of Shear-PCA on 206 Black Abalone[a]

	Eigenvectors		
Variable	PC1	PC2	PC3
Length	0.42	0.09	−0.05
Width	0.42	0.04	−0.08
Height	0.41	0.20	−0.10
Tremata number	0.23	0.74	0.16
Tremata length	0.42	0.14	−0.13
Tremata spacing	0.36	−0.31	−0.48
Tremata diameter	0.34	−0.52	0.84
Eigenvalues	5.49	0.88	0.34
Percent variation	78.4	12.6	4.9

[a] From Tissot (1988).

and higher axes can be fit as long as there is sufficient variation to be explained. In general, axes fitted after roughly 90–95% of the variation has been accounted for, are picking up biologically uninterpretable "noise."

The relationship of each trait to the axes is given by the eigenvectors or "loadings." These will vary from 1 to −1 because they represent the cosine of the angle between the fitted axis (e.g., first axis) and the vector of each trait. Thus, if the first axis explains most of the variation, as in Fig. 2-16, the loadings will all be near 1 since each trait deviates little from that axis (1 being the cosine of $0°$, indicating little deviation; 0 indicating $90°$ deviation). Importantly, even where traits are not so tightly associated, the loadings can indicate which traits covary since similar loadings will indicate similar degrees of deviation. These ideas are illustrated in Table 2-2, showing data from abalone measurements (from Tissot, 1988). The first three axes account for over 95% of the variation. In addition, we see that the basic shell measurements (length, width, height) have similar loadings and so covary considerably. In contrast, the tremata (respiratory pores) traits show much deviation from the shell growth patterns, especially on the second and third axes.

Loadings also point out a crucial link between multivariate and bivariate analyses (aside from the geometric one seen in Fig. 2-16). As demonstrated by Shea (1985; confirming the original work of Brown and Davies, 1972), the loadings of the first axis (or component) in PCA will directly reflect the specific growth ratio, k, when normalized (i.e., the vector of the traits are standardized so that sum of the squared elements equals one). That is, the ratio, k (as defined in the bivariate section), between any two traits is simply the normalized loading of one divided by the other. Thus, the ratio between length and width in the abalone (Table 2-2) is about 1, indicating isometry. Similarly shown (for example) is the high negative allometry of tremata number relative to length, as calculated from the loadings.

It may now be evident that the first axis in morphological analyses is concerned with explaining variation among traits associated with overall increase in "size." Graphically, this is seen in the longitudinal axis of Fig. 2-16 such that "body size" is represented by that vector of covariance among all traits analyzed. As formally defined by Bookstein *et al.* (1985), *size* is "*a general factor* which best accounts for all observed covariances among a set of distance measures taken on individuals of varying size." Thus, it is the first factor as estimated by the "best-fit" line (minimum deviations) by whatever algorithm is used to fit the line. The crucial benefit is that it allows the worker to roughly quantify the otherwise qualitative notion of body size. Secondary factors (axes) may then represent "shapes," allowing roughly quantitative assessment of that qualitative notion. A recent, major debate on multivariate size and shape analysis is found in *Systematic Zoology* (1989, pp. 166–180). Many of the following points are expanded upon there.

There is a common misconception the first axis is a "size-only" axis. This is not true. As shown by Shea (1985), the first axis does represent variation caused by size change, as already noted; but this change need not be isometric. For example, consider the point cloud in Fig. 2-16. If the longitudinal (first) axis has a slope of one among all axes, then it truly is multivariately isometric and there is no shape change with size. However, very often the slope will deviate from one among some axes such that "size" (first axis) increase will be accompanied by shape change (growth ratios will differ). For this reason, the first axis is more appropriately called a *size-determined* shape axis, not size-only. Biologically, it represents consistent, proportional changes that occur with overall growth. In terms of allometric heterochrony, this amounts to saying that the first axis summarizes offset changes (i.e., allometric progenesis and hypermorphosis) where shape change results from changes in body size alone.

Succeeding axes focus on trait changes that deviate from the general "size-determined" proportional patterns of the first axis. For example, Table 2-2 shows that tremata growth deviates quite a bit from the shell dimensional characters, as deduced from their loadings on the second and third axes. Such "shape" changes account for about 22% of the differences among individuals. This variation is far from randomly distributed among the geographic groups analyzed by Tissot (1988). Graphically, this is shown in Fig. 2-17 where the different groups vary distinctively in the ontogeny of these tremata traits. The second axis (in particular, tremata number has a high loading on this axis so it is important in defining this axis) serves to separate the groups as "body size" (first axis) increases ("grows"). In terms of allometric heterochrony, the second axis (and potentially the third and so on, until only "noise" is left) describes changes attributable to allometric rate (neoteny, acceleration) and onset (pre-, postdisplacement). That is, they describe changes out of proportion with the general size-determined pattern. More basically, they represent growth rates of morphological subsystems (growth fields) that deviate from the overall global pattern of the collective whole. Indeed the holistic view of ontogeny is that of primary, secondary, and tertiary (and higher) growth gradients that occur among multiplying cells (Chapter 3).

In sum, in the comparison of ontogenies of different species, the more variation the first axis accounts for, the more the species differ only in size and size-linked allometric changes (i.e., the more the traits share proportional growth rates). Later axes show deviations from this basal size–shape association. In the case of the abalone, the deviations in growth rate of the tremata are clinal and subspecific, and manifested in allometric rate changes. While the multivariate approach can offer a simultaneous view of trait growth, it is not qualitatively different from the bivariate view and is still limited by the

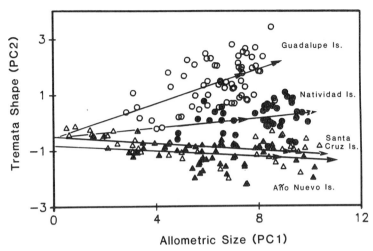

Figure 2-17. Geographic heterochronic variation in the black abalone. Ontogenetic trajectories for four populations, based on first ("size-determined" shape) and second ("tremata shape") axis PCA scores. From Tissot (1988).

lack of age-free data. A possible qualification of this is Shea's (1985) report that nonnormalized loadings of traits on the first PCA axis are proportional to absolute growth rates. It is difficult to theoretically visualize why this might be so, but this suggestion is very intriguing and therefore badly needs further testing.

Chapter *3*

Producing Heterochrony
Ontogeny and Mechanisms of Change

Time is the fire that consumes me, but I am the fire.
Jorge Luis Borges

1. INTRODUCTION

The highly orchestrated cascade from which a complex individual is created from a single fertilized cell is perhaps the greatest wonder of Nature. Here is a self-directed process, wherein the building materials, the builders, and the supervisors are all produced and used at the correct time and place, in blatant circumvention of the most basic law of the universe: entropy, the second law of thermodynamics. Obviously, such a process must consist of a very tightly controlled sequence of nested causal processes. Yet the most intriguing aspect is the ease of creativity: the abundance of ways in which this process can be successfully perturbed to produce different, but viable, individuals (and thus evolution, after the winnowing by selection).

In this chapter, we take a general look at the ontogenetic process and how it can be altered. We will promote the views that (1) heterochrony is the cause of most developmental alterations and (2) heterochrony can cause major novelties. The former view is held because most intra- and interspecific variation arises from minor heterochronic size and shape changes late in ontogeny. The latter view is held because heterochronic changes in local cell populations in early ontogeny can produce substantial changes in cellular interactions. In regulative development this can give rise to new cellular juxtapositions and even new tissues because cell fates are determined by spatial context. These views, especially the latter, are in contrast to the traditional concept of heterochrony as an important but "minor" subset of developmental changes (e.g., Raff and Kaufman, 1983; Bonner, 1988; Thomson, 1988). This is because the concept of heterochrony arose as a way to classify simple regularities between the ontogenies of groups. As a result, it focused on observable (large-scale), relatively late-acting developmental changes, without consideration of the biochemical and cellular processes underlying them. However, as we shall see, the same rate and timing changes that cause such simple late-stage allometric heterochronies (in "contiguous morphospace") can also cause major "jumps" in morphospace when acting in early ontogeny.

1.1. Development as Self-Assembly of Cells

Heterochrony becomes much clearer when development is viewed from the cellular level, as a process of self-assembly by cells. Heterochrony is then seen as change in rate or timing in the cascade of cell assembly. At a more ultimate level, this change is, in turn, caused by changes in the flow of information among cells, regulating that assembly. At the most ultimate level are changes in genes, creating and responding to that information. In short, to echo Oster and Alberch (1982), if development is "an increasingly complex dialogue between cell populations," then heterochrony is seen in rate or timing changes in that dialogue. In the past, workers have focused only on the dialogues of late ontogeny (mostly a time of simple growth, or mitosis from the cell view). This was because it was convenient to their own scale of reference. Our cellular view complements that of Buss (1987) who sees evolution itself as a process of cellular self-assembly. In these terms, heterochrony is an inevitable, common mechanism of change as mistakes in assembly have led to sometimes successful (and "advanced") permutations into new configurations of cell assemblages.

The cell viewpoint also clarifies yet another fuzzy point, that of the "regulatory gene" which is so glibly invoked in discussions of heterochrony. As increasing numbers of workers (e.g., see Levinton, 1988, for general discussion) have noted, "regulatory gene" is a concept that includes a wide variety of disparate genes, coding for many different products and functions. In the cellular view, any gene that affects the timing of cell dialogue is "regulatory." Thus, any gene that affects the timing of a chemical signal's: (1) transmission (rate of chemical signal synthesis), (2) reception (i.e., threshold of gene sensitivity), or (3) travel time (properties of travel medium) between cells would be included. Such genes would include those affecting such diverse properties as those of the cellular matrix, cell membrane permeability, receptor molecule function, among many others (Oster *et al.*, 1988).

Most importantly, the cellular view lifts the study of heterochrony out of being simply a taxonomy of patterns. Much confusion over heterochrony has arisen because, while it is defined as a process—"change in timing of development"—heterochrony is usually described and discussed as a pattern. It has been a taxonomy of patterns diagnosed from the similarities of ontogenetic morphologies (this problem was noted by de Beer, 1958). This has led some to comment that heterochrony is not worth much as an explanatory concept (Kauffman, 1983) or that it is hard to relate heterochrony to major developmental processes such as pattern formation (Wolpert, 1982).

Lest the reader accuse us of "pan-heterochrony," we will not argue that *all* developmental alterations are most fruitfully viewed in heterochronic terms. Our main argument is simply that traditional views of heterochrony as an interesting but minor subset of developmental change are too narrow. Our major evidence is that the developmental causes of heterochrony, as traditionally viewed (the late-acting, mainly allometric changes), are often the same as more novel, "drastic" changes. It is mainly a matter of beginning to look at change in rate and timing at finer spatial and temporal scales.

Heterochrony can then become an organizing principle at these scales as we compare (for example) the slower rate of cell migration in certain parts of the early embryo of a descendant relative to the ancestor. Incorporation of this view has already begun. Hall (1984a,b) and Thomson (1988) for instance have written of heterochrony in terms of timing of tissue induction while Minelli and Bortoletto (1988) refer to timing of diffusion flow.

Here we want to formalize this view into the realization that changes in rate and timing can occur at any scale (molecular, cellular, tissue, organism) and that one can sometimes usefully apply the heterochronic perspective to these changes. This requires explicit discussion of the scales at which phylogenetic comparisons are made. Because development is a highly orchestrated cascade, small rate or timing changes at the lower (finer) levels (e.g., diffusion flow) will often translate into complex results at higher levels (e.g., tissue grade). The nonlinearity of the system will amplify some changes (positive feedback) and dampen others (negative feedback) as they cascade upward. Heterochronic changes in later ontogeny are more easily related to lower level changes because they are simpler (mainly just cell growth). However, these changes ("inputs") are often no more "heterochronic" than the early ontogeny changes that happen to result in more complex tissue-level changes ("outputs") because more is going on then (cell differentiation, inductive interactions).

1.2. Preview and Outline of Chapter

First we present a very basic outline of the developmental process, mainly in terms of cellular assembly. Three phases of development are acknowledged: neofertilization (influence of maternal DNA on cell assembly), differentiation (cell differentiation, migration, and mitosis), and growth (mainly mitosis and some migration). Next we discuss how alterations in rate and timing of these cell assembly processes can result in ontogenetic changes, discussing a variety of examples. For simplicity we divide these heterochronic changes into two basic kinds: **growth heterochronies**, of late ontogeny, and **differentiative heterochronies**, occurring earlier in ontogeny. The dividing criterion is whether the timing or rate change occurred after differentiation of the affected part(s).

The fourth section discusses the mediators of cell dialogue, the biochemical messengers such as morphogens and hormones. A knowledge of these is important for understanding a key step in rate and timing change: information transfer among cells. The fifth section discusses the ultimate cause of change, the genetic level. As might be expected, our knowledge here is skimpiest of all, although some important modifications to the basic "regulatory gene" concept can be made. We close the chapter with sections which summarize the major concepts using a simple heuristic "tree" or bifurcation model. This includes an attempt to unite these concepts with heterochronic and allometric classifications, as presented in the last chapter, and the concept of hierarchy.

2. ONTOGENY: A BRIEF SKETCH

The intricate network of genetic, biochemical, and cellular interactions which lead to the assembly of an organism can hardly be covered here. For that, the reader is referred to any of a number of texts on developmental biology (Davenport, 1979, and Gilbert, 1985, are especially good). Instead, we seek to outline the basic processes only to the extent needed to understand how general heterochronic changes in those processes produce evolutionary change. Epigenetics, the name given to this network of assembly, has been defined in a number of ways, but generally refers to the control of gene expression by the microenvironment(s) encountered by cells during development (see Hall, 1983, for discussion). This control is mediated by a large number of biochemical interactions, such as hormones and morphogens, and even physical phenomena such as mechanical and electrical stimuli. Since virtually all cells in the adult organism contain the complete gene complement of the original zygote, control of gene expression must occur by differential transcription or translation of those genes. Waddington's (1940) famous analogy of the epigenetic landscape aptly summarizes this process.

Bonner (1988) has recently emphasized the important point that much of development is simply driven by the "inertia" of biochemical reactions which drive themselves and have their own characteristic rates. Genes can be seen as producing input (synthesizing products) at critical times (to affect the cascade) but they are often remote from much of the "action." This is necessary because all of the information for every step cannot possibly be encoded in the limited space available, given the size of the storage units (DNA molecules). The interesting result is that the adult human body contains about 5 $\times 10^{28}$ bits of information (more than 10^{19} printed pages) in its molecular arrangement but only about 1 $\times 10^9$ bits in the genome (about 10^4 printed pages), according to Calow (1976). This only makes sense in light of Bonner's insight. Indeed, at least some genes (especially "regulatory" genes) can be usefully seen as simple statements in the genomic "computer program," such as STOP, START, or GOTO statements. It takes very little information to encode each statement *per se,* but they have powerful control over what happens at each time and place and therefore potentially much of the whole cascade. For instance, complex events will continue until an inhibitor ("STOP" statement) is produced. In this view, at least some heterochrony is caused by changes in these statements, changing the time when events occur (onset/offset). (Again, the complexity of interpretation arises when these simple onset/offset changes at fine, e.g., cellular, scales do not extrapolate into simple onset/offset large-scale tissue changes.) This perspective also provides some measure of the complexity of development in general: the roughly 100 trillion cells in the adult human body could in theory be generated by just 45 binary divisions from the zygote. The information needed to direct this would be very much less than the billion or so bits actually encoded in the genome: one simple START and one STOP statement would provide the ultimate global control of the entire fission process. If there was a "dilution" process of an initial substance with each fission, it could even be carefully timed (i.e., the STOP statement would be initiated at some threshold in each cell at the same time).

2.1. Three Stages of Development

To simplify discussion of development (and epigenesis in particular) and its alteration, we will subdivide development into three basic phases: **neofertilization, differen-**

tiation, and **growth.** In theory at least, change in rate and timing of cellular interaction can occur at any of the three phases. In general, as von Baer's famous observation noted, changes in the earlier phases will have greater effect. Earlier changes generally have less chance of creating viable products because they more greatly alter a functioning, integrated system (although they can sometimes succeed, as discussed later). While each species obviously goes about its epigenesis somewhat differently, the basics are the same at the cellular level: growth (mitosis in multicellular forms), to increase the number of cells; differentiation, to increase the kinds of cells; and cell movement, to alter the spatial orientation of cells as growth and differentiation proceed. The directionality that results is of course toward increasing number and kinds of cells but it is also toward progressive restriction of cell fates. Thus, the last phase is primarily cell replication and movement.

2.1.1. Neofertilization Phase

The neofertilization phase includes the processes just after the egg is fertilized. At first, maternal information (from the egg alone, passed on by nurse cells, among other modes) directs all activity in most animals. As cell cleavage progresses, the embryonic (zygotic) genome exercises progressively more control. In humans the embryonic genome begins expression between the four- and eight-cell stage (Braude *et al.,* 1988). One of the main kinds of information transmitted during this and all phases is **positional information** which tells each cell where it is (spatially) and thus how (and when) it should "behave," i.e., differentiate, divide, and migrate. Since 1969, Wolpert has been one of the major researchers in this area (Wolpert, 1969; 1982 for a more recent overview). The distribution of this information is often called **pattern formation.** As will be discussed later, this information distribution often involves sequential transcription of genes, stimulated by products of previously transcribed genes. Many readers will know that both positional information and pattern formation are "loaded" phrases that have caused many arguments. We are therefore somewhat hesitant to use them here but feel that if the reader limits the meaning of these terms to our specific, explicit definitions just outlined, their meaning in the discussions that follow will be clear.

Maternal positional information seems to be based on one of a number of possible heterogeneities in the newly fertilized egg, as an initial point of reference. Some mollusks for instance use point of entry of the sperm as an initial reference (first clearly shown by Morgan and Tyler, 1938). The dominant vector of gravity is another reference point used (Smith, 1985). In many animals, formation of the basic body pattern occurs while the zygote is dividing into a population of some 10,000 cells. During this time, positional information is "distributed" and "founder" regions of cell groups are set up accordingly. Impressively, this basic organization is complete within 1 or 2 days after fertilization in most animals although complex organisms such as humans take over 30 days (Cooke, 1988). After this, the founder regions are sequentially subdivided into progressively smaller "modules" (i.e., growth fields). Driever *et al.* (1989) have shown that the maternal *bicoid* gene produces a protein gradient binding to DNA, and playing a key role in this process. This "modularization" process is shown in Fig. 3-1 for the widely studied *Drosophila.*

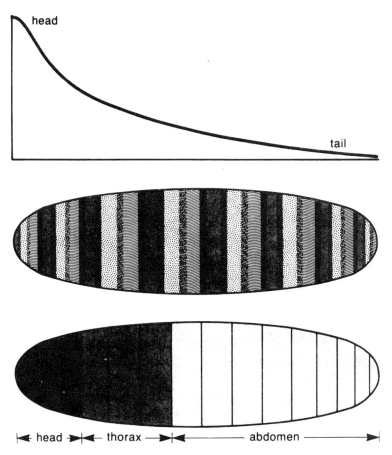

Figure 3-1. Diagrammatic illustration of a gradient of information about body position in developing *Drosophila* embryo. As development proceeds, genes and gene products interact to produce a progressively more refined "modularization" of the individual. From Cooke (1988).

2.1.2. Differentiative Phase

In the second, differentiative phase, progeny cells of the founder groups undergo a complex series of mitosis, cytodifferentiation, and cell movement characteristic of cells in each initial founder region. During this time the zygotic genome comes to dominate the rate and timing regulation of events. This phase culminates in tissue differentiation and organogenesis. In some groups of organisms, such as nematodes and mollusks, cell fates are determined relatively early. In this **mosaic** development, positional information from the neofertilization phase is used to somehow "program" the fates of entire cell lineages. Cell activity is only slightly, if at all, context-dependent as development proceeds. However, in other groups such as echinoderms and vertebrates, development is more **regulative** such that cell differentiation is not determined until later. In such cases, cells within each founder region are **pluripotent** (i.e., of indeterminate type) until they come into contact later with positional information which causes them to differentiate.

They may multiply and migrate in the pluripotent state and become committed ("switched on") long before cytodifferentiation actually occurs.

In the past, the dichotomy between mosaic and regulative development has been overdrawn, with the former being associated with protostomes and the latter with deuterostomes. However, this dichotomy is oversimplified and even if it were not, protostomes and deuterostomes do not all fall into one or the other category (see Raff and Kaufman, 1983, for further discussion). Nevertheless, if not taken too far, the concepts are clearly useful. Thomson (1988) has pointed out that to ontogenetically "build" a complex organism, regulative processes are probably necessary. It would simply take too much information to "program" each step. A key point is that regulative processes are therefore relatively plastic in their responses to perturbation (e.g., timing), giving them much "creativity." An example is the stimulation of differentiation by a wide variety of substances (Thomson, 1988), as discussed below. Unfortunately, the study of heterochronic causes at finer levels (e.g., genetic, cellular) has focused almost entirely on organisms with highly mosaic development (see Ambros, 1988, for review) so that we are not getting a complete view. However, for now this focus is a practical necessity because the determinate nature of mosaic development makes it much easier to analyze rate and timing perturbations.

The best known aspect of regulative development is **induction**. This has many definitions but generally refers to the evocative action of one cell (or tissue when referring to a cell aggregate) upon another (Berrill, 1971). One cell type, the inductor, often acts upon another cell type to produce a third cell type. The discovery of this process is associated with Hans Spemann who made some of the earliest use of tissue transplantation experiments. Much recent research has dealt with determining the mechanisms behind this evocative action: how do cells communicate? In general, Wolpert's ideas cited above focus on the existence of some kind of gradient of positional information. The signal which transmits the positional information is called a **morphogen**, a general term for anything that causes a cell to differentiate, divide, or move (**mitogen** is often used to specifically refer to that subset of morphogens that stimulates mitosis). These signals no doubt vary in kind, depending in part on the distance of information transfer: long-range, short-range, and direct contact (see Hall, 1988, for an excellent review of induction). These are shown in Fig. 3-2.

In long-range interactions the inductor cells are thought to produce molecules that are relatively small but specific in their action. Proteins are too large to diffuse readily and small ions are not specific enough in their action (Crick, 1970). The first positively identified morphogen, retinoic acid, occurs in a gradient in chick embryo limb buds, and fits this description quite well (Thaller and Eichele, 1987). As expected, the gradient is exponentially decreasing from the source cells and levels of concentration seem to determine cellular response (often a threshold response), organizing limb bud growth. Examples of this type of interaction can be found in Maclean and Hall (1987).

In short-range interactions the cells involved are no more than 20 to 40 μm apart. Often these are mediated by an extracellular matrix. As discussed below, bone growth is thought to occur in this way (see Kemp and Hinchliffe, 1984, for a review of this mechanism). Direct cell-to-cell contact is sometimes necessary for information transfer. This may occur via direct interaction with the nucleus and genome of the inducing cell or via interaction with the cytoplasm (Fig. 3-2).

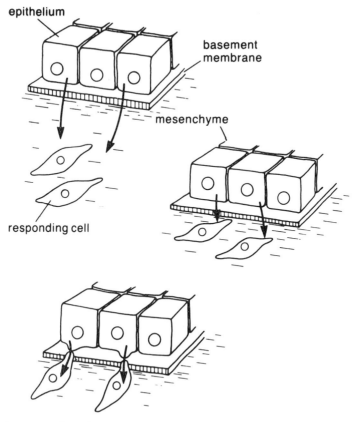

Figure 3-2. Three types of signal interactions among cells. Top: long-range interactions, where no direct contact is needed. Middle: short-range interactions, where receiving cells (mesenchyme) must be in contact with extracellular basement membrane secreted by sending cells (epithelium). Bottom: direct-contact interactions, where cells must be in direct contact. From Hall (1988).

In the receiving cells of any of these interactions are receptor molecules that ultimately transmit signals to the genome. These "pull a switch" and initiate the sequence leading to cytodifferentiation.

2.1.3. Growth Phase

The third phase in development is the growth phase. This refers to that time after cell and tissue differentiation are complete and the developing organism undergoes mainly an increase in cell number of the tissues. Integrative mechanisms become increasingly important at this time, in particular the hormonal system controls overall growth. On a local level, mitogens, described in detail in Section 4, are very active. In humans, this final phase begins at about 2 months after conception (Harrison *et al.*, 1988). As we shall see, there is much interplay between global and local control at this time, with much differential cessation and rate of growth among tissues. Thus, there is great opportunity for heterochrony.

3. HETEROCHRONY: CHANGE IN RATE OR TIMING OF ONTOGENY

The vast majority of differences among closely related organisms are due to simple changes that are manifested relatively late in ontogeny. Usually these are allometric modifications of size and shape. Thus, they amount mainly to changes in cell number and/or arrangement. Even "major" evolutionary changes largely involve such modifications. The 100 trillion or so cells in the human body contain about the same 200 cell types found in frogs, reptiles, and other mammals (Wolpert, 1978), making the main differences ones of cell number and configuration. In terms of the above outline, this would seem to imply that much evolution occurs via changes in the growth phase. This would be in keeping with the commonly held principle that more lately occurring changes are less disruptive and therefore more common. However, this implication may be more apparent than real. A nonobvious point is that many (perhaps most?) of these late-manifested differences are actually caused by changes (often rate and timing changes) in the preceding, differentiative phase. Thus, as we discuss below, changes in limb size may have more to do with differences in (pregrowth) limb bud formation than limb bud growth.

We subdivide our discussion of heterochronic changes into two major groups. **Growth heterochronies** affect tissue after differentiation and can act either globally or locally, during the growth phase. Sequential and organizational heterochronies (Chapters 2, 7, and Glossary) usually fall in this category. In the latter part of this section we deal with the second type, **differentiative heterochronies**. These act during the earlier, neofertilization or, especially, the differentiative phase discussed above, when changes may alter rate or timing of inductive and other pregrowth interactions. We will show that the most "creative" heterochronies are those local differentiative changes which create new kinds of inductive interactions. However, many differentiative changes affect only the amount rather than kind of induction. This only alters the size of the resulting tissues. The following examples are not even remotely exhaustive, but are meant to be illustrative of the principles involved. Nor do we consider this classification to be anything but a first step toward assembling and formalizing heretofore loosely organized concepts.

3.1. Growth Heterochronies

Global growth heterochronies are well known as instances of body size change in evolution and domestic breeding. These are discussed first, followed by a discussion of local growth heterochronies which affect only some growth fields.

3.1.1. Global Growth Heterochronies

Size change is the most common evolutionary change (Chapters 6 and 8) and the developmental cause of size change is usually "global" late stage change in rate or timing. Many fossil examples of Cope's rule of size increase (e.g., "Irish Elk," titanotheres; see Chapters 6 and 8) clearly involve such simple changes where body growth is accelerated or prolonged, and allometric correlates "tag along" (McKinney, 1984). However, a main point is that even in these "global" cases, the changes are rarely if ever "across-the-board." That is, at least some growth fields lag or are accelerated relative to the ancestor, out of proportion to the "overall" change. They are "dissociated" in terms of Chapter 2. This is probably related to the practical inability to completely integrate such a complex assembly. In some cases there is also often direct external selection on these fields to deviate allometrically as a means of "fine-tuning." In spite of some claims to the opposite, completely isometric changes in size are rare. This is because size changes almost always involve new demands (physical, behavioral) that are not linear extrapolations of other sizes.

In mammals, the most important overall mechanism for control of postnatal body growth is somatotropin or growth hormone (GH). This polypeptide is produced by the pituitary gland which is in turn stimulated by hypothalamic production of the neurotransmitter growth hormone releasing factor (GHRF), and inhibited by the production of somatostatin (Harrison *et al.,* 1988). Circulating through the body, GH stimulates the liver and other cells (e.g., bone epiphyses) to produce the insulinlike growth factor I (IGF-I) which acts as a mitogen in a variety of tissues (Nilsson *et al.,* 1986). Interestingly, GH itself undergoes a fairly complex (chemical) ontogeny which has been recently analyzed in transgenic mice (Borrelli *et al.,* 1989), and is probably quite similar to that of our own. Clearly there is much room in this chain of events for alteration of growth rate or timing by changing the rate or time of activity of GH and the factors it stimulates. Even more, another potentially changeable link in the chain has recently been discovered in that retinoic acid now appears to control GH production in pituitary cells by acting on GH gene expression (Bedo *et al.,* 1989). This is yet another remarkable case of evolutionary parsimony in that retinoic acid, noted above, is the first identified morphogen, acting in chick limb growth. Rate or timing change in the production sequence of one or more of these peptides is not the only way to produce heterochronies among the tissues. Changes in threshold responses (in target tissue) to the same amount of stimulus would have the same effect.

Two well-documented examples of body growth rate changes, in rodents and primates, have been discussed extensively by Brian Shea. In the first case (Shea *et al.,* in press), Snell dwarf mice growth was compared to that of transgenic giant mice. The dwarf mice produce almost no GH due to a pituitary malfunction, while the transgenic mice (created by microinjection of fusion genes coding for increased production of rat GH and IGF-I) produce abnormally high amounts. Shea found that while the duration of growth was not significantly altered, the rate of growth was. The transgenic mice grew much faster (body acceleration), reaching a much larger adult size (Fig. 3-3). Also of note in Fig. 3-3 is that, at least for long bones, the change truly was "global" with allometric "extension" (allometric hypermorphosis of Chapter 2).

In the primate case, Shea (1983, 1988) showed that gorillas do not grow for a longer time than chimpanzees, but grow faster to attain their roughly threefold greater adult body size (discussed in Chapter 2; see also Chapter 7). This is also true for the two chimpanzee species; the pygmy chimpanzee simply grows slower than the larger form.

Much of the size (and size-determined allometric) polymorphism exhibited by dogs (see Wayne, 1986, for definitive study) seems to be due to a similar process. There is a strong correlation between adult size and levels of circulating IGF-I in various dog breeds

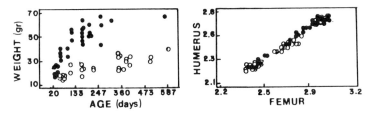

Figure 3-3. Allometric plots of Snell dwarf mice (open circles) and transgenic giant mice (dark circles). Modified from Shea (in press).

(Eigenmann *et al.,* 1984). The same is true of size and IGF-I levels in rabbits (Masoud *et al.,* 1985). Less direct evidence implicates the same mechanism in horses, cattle, and pigs in that they show many of the same allometric extrapolations as Shea finds in mice (Wayne, 1986). While the hormonal mechanism is somewhat different than in mammals, some frogs also grow larger via increased growth rate (Emerson, 1986).

In contrast to the above, other global heterochronies reflect changes in timing rather than rate. Historically, most workers seem to have assumed that larger size is always attained via longer growth duration. As discussed in Chapter 2, this is based on the gross "bacterium to whale" relation and may not apply at all to species or generic level changes. In any case, this is an empirical question that deserves much more study than it has gotten. Three crucial questions need to be addressed regarding how large size is attained: (1) Which is more common, rate increase or offset delay? (2) What environmental conditions favor rate increase, as opposed to offset delay as a mechanism of size increase? (3) Most relevant to this chapter, is it genetically or developmentally "easier" to become larger through growing faster or longer? Since size change is the most common evolutionary change, a true understanding of evolution must address these. In any case, as a counterpoint to the rate mechanisms of size increase above, we note the following examples of increased size via delayed growth offset.

The first is an interesting comparison to Shea's mice example, involving two strains of musk shrews which likewise vary greatly in size. Ishikawa and Namikawa (1987) showed that, unlike the mice, the size difference in the shrews was based on difference in global growth duration instead of rate. The larger strain grows longer, with the male's rapid, juvenile growth phase slowing down at 34 days versus only 15 days for the smaller strain; similar differences are seen when comparing females, which are smaller in these sexually dimorphic strains (Fig. 3-4). Note that this graph might initially be diagnosed as acceleration since the larger strain grows faster. Yet this increase is ultimately caused by prolonged juvenile and prenatal growth phases (the larger strain is born larger). The larger size at birth means that there are a larger proportion of cells multiplying so "acceleration" is probably a derived feature of the hypermorphosis. Contrast this with the true global acceleration of Shea's mice example (Fig. 3-3) where growth terminates for both strains at about the same time and they start at about the same neonatal size. This subtlety, that delay in rapid growth phases can cause "rapid" growth, is often confused with "pure" acceleration where the timing of the phase is not altered. Rather, growth during that phase is somehow increased. This suggests different underlying genetic and developmental mechanisms, one involving timing (e.g., some kind of "clock," perhaps using number of multiplicative cell cycles), and the other involving rate (e.g.,

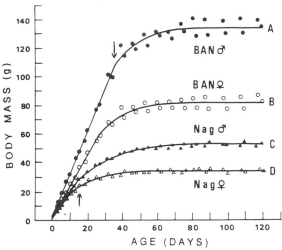

EYES OPEN

BELLY UNDERFUR

EARS OPEN

BELLY PIGMENTED

DORSAL UNDERFUR

UNVISUALIZATION OF
VISCERA THROUGH
THE BELLY WALL

DIGITS FREE

OSTIUM
UROGENITOANALIS
SEPARATED

DORSAL PIGMENTED

DORSAL COLORLESS
GUARD HAIRS

BELLY COLORLESS
GUARD HAIRS

GUARD HAIRS ON THE
TAIL AND FACIAL
VIBRISSAE

CLAWS

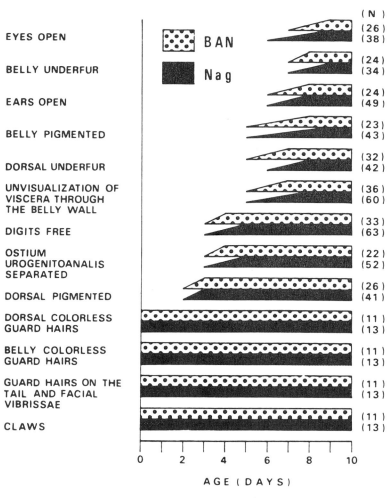

increased mitogen production or sensitivity). The exact mechanisms are poorly understood (more discussion of "rate" and "timing" genes is presented below). A final point of interest is the lack of differences between the shrews in all major developmental events (Fig. 3-4) illustrating a dissociation of these from body size growth itself.

Humans are another good example of size increase through delayed timing, as discussed in Chapter 2. When compared to the chimpanzee and our hominid ancestors, our evolution shows a progressive extension of all growth phases, from fetal phases to maturity and death. We are also a good example of the problem of distinguishing rate from time heterochronies: we grow bigger by delaying sexual maturation (hypermorphosis), yet one reads endlessly that we are neotenic and grow "slower." In fact, we do not grow more slowly, only longer. We extend both slow and rapid growth phases. This is all discussed in detail in Chapter 7. Interestingly, humans also provide an example of size differences from rate alone, at the subspecies level. The African pygmy apparently grows as long as "normal," larger humans but grows at a slower pace (Merimee and Rimoin, 1986; Shea, 1988). The proximal mechanism for this seems to be low IGF-I levels (Shea et al., in press; see also Chapters 7 and 9).

3.1.2. Local Growth Heterochronies

Local growth heterochronies are those in which rate or timing changes in growth occur after the tissue or organ has differentiated. For example, a limb would differentiate as the same in size and kind in the descendant but alterations in mitosis (and possibly cell migration) occur afterwards. The "dissociated" heterochronies of Chapter 2 would include these kinds. Just how these changes occur depends on the signals used at the local level to transfer information. We note two of perhaps many kinds: those involving changes in local sensitivity to "global" (hormonal) mitogens and those involving changes in local mitogens themselves.

In one of the best documented examples of the first type, Riska and Atchley (1985) showed that hormonal influence diminishes in its effect on certain target tissues as development proceeds. IGF-I has little effect on brain growth although it can greatly affect body growth, as discussed above. In contrast, IGF-II, which acts earlier, during fetal development, can increase mitosis in both brain and body because brain tissue is responsive to it then. Later, brain responsiveness to it is lost (see also Chapter 7). A similar process occurs during puberty when only certain tissues grow in response to the secretion of sex hormones and only during certain time "windows" (Ritzen et al., 1980). The point is that any changes in threshold sensitivity to the signals will produce corresponding changes in rate or duration of growth in these local areas. Such local changes are also known in salamanders; they are discussed below in differentiative heterochronies because they often affect more than just growth of the local tissues.

Figure 3-4. Top: Postnatal body mass gains in two strains (BAN, Nag) of musk shrews. Arrows denote approximate age (and size) when rapid juvenile growth phase begins to plateau. Bottom: Comparison of developmental events between the two strains, showing no appreciable difference. Modified from Ishikawa and Namikawa (1987).

The second type of local growth change, where widely acting hormones are not directly involved, may be illustrated in long bone growth, one of the best studied of all developmental systems. Thorngren and Hansson (1981) have argued that differences in long bone growth rates are due to variations in the relative frequencies of cell division at the epiphyses. The exact cause of such variation is uncertain but could relate to local mitogen sensitivity or the amount of mitogen produced locally, or even perhaps some intrinsic traits of the proliferative cells themselves. The second case, diffusional determinants of heterochrony, is discussed below.

In contrast to the above, which showed how changes in local growth rate may arise, other changes may reflect alterations in offset timing of local growth. For instance, cartilage is largely avascular so that it must receive nutrients by diffusion (Hall, 1984a). Thus, growth offset begins when the cartilage becomes so thick that it outgrows its blood supply (Moss and Moss-Salentijn, 1983). Mutations which increase the supply of nutrients, either by changing the nutrient diffusion gradient or by increasing the size of small blood vessels that may feed parts of the cartilage, could prolong growth and increase size of the cartilaginous structures or organ(s) involved. Note that such a mutation does not involve a gene in the traditional "rate" or "time" gene sense. A similar argument could be made for changes in rate and timing of cell death, which can be an important mechanism of morphogenesis in many organisms. For example, digit formation in reptiles is by interdigital cell death (Fallon and Cameron, 1977). Delay of cell death can lead to interdigital webbing. Of particular interest is the large role of cell death (and changes in it) in brain development, discussed in Chapter 7.

3.1.3. Overview of Growth Heterochronies

Changes in growth rate and onset/offset of growth, occurring after tissue differentiation, can be either global (whole organism) or local. Global changes, such as illustrated in primates or rodents, can provide a powerful means of changing body size through rapid yet integrated changes. However, even global changes are rarely, if ever, truly "global": local allometric deviations occur because the causal hormonal changes may have differential effects depending on when they act, and target tissue responsiveness. Local changes, sometimes related to mitogenic stimuli or removal of constraints, provide a powerful means of organ/tissue adaptation. Being decoupled from other growth fields, local changes can result in major shape changes. Both global and local growth changes are late-occurring in ontogeny and thus minimize many logistical problems.

A crucial but very poorly understood aspect of growth control relevant here is the interplay between local and global factors. In their excellent review, Bryant and Simpson (1984) discuss the complexity of this interplay. In general, there is considerable change in the coupling (partial and complete) and decoupling of growth fields during ontogeny. Notable among the informative studies cited are the tissue transplant experiments (Twitty and Schwind, 1931, is perhaps the classic pioneer work) which show that transplanted tissues grow at the rate and to the final size of tissues characteristic of the donor organism. Thus, the new hormonal environment of the receptor individuals does not alter the intrinsic tissue controls.

3.2. Differentiative Heterochronies

As already noted, cellular processes during the neofertilization and differentiative phase involve not only mitosis and migration, but also cell differentiation. Therefore, changes in the time and rate of cell activities here can have *much more effect* on the final morphology *by changing patterns of differentiation* (new configurations and even tissues). Hall (1983, 1984a,b) in particular has been a cogent proponent of such heterochronic changes. However, most rate or timing changes during or before differentiation probably do not have such novel or even major effects. This, in part, is because development is very integrated and can compensate in many ways. In fact, it is likely that the large majority of such changes (where they have any effect at all) mainly act to alter the ultimate size of the affected tissues or organs via change in the number of proliferative (stem) cells allocated. Yet the potential for great change is there, as we shall see. As with growth heterochronies, differentiative heterochronies are subdivided into global and local categories.

3.2.1. Global Differentiative Heterochronies

At first thought, it is perhaps difficult to accept that changes could occur in any but the latest stages of ontogeny since earlier changes tend to have much greater effects, ostensibly reducing the chances for a viable organism. However, it is now known that closely related organisms can in fact differ in their early ontogenies, indicating that early changes can successfully occur (Raff, 1989; and especially Thomson, 1988, for discussion). As we discuss shortly, echinoids and frogs show this most clearly in studies done to date. A rather different kind of global differentiative change can also occur in later development, just before metamorphosis. It involves much less complex changes, being caused primarily by changes in the timing of metamorphosis. This is best known in arthropods and amphibians as discussed below. In all cases, we reiterate that our use of "global" is strictly relative. We simply mean that large numbers of growth fields are affected, not every last one. In addition, note that, perhaps nonintuitively, even these larger changes are often not very drastic in altering the adult phenotype.

Probably the best documented early global changes are those of the regular echinoid *Heliocidaris* as recently documented so well by Raff and his co-workers (Raff, 1987; Raff *et al.,* 1987; Parks *et al.,* 1988). *H. tuberculata* has the typical mode of development wherein a feeding larva (pluteus) is produced that ultimately metamorphoses into an adult. Yet there is a closely related species, *H. erythrogramma,* that shows direct development, producing adults directly from the egg in much less time, by omitting the larval stage. This change first appears very early in development; *H. erythrogramma* has a larger egg, a different cleavage pattern, and early appearance of the adult skeleton and other adult structures. Importantly, the direct development involves more than just omission of developmental pathways, although there is much of that. Aside from the larger egg, the first major change is the much larger number of mesenchyme cells, from 32 in the larval developer to over 1700 in the direct developer. This is accomplished by prolonged mitosis of these cells, made possible by suppressing micromere/macromere asymmetry of unequal cleavage to achieve equal cleavage. The greater number of mesenchyme cells

Table 3-1. Order of Appearance of Developmental Features and Heterochronies in Typical and Direct Developing Sea Urchin Species[a]

Feature	Typical (euechinoids)	Intermediate (*Peronella japonica*)	Direct (*Heliocidaris erythrogramma*)
Egg size (μm)	60–345	ca. 300	ca. 450
Fertilization membrane	Yes	Yes	Yes
Radial cleavage	Yes	Yes	Yes
4th cleavage	Small micromeres	Large micromeres	Equal cleavage
Blatula	Smooth	Convoluted	Convoluted
Active hatching	Yes	Yes	Yes
Ciliated	Yes	Yes	Yes
Primary mesenchyme	32 cells	Large number of cells	Large number of cells
Archenteron invagination	Yes	Yes	Yes
Archenteron elongation	Yes	No	No
Next step	Initiation of larval skeleton	Initiation of larval skeleton	No
Next step		Invagination of vestibule	
Next step	Morphogenesis of typical pluteus arms	Partial pluteus with variable skeleton	No larval skeleton
Next step	Formation of gut	Not formed	Not formed
Next step	Formation of coelom	Formation of coelom	Formation of coelom
Next step	Formation of hydrocoel	Formation of hydrocoel	Formation of hydrocoel
Next step	Invagination of vestibule		Invagination of vestibule
Next step	Development of echinus rudiment	Development of echinus rudiment	Development of echinus rudiment

[a] From Raff (1987).

permits accelerated formation of adult structures. The omission of stages begins just after gastrulation. These are shown in Table 3-1 along with the developmental patterns of an echinoid which is somewhat intermediate between the direct and feeding larval modes. In this intermediate case, we find not only omission and acceleration of stages but also inversion of order of some of the stages.

Thus, the heterochronic alterations leading to direct development involve more than simple acceleration of all pathways or omission of some. Even the simplest case, the direct development of *H. erythrogramma*, while involving mainly omission also shows an increase in mesenchyme production (acceleration), necessary to accommodate the early forming adult structures, among other changes. In light of this, it is surprising how easily such early changes are apparently carried out: direct development has evolved independently in at least six orders of sea urchins (Raff, 1987). Indeed, about 20% of all species are direct developers (Emlet *et al.*, 1986). Most impressive in Raff's studies is that he and his co-workers have been able not only to document the changes but also to identify the cell level causes. This much-neglected aspect is essential if heterochrony is ever to be understood as a process rather than a pattern, as discussed elsewhere. In this case, the changes come about because, in addition to mesenchyme acceleration, different cell lineages within the embryo undergo different within-lineage heterochronies (Parks *et al.*, 1988). The extreme flexibility here arises from the relative independence of these lineages while maintaining a surprising ability to successfully interact with each other. This would seem to agree with Buss's (1987) view of cell selection acting to promote cell competition and/or cooperation as the basic mode of metazoan evolution.

Frequent early heterochronies are not limited to echinoids: direct development is found in hundreds of amphibian species and has probably evolved repeatedly in each extant order (Duellman and Trueb, 1986). Cases of direct development via heterochrony in frogs are being investigated by a number of workers (Raff and Kaufman, 1983; Wassersug and Duellman, 1984). While the specifics are not as well worked out, the mechanisms appear at least roughly similar at this point. It is also true that nonmetamorphic organisms can successfully experience very early changes in development. Berrill (1955) argued that "higher" chordates arose from tunicates via neotenous retention of the larval body plan. Whether or not he was right about the specifics, there is a clear pattern of timing shifts during very early cell cleavage (Table 3-2). More complex organisms have progressively higher cleavage numbers (delayed offset of cleavage) and therefore more cells during gastrulation (and more importantly, in later development as well).

In addition to global change in very early ontogeny, later changes may also be global through integrative (hormonal) mechanisms. In particular are organisms which undergo metamorphosis such as arthropods, amphibians, and many marine invertebrates. Matsuda (1987) notes that some 18 phyla of marine invertebrates have pelagic larvae which undergo metamorphosis and he lists many examples where global heterochronies have caused evolutionary change through change in timing of metamorphic events.

Probably the most familiar metamorphic heterochronies are those of salamanders, especially the axolotl. In one of the simplest changes, (indefinitely) delayed secretion of the thyroid hormone thyroxine will result in delayed metamorphosis. The result is that larval somatic traits are never lost although sexual maturation and large size will be attained (see Raff and Kaufman, 1983, for review). Similar delays are common in frogs (e.g., Emerson, 1986) and insects (Matsuda, 1987).

At least numerically, the most successful group to exploit this type of heterochrony has been the arthropods, especially the insects (see French, 1983, for review in developmental context). This success is especially impressive given the relative simplicity of the mechanisms involved. Very generally, two hormones are the proximate motivators of change, ecdysone and juvenile hormone, produced by the prothoracic glands and corpora allata, respectively. As long as both hormones are circulating, the larval stage is

Table 3-2. Relationship between Timing of Cell Determination and Ultimate Larval Cell Numbers in Tunicates and Chordates[a]

Animal	Gastrulation		Notochord approx. no. of cells	Tail muscle approx. no. of cells	Diameter of egg (mm)
	Cleavage no.	Approx. no. of cells			
Oikopleura	5–6	38	20	20	0.09
Styela	6–7	76	40	36	0.13
Amphioxus	9–10	780	330	400	0.12
Petromyzon	11	2,200	500	—	1.0
Triturus	14	16,000	1200	—	2.6

[a] From Raff and Kaufman (1983).

maintained and the organism feeds and grows in that form. When the juvenile hormone is no longer secreted, metamorphosis begins. Ecdysone, acting alone, functions to promote formation of "adult" tissues. Prolonged secretion of the juvenile hormone leads to larger size and "overdevelopment"; premature termination of secretion leads to smaller size and "underdevelopment." A particularly obvious sign of the latter is wing reduction. Depending on degree of prematurity, this can cause changes ranging from microptery through aptery. (For a more complete description, see Matsuda, 1987; for example, there are actually a number of juvenile hormones.)

Global metamorphic heterochronies in social insects are especially interesting because size and shape changes of individuals can have important effects on societal roles. For example, Wheeler and Nijhout (1981) applied juvenile hormone to the last larval instar stage of an ant species, delaying metamorphosis and causing a longer period of growth. Instead of a worker ant, a soldier ant, characterized by its larger size and hypertrophied jaws, was produced. Further, it seems likely that behavior is also affected in such instances, e.g., "overdevelopment" (peramorphosis) of aggression being produced (Chapter 7). The evolutionary importance of this process is shown nowhere more clearly than by Roisin (1988) who documents an entire termite caste system based upon individual polymorphism produced by different times of metamorphosis. In this case, the heterochronic mechanisms have been incorporated into the species' genome to produce a functional society. We suggest that this is true for many social insects.

3.2.2. Overview of Global Differentiative Heterochronies

The adaptive utility of global differentiative heterochronies is apparent: they provide a way of producing coadapted traits in morphologies at the times dictated by their environment via life history selection. For instance, in the echinoid example, larval feeders take several weeks to become adults in contrast to just 4–5 days for direct developers. Whether global or only partly global, a main point is the greater morphological variation available to organisms with larval stages radically different from their adult forms. Through heterochronic truncations of the pathways of all or only some traits, these radically distinct larval morphologies can be incorporated into the adult form. In this view, metamorphosis is very rapid, "compressed," and often extreme differentiation. The very abundance of metamorphic organisms (of course insects in particular, making up the largest group of animals) is in part (perhaps largely so) due to the ease and success of such heterochronic changes.

As with nonmetamorphic heterochronies, even apparently straightforward cases of global change often entail some local alterations. For example, while the gross morphology of the axolotl is larval (i.e., neotenic since development is retarded), red blood cell properties, serum proteins, and other biochemical changes occur that are characteristic of adults in nonaffected "normal" individuals (Ducibella, 1974). Similar "dissociations" in tissue morphology are also common (e.g., Alberch and Alberch, 1981) and will be discussed below under local changes.

3.2.3. Local Differentiative Heterochronies

Where rate or timing of only some cell activities are affected, major novelties can occur because more than just local changes in number of the same kinds of cells are involved (in contrast to local growth alterations). Here, during the neofertilization and differentiative phases, disjunctions of rate and timing can significantly rearrange cells, changing spatial juxtapositions of tissues, and even create new tissues as new inductive interactions are made. However, it is also true that most of the time no new interactions are created; rather, only the duration of old interactions is altered, changing only the size of tissues or organs after differentiation. For this reason, we propose two distinct categories of local differentiative heterochronies: **size differentiative** and **novel differentiative**.

3.2.3.1. Size Differentiative Heterochronies. In spite of the existence of purely post-differentiation changes in rate and offset timing of cell mitosis, there is much evidence that ultimate organ size is often strongly determined by rate and timing changes which affect cellular pathways *before* cellular and tissue differentiation is finished. Hall (1984a) enumerates well-documented examples. The general picture that emerges is that the number of stem cells, the proportion of those that later divide, their intrinsic rate of division, and even the amount of cell death (among still other factors) are determined by pathways (e.g., cellular interactions) before or during the condensation of an organ or tissue. These largely determine the rate and duration of growth of that organ, and ultimately its size because specific growth rate of tissue declines exponentially as it progressively loses its ability to replicate cells (Laird, 1965; Kowalski and Guire, 1974). In terms of the allometric equation discussed in Chapter 2, the allometric exponent, k, represents this diminishing rate of replication as compared between two organs and b represents the relative number of initially replicating cells (Katz, 1980).

For example, adult long bone size seems to be at least partly fixed by the number of mitotically active cells in the original condensation (Moss-Salentijn, 1974; Kember, 1978). This number can often, in turn, be related to timing of the condensation process. A delay of tissue induction, for example, means that cells that would have been committed to that condensation may in the meantime be induced by other interactions to form other tissues. When the delayed induction and subsequent condensation does occur (e.g., bone formation), there will be fewer cells available (Hall, 1984a). It is true that the impressive compensatory mechanisms of development often correct for such changes, but this may not always be so. In this case, the ability to compensate for lost cells is often lost just before condensation occurs (Kieny and Pautou, 1976).

Alberch (1985b) has pointed out that the number of digits on a dog's foot may be related to the size of the limb bud. This often seems to be a matter of body size: small dogs often lose a digit while large ones add one because larger dogs have larger limb buds. Thus, this example might be cited as one of global differentiative change. However, we include it here because the limb bud itself is the focus and, as Alberch notes, mutations affecting only the limb bud can occur which would alter digit number independent of body size. Such a mutation could affect the rate or timing of any number of processes (e.g., mitotic rate of precursor cells) that would alter the size of the resultant limb bud.

Aside from affecting the size of the initial, differentiating organ (e.g., bone, limb bud just noted), differentiative heterochronies can occur later, while the organ is in the process of differentiation. This broaches the difficult subject of trying to relate heterochrony to pattern formation [Wolpert (1982) noted this gap in our understanding]. While a great deal of work obviously remains to be done, we suggest that a useful way to view this relationship (that is consistent with our whole approach) is one of changes in

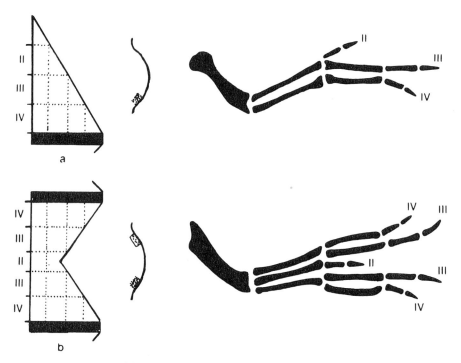

Figure 3-5. Determination of the anterior–posterior axis of the chick limb by the zone of polarizing activity (ZPA). (a) Normal limb with normal gradient of positional information; (b) results of tissue transplantation experiment, with altered information gradient. Modified from Raff and Kaufman (1983).

the rate or timing of local morphogen diffusion, i.e., change in information transfer at the cellular level. A good example is the chick limb, one of the best-studied systems in development. Here, positional information is thought to diffuse from a localized source on the posterior edge of the limb bud. This so-called ZPA [zone of polarizing activity, by Tickle *et al.* (1975), who first presented this model] may thus form a concentration gradient along the anterior–posterior axis (Fig. 3-5). In theory, this information will instruct the cells of the bud on differential mitotic activity and migration, and so determine the eventual limb size, and number of digits on the limb (see Raff and Kaufman, 1983, for discussion). Decreases in the concentration gradient, caused by late or retarded production of the morphogen(s), could result in a reduction in limb size and/or digit number. Phenotypically, such a limb could be classed as "underdeveloped" relative to the ancestor. If this scenario is correct, it could give us a way to mathematically describe heterochrony at a chemo-developmental level earlier than the usual morphometric level since diffusion equations can be employed. Thus, the concentration for a one-dimensional gradient as this is:

$$C = C_0 e^{-\lambda x}$$

where C is the concentration gradient of the morphogen at distance x from the source and C_0 is the initial concentration. λ is the diffusion constant representing the properties of the morphogen under examination (Holder, 1983). Heterochronic changes in rate or timing of morphogen production (or medium transmissivity, or any number of factors) can thus be modeled to produce limbs of varying sizes and "development" (Holder, 1983). While the use of diffusion timing may at first seem to be "stretching" the definition of heterochrony, we will see later that such processes are often the ultimate cause of many heterochronies. Hanken (1989 for review) has related heterochrony to patterning, as discussed below.

The processes which determine proximal–distal growth seem to be somewhat different from those of anterior–posterior growth. As the limb grows in the distal direction, more and more distal structures are added to the terminus. Wolpert and others (1975) have inferred that this is because cell instructions are determined by the time at which cells form at the terminus. Since the relative size of the primordia for each limb bone is about the same size (Raff and Kaufman, 1983) when they forms at this terminus, their different sizes (e.g., ulna versus wrist bone) are due to proliferation dynamics after the primordia are formed. All of this means that while the precise mechanisms of proximal–distal size determination are complex and still somewhat speculative, there is much opportunity for changes in timing and rate to be a major factor in the ultimate size of both the limb and even its component bones. Changes in the timing of cell formation at the terminus will affect cellular activity while changes in proliferation rates (and timing) could affect individual bones. Again we see a hierarchy of possible effects, although the mechanism (and thus feasibility) of such change is obscure. One possibility for at least some of the proximal–distal growth is determination via localized cell interactions. Here cells transfer information only on a very short-range (cell–cell) basis, using a shared polar coordinate system as a frame of reference. Holder (1983) discusses this model at length and suggests that it accounts for much of limb growth.

Differentiative changes before or during metamorphosis are also a common cause of size changes. Hanken and Hall (1988) show how timing differences of skull bone ossification in toads account for species distinctions. Indeed, as they note (and document), cranial bone rate and timing changes are very important in amphibian evolution in general (e.g., Alberch and Alberch, 1981). In addition, Hanken and Hall (1988) make the key observation that premetamorphosis is not simply a growth phase characterized by little morphological change, as is often stated (e.g., Etkin, 1968). Rather it is a time of differentiation, e.g., of osteogenic sites, whereas postmetamorphosis is a time of cell proliferation and calcium deposition. This is in accord with our classification of later (mainly growth) versus earlier (differentiative) heterochronies. However, the dichotomy should not obscure the fact that the end result of even these pregrowth changes often mimics later growth changes in changing mainly size of the organism or organ. Indeed, as noted above, global premetamorphic delays or rate increases are major mechanisms of body size increase in insects and amphibians. Given the massive restructuring of morphology during metamorphosis, this is perhaps not as obvious a conclusion as one might first think. As with local growth changes, the cause of local metamorphic differentiative changes can sometimes be differential tissue response to a global signal, i.e., hormone, as discussed further shortly.

3.2.3.2. Novel Differentiative Heterochronies. Rate or timing changes in the migration and differentiation of cells can lead to new tissues, or at least major new cell spatial configurations within a tissue, as opposed to simple changes in size of tissues, as have been discussed so far. These changes can be especially "creative" in organisms with regulative development in that changes in time or rate of induction can occur if, for example, some cells delay migration and others do not so that new cell interactions occur. Since cells can be experimentally induced by a large number of stimuli (Thomson, 1988), these new interactions can still lead to induction. Thus, uncommitted cells can be stimulated by adjacent cells not normally in proximity, creating new cellular juxtapositions. In the words of Hall (1983, p. 374): "Changing elements of epigenetic control, such as position of epithelia and mesenchyme within the embryo, timing of cell migration, length of time inducers retain inductive ability, etc., without any necessary change in the nature of the inductive signals or in the competence of the responding cells . . . provides a fundamental mechanism for generating evolutionary change." Or, in information terms, as Thomson (1988) has put it, a change may lead to only some cells being affected by a signal, leaving others available to acquire new properties later in the cascade. Such changes are most obviously made in regulative development but mosaic changes are not excluded since new cell juxtapositions can form even where intrinsic changes occur. We believe that such new local interactions are a major source of "creativity" in differentiative heterochronies. Most, if not all, global changes do not "create" much outside what is already in the developmental repertoire. To change too much at once would indeed create a truly hopeless monster. Instead, innovations can be made on a local scale and modified thereafter.

One of the best documented cases of this kind of change is the mammalian jaw (Hall, 1983, 1984a). Much development in mammals occurs from interaction between the two major kinds of cells, mesenchyme and the sheetlike epithelium. An important source of mesenchyme cells is the neural crest. These cells are extremely versatile because when they migrate out of the crest they interact with various epithelia to form nerve cells, pigment cells, and other types, including cartilage and bone, depending on the epithelium involved. Thus, there is a great deal of creative potential where changes in timing of migration, for instance, would alter the mesenchyme–epithelium interaction (obviously not limited to the jaw). Hall (1975) and Maderson (1975) explore the potential of temporal shifting of the mesenchyme–epithelial boundaries in detail (especially many other aspects of craniofacial formation).

In the case of the jaw, many changes have occurred among the major vertebrate groups. As summarized in Table 3-3 and Fig. 3-6, both the site (the particular epithelium) and the time of migration have changed in the epigenetic initiation of chondrogenesis of Meckel's cartilage, an important component of mandibular development. This example of timing shift goes beyond a simple illustration of the ontogeny of one part because the formation of Meckel's (i.e., branchial) cartilage helps initiate a whole sequence of interactions leading to the formation of bone, teeth, and the whole mouth cavity (Cassin and Capuron, 1979). Indeed, in the jaw alone, the whole series of events leading to the mandible is potentially changeable such that Atchley (1987) has suggested that each part of the jaw be viewed as having a separate ontogenetic history with its own intrinsic and extrinsic timing and rate controls. Another key ramification of just this one change is that the evolution of ossicles in the mammalian middle ear requires the late induction seen in mammals, to allow some mesenchyme cells to shift their site of chondrogenesis to the middle ear (Hall, 1983). Oster and Alberch (1982) have modeled

Table 3-3. The Timing of Epigenetic Initiation of Chrondrogenesis in Forming Meckel's Cartilage[a]

Group	Site of inductively active epithelium	Relationship to neural crest cell migration
Cyclostomes		
Lampetra fluviatilis	Branchial ectoderm	
	Branchial endoderm	Mid migration
Urodele amphibians		
Triturus alpestris		
Ambystoma mexicanum	Pharyngeal endoderm	Mid migration
Pleurodeles waltlii	Pharyngeal endoderm	
	Dorsal mesoderm	Mid migration
Anuran amphibians		
Discoglossus pictus	Pharyngeal endoderm	Mid migration
Bombina spp.		
Birds		
Gallus domesticus	Cranial ectoderm	Early migration
Mammals		
Mus musculus	Mandibular epithelium	After migration

[a] From Hall (1983). See Fig. 3-6.

many other permutations of epithelio-mesenchymal interactions, with dermal tissues, to form hair, scales, and glands among other tissues.

Another aspect of interactive timing change is the redundancy inherent in the multiple induction of some tissues. For instance, for the amphibian eye lens to develop, sequential interactions with endoderm, heart mesoderm, and retinal tissues occur. These

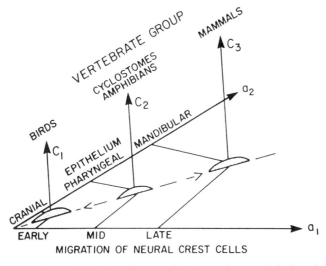

Figure 3-6. Meckel's cartilage, which is homologous within the vertebrates and the branchial cartilage of the cyclostomes, develops from neural crest-derived cells after undergoing an inductive tissue interaction with an embryonic epithelium. Both the inductively active epithelium (cranial, pharyngeal, or mandibular, a_1) and time during neural crest cell migration when the interaction occurs (early, middle, late, a_2), vary among groups. The interaction occurs earlier in bird development (C_1) than in amphibians (C_2) and latest of all in mammals (C_3). From Hall (1984a). See also Table 3-3.

act as inductors on the response epidermis. However, when the retinal inductors were removed from the sequence, lens formation still occurred in 42% of the trials (Jacobson, 1966). This increases the flexibility of the epigenetic cascade by permitting one of the participants of an interaction to be eliminated (and thus perhaps end up elsewhere, to participate in another interaction).

The complete loss of interactions (instead of new ones as above) is another mode of local differentiative novelty. Many of the well-known atavisms illustrate how this could occur. For instance, the induced production of teeth in chicks by combining dental ectoderm from the chick with mesenchyme from a mouse (Kollar and Fisher, 1980) is a revived interaction enabled by humans, originally lost through timing change of tissue relations. When lost, this would have led to the "origin" of the bird's beak. Just why the old pathways should be retained is not necessary to the point but in many cases it is probably because, while not phenotypically expressed in adult morphology, many of them are necessary links in the developmental chain of events. This example also serves to illustrate again the flexibility of regulative epigenesis because if the mouse mesenchyme is combined with chick limb ectoderm instead of dental ectoderm, limb tissue (cartilage) results (Hata and Slavkin, 1978).

In his critique of evolutionary saltations, Levinton (1988) criticizes the notion of atavisms as indicators of developmental creativity by emphasizing that they are not new creations but simply revived preexisting pathways. We believe that this misses a main point: that the novelty is not the atavism itself, but the *new context* in which it may be placed. New interactions are created by reviving old pathways in new places. A good example is the external shell of the pelagic octopus *Argonauta*. As suggested by Raff and Kaufman (1983), this shell seems to result from the reactivation of the old shell secreting glands of the ancestral mollusk. However, in this animal the shell is secreted by two specialized arms. The old pathways are reactivated not in the old location, the mantle, but in a new one, two of the arms. Such a reactivation is surely much easier than creating new pathways of secretion from nothing, and it is certainly a novelty given the shell-less (and very old) state of other octopi.

Differentiative novelties need not be as drastic as the ones discussed so far ("novelty" being a highly subjective term at any rate). Bleiweiss (1987) has elegantly outlined a heterochronic pattern of timing changes in the differentiation of bird feathers. Sometimes the changes are simple arrested development of barb formation; at other times, disrupted development occurs. Brylski and Hall (1988) record what seems to be an excellent example in the origin of external cheek pouches in geomyoid rodents. During development the corner of the mouth participates in the evagination of the buccal epithelium to form the external pouch. In rodents with internal cheek pouches, the mouth epithelium is not continuous with the buccal epithelium and so does not participate in the evagination. Thus, Brylski and Hall suggest that the external condition is a trait derived from ancestors with internal pouches wherein the location or direction of the buccal evagination has changed (anteriorly) causing the mouth to participate in it. Such a change may be attributable to simple alterations of rate or timing parameters in some part of the local developmental sequence (e.g., a prolonged migration of cells in the anterior direction leading to the shifted merging of epithelia) although the exact mechanism is speculative for now. This change is particularly interesting because even though it has had a major effect on life-style and adaptation, it results from a very small alteration in development.

There are a large number of chick and mouse mutations which have pleiotropic heterochronic effects on skeletal growth (review in Hinchliffe and Johnson, 1980). In particular, they affect size and shape of the limb buds and cause abnormal differential redistribution of mesenchyme to different regions. In creeper in chicks there is differential growth rates of cartilage rudiments.

Other examples of nondrastic differentiative changes would include a large group of alterations involving "partial" metamorphosis. We have already discussed many local deviations from "global" change in earlier sections (e.g., axolotl). These are novel in the sense that "mosaic" morphologies are created which combine the often radically different pre- and postmetamorphic traits. The proximate mechanisms of change vary, but they are generally related to the hormone production–tissue response interaction. In the blind cave salamander (*Typhlomolge*), for example, the thyroid is nearly absent. In *Proteus anguinus,* another blind salamander, hormone production is normal but the target organs cannot respond (Levinton, 1988). An especially good example of a mosaic has been found in the salamander *Notophthalmus* (Reilly and Lauder, 1988), which shows neoteny in the variable retention of larval ceratobranch and external gills. Of importance to the subject of novelty with potential for evolutionary "advance" is that these gills appear to be comparable in all respects (and origin) to those found in ray-finned fishes and lungfish.

Finally, we return to the problem of diffusional rate and timing (and pattern formation) and attempt to relate it to cell and tissue level heterochronies. We discussed earlier how this might work with size differentiative heterochronies and turn now to cases where more novel effects may be observed. One of the simplest and well known are the diffusion models of pigmentation begun by Turing (1952) and elaborated on by Murray (1981; Oster *et al.,* 1988). These provide good evidence that coat markings, ranging from stripes to spots, arise as a consequence of changes in diffusion parameters. There are many ways to alter such parameters: changing initial concentration, medium transmissivity, and so on; among the most interesting is the changing of the growth rate of the organ or animal (spatial gradient) across which diffusion occurs. Thus, Murray has shown that the size of the growing field can determine whether an animal is spotted or striped.

While such alterations are hardly major novelties, they are useful for their view of the elegant simplicity by which "new" traits can be created (e.g., spots from stripes). The clarity of view here is because the biochemical dynamics involved are directly relatable to the product. Usually of course the "translation" of scaling upward through the cellular level is much more complex, such as in many of the earlier examples where more complex changes, in cellular interaction, resulted from changes in time and rate of information diffusion. However, there has been some important work that has helped relate even some complex cellular interactions to the notion of pattern formation. At least some of this may be related to diffusional changes of morphogens in the same manner as the pigmentation changes. For example, Alberch (1985a) discusses how "new" limb patterns are produced in *Ambystoma* that are not comparable to ancestral patterns. In an excellent summary article, Hanken (1989) has extended this to show how novel skeletal patterns result from scaling effects of miniaturization. Specifically, size change alters spatial patterning (i.e., causes "repatterning"), leading to novel arrangements in limb morphology within a single group of salamanders. Hanken might object to our attempt to precisely relate heterochrony to this repatterning but, as he discusses, there is clearly some kind of relationship between spatial and temporal repatterning. He provides numerous examples showing that such repatterning may be an important source of novelty in evolution. Finally, Minelli and Bortoletto (1988) discuss timing of diffusion flow in arthropods. Driever *et al.* (1989) demonstrate how the *bicoid* morphogen determines the embryonic spatial domains of *Drosophila*. Might not change in diffusion timing lead to major changes here as well? Is this "heterochrony"?

4. BIOCHEMICAL MEDIATORS OF HETEROCHRONY

To pursue the metaphor of heterochrony as *change in timing of cell dialogue,* it is important to look at the media of the dialogue: the morphogens, hormones, and other messengers which are sometimes altered and always participate in changes (at least in regulative processes). In general, they serve as inductors, inhibitors, and receptors by which the genes between cells communicate. They may thus be conceived as the intermediate link between the higher cellular level processes of mitosis, differentiation, and migration, and the lower level processes of the gene. This area is one of the fastest growing and most complex fields in science and we cannot hope to do it justice in a book with as broad a scope as this. Instead, our goal is to provide a sampling of some of the most recent advances and integrate them with some general principles that shed light on the operation of heterochrony. We see this section, even more than the others, as a tentative effort to give form and pattern in an evolutionarily meaningful way to the many facts emerging from developmental biology.

4.1. Extracellular Mediators

One of most important realizations to emerge from the burgeoning data in this area is that many of the major morphogens are closely related in chemical structure, and are often clearly derived from one another. What was once a large number of disparate polypeptide "growth factors" can now be assigned to a fairly small superfamily of categories (Table 3-4) that are shared by many groups, from *Drosophila* to mammals, and often serve the same function. Obviously there has been much evolutionary conservation. Of equal interest is that in some cases the same morphogen helps control the timing of many processes: mitosis, cytodifferentiation, and migration, often in conjunction with only one or two other factors or determinants (see Mercola and Stiles, 1988, for thorough review). In all cases, there is much potential for heterochrony in these cellular processes. Changes in the rate or timing of production of these mediators, sensitivity to them, or transmissivity of any intervening matrix can change the timing of intercellular communication.

The following is a brief discussion of the major growth factors, generally in a heterochronic context. Growth factor families (Table 3-4) are grouped according to nucleotide and amino acid sequence homology and receptor-binding activity.

Platelet-derived growth factor (PDGF) is the major mitogen found in serum and is a potent growth factor for the many cells of mesenchymal origin (Beckmann *et al.,* 1988).

Table 3-4. Growth Factor Families[a]

Group	Members
Epidermal growth factor	EGF
	Transforming growth factor-α (TGA-α)
	Vaccinia growth factor (VGF)
Insulinlike growth factor	IGF-I
	IGF-II [somatomedin C; multiplication stimulating activity (MSA)]
Transforming growth	TGF-β_1
factor-β	TGF-β_2
	TGF-$\beta_{1.2}$
	Inhibin-A
	Inhibin B
	Activin-A
	Activin AB
	Müllerian inhibiting substance
Heparin-binding growth	Acidic HBGF [acidic fibroblast growth factor (aFGF), endothelial cell growth
factors (HBGF)	factor (ECGF)]
	Basic HBGF (bFGF)
	Products of the *int-2*, *hst*, and Kaposi's sarcoma proto-oncogenes
Platelet-derived growth	PDGF-A
factor	PDGF-B (*cis* product)
	PDGF-AB

[a] From Mercola and Stiles (1988).

A good example is its role in promoting mitosis and migration in rat optic nerve development and in the timing of cytodifferentiation. Secreted by nearby astrocytes, PDGF will cause progenitor cells to proliferate, migrate, and differentiate into oligodendrocytes over several weeks. Without PDGF, the progenitor cells stop dividing and differentiate in a few days (Noble *et al.*, 1988). In addition, Raff *et al.* (1988) have shown that there is an internal mitotic clock within the progenitor cells which sets a maximum number of divisions a cell (and its progeny) can undergo before differentiating. Apparently, after the maximum number is reached, the cells lose their mitogenic responsiveness to the PDGF and cease division. This withdrawal from the cell cycle then stimulates cytodifferentiation. This also explains the fact that mitosis stops and differentiation begins when no PDGF is provided. In addition to promoting mitosis, PDGF may also act as a chemotactic agent, providing a gradient which the migrating progenitor cells follow (Noble *et al.*, 1988). Further "fine-tuning" of PDGF action can be made by antibodies which inhibit its effects. This growth factor acts at the earliest stages of embryogenesis (Mercola *et al.*, 1988) so that timing changes can theoretically occur very early in development. Mutations have recently been discovered in mice and humans which disable receptors of PDGF so that affected cells do not proliferate in response to it (Escobedo and Williams, 1988). Here then is a precise mechanism whereby timing change in mitosis can occur.

The epidermal growth factor (EGF) family is one of the best characterized. It also shows diverse activity as a mitogenic and morphogenic promoter. Adamson and Meek (1984) showed that while the number of EGF receptors increases in fetal mouse tissues, the affinity for EGF decreases two- or threefold during gestation. This is thought to reflect

the changing roles of the receptors and resultant changes in tissue growth and differentiation. Thus, EGF may stimulate proliferation of stem cells and differentiation in more
mature cells (Sporn and Roberts, 1986). EGF precursors are thought to encode trans-
membrane peptides which may act on the cell surface to mediate cell–cell interactions
by recognizing a complementary receptor on another cell (Teixidó *et al.*, 1987).

EGF is particularly important as a mitogen in epithelial and connective tissue (review in Carpenter and Cohen, 1979). It is thought to be a regulator of local growth fields
(Rall *et al.*, 1985) and has wide-ranging systemic effects as well (Coleman *et al.*, 1988).
Of direct heterochronic interest is its ability to promote early onset of such ontogenetic
events as tooth eruption and eyelid opening in the mouse (Cohen, 1962). Nieto-Sampedro (1988) has recently presented evidence of the importance of EGF in glial cell
mitosis in the brain. Some "fine-tuning" may be carried out by retinoic acid which is
known to enhance EGF binding and synthesis of EGF receptors (Kukita *et al.*, 1987).
Thus, Abbott *et al.* (1988) have shown that retinoic acid will cause continued cell proliferation in mouse palate formation beyond the time it normally does, because EGF has
continued effect. DNA synthesis continues and programmed cell death does not occur
(indicating that EGF also affects timing of that important morphogenetic event). Recall
the discussions above of retinoic acid as the first proven morphogen, and its key role in
limb development.

The heparin-binding growth factors include nearly a dozen previously identified
factors (Lobb *et al.*, 1986). Fibroblast growth factor (FGF) is one of the best known.
Among its activities is its role in muscle growth in stimulating myoblast proliferation and
inhibiting the onset of muscle differentiation (Kardami *et al.*, 1988). It is also a potent
mitogen for a number of mesodermal and neural crest (see below)-derived cells. Recent
studies reveal two forms, aFGF and bFGF, which show 55% sequence similarity (Seed *et
al.*, 1988). The first form is found in neural and kidney tissue while the second is found in
pituitary, brain, placental, cartilage, bone, and muscle tissue (references in Seed *et al.*,
1988). It seems to be vital to chick limb bud development (Seed *et al.*, 1988), showing
position-dependent growth. Apparently, cells in the posterior region of the bud control
normal limb development (Aono and Ide, 1988).

Insulinlike growth factors (IGF) are small single-chain polypeptides that bear a
remarkable amino acid sequence resemblance to human proinsulin (review in Blundell
and Humbel, 1980). There are two major forms, I and II, which are closely homologous.
As noted earlier, they are especially important in mammalian development. IGF-II is
mainly a fetal mitogen active in embryonic tissues and declining during gestation. However, high levels are found in the adult mammalian brain where it apparently continues
to stimulate axon outgrowth of existing neurons (Haselbacher *et al.*, 1985). IGF-I shows
an opposite overall pattern. Instead of decreasing within a few weeks after birth, as does
IGF-II, IGF-I shows increasing levels of tissue concentration. Both forms may operate
over large areas, circulating throughout the organism bound to protein carriers, and
locally in an autocrine fashion (Gluckman, 1986). This pattern can help explain the
mammalian brain–body allometries discussed elsewhere. Heterochronic changes in
IGF-II would be expected to have a more global effect, on both brain and body, since
they would generally act earlier. Also, IGF-II seems especially active in neural tissue.
Late-acting IGF-I would affect mainly body size alone because most brain growth is
finished early on (Chapter 7).

Finally (in this superfamily on Table 3-4), there is the transforming growth factor family (TGF). This factor elicits a wide variety of responses, depending on cell type and other growth factors present. Thus, it may either promote or inhibit cell division in culture (Roberts et al., 1985), and show variable effects on differentiation; e.g., chondrogenesis of epithelial cells is stimulated while myogenesis is inhibited (see Sporn and Roberts, 1986).

An important growth factor not shown in Table 3-4 is nerve growth factor (NGF) which plays a crucial role in the development of the central nervous system, especially sensory and sympathetic neurons. For example, rat fetuses subjected to NGF deprivation (injected with NGF antibodies) develop into adults with greatly reduced numbers of these neurons (Johnson et al., 1980). Neurons have been shown to bind NGF directly and fetal neurons require NGF for survival (Godfrey and Shooter, 1986). Compared to the other factors discussed, the main source of NGF is often very localized, since it is not produced until the sensory nerve fibers grow into their target organs. These organs (or fields, e.g., epithelial cells) then begin to synthesize the substance (Davies, 1987; Korsching and Thoenen, 1988; Scarpini et al., 1988, for further discussion). Cell–cell or cell–matrix interactions modulate biochemical response to NGF via receptor molecules. Of particular interest here is that cell response can be altered by these molecular mechanisms (Doherty et al., 1987). This indicates a possible mode of changing cell sensitivity and therefore causing heterochronic alteration of subsequent cell activity.

The molecular aspects of the all-important neural crest cell migration and morphogenesis, discussed in the context of differentiative heterochronies, are also becoming clearer. Perris et al. (1988) have shown that instructions to these cells are received during their migration through the extracellular matrix. They seem to be fairly unrestricted in phenotypic expression until parts of the matrix provide specific information, after which they become committed. Thus, change in timing of migration can change time of commitment. The glycoprotein, tenascin, coincides with the pathways of migration and seems to be an important player in this morphogenetic determination, although the exact role is unknown (Mackie et al., 1988).

4.2. Intracellular Mediation

Among the organisms with mostly invariant (mosaic) cell development is the nematode, *Caenorhabditis elegans*, which has been used extensively for heterochronic studies because of its small size, transparency, ease of culture, and simple genetics (see Hodgkin, 1986, for review). In such organisms, instructions for cell division and differentiation do not come from cellular neighbors (e.g., via growth factors discussed above, as used in regulative development). Rather the regulatory proteins are expressed and act within each cell alone. Not only do mitosis, differentiation, and cell migration normally follow a precisely programmed sequence, but there also is extensive programmed cell death (followed by phagocytosis). In such cases, laser ablation of neighboring cells has no effect on development of remaining cells. This invariance allows cell lineages to be thoroughly mapped and heterochronic changes in expression can be easily identified as temporal deviations. [A minority of cells are regulative, mostly in the postembryonic stage and in only a few parts of a lineage (Hodgkin, 1986).] As a result, many current findings on cell-level heterochrony come from this organism.

Chalfie *et al.* (1981) reported one of the earliest known heterochronic mutations, *lin-4* (*lin* for lineage abnormal). This perturbs the timing of cell divisions, causing some somatic cells to repeat larval division patterns at a time when they should have ceased dividing and begun differentiation into adult cells. Terminologically, this might be called cell-level mitotic hypermorphosis since offset timing is delayed. Alternatively, it is also postdisplacement of differentiation, since the onset of that event (usually contingent upon mitotic cessation) is delayed. We note this not to encourage the merciless use of such jargon, but for the opposite reason: to illustrate its strictly contextual utility. Usage of such terms must specify the exact events considered. There is also the question of scale. Where in the upward causation of heterochrony, from genes to tissues, is the reference made? This is discussed further, in the last section.

Ambros and Horvitz (1984; Ambros, 1988, for review) have subsequently reported three more heterochronic mutations: *lin-14, lin-28,* and *lin-29. Lin-14* is a widely acting mutation, causing a number of cell lineages to either repeat the fates of their progenitors or express the fates of their descendants (a kind of post- or predisplacement of differentiation, respectively). Which of the two heterochronic events occurs depends on the genes involved, as discussed in the next section on genetic aspects. In contrast, the *lin-29* mutation is much more restricted in scope. Only certain cuticle-forming cells are affected. This contrast is some of the first direct evidence of a hierarchy of heterochronic effects at this level.

4.3. Comparison of Extra- and Intracellular Mediation

While the distinction between rate and timing changes initiated by information outside the cell versus inside seems clear enough, it is important to note that the mechanisms are often similar. In both cases a regulatory protein stimulates genes which promote the change, as discussed in more detail below. Further, these proteins seem to be similar in all cases. For example, TGF, used extensively in extracellular mediation of morphogenesis, is homologous (has sequence similarity) to a morphogenetic protein of *Drosophila* (Padgett *et al.,* 1987). Similarly, the gene encoding EGF shares domains of homology to homeotic loci in *C. elegans* and *Drosophila* (Mercola and Stiles, 1988). Again we see much evolutionary conservation in the signaling agents, regardless of whether the signal is produced and acts entirely within the cell or is produced in neighboring cells and traverses short or long (e.g., circulating growth hormones) distances.

Perhaps most arresting of all in their similarity among many organisms are the basic regulatory mechanisms of mitosis that are triggered, whether it be externally or internally. From yeasts to humans, control of the cell cycle has been recently found to be simple, involving just two major regulatory proteins [overview in Marx (1989) from which this immediate discussion derives]. Kinase is the proximate enzyme that triggers the biochemical mitotic sequence. The other protein is cyclin, which triggers the kinase. The all-important timing appears to be regulated by the synthesis of cyclin, which is made constantly in the cell. When some threshold concentration is reached, cell division begins and cyclin concentration drops, and its accumulation begins again. Here is a (perhaps the) major mechanism which can be altered to change the timing of cell behavior in development. Mutations affecting (directly or indirectly) rate of cyclin synthesis, or threshold sensitivity of kinase to cyclin concentration, are among a number of possible changes that could alter timing of mitosis. Indeed, a number of mutations have been isolated that do affect such timing, such as mutations in the "*string*" gene of *Drosophila*. Strong conservatism is seen in human kinase which is 63% identical to that of the yeast enzyme. Given an evolutionary divergence of over a billion years, this is striking, and no doubt a testimony to the extremely critical role it plays in development and maintenance of the organism.

In spite of these similar mechanisms it is apparent that some major differences in heterochronic effects exist between extra- and intracellular mediators. Because regulative development consists of so many contingent events which are plastic enough to adjust to new contingencies, rate or timing changes in regulative cells can be much more "creative" by causing new interactions and cell rearrangements to occur, as already discussed. Changes in autonomous cells do not spread, although of course they may be large-scale if many cell lineages are "reprogrammed" by mutation. We think the point is important because most cell-level heterochronic studies have focused (for good practical reasons) on mostly invariant systems like the nematode. Care must be exercised in extrapolating to cellular heterochronies in regulative systems.

4.4. Cell Adhesion and Heterochrony

The fine-level changes that can occur in cellular migration, mitosis, cell death, and cytodifferentiation obviously subsume a large number of the steps leading to tissue formation and organogenesis. However, for the process to be complete, the cells must be spatially fixed and integrated. In particular, cell types must be able to distinguish and adhere to one another. Such selectivity and adhesion must have a molecular basis, namely cell adhesion molecules (CAMs) many of which have been identified (review by Damsky *et al.*, 1984). These surface glycoproteins were first isolated from chick embryos (Thiery *et al.*, 1977) and have become a major area of research on tissue organization.

The relationship of CAMs to heterochrony has been most directly argued by Edelman (1986) in evidence for his regulator hypothesis of cell positioning. Basically, this asserts that the expression of CAMs are spatiotemporally regulated and promote a variety of organizational events. A study by Takeichi (1988) is one of the clearer examples, showing how the onset or offset of expression of subclasses of CAMs correlates with collective cellular activity. Inhibition of CAM expression induces dissociation of cell layers.

Changes such as this, e.g., affecting the timing of expression of cell adhesion properties, could often induce phenotypic changes that are not clearly comparable to the unchanged (ancestral) state. Thus, as argued by Oster and Alberch (1982), small changes in cellular mechanico-chemical properties could have drastic, "nonlinear" effects, e.g., a drastic rearrangement of cells. A good example is found in chick feather patterns, which are highly disrupted during development by altered synthesis of CAMs (Goetinck and

Carlone, 1988). However, not all such changes need be manifested this way. As argued by Edelman (1986), many changes in the timing of CAM gene expression may be covariant, so that change will be more comparable to the ancestral pattern. It would show a traceable vector (Chapter 2) that would be more continuous at the tissue grade scale.

5. GENETICS OF HETEROCHRONY

There are about 10,000–100,000 genes in each eukaryotic cell. In a multicellular organism, most cells have the same gene complement, that of the original zygote (exceptions include red blood cells). Thus, it is clear that only some genes are expressed in each cell, given the wide variety of cell type, cell activity, spatial position, and so on that occur during development and maturity. A distinction is often made between "**developmental genes**" and "**maintenance genes**," wherein the former operate only during ontogenetic processes. However, like the other dichotomous gene distinctions discussed shortly, this seems to be an oversimplification. As one of many examples, mature liver cells can be "switched on" to re-create new cells in cases of widespread destruction.

The first question then is how gene expression is controlled. In bacteria, this occurs mainly at the level of transcription such that only a small number of genes are actively transcribed at any one time. A typical transcription control interaction is where repressor proteins bind to loci to prevent transcription and an inducer substance binds to the repressor (leaving the loci free) when transcription is needed. Unfortunately (for our understanding), this simple model found in prokaryotes has proven much too inadequate to explain eukaryotic gene expression. Some idea of the differing complexity is conveyed in the facts that at any given time, only about 4% of the genome of any eukaryotic cell is actively expressed while at least 75% and up to 100% of the genome of a prokaryote is being expressed (Davenport, 1979).

An effort to generate a more complex model, based on the prokaryotic model, for eukaryotes was first attempted by Britten and Davidson (1969). It has drawn much attention but there is little direct evidence for it so far. One of the most contentious aspects is their emphasis on the importance of repeated DNA, which role now seems improbable (Levinton, 1988). Nevertheless, the notion of hierarchies of gene "batteries," which are sequentially switched on/off in binary fashion, interacting with positive and negative feedback loops, seems to be valid, as evidence emerges (Bonner, 1982). In addition, while the exact mechanics are unclear, it is evident that much, if not most, control is indeed accomplished by *selective transcription,* in eukaryotes. Thereafter, the copying mRNA transfers the transcribed information outside the nucleus where it is translated into one of a myriad of enzymes or other peptide products (although there may be considerable delay between transcription and translation in eukaryotes). This is not to say that all gene expression is transcription controlled as there is definite evidence of translational and other controls in some organisms (Raff and Kaufman, 1983; Gilbert, 1985). An excellent review of DNA-binding transcriptional control and rate-limiting steps in mammals is by Mitchell and Tjian (1989).

The importance of all this to heterochrony is that development can be characterized as differential transcription (and to an unknown extent, translation) of genes in different cells and tissues at different times and rates, with each step ultimately initiated by the transcription and translation products of the previous step. The net result of these steps is the orchestrated assembly of cells. Therefore, an understanding of genetic aspects of heterochrony must focus on changes in the timing and rates of differential transcription, translation, and the other forms of selective gene expression.

5.1. A Proposed Hierarchy of Gene Regulation: Three Basic Levels

For far too long, discussion of genetic regulation has centered around the overly simple regulatory/structural gene dichotomy. **Structural genes** are generally defined as those which code for proteins used in cellular machinery (during growth and maintenance). **Regulatory genes** generally refer to genes which produce substances controlling structural genes, perhaps by binding to sites "upstream" of the structural transcriptional unit. The typical discussion cites the abundant evidence that morphological differences among species rarely correspond to differences in structural proteins, as determined by sequencing. The classic examples are that humans and chimpanzees are about 99% similar in their protein sequences in spite of great morphological differences while frogs may be very similar in morphology but very different in protein sequences (King and Wilson, 1975; Cherry *et al.*, 1978). Or, as Britten and Davidson (1971) pointed out, higher and lower organisms have some 90% of their enzymes in common. The conclusion is that if protein sequences, as produced by structural genes, do not determine evolutionary (morphological) changes, then it must be something else. The problem of course is that "something else," the "regulatory gene" processes usually implicated, includes a great variety of different kinds of activities.

This is not to say the concepts of regulatory and structural genes are of no use. To the contrary, we believe that Raff and Kaufman's (1983) observation that "structural genes supply the materials for development and regulatory genes both provide and interpret the blueprint" is very useful. Our main point is that sufficient knowledge is now available to refine the concepts. Certainly, we are not the first to note the oversimplicity of the structural/regulatory dichotomy (e.g., Grant, 1985; Levinton, 1988). Our goal is to synthesize some of these other discussions in a heterochronic context. This will shed light on the genetics of heterochrony beyond the crude "rate" and "timing" gene concepts of the past. In addition, we point out that, for now, our use of the term "hierarchy" is strictly pragmatic, and its meaning should be clear from the context of the discussion that immediately follows. A much broader discussion of this critical but much abused concept is presented in the last sections of this chapter, where we try to draw many ideas of this chapter together.

5.1.1. D-genes, S-genes, and R-genes

Arthur (1984) has made (what we feel is) an important distinction among three mutually exclusive types of genes. Consider the basic genetic situation in Fig. 3-7 wherein a regulatory gene X regulates structural gene Y in its production of E, an enzyme

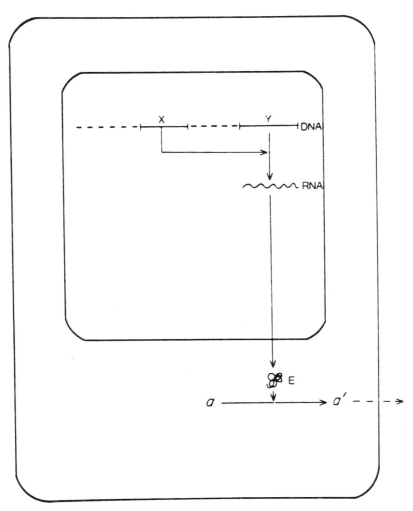

Figure 3-7. A gene control system in which a structural gene, Y, which produces an enzyme, E, is regulated in an unspecified way by a controlling gene, X. Whether X and Y regulate development depends on whether the enzyme's product, a', is a morphogen. From Arthur (1984).

(from a precursor). If the enzyme is a morphogen, e.g., one of the many growth factors discussed above, or any peptide involved in the stimulation of mitosis, cytodifferentiation, migration, cell death, "pattern formation," or any other morphogenetic (or developmental) process, then both genes are developmental or **D-genes**. In contrast, if the enzyme has a nondevelopmental "maintenance" function (e.g., digestive enzyme in an adult), the X gene is a regulatory or **R-gene** and the Y gene is simply a structural or **S-gene**. The main value here is that we can clearly distinguish between "regulatory" gene in the sense of regulating development and in the other sense of regulating another gene. The drawback of this triad is that it does not explicitly distinguish between the regulating D-gene and the structural one. Hence, we will use the terms regulator D-gene and structural D-gene.

In earlier sections we have promoted the view that heterochrony can be seen as a change in the rate or timing of the flow of developmental information affecting timing or rates of cell assembly. In the current context, we can further extend this view to that of a change in a D-gene which affects that flow, e.g., the rate and timing of production of a morphogen (or, conversely, a change in sensitivity to a morphogen). The potential for further modification of timing or rate is seen in that a morphogen's properties can be altered by other morphogens and inhibitors, which are produced by still other D-genes.

As an example, consider the flow of information in the action of the growth factors just discussed (much of this, and the next, paragraph is based on the excellent review by Mercola and Stiles, 1988). The sequence is initiated when the morphogen is produced. After diffusion and upon contact with the target cell, the growth factor binds to a specific, high-affinity receptor (usually a tyrosine-specific protein kinase, although IGF-II has a different receptor) which spans the outer cell membrane. This binding action is thought to induce soluble, intracellular messengers which transmit a signal to the nucleus (review in Rozengurt, 1986). The identity of the messengers is uncertain but may be phospho-proteins, inositol phosphates, diacylglycerol, or simply ions, among many possibilities. Very soon thereafter, changes in gene expression can be detected, so that the whole intracellular sequence, from initial binding to genetic response, occurs in only a few minutes.

A main point with respect to differential heterochronic effects is that some growth factor effects are cell-lineage specific. For example, NGF will stimulate offset of mitosis and onset of differentiation in PC12 cells (derived from rat pheochromocytoma) into the mature neuronal phenotype. Yet the same morphogen will promote *onset* of mitosis in fibroblast cells. Even more interesting is that the same D-genes are stimulated ("switched on") by the messengers in both cases, the *c-fos* and *c-myc* genes. The key determinant in the differing responses seems to be their current status in their cell cycle. The undifferentiated state of the PC12 cells compared to the differentiated status of the fibroblast cells seems to explain their differing responses to the *c-fos* and *c-myc* gene products. Thus, changes in timing or rate of production of the same morphogen can affect both differentiation and mitosis in different cells. And since mitotic offset often correlates with differentiation onset, it can affect timing of both processes in the same cell as well.

5.1.2. A Hierarchy of D-genes

The discussion above indicates a hierarchy of D-genes, divisible into two gross categories: (I) those regulating activity among cells (*intercellular* processes), and (II) those regulating the *intracellular* responses to the higher level information. This second level is further divisible into a hierarchy of those genes regulating progressively more refined (specific) intracellular activity.

I. These are the D-genes responsible for the production of the many hormones and other morphogens, such as growth factors, which serve in "regional" extracellular communication. Mutations in such genes seem likely to be one of the major forms of heterochronic change due to the fact, noted earlier, that most morphological (and evolutionary) change involves only change in arrangement and numbers of cells, not in intracellular processes (producing new kinds of cells). Changes in the rate or timing of mitosis in the growth phase, or in mitosis, differentiation, and migration of cells in the differentiative phase would seem to often be rooted in these genes which control the timing and rate of cellular production of the morphogens involved. Greater production of a mitogen for instance would result in earlier attainment of stimulus threshold levels and thus earlier stimulation of mitosis. More cells may also be subjected to threshold levels (due to the greater initial concentration) and so more cell growth would occur in the growth field as a whole. Zuckerkandl (1976) is among those who have postulated the existence of just such genes. He has specified genes which code for: (1) onset of synthesis of gene product, (2) duration of synthesis, (3) rate of synthesis, and (4) total output of synthesis.

The spatial scale of any such change will in large part be determined by the stage of development when it occurs. A small embryo can be subdivided into only a few "diffusional domains" as determined by the physicochemical constraints of diffusion [e.g., Crick (1970) has estimated that morphogenetic fields are limited to 50–100 cells maximum]. Obviously, as body size increases, more and more domains are created beyond which communication by diffusion alone is unattainable or impaired. Arthur (1984) has extrapolated this view to the ultimate conclusion that the adult phenotype is finalized by a large number of D-genes having minor or negligible morphogenetic effects, many of which are the polygenes of quantitative genetics. A good example of a very late, local growth field change, with a simple polyallelic basis, seems to be the classic heterochronic case of the amphipod *Gammarus chevreuxi* (Ford and Huxley, 1929). Here the homozygous recessive genotype, *rr*, results in delay in the onset and decrease in the rate (neoteny) of melanin deposition in eye pigmentation. While such traits in populations are rarely couched, or thought of, in heterochronic terms, that is what they are. The vast majority of subspecies polymorphism, from height to shape variation, may be considered as changes in rate and timing of cell activity among individuals (Chapter 4).

While generally agreeing that D-gene effects decrease in magnitude as ontogeny progresses (following von Baer), we would alter this oversimplistic view of monotonically diminishing effects to include the endocrine and other hormonal processes, discussed earlier, that involve circulation through the body's vascular and other systems. These "integrative" systems counteract the diminishing control of morphogenetic fields by providing late-acting regulation of large-scale effects. Thus, mutations affecting body growth and maturation in the late growth stage discussed in an earlier section are not only possible but are probably an abundant form of heterochrony. These integrative systems do not play an important role until late in development because the circulatory/vascular system needed for dispersal have not developed until then (Davenport, 1979).

II. The second level in the hierarchy represents D-genes in charge of intracellular responses to the morphogen information from D-genes of level I. This second level includes two further levels: A, the "switch" genes which make decisions based on the morphogen information; and B, the genes which carry out the decisions. In brief, the D-genes of A tell the D-genes of B what to build (e.g., a new cell in the case of mitosis, new structures in the case of cytodifferentiation) and the D-genes of B regulate the structural genes in building it. No doubt a further hierarchy exists within B as regulator D-genes coordinate activity with one another and their products.

To summarize briefly: IIA D-genes are expressed when the intracellular messenger enters the nucleus and stimulates them. To use the previous example, these are the D-genes which, upon exposure to the *c-fos* and *c-myc* products, initiate either mitosis or differentiation, based upon the cell's current status. Presumably these products cause the regulator D-gene to stimulate production of some message by a structural gene which in turn stimulates other regulator D-genes. These last (IIB) control structural D-genes which contribute directly to cellular change. For example, if the mitogenic sequence is initiated, the IIB D-genes stimulated are those involved in the production (e.g., of various proteins) of more of the same kind of cells. If the differentiative sequence is set off, D-genes involved in production of proteins used in the metamorphosis of the affected cell to the more mature stage are stimulated.

5.2. Genetics and Heterochronies of Mosaic Development

While the above scenario may apply to the many cases of heterochrony in regulative development, there remains the problem of invariant cell development, such as found in *Caenorhabditis*. As extracellular communication is not a determinant in morphogenetic cell activity, the rate and timing of changes must come from intracellular D-genes. This is not as qualitatively distinct from regulative behavior as it might seem at first. It simply means that the "switch" D-genes (IIA) are responding to intracellular rather than extracellular information. The cascade of events regulated by lower level (IIB) D-genes after the sequence is initiated (e.g., differentiation) is probably much the same, given the similarity of cells and subcellular processes among organisms.

The main question with respect to heterochrony is thus: what internal mechanisms determine the timing (and possibly rate) of information transmission within each invariant cell and allows them to be at least roughly synchronized? The answer is unknown at this time. One possibility in at least some cases is that an initial substance (e.g., macromolecular) is progressively "diluted" as mitosis proceeds and the substance is more widely distributed. After a certain number of cycles, a threshold level is reached which terminates mitosis and begins differentiation. Mitotic offset and differentiative onset would be covariables, as noted earlier.

While the exact mechanisms are not understood, some clues are known. The *lin-14* dominant mutation of *Caenorhabditis* will cause cells to repeat the lineages of their progenitors. However, recessive mutations will lead to the opposite: early onset of ancestral cell fates. This has led to speculation that the mechanism involves dosage of the *lin-14* "gene product" (an intracellular morphogen similar to extracellular ones?). Higher levels specify earlier fates; as levels of the product diminish with time, more mature fate thresholds are reached. The dominant mutation increases the *lin-14* activity causing repetition of earlier fates while the recessive does the opposite by diminishing gene product levels earlier (Ambros and Horvitz, 1984). In this case the "clock" may involve some kind of programmed decay in morphogen productivity of the *lin-14* (and perhaps other) gene(s). Indeed, the whole sequence of invariant cell ontogeny may operate via a series of such decays, each reaching a threshold level which stimulates the next D-gene(s) into activity. (Compare to control of cell cycle via external morphogens, Section 4.)

An interesting counterpart to this heterochronic change is the *lin-12* mutation which represents a spatial change. Here the dominant mutations, probably causing overproduction of the *lin-12* product, cause cells to adopt fates normally expressed at other locations (Hodgkin, 1986). Similar homeotic mutations are well known in *Drosophila,* which also act in an intrinsic manner (Garcia-Bellido, 1977). In such cases a given gene operates in strictly binary fashion (Hodgkin, 1986), choosing one of two cascades of differentiation. In the case of the temporal (heterochronic) genes the binary choice seems to be on/off of only one cascade, the only variation being when the choice is made. (For further discussion, see Ambros, 1988.)

5.3. General Discussion of Genes and Heterochrony

Since the intracellular master "switch" genes (IIA) determine when a cell undergoes mitosis, differentiation, or migration, we suspect that these should play a major role in heterochronic change. It is unlikely that D-genes "beneath" this level show much alteration since most evolution is by change in number and position of cells. Probably for basic logistic reasons, there is little tampering with kinds of cells produced, aside from the timing of differentiation. As these master genes are responding to stimuli, there would seem to be only two basic ways for D-genes to be involved in the timing of these cellular processes. One is to alter the time at which the threshold stimulus reaches these genes. In the case of extracellular morphogens, this is clear enough. Change in onset time or rate of morphogen production by regional D-genes (group I) would alter timing of threshold stimulus (and perhaps the number of cells affected). A number of examples are discussed in Section 3, as in changes resulting in increased and prolonged productivity of growth hormones in hominoid primates. The ever-cited axolotl would be another example, in those cases where neoteny is induced by failure of the hypothalamus to produce thyroxine (i.e., onset time is permanently delayed; Blount, 1950). This has a very simple genetic base: a single gene with two alleles (Tompkins, 1978).

The timing of intracellular stimuli was just discussed and although the information is not from a regionalized central source, the principle is the same in that time and rate of information ("gene product") release is the key determinant. The main qualitative difference is that in extrinsically mediated change there are more factors in modulation: rate and timing of other morphogens and inhibitors which interact with the signal morphogen may be a factor. Intrinsically moderated information is transmitted much more directly to the master D-gene(s). In short, extrinsic regional signals are subject to other cascades which have rate and timing parameters of their own, introducing more potentially alterable control parameters. In terms of information theory, this would give regulative heterochronies more potential for "fine-tuning" the process and morphological end product, and more variety in general.

The second way for D-genes to alter the timing of cellular morphogenetic activity is to change the stimulus effect (threshold) on the receiving D-gene upon reaching it. Even if the rate or time of stimulus transmission is the same, the time of onset will obviously now change since the stimulus will "register" sooner or later. A possible mechanism for this would be modification of DNA binding sites. The ability of differential alteration of response among tissues and cells is an important result of this. For instance, in the axolotl, target tissues can differentially undergo altered response thresholds to thyroxine, resulting in heterochronic mosaics (Tompkins, 1978).

There are also changes in genes, that are not normally "D-genes," that could affect this information flow. Changes in the medium between the morphogen producer and receiver could alter timing. Thomson (1988) has suggested this for changes in the properties of the extracellular matrix. Changes in the cell membrane or the messengers (receptor molecules) which carry the signal to the nucleus would have similar effects. Similarly, researchers have recently experimentally blocked gap junctions between cells. This affected the timing of intercellular communication by small molecules and produced developmental abnormalities (Marx, 1989). Hence, while the D-gene concept is useful, it is important to realize that "structural" and other properties encoded in genes not actively involved in directing development may play a role in producing heterochronic changes.

Not only is the regulatory/structural contrast too simple but mention should be made of the *"rate gene"* and *"timing gene"* terms often bandied about. [Rate genes made their appearance at least as early as 1918 in Goldschmidt's work (Gould, 1977); temporal genes have been reviewed by Paigen (1980).] Given the discussion above, many, if not most or even all, cases of heterochrony result from a battery of genetic interactions which seem rarely initiated by a simple "rate" or "time" gene. Regarding extracellularly mediated morphogenesis and growth, we have seen that changes in: output of a morphogen, media properties, or any number of inhibitors or enhancers can alter rate or timing of cellular response and activity. The D-genes which initiate such output are often simply responding to environmental stimuli of their own and so do not fairly seem to warrant "time" gene designation. Even less justifiable are cases where response time is changed due to mutations affecting response to stimuli (e.g., cell membrane transmissivity to morphogens or gene threshold changes). "Time" gene may be useful where it affects the initial amount of a substance determining the number of mitotic cycles, leading to timing alterations of mitotic offset in invariant cell lineages. Perhaps the most justifiable application of the terms is that of "rate" gene to whatever genes might be controlling the rate of morphogen production, which could translate into rate changes in cell activity.

One of the clearest views of the whole process is offered by Bonner (1988) who discusses the concept of **gene nets**. These are the networks of genes and their products which act to integrate yet "modularize" development. Most importantly, he points out that the vast network of biochemical processes and interactions have intrinsic regulatory properties of their own. Heterochronically, this means that much of the timing and rates are determined by their own intrinsic rates of reactions. The genes themselves are removed, sometimes far removed, from this biochemical cascade, maintaining control only by carefully timed "injections" of their products into crucial "branch points" where small inputs have big effects. Recall our discussion at the beginning of the chapter that this is why so few genes are required to create so much organization: most of the order is created by the "inertial flow" of the gene products. The (structural) genes need code only for the small inputs, that must be carefully timed via regulator genes responding to stimuli of their own (previously constructed gene products). While we have discussed the D-genes as those which are most directly involved in these "injections," other genes may affect the cascade through less direct effects, e.g., physicochemical properties affecting flow. Finally, above the biochemical level, and thus even farther above the direct control of the gene level itself, is the cell level. Here is where the process of self-assembly becomes manifested and selection is therefore also strongly manifested.

6. SUMMARY: BRANCHING MORPHOGENETIC TREES

To illustrate the concepts of this chapter, we will use **morphogenetic "trees."** These have a long history in developmental biology and continue to be used (Smith, 1983; Arthur, 1984). They have major advantages over other heuristic, conceptual devices used to visualize ontogeny. The epigenetic "landscape" of Waddington (1940) is totally metaphorical, and has no potential for quantification. The topologies of Thom's (1975) catastrophe theory are more amenable to quantification but are analytically very cumbersome. Even where they can be made tractable, they are not readily interpretable in terms of real morphogenetic processes. Obviously the tree model, as does any partial representation of reality, has its shortcomings too, as we will see shortly.

Shown in Fig. 3-8, the morphogenetic tree can be viewed as modeling development as cellular assembly, focused on earlier. As discussed in more detail by Alberch *et al.* (1979), Slatkin (1987), and Thomson (1988), traits develop from the same tissue until a transition occurs, after which they develop at least partially independently. *Cells thus have a shared history (time axis) until some event acts to spatially separate them (space axis).* Only one spatial dimension is shown for simplicity, though of course three are really involved. The advantages of this model are that it is quantifiable, as we will see, and is readily related to actual morphogenetic events. Everything from cellular activities (mitosis, migration, differentiation) to biochemical diffusion can be visualized. Such trees are not limited to these cellular- and subcellular-level events; indeed they are often used to demarcate branching of "traits" or tissue level changes (Smith, 1983). Also in Fig. 3-8, we see that the tree approach can supplement the "closed loop" model of Oster *et al.* (1988) by giving it a directional arrow of time (an "open" system). Conversely, the closed loop model supplements the tree by emphasizing the role of feedbacks and contingencies in the system, such as the critically timed "injections" discussed above.

As always, simplicity is bought at a price and we do not pretend that this view is analytically powerful. For more rigorous treatments of bifurcations the reader is referred to the excellent works of Oster (e.g., Oster and Alberch, 1982; Oster *et al.*, 1988). Nevertheless, the conceptual basis of developmental bifurcations is the same. Our purpose here is mainly to provide a graphical, conceptual framework. The tree directly depicts the spatiotemporal contingencies of cellular assembly. It reifies the usually vague (but often stated) idea of a cellular or developmental "pathway."

A serious drawback of the tree approach, noted by Raff (1989), is that, unmodified, it naively assumes that development is entirely nested in space–time: early changes always affect more cells than later changes. As a very broad generalization, this is of course true, as canonized by von Baer (Raff and Kaufman, 1983); e.g., changes in the zygote or early embryonic stages will have much more effect on the whole organism than a change in one or more limb buds. It is also reflected in a less well-known generaliza-

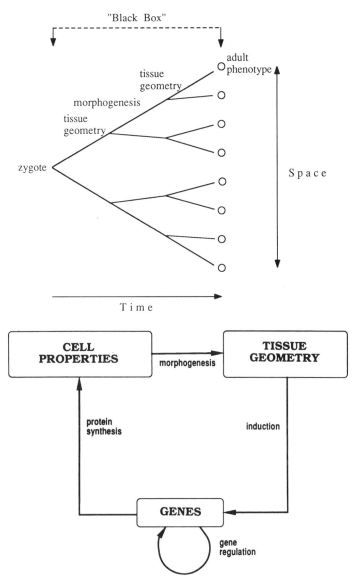

Figure 3-8. Top: a morphogenetic tree showing time–space relations among cells, as they divide, move, and/or differentiate during the "black box" of development, from zygote to adult phenotype ("organ topology"). Bottom: closed loop gene control and relations among gene, cell, and tissue levels. Note how this loop depiction can be incorporated into the morphogenetic tree at the top, pointing out how development is a sequence of interactive loops. Bottom figure modified from Oster *et al.* (1988).

tion, **Pearson's rule**, which says that the correlation (covariation) among body parts increases with their proximity to one another (Levinton, 1988). In terms of earlier discussions (Chapter 2), this is reflected in spatial serial correlation across cell ("growth") gradients that make fitting curves to growth trajectories so difficult. This is diagrammatically shown in Fig. 3-9 in the concept of the **growth field**, or "modules" of

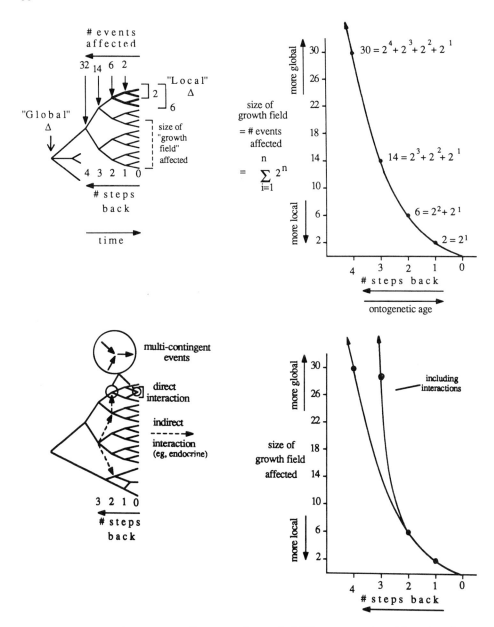

Figure 3-9. Top: morphogenetic tree illustration of "growth field" concept and spatial nesting of events caused as related to temporal separation. In general, later heterochronic alterations will have smaller effects because of this space–time nesting. For example, as shown on the right, there is an exponential decrease in effect as the change occurs later. Put another way, because development is a multiplicative process (in many ways, not just mitosis), earlier events cascade through the system with progressively greater impact the earlier they occur. Bottom: a more realistic model which accounts for the fact that spatial units (e.g., cells, tissues) may interact again after they have separated (branched). This can actually increase the effect of early events, as shown. This is not to say that earlier effects are invariably greater: this simple model does not take into account the many homeostatic mechanisms which can "dampen" or contain such changes by channelization (e.g., in the frog and echinoid examples in text).

cells, controlled by Bonner's (1988) modularized "gene nets." Such covariation is perhaps the key interface linking ontogeny to evolution because it lies at the heart of how much development "constrains" variation, and therefore controls evolution. For example, selection operating on one growth field may "drag along" another that covaries with it. This is a recurrent theme of the chapters that follow and discussion of its effects is reserved until then. In this chapter, we are concerned mainly with the mechanics of how such covariation is produced.

As discussed above, the idea that effects unerringly diminish monotonically is naive. This can seen in two ways. First, early changes may not have big effects. This is shown in direct-developing sea urchins and frogs (work of Raff discussed above; see also Raff and Kaufman, 1983) wherein drastic differences in the early development of related species result in similar adults. Second, late changes do not always have small effects. Changes in hormonal processes occur very late in ontogeny, only after the integrative systems are well-developed, yet have major impacts. Dwarfism and related syndromes as illustrated by the human pygmy discussed above are good examples. However, these caveats do not necessarily negate the tree approach because they can be incorporated into it (Fig. 3-9 of late but "big" effect). Nevertheless, the main point of Raff (1989), that models can never supplant empirical information, is undoubtedly true. Such issues as just how much development has constrained or "channeled" evolution can never be resolved by rhetoric or diagrams with arrows. We repeat that our goal here is strictly to provide a conceptual framework for the organization, presentation, and understanding of such information.

6.1. Trees and Growth Heterochronies (Late Ontogeny)

The tree concept is useful in modeling and classifying heterochronic processes: the major cell processes (and hence changes in time and rate) of late ontogeny are mitosis and migration. The third activity of cell differentiation mainly occurs earlier. A branching model of late ontogenetic mitosis is illustrated in Fig. 3-10. Also shown is how the space–time relations can be translated into the standard heterochronic size (i.e., space)–time plots discussed in Chapter 2. For instance, progenesis is seen as early termination of mitosis while neoteny is slower mitosis (branching). Similar schematics can be visualized which show changes in migration; e.g., early onset of spatial movement. Obviously many more cells are involved in real growth but the principle is unchanged. Also shown in Fig. 3-10 is the role of information flow in the rate and timing of bifurcations. Thus, one might refer to retarded rates of morphogen diffusion as an ultimate cause of tissue grade timing or rate changes, at least where regulative growth is concerned. This is not to say that heterochronic terminology employed at this scale is especially desirable or useful in all circumstances. No doubt most phylogenetic comparisons will continue to be made at tissue and organismic scales. However, a truly complete understanding of developmental changes must at least be aware of the small-scale changes underlying the larger-scale ones. This is essential if the study of heterochrony is ever to rise above the level of descriptive morphological pattern taxonomy (further discussion in McKinney *et al.*, 1990).

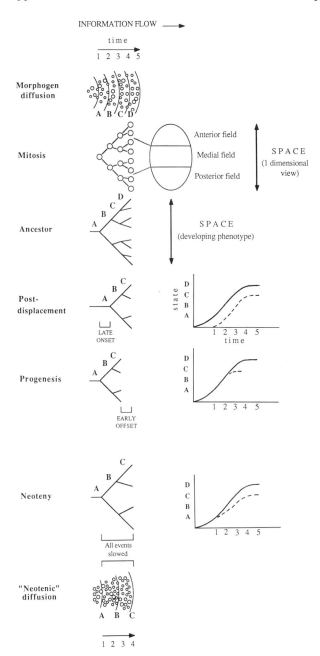

Figure 3-10. Heterochronic classification using a branching diagram. In this case, cell-level events are shown (mediated by molecular-level, morphogens, events) but other levels (e.g., tissue) are possible. Ancestral ontogeny shows four events (A–D) occurring in space (one dimension for ease of viewing). Changes in onset time, offset time, and rate of spatial events are shown below the ancestral ontogeny (these are paedomorphic events; the opposite would occur in peramorphosis). Bivariate classification of these events (sensu Chapter 2) is also shown. This depiction is not limited to mitosis or morphogen diffusion; any spatiotemporal events may be classified in this way, although change in rate or timing of cell dialogue is certainly a major root cause of heterochrony.

Two key concepts usefully viewed in the tree approach are those of depth and breadth of ontogenetic change. These terms are often used loosely, but in the tree model, **breadth** of change can be viewed as the number of cells (or traits) affected; it is the size of the growth field (sensu Fig. 3-9). **Depth** of change can be viewed as the degree to which the cells (fields) are affected. For instance, if growth is accelerated in a trait or growth

field, breadth would be the number of ancestral cells (or traits) affected by the acceleration and depth would be the amount of acceleration (e.g., twofold, or fivefold increase).
For timing change, depth would be how early onset was (much depth if very early) or
how late the offset was (much depth if very late). This is essentially the number of steps
back in Fig. 3-9. In both cases, the greater the depth, the greater the net effect on the
original cells generally is (see Chapter 8; see McKinney, 1988a,b, for evolutionary context of breadth and depth).

Slatkin (1987) has pioneered the necessary incorporation of quantitative genetics to
the developmental bifurcation view. As above, traits are seen as sharing the same history
until a branching occurs, after which they develop in (partial) independence. Each
branch grows for a characteristic duration (t) and rate (r), determined by genetic parameters. As shown in Fig. 3-11, the end product is a "topology," which essentially refers to
the final adult growth field configuration in our model, with which it can be homolo-

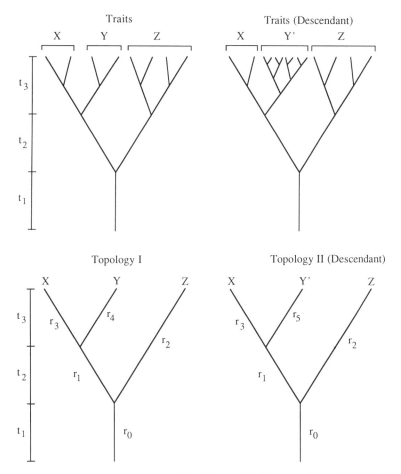

Figure 3-11. Topologies and traits (X, Y, Z) produced during development, illustrated by branching tree
diagrams. Each tissue grows for a characteristic time (t) and at a given rate (r), after which it branches
(differentiates) into other tissues. Organ Y is accelerated in the descendant, shown on right. See text.

gized. For instance, trait X (Fig. 3-11), growing during the third phase (t_3) at rate r_3, will increase in size by t_3 times r_3 (e.g., 5 hr $\times$ 1 multiple/hr = 5 multiples or 5-fold increase from its size at the end of t_2). The final size of each trait then will be the cumulative result of each rate and duration of growth, some shared with other traits, others not, depending on whether or not the rate/timing change occurred after branching. Thus:

$$X = r_0 \, (t_1) + r_1 \, (t_2) + r_3 \, (t_3)$$

The rate and duration of growth of each branch are under genetic control and we thus can model heterochronic changes in rate (e.g., r_4) and duration (e.g., onset of t_3, offset of t_3). This is shown in Fig. 3-11 where a descendant has undergone acceleration of trait Y (in this case with no change in duration). Slatkin's topological and our cell-multiplication depictions of this acceleration are both shown for comparison. As noted, earlier changes will usually affect more traits. Notice, however, that this view is not as simplistic as it may seem since late but "big" changes can be modeled by changing the rate or duration of any number of later branches. Slatkin explores a number of interesting implications derived from the model by assuming that the rate and time parameters have a polygenic basis, and can therefore each be treated as a quantitative-genetic trait in its own right. This takes us to the issue of selection and environment, which is examined in subsequent chapters.

6.2. Trees and Differentiative Heterochronies (Early Ontogeny)

Recall that early ontogenetic heterochronies, occurring before differentiation, can result in either simple size changes or novelties. The former can be illustrated above. For instance, if Y' (Fig. 3-11) was a larger stem cell population instead of an adult phenotype, then later multiplication (trait multiples) could be affected accordingly. However, for novelties we must consider the third cell process, differentiation. Here rate or timing changes result in spatial arrangements that can lead to new cellular juxtapositions (configurations) and sometimes even alteration of induction. In both cases, "novelties" can occur, as illustrated in branching form in Fig. 3-12. While the term "novelty" is somewhat subjective, we discussed earlier how, in developmental novelties, it essentially connotes a trait (or trait suite) that is "noncomparable" to a trait in the ancestral ontogeny.

6.3. Summary: Allometric and Disjunctive Heterochronies

A very diagrammatic view of all possible kinds of change in an ontogeny is shown in Fig. 3-13. In this view, a single adult phenotype (or topology to use Slatkin's term) is not modeled. Rather the figure shows the collective total of potential morphs producible by modification of the ancestral species's ontogeny. (To limit it, let us say only the physiologically viable potential morphs are shown.) Most of these are simple allometric changes from late ontogenetic perturbations (growth and size differentiative heter-

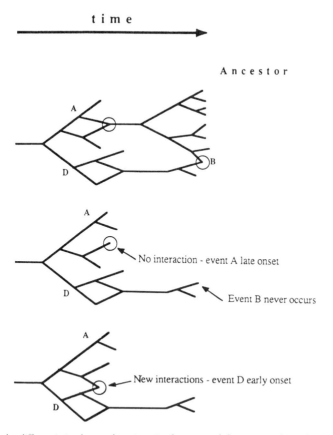

Figure 3-12. Early, differentiative heterochronies arise from spatial disjunctions (y axis) among tissues/cells. New interactions occur and/or old ones do not when rate or timing changes (of information flow) happen.

ochronies). These are, very generally, nested and there is a great deal of contiguity (and continuity) in the morphospace. Alberch (1982) has also discussed this, in terms of "clumped" morphospace. In the jargon of our tentative terminology of this chapter, such allometric (contiguous) changes would specifically result from late ontogenetic, growth heterochronies, or earlier-acting size differentiative heterochronies. These are often controlled by a number of rate and timing genes (polyallelic, multiple loci), affecting a number of traits.

More controversial is the notion that early rate/timing perturbations of the trajectory can produce the viable morphs outside of contiguous space. As discussed, these can arise from any of a large number of alterations in cell properties, cell movements, and so on as mathematically analyzed by Oster's work cited above, and just noted in the above section. Such a threshold effect is very common in nonlinear systems. Indeed, with all of its tightly woven, intricate local and global feedbacks (both positive and negative) and exponential, multiplicative interactions, a complex ontogeny is arguably the most "nonlinear" natural process in the known universe. This applies not only to cell-level behavior, but also accounts for the problems of classifying heterochrony as well. Because of the

Potential Descendant Morphospace

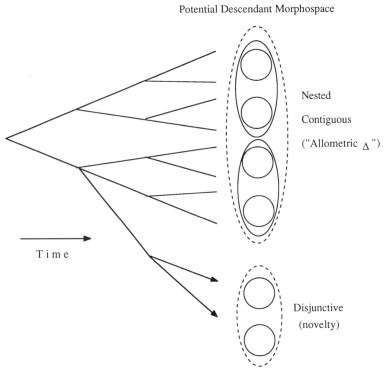

Figure 3-13. Branching tree diagram of the morphospace available to a species's collective genome: nested, contiguous morphospace, determined by a number of rate and timing genes at various loci (and often polyallelic) affecting various traits. Also illustrated are the much rarer "disjunctive" heterochronies, which generally act earlier and can cause major tissue rearrangements ("novelty") outside of the ancestral region of morphospace.

enormous spatial connectance and serial (temporal) contingency of the system, minor changes at one scale (the molecular level) can cascade to a variety of amplifications to higher levels, from minor to profound. Thus, changes in rate or timing at molecular levels can result in simple, "linear" extrapolations in rate or timing at the tissue level (i.e., allometric), can have saltatory effects, or no effect at all.

The case of "nonlinear" creations of novelty (defined above as noncomparable to ancestral traits) is as important and fascinating as it is problematic. This calls to mind Dobzhansky's (1969) insight when faced with the similar problem of the novel emergent (he called them transcendent) properties of the human brain: "novelty" must be judged on the basis of predictability rather than analysis of component differences between the novelty and the ancestral state. *Novelty* is thus defined on the basis of how well we can predict the properties of a structure given thorough knowledge of the ancestral structure. In the context here, we see that developmental changes can create (viable) descendant structures that would be very difficult to predict (novel, noncomparable), given the highly nonlinear, interactive complexity of the developmental process. To continue the "morphospace" metaphor, the resultant morphs are *disjunctive,* being discretely different from ancestral morphs.

The scale at which heterochronic terms are useful (if at all) in describing and understanding developmental changes depends on the investigator and his goals. This brings us to the final topic of the chapter: the integral but knotty issue of hierarchy.

7. HIERARCHIES AND HETEROCHRONY

The concept of hierarchy has been implicit throughout this chapter, as indeed it must be since hierarchy springs from complexity and no process rivals the complexity seen in constructing a complex organism. Indeed, hierarchical descriptions help manage complexity by isolating dynamics at a single level (e.g., cell level) and ignoring details at lower organizational levels. In spite of this potential as a conceptual aid, hierarchical reasoning can be very confusing, especially when discussed in the abstract. For that reason, we begin with a basic discussion of hierarchical theory as applied to development.

7.1. Basic Hierarchical Principles as Applied to Ontogeny

A growing number of workers are becoming interested in hierarchy in biological systems. Thomson (1988) has discussed a hierarchical view of development, and includes all major references of on-going work, such as that of Salthe (especially Salthe, 1985) and Eldredge. Coming at hierarchy from another "level," ecologists have also written extensively on the subject. Allen and Starr (1982) and, particularly, O'Neill *et al.* (1986) provide excellent reviews from this perspective. Our primary concern here is with the use of hierarchy as an aid to understanding the mechanics of heterochronic changes. We do not dwell on, but will briefly mention, the currently topical issue of hierarchies of selection.

Loosely used, a hierarchy is simply a system that can be organized into levels or ranks. As there are many kinds of phenomena that can be ranked in biological systems, many hierarchical arrangements are possible. This has led to much obfuscation, when a worker tries to isolate "the" hierarchy in a living system. In addition, there are different kinds of hierarchies. O'Neill *et al.* (1986) make a number of useful distinctions. A **nested** hierarchy occurs when the higher levels are composed of, and contain, the lower levels. For example, an adult organism is composed of organs, composed of cells, composed of molecules, and so on. This particular example is a nested **vertical** hierarchy in that there are higher and lower levels. In contrast, we may also speak of a nested **horizontal** hierarchy where the nested elements are composed of units on only one level. For example, if we focus on the cell level, we can identify cell systems (such as organs) that can be broken down into more refined subsystems (anatomical structures or cell aggregates, within the organ). Koestler (1969) used the term *holon* for horizontal units in which the components interact more frequently than with other units, and the term has become common.

The relevance of all this to ontogeny is clearer when we make one final distinction put forth by O'Neill *et al.* (1986), that of hierarchies of *static* versus *dynamic* phenomena.

In the example above, the adult phenotype has been described as a nested vertical hierarchy with nested horizontal structures. Since we focused on matter (atoms, cells, and so on) at one point in time, this is a static hierarchy. However, it is obviously more enlightening to the study of ontogeny (and its changes) if we can find a dynamic hierarchy that captures what is going on: cell assembly through time, orchestrated by smaller units within the cells. Thus, we might generate the following dynamic vertical hierarchy: DNA synthesis–gene (information stimulus)–morphogen (information transfer)–cell (construction, movement)–organ (construction). Vertical changes in the lowest level (DNA, i.e., molecular) can cascade upward, causing rate or timing changes at higher levels. In turn, such vertical changes can also cascade horizontally. At the cell level, for example, holons during growth are not the static organs discussed above but interacting growth fields. Where rate or timing of interactions is affected, major (or minor) changes can occur among the growth fields. If ontogeny were a simple process, without much interactive complexity, e.g., simply multiplicative growth of cells, then the nested static hierarchies of each ontogenetic point in time would closely match the nested dynamic hierarchy of cell assembly. Temporal and spatial nesting would be perfectly or at least highly concordant. In such a process, early changes would always have proportionately greater effects (e.g., the nesting of "contiguous" morphospace in Fig. 3-13). "Progressive" ontogenetic evolution would always lead to terminal addition (and recapitulation). The human desire for such simple patterns has often led to the imposition of this perfect space–time nesting on ontogeny in the past, but, as we have said, at some variance to the truth.

Evolution of Ontogenetic Hierarchy

The current hierarchical complexity of complex organisms is of course the result of a long period of selection on units at lower levels. Intermittently, "random" changes sometimes add to the complexity, and are conserved if beneficial. Since selection continues to act on lower levels as well as later-evolving, higher ones, a hierarchy of selection (see especially Buss, 1987, and Thomson, 1988) evolves in step with the hierarchy of structure. This would agree with two recent review articles on the hierarchy of selection which conclude that "there is no clear consensus concerning the relative importance of the various phenotypic levels in the selective interactions with the environment" (Tuomi and Vuorisalo, 1989), and that hierarchy "renders sterile the debate concerning 'what is THE unit of selection?' . . . It is likely that, for major traits, several levels are often involved" (Gliddon and Gouyon, 1989). Molecules, genes, cells, tissues, and individuals are all replicated during ontogeny and all have their own environmental variables to contend with, at their respective scales. It seems likely that, as with heterochronic diagnosis, our perception has been biased by our own scale of reference. The individual organism (especially multicellular) is closest to our own spatial scale and so has historically absorbed most of our attention as the focal level of selection.

A perhaps nonobvious process is that this Darwinian drive of random but sometimes conserved change leads to increasing complexity (i.e., more levels) in both vertical and horizontal hierarchies. If we posit a nonperfectly replicating molecule (e.g., DNA) as the ultimate ancestor, there is a resultant increase in the vertical hierarchy: DNA–gene–

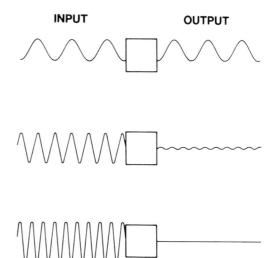

Figure 3-14. Attenuation of input signal by a "black box." Low-frequency signals (top) pass unmodified. However, even high-frequency signals at low levels (e.g., cell) are "dampened" (middle) or completely isolated (bottom) at higher levels (e.g., tissue, individual) in the hierarchy (output on right). From O'Neill *et al.* (1986).

morphogen–cell–organ–metazoan and so on, as is well known (this all-important "nowhere but up" process is discussed in Chapter 8). However, there is also an increasing hierarchy in each horizontal level: more and different kinds of genes evolve (hierarchy of regulator genes, structural genes), different morphogens with variable degrees of control, and more and different kinds of cells arranged into more complexly interacting growth fields. In brief, the holons also become arranged into more complex hierarchies. In many cases, the driving force behind this vertical and horizontal hierarchical expansion is "random" change in rate or timing of the assembly process (heterochrony).

7.2. Hierarchy and Process Rates

O'Neill *et al.* (1986) discuss the fundamental point that organization results from differences in process rates (a point originally made by Simon, 1962). Even when viewed as a static hierarchy, a complex system, such as a metazoan, operates over a wide spectrum of rates. Physiological processes at the molecular level are obviously much faster than those at organ levels, for instance. In terms of information theory, this means that signals originating at lower levels are filtered when passed through higher levels. Generally speaking, each level is well buffered from signals from lower levels (Fig. 3-14). For example, individual cells are hardly sensitive to every change at the atomic level in their internal processes; similarly, huge numbers of cells may die in a metazoan with little or no effect on the organ or individual level. Of course, being buffered is not the same as being isolated and if lower-level change is sustained and drastic enough, it will become felt at higher levels, even though the effects are "averaged" out. If enough liver cells are killed, for instance, organismic-level dysfunction will appear. Further, some "unusual" low-level changes can cascade upward (become amplified) quite rapidly to cause relatively rapid changes at higher levels: for example, a gene ("molecular") mutation causing a cancerous cell can have obvious effects on the organ and organism.

Having used (again) the static view as the simplest example, let us turn to the dynamic view of rate hierarchies in ontogeny. In this case, we see the same buffering of rates: DNA changes often have no major effects at the cell and higher levels (e.g., repeated DNA). Even some fairly dramatic genetic and cell-level changes can have little effect on adult phenotypes (as in the frog and echinoid examples above). This would also agree with the observation that most genetic changes that do have effects are relatively minor, being mainly the allometric/size changes of the growth and differentiative phases. On the other hand, "unusual" lower-level changes can cascade upward: these are the novel differentiative heterochronies, where small changes have profound "rapid" (macroscale) effects compared to the ancestral ontogenetic process. Thus, a small genetic alteration of timing of cell migration can have large effects on early (and later) ontogeny.

7.3. Hierarchy: Overview

The hierarchical view of ontogeny, at least so far, is more of an organizing perspective than a device for yielding profound insights. That levels of organization appear in complex entities is virtually a tautology, and most of us would intuitively expect that higher levels ("larger" phenomena) would have slower process rates: higher levels are composed of progressively less numerous, larger components and things will naturally take longer (on average) to happen on that scale. The most important explanatory aspects are what determines the dynamics at any particular level, i.e., what causes that set of interactions. This critical information, the determinants ("laws") of vertical and horizontal interactions, is something that the hierarchical approach in itself cannot tell us. It can identify the levels, but cannot tell us why they exist. Why, for example, does one low-level change in ontogeny cascade upward to produce "saltations" while others have no effect whatever? For the answers to these kinds of questions, we must turn to empirical investigation of the ontogenetic processes themselves, learning the specific dynamics at each level. As Raff (1989) has said, there is no substitute for the raw data of experimental developmental biology. To that end, we suggest that the role of heterochronic changes in development will prove to be very important at a number of levels. This is perhaps the central message pervading this chapter.

Chapter *4*

Heterochronic Variation and Environmental Selection

> *Slow coming to perfection, both for growth and ripeness signifies long life in all creatures. Concerning the length and shortness of the life of living creatures, hitherto negligently observed, and proceeding from divers causes, instead of certain rules hard to find, these notes following may be added.*
>
> Francis Bacon, 1638, *The Historie of Life and Death*

1. INTRODUCTION

Classically, heterochrony has been mainly concerned with evolutionary change at the species level. It has been viewed as little more than an ancillary part of evolutionary theory that deals largely with patterns of morphogenesis. But as discussed in Chapter 3, in our view heterochrony is a fundamental process of morphogenesis. Biological evolution, after all, arises from the complex interplay between extrinsic factors, such as natural selection, and intrinsic factors, such as the nature of an organism's developmental system. This has been shown to embrace two fundamental concepts (Gould, 1989): *historicism*, that is, form as dictated by past events and past relationships: and *formalism*, interpreted as the "rules of structure," wherein an organism's properties are constrained to some degree by the physical consequences of its inherent structure.

Whether or not a new phenotype survives is not simply a question of extrinsic factors, but also depends upon what the phenotype is (based on historical and formal factors) and what it does (Thomson, 1988). If it succeeds, then it effectively defines its own selective regime. It is this "tug-of-war" between intrinsic and extrinsic factors that lies at the very heart of evolutionary change (McKinney, 1988b). Extrinsic factors may sometimes exert a direct perturbing influence on an organism's developmental system. Alternatively, intrinsic perturbations to the developmental system may occur and gener-

99

ate a wide variety of heterochronic morphotypes in a relatively passive extrinsic setting.

Whether or not speciation occurs is dependent upon a wide range of factors. One of the most important is the availability of niche space for the newly evolved morphotypes. But it is this generation of what may be termed **heterochronic morphotypes**, i.e., morphologies produced by heterochronic processes within species, that is the subject of this chapter. It is necessary to view the whole gamut of intraspecific variation firmly in a basic heterochronic framework before we can progress to detailing the role of heterochrony in speciation: its influence on the directions, rates, and targets of evolutionary change. It is our contention that the vast majority of cases of intraspecific morphogenetic variation arise from heterochrony, including such aspects as polymorphism and sexual dimorphism. If heterochrony has a profound influence on the regulation of morphological variation and on rates and directions of evolution at the species level and above (Alberch, 1980, 1982; Alberch and Gale, 1985; Eldredge and Gould, 1972; Foster, 1985; Hoffman, 1981, 1982; Rachootin and Thompson, 1981; Williamson, 1981; McNamara, 1982; Maderson et al., 1982; Levinton, 1986; Maynard Smith et al., 1985; Oster et al., 1988; McKinney, 1988b), then its role in inducing intraspecific morphological variation must form an integral part of analyses of the significance of such variation in evolution. In this chapter, we extend our discussion in Chapter 3, moving from the cellular and genetic levels to the level of the individual organism. In particular, we wish to focus on the interplay between an organism's development and its environment. This will serve as a bridge connecting development with selection, and in turn both of them with the evolutionary topics in the chapters to follow.

Phenotypic Plasticity

An organism's developmental system can be considered to be a constrained system that protects the "integrity" of the species and confines it within certain morphogenetic bounds. This has been variously termed "canalization" (Waddington, 1957, 1962; Rendel, 1968, 1979); "developmental homeostasis" (Lerner, 1954); and "developmental constraint" (Maynard Smith et al., 1985). Waddington (1975) defined **canalization** as a property of an organism's developmental system which buffers the course of ontogeny from environmental and genetic perturbations, thus ensuring the production of optimal adult phenotypes with minimal variance. The ability of the phenotype to resist both extrinsic, environmental and intrinsic, genetic pressures indicates regulation of the developmental program (Alberch, 1980). However, where high levels of phenotypic plasticity occur, this is said to be evidence that the developmental system is poorly canalized (Bradshaw, 1965; Pachut and Anstey, 1979; Hickey, 1987). Such phenotypic plasticity produced by a less constrained developmental system is caused by heterochrony. A wide variation from the phenotypic "norm" may be caused by any of the heterochronic processes, as the changes may occur by perturbations to timings of onset, offset, or rates of growth.

The whole concept of **developmental constraint** and its role in evolution has received a great deal of attention in recent times. A major problem has been that the term "constraint" has been used in different ways and continues to be used with more than one meaning (Gould, 1989). One inappropriate way in which it has been used is to describe any causal biological change. But as Stearns (1986) has pointed out, this is tantamount to describing all biology as the consequence of constraints. Unfortunately, the term "constraint" is used in two ways, in both a positive and a negative sense. In the negative sense, developmental constraints are the limits imposed upon natural selection to move in a particular direction (Gould, 1989). In the positive sense, it is used to confer the sense of channeling phenotypic change in a direction that is influenced by past historical factors and formal structure (these are combined in ontogeny and are the intrinsic components of evolution), rather than by any exisiting adaptations. As Gould (1989) has pointed out, Maynard Smith *et al.* (1985) interpreted developmental constraint in the "negative" sense, i.e., limitations on phenotypic variability. Yet, as they also point out, constraint will have a strong influence on the evolutionary pathways that are followed. Essentially, by constraining phenotypic change, due to the influence of intrinsic factors produced by past historical effects and the formal structural rules, developmental constraints will restrict many of the options open to the organism. This will cause a channeling along a certain morphological, and thus evolutionary, pathway.

Localized populations within species with relatively high degrees of phenotypic plasticity induced by intraspecific heterochrony, are those most likely to be the targets of selection at the species level. As Guerrant (1988) has pointed out, phenotypic plasticity is not just bothersome noise, but is a phenomenon of considerable evolutionary and ecological significance in both animals and plants. However, it does not follow that high phenotypic plasticity should be a precursor of speciation. Phenotypic plasticity itself may be of particular adaptive significance. It may only be because of such intraspecific heterochrony, which induces wide phenotypic variation, that certain species were able to evolve. This, as we shall demonstrate, arises from the innate ability of such species to survive greater degrees of environmental perturbations than their sister species that have less extensive plasticity. From this it might be supposed that species in more highly disturbed, variable (i.e., unstable) environments (r-selected) show a greater degree of phenotypic plasticity in the form of intraspecific heterochrony. Likewise, species with more highly canalized developmental systems will preferentially be associated with more stable, K-selected environments. This heterochronically induced phenotypic plasticity may be expressed as general morphological variation; alternatively, it may only be expressed at certain times, as polymorphisms, and under the influence of specific extrinsic factors that trigger it.

One way to recognize phenotypic plasticity and canalization is to compare ontogenetic trajectories among individuals and among populations (Johnson, 1981). This can be carried out on populations in different habitats. Phenotypic plasticity ensues when ontogeny among populations diverges from initially similar early ontogenetic morphologies. Organisms with canalized developmental programs will have similar ontogenetic trajectories between populations, and so exhibit minimal morphological divergence during ontogeny. The basic question raised by studying the role of perturbations on developmental programs in affecting morphological plasticity is whether or not high degrees of phenotypic variability have a genetic basis, and if so to what extent the regulation of developmental programs affects such variability.

2. THE NATURE OF PHENOTYPIC VARIATION

To understand how heterochrony works in evolution, both at the species level, as well as at higher taxonomic levels, such as in the generation of evolutionary novelties, we consider first the nature of phenotypic variation and the development of polymorphism. Mayr (1970, p. 89) has defined **polymorphism** as "the occurrence of several strikingly different discontinuous phenotypes within a single interbreeding population." To Ford (1940), polymorphism is the occurrence together in the same locality of two or more discontinuous forms of a species in such proportions that the rarest of them cannot be maintained by recurrent mutation. While the discontinuous nature of polymorphisms seems to be generally accepted at the phenotypic level, Franklin (1987) has queried the exclusion of continuous variation from the concept of polymorphism. Pure semantics are not the question here. Any discussion of the rates at which evolution proceeds (see Chapter 5), particularly if it involves preferential selection of certain phenotypes, is contingent upon the nature of phenotypic variation. If there is a continuum between extreme end members, then subtle shifts in phenotypic means over time have the potential to produce a gradual response. However, if the polymorphs are discontinuous, then selection of any of these polymorphs can result in a rapid morphological transition. While there may be some merit in using the term "polymorphism" in its broad sense, as advocated by Franklin, we follow accepted practice and restrict "polymorphism" to the concept of discrete phenotypes.

Polymorphism is generally considered to be distinct from phenotypic plasticity *per se,* in that the former has a genetic basis. **Ecophenotypic plasticity** is the degree to which a single genotype may show phenotypic variation under fluctuating environmental conditions. Much intraspecific variation attributable to phenotypic plasticity occurs by slight changes in developmental rates, producing a continuum of morphotypes. We would argue that much of this sort of variation is not merely induced by environmental perturbations, but arises from intrinsic alterations to the developmental program. This we describe as **phenotypic variation** (as opposed to ecophenotypic variation). Thus, the morphogenetic distinction between "polymorphism" and "phenotypic variation" is merely that the former produces discontinuous phenotypic variation, while the latter can form a continuous array of traits within a population. Both discontinuous and continuous phenotypic variation can be induced by intrinsic perturbations of the developmental system.

Many cases of environmentally induced phenotypic variation show that it is the organism's developmental program that is affected, inducing heterochronies. The balance of effects of intrinsic and extrinsic agents determines the evolutionary potential of heterochrony. When extrinsic agents (e.g., temperature perturbations) affect the developmental program of an organism, the heterochronic changes producing phenotypic variation may not often have great potential for being the springboard for the evolution of a new species. However, it is this very lability, the power of the organism to be externally "manipulated," that may be the important target of selection. By responding morphologically or behaviorally to extrinsic agents in a "positive" manner, the species thrives and prospers; it moves with the external changes; it sways in the environmental breeze. If its developmental system was not able to respond in such a way, then the species would snap off the evolutionary tree, as readily as a rigid branch in a gale. As Tomlinson (1987) has observed in plants, "plasticity of organization rather than initial architecture may be the more significant adaptive mechanism." However, from an evolutionary viewpoint, internally generated phenotypic variation, with its basis in heterochrony, is far more potent in the generation of evolutionary novelties.

If it can be demonstrated that both polymorphism and phenotypic variation have an underlying regulatory genetic basis, then both can provide the raw material for extrinsic selection pressures to operate on. We will argue that many cases of polymorphism are due to heterochronically initiated perturbations to the developmental program. This is not to deny the role of extrinsic factors actually inducing phenotypic changes. For instance, homeotic transformations in *Drosophila melanogaster* produce discrete, nonrandom sets of developmental trajectories which may be induced by environmental perturbations during development (Wake and Larson, 1987). We shall demonstrate how such extrinsic factors have the ability to alter developmental programs. Subsequent geographical or ecological isolation can result in such changes becoming genetically fixed in these descendants. Thus, in terms of the duality of extrinsic and intrinsic factors, extrinsic factors may induce intrinsic changes, or alternatively intrinsic factors may operate initially, and are then subsequently selected by extrinsic factors.

It is rather surprising that polymorphism has rarely been looked at from a developmental perspective. Generally speaking, polymorphisms have been seen as arising from the simultaneous occurrence in a population of several genetic factors (alleles or gene arrangements) producing discontinuous phenotypic effects. Opinions on the significance of polymorphisms to evolution have been highly variable. It used to be said that much polymorphism was evolutionarily irrelevant. However, as Mayr (1970) has observed, such a view does little more than obfuscate the complex relationship that exists between genes and phenotypes. But just how much of the variability arising from polymorphism has adaptive significance and the potential to be genetically fixed in descendant populations by barriers to breeding? And how does this affect rates of evolution?

3. EXTRINSIC FACTORS INDUCING HETEROCHRONIC CHANGE

First we look at the effect of extrinsic perturbations on the developmental program, and whether the susceptibility of organisms to such stimuli has a genetic basis. Smith-Gill (1983) considers that developmental plasticity (which results in morphological plasticity) may be of two types: (1) *phenotypic modulation* and (2) *developmental conversion*. In both of these models, extrinsic factors trigger developmental change.

1. In **phenotypic modulation**, generalized phenotypic variation results from environmental factors affecting rates or degrees of expression of the developmental program. However, under this scheme, genetic programs controlling development are not thought to be altered.

2. In **developmental conversion**, specific environmental cues activate alternative genetic programs controlling development. Such developmental conversion may result in the creation of alternative morphotypes (polymorphs), or lead to a premature cessation of the developmental program.

3.1. Phenotypic Modulation

Phenotypic modulation (as defined by Smith-Gill, 1983) occurs when general phenotypic variation arises from the susceptibility of different parts of a developing organism to direct environmental stimuli. While the genetic program of the developmental system does not suffer any changes, heterochrony may occur as rates, or degrees, of expression of the developmental program may be altered. Extrinsic factors, be they biotic or abiotic, principally act directly on the organism's hormonal system. For instance, determinants of sexual dimorphism in mammals (Wade, 1976), birds (Whittow, 1976), and reptiles (Crews *et al.*, 1985; Shine and Crews, 1988) are gonadal hormones. Perturbations to hormonal production, such as male testicular hormone in the male red-sided garter snake, result in smaller body size in males (Crews *et al.*, 1985) (see Chapter 6). The hormonal environment during early ontogeny is important as a determinant of shape and size in adults. Thus, extrinsic perturbations to hormones can induce major phenotypic changes, which may induce differential niche utilization if trophic apparatuses are involved. This applies to both intra- and intersexual polymorphs. In heterochronic terms, changes in the timing of hormonal inductions related to maturation will produce either global progenesis or global hypermorphosis.

Rates of existing developmental programs may also be altered: they may be either slowed down (neoteny), producing paedomorphic traits, or speeded up (acceleration), producing peramorphic traits. Smith-Gill (1983) has stressed the role of environmental modulations affecting the developing organism at all stages of its life history. The earlier that the changes occur in the developmental program, the more strident will be the phenotypic change. Organisms are considered to be most sensitive to environmental perturbations during periods of high morphogenetic activity, or alternatively when the neuroendocrine system is not sufficiently mature to deal with physiological homeostasis.

The variable phenotypes arising from phenotypic modulation range, at one extreme, from teratologies, that may be induced by extreme environmental perturbations, through to the normal range of phenotypic variation induced by minor environmental effects. However, while just minor perturbations to the timing and rate of growth during morphogenesis may induce major phenotypic abnormalities, in the form of teratologies (Cock, 1966), the range of such abnormalities may be subject to some degree of control. This has been demonstrated in studies by Edmonds and Sawin (1936) and Sawin and Edmonds (1949) (see also Alberch, 1980) on variations in the aortic arch of domestic rabbits (Fig. 4-1). Essentially, they documented only six major types of structural arrangements of the aortic arch. The frequencies of these groups differed substantially (Fig. 4-1). This they attributed to differential growth rates between the anterior thoracic region and the axial skeleton, and not to any effect of genetic inheritance.

As we have pointed out, an organism's development may theoretically be prone to alteration by environmental perturbations at any stage of development. However, there are particular times when structures are more liable to change. Tissues that are metabolically active at the same time as an environmental perturbation are likely to be more sensitive and thus susceptible to change (Smith-Gill, 1983; Zwilling, 1955). This differential sensitivity will therefore directly influence the relative timing of growth and maturation of particular structures. In their study of the green frog *Rana clamitans*, Berven *et al.* (1979) showed how environmentally induced acceleration or retardation in development of particular structures relative to one another resulted in substantial alterations to the phenotype. Populations of this frog that inhabit a montane environment take longer to complete metamorphosis and are substantially larger than their lowland counterparts.

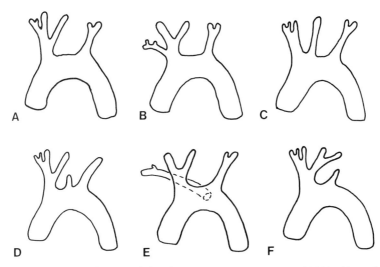

Figure 4-1. Variation in types of aortic arch found in postmortem examination of 3000 rabbits. There are six major types that account for 99.7% of the total variation: A, 81.6%: B, 15.1%: C, 0.7%: D, 1.0%: E, 0.06%: F, 1.23%. Type A is the normal phenotype. Redrawn and modified from Sawin and Edmonds (1949) and Alberch (1980).

However, when transferred to the lowland environment, their developmental rates are accelerated to an even greater extent than their lowland counterparts; metamorphosis is achieved more rapidly, and at a smaller size than the lowland larvae. Much of this variation in development is directly attributable to variations in temperature, affecting both growth and differentiation, and producing continuous variation of phenotypes. The significant genetic difference between montane and lowland frogs was to reduce the range of variation produced by environmental changes in the montane frogs. Not only is metamorphosis affected by environmental modulations during amphibian development (Smith-Gill and Berven, 1979). Environmental modulation can also affect age and size of onset of maturity in anurans (Berven, 1982a,b).

However, other environmental factors can produce dramatic phenotypic effects. Bernays (1986) demonstrated how differences in diet in the grass-feeding caterpillar *Pseudoaletia unipuncta* can have a pronounced effect on head size, changes in diet inducing changes in head allometry. Individuals reared on hard grass developed heads with twice the mass of those fed on soft, artificial diet, even though body masses were the same. Individuals fed on an intermediate diet (soft wheat seedlings) had intermediate head masses (Fig. 4-2). Bernays attributed these allometric differences to an increase in muscular development, which resulted in a significant morphogenetic effect on head size. Size differences in the heads, with correlated differences in mandibular strength, directly affect the insect's ability to cope with foods of different hardnesses: those with large heads are adaptively more suited to dealing with hard grasses. The same pattern was observed by Bernays (1986) in 82 grasshoppers and 76 caterpillars from North America and Australia. Grass eaters consistently had relatively larger heads than herbaceous-plant eaters (Fig. 4-3).

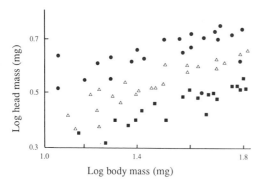

Figure 4-2. Relationship between dry head mass and total dry body mass in fifth-instar caterpillars of *Pseudoaletia unipuncta* reared from hatching on three different diets: the hard C_4 grass *Cynodon* (circles): the relatively soft C_3 grass *Triticum* (triangles): and artificial diet (squares). Redrawn from Bernays (1986, Fig. 2).

3.2. Temperature as an Extrinsic Factor

3.2.1. Effect on Arthropod Development

Among abiotic factors that can induce heterochrony by acting on the organism's hormonal system, probably the most important in ectotherms is temperature (Moore, 1942; Madge, 1956; Howe, 1967; Loch and McLaren, 1970; Smith-Gill, 1983). This generally affects the induction and/or inhibition of maturation (Singh-Pruthi, 1924; Newell, 1949; Wigglesworth, 1954; Bonner, 1968; Matsuda, 1979; Zuev *et al.*, 1979), by the impact of temperature changes on the hormones associated with growth and development. Research on the functioning of endocrine systems in arthropods, principally

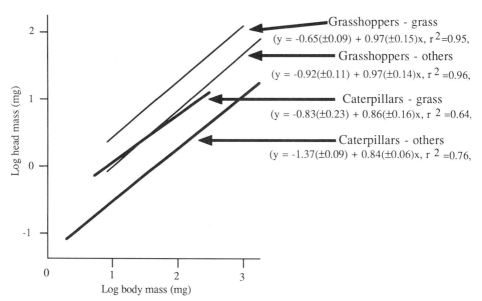

Grasshoppers - grass
$(y = -0.65(\pm0.09) + 0.97(\pm0.15)x, r^2 = 0.95,$

Grasshoppers - others
$(y = -0.92(\pm0.11) + 0.97(\pm0.14)x, r^2 = 0.96,$

Caterpillars - grass
$(y = -0.83(\pm0.23) + 0.86(\pm0.16)x, r^2 = 0.64,$

Caterpillars - others
$(y = -1.37(\pm0.09) + 0.84(\pm0.06)x, r^2 = 0.76,$

Figure 4-3. Relation between dry head mass and total dry body mass in grass specialists and other foilage feeders among grasshoppers and caterpillars, showing growth of relatively larger head in grass feeders. Redrawn from Bernays (1986, Fig. 1).

insects and crustaceans, apart from unraveling the mechanisms of the various hormonal systems, has revealed a close similarity in the methods of control of molting and maturation in different groups of arthropods (Wigglesworth, 1936, 1965; Wyatt, 1972; Doane, 1973; Gilbert and King, 1973; Willis, 1974; Gilbert, 1974; Ashburner, 1980; Aiken, 1980). In particular, the molting hormone ecdysone has been found to induce molting in a number of groups of arthropods, including insects, crustaceans, arachnids, and xiphosurans (Wyatt, 1972). Similarly, the other major hormone controlling morphogenesis, the juvenile hormone, acts in a similar manner in both insects and crustaceans.

It is thus probable that ecdysones were also used to regulate molting in arthropods prior to the divergence of the major arthropod groups in the Precambrian (Wyatt, 1972). There is probably a closer relationship between trilobites and crustaceans than between crustaceans and insects (Eldredge, 1977). Yet the close similarity in endocrine systems between these latter two groups suggests that morphogenesis in extinct groups of arthropods, such as trilobites, is likely to have been controlled by hormonal systems analogous to those found in living arthropods.

The molting sequence and the morphological expression of the molt in arthropods is under endocrine control. Neurosecretory cells in the brain produce a hormone that stimulates the synthesis and release of the molting hormone ecdysone. This molting hormone induces the molting process. Independent secretion of the juvenile hormone by specific glands modifies the expression of the molt and acts in conjunction with the molting hormone. It is possible that the juvenile hormone may act directly on regulatory genes, repressing their activity (Williams and Kafatos, 1971; Ashburner, 1980). If this is the case, then the distinction between phenotypic modulation and developmental conversion becomes somewhat clouded. While the juvenile hormone is present at a sufficiently high titer, the organism will display juvenile morphological characteristics. When this hormone ceases to be produced, metamorphosis to the adult phase occurs.

Extrinsic perturbations, such as temperature changes, can induce heterochronic changes by altering the molting sequence. In arthropods, molting is initiated by the secretion of ecdysone. This activates the epidermal cells to secrete a new cuticle. Changes in the timing of induction of deposition of the new cuticle will affect the degree of intermolt morphological development. Thus, if the cuticle deposition occurs soon after the previous molt, there will be little morphological change between instars. If, however, the onset of deposition of new cuticle is inhibited by delay in ecdysone secretion, then there will be a greater degree of morphological development between instars (Wigglesworth, 1965). Similarly, it has been shown (Wigglesworth, 1933) that artificially inducing the bud-bug *Cimex* to precociously deposit its cuticle results in the appearance of certain morphological features in an intermediate stage of development. Similar effects have been obtained by accelerating onset of molting in *Rhodnius* (Wigglesworth, 1961) and in *Hyalophora* (Staal, 1968; Willis, 1974).

The classic work of Wigglesworth (1936) on the effects of experimentally inducing precocious maturation in the assassin bug *Rhodnius* by artificially excising the source of the juvenile hormone, revealed the great importance of this hormone to the timing of maturation and the morphological character of the adult. Similar experiments on the silkworm *Bombyx* (Bounhiol, 1936, 1937) and the phasmid *Dixippus* (Pflugfelder, 1937) have shown how paedomorphs could be produced by artificially reducing the number of juvenile molts. A similar effect is seen in trilobites: precocious onset of maturation

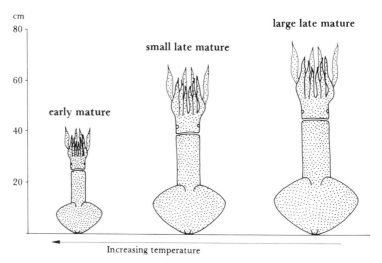

Figure 4-4. Effect of surface water temperature on onset of maturity in the Atlantic squid *Sthenoteuthis pteropus*. Modified and redrawn from Matyja (1986, Fig. 9).

caused by premature inactivation of the juvenile hormone resulted in the generation of an incomplete number of thoracic segments and retention of ancestral juvenile features in the descendants. In trilobites this has been termed *terminal progenesis* (McNamara, 1983a, 1986b). The effect of temperature in controlling terminal progenesis in the mealworm *Tenebrio mollitor* (Singh-Pruthi, 1924) and *Rhodnius* (Wigglesworth, 1965) lends support to the suggestion that temperature may have been influential in controlling progenesis in olenellid trilobites (McNamara, 1978). This is discussed in Section 3.4.

3.2.2. Effect of Temperature on Molluskan Development

Temperature changes can also affect the timing of onset of maturation in mollusks by directly targeting the organism's hormonal systems. For instance, in the pteropod *Clione limacina* (Lalli and Conover, 1973), a northern cold-water race grows up to 70 mm in length, whereas a southern, warmer-water race attains a maximum length of only 12 mm. This form attains sexual maturity at a body length of less than 3 mm, whereas the larger, northern form is still an unmetamorphosed larva at a body length of 15 mm. Similar patterns are seen in cephalopods. Zuev *et al.* (1979) found three groups of mature females of the Atlantic squid *Sthenoteuthis pteropus* of different body size (Fig. 4-4): an early maturing form, maturing at a mantle length of 17–30 cm, with a final length of 35 cm; a small late-maturing form, maturing at a mantle length of 33–40 cm, with a final length of 45 cm; and a large late-maturing form, maturing at a mantle length of about 40 cm, and attaining a maximum length of 50–60 cm. Onset of maturation in the squid was found to correspond with surface water temperatures, the smallest, earliest maturing form occurring in water temperatures of 26–30°C, the larger maturing form, in water 18–22°C. The same inverse relationship between body size [which in cephalopods is known to equate well with timing of onset of maturity (Wells and Wells, 1977)] and temperature is evident in the tropical Atlantic squid *Ornithoteuthis antillarum* (Nesis and Nigmatullin, 1977) and the Indo-Pacific squid *Sthenoteuthis oyalaniensis* (Nesis, 1977). In this species, early and late maturing males and females differ in final body size, the earlier maturing forms being restricted to the warmest water zones of the geographic range of the species.

The adaptive significance of such disparate progenetic and hypermorphic morphotypes is likely to be that it allows occupation of a wide range of habitats, even though phenotypically the species differs in each. The evolutionary significance of such adaptations produced by extrinsic perturbations to the developmental system should not be underestimated. There is evidence from the fossil record that such extrinsically orchestrated phenotypes have a long evolutionary history (see Section 3.4).

3.3. Developmental Conversion

The second form of extrinsically induced developmental plasticity is known as "developmental conversion." This principally involves changes in the timing of onset of maturity, or of major periods of morphogenetic transition, and results in the development of polymorphs. Developmental conversion will only have a phenotypic effect, however, if growth parameters are allometric. Timing of maturation will have no phenotypic effect if growth is isometric. The greater the allometry, the greater the potential for expression of a wide range of phenotypes brought about by variation in the timing of maturity or metamorphosis. Only certain stages of development tend to be affected by developmental conversion. This occurs in two ways (Smith-Gill, 1983): by the activation of an "on–off" switch; or by the expression of quite different morphotypes. Either way, such changes induce morphological saltations. The role of extrinsic factors is not so much to cause changes, but to trigger nascent developmental pathways that are already genetically programmed.

One such way is by triggering "on–off" switches which can cause cessation or initiation of the developmental program. Such a situation occurs in organisms which show diapause. Here an environmental factor causes cessation of the developmental program. However, because the diapause is not permanent, further environmental cues can produce a reversal, with reinstigation of the original developmental program. Such developmental conversion may be termed *reversible developmental conversion*. In insects, this may be induced either by abiotic factors (such as temperature changes) or by biotic factors [such as queen pheromone diapause or metamorphosis of ant larvae (Brian, 1965), or synchronization of sexual maturity, by relative acceleration or retardation of the onset of maturity by other members of the colony (Butler, 1967)]. Whether or not environmental factors produce diapause is partly dependent upon the stage of morphological development that the organism has reached. Thus, the same environmental cue that initiated diapause at one stage of the organism's development may have no effect at all at another stage. Smith-Gill has recognized three constraints on the initiation of diapause: (1) a physical protection from the environment is necessary (such as an impermeable case); (2) the developmental period must have low morphogenetic activity; (3) the environmental cue has to be able to be received and interpreted by the organism's central nervous system. To produce this facultative diapause a specific set of genetically controlled events has to be initiated. Diapause does not affect the main developmental program, for when terminated, the normal program is reestablished.

The second form of developmental conversion, which may be termed *irreversible developmental conversion,* is permanent in its effects as environmental cues determine which alternative developmental pathways can be exploited. Two factors constrain which particular polymorphism is developed (i.e., which developmental pathway if followed): the switch from one pathway to another must precede morphogenetic activity; and the environmental cues must be physiologically translatable (Smith-Gill, 1983). If the environmental cue does not occur at the critical time during the organism's development, no change will occur. Many examples of such irreversible developmental conversion may have been described, for example in Hymenoptera; migratory locusts (alternative adult phases); and aphids (alternative modes of adult reproduction, i.e., parthenogenesis or bisexuality).

3.3.1. Pheromones and Population Density as Factors in Intraspecific Heterochrony

There is some evidence that in many social insects one extrinsic factor that greatly influences the timing of onset of sexual maturity is the production of primer pheromones by certain members of the community. These pheromones trigger a chain of physiological changes that either retard the onset of maturity (hypermorphosis), as in adult honeybees, or accelerate it (progenesis), as in some desert locusts (Butler, 1967). Sometimes, both hypermorphosis and progenesis may be induced within the same species. The presence of adult males within high-population-density colonies of the desert locust (*Schistocerca gregaria*) induces progenesis in any immature males and females with whom they are living (Norris, 1954; Butler, 1967). Mature females also have the ability to stimulate maturation, but not as effectively as males. However, females less than 8 days old have the reverse effect: they retard male maturation. It is high population density that results in the attainment of synchronous maturation by acceleration of laggards. It also retards potentially precocious forms from attaining maturity at too small a size. The acceleration pheromone in mature desert locusts is a volatile substance that covers their bodies. Just 1/5000th of the amount present on an individual at any one moment is enough to elicit a response in an immature male (Loher, 1961).

Experimental work has shown that removal of the corpora allata (the endocrine glands that secrete juvenile hormones) of immature male desert locusts, which prevents them from reaching maturity, not only inhibits production of the accelerating pheromone, and thus the potential for induction of maturity in other males, but seems also to result in a pronounced delay in the onset of normal maturity in others (Norris and Pener, 1965). It has been suggested (Butler, 1967) that similar accelerating pheromones may also be present in ladybirds (Coccinellidae), coming into effect when immature adults congregate. The pheromone induces synchronous sexual maturity, allowing mating, followed by dispersal.

In Hymenoptera, similar olfactory/gustatory pheromones have the effect of retarding the onset of maturity. The presence of a sexually mature queen often inhibits development of sexual maturity in workers. If the queen is removed, then increased ovary development occurs in the workers, and eggs are laid. This happens in ants [*Formica rufa pratensis* (Bier, 1956), *Leptothorax tuberum unifasciatus* (Bier, 1954)], bees [*Apis cerana* (Sakagami, 1954), *A. mellifera* (Butler, 1959; Pain, 1961), *Bombus* spp. (Cumber, 1949), *Lasioglossum* spp. (Ordway, 1965), *Trigona frieremaiai* (Sakagami *et al.*, 1963)], wasps [*Polistes* spp. (Deleurance, 1946) and *Vespula* spp. (Imms, 1957)]. In *Apis mellifera*, the honeybee, the pheromone that inhibits development of ovaries has been isolated. It is secreted by the queen from mandibular glands and spread over her body.

Population densities can also have a pronounced effect on developmental conversion in salamanders. In the classic case of the axolotl, this paedomorphic salamander can, under the right environmental conditions, be induced to metamorphose. This example of facultative paedomorphosis is an example of reversible developmental conversion. It has been argued (Semlitsch and Gibbons, 1985) that in many natural populations of paedomorphic salamanders, natural selection is likely to favor such facultative, rather than obligate, paedomorphosis, particularly where the drying up of ephemeral ponds is occurring. Being able to switch back to the ability to metamorphose is selectively more advantageous in such a situation. An individual undergoing facultative paedomorphosis (by reversible developmental conversion) would therefore have the best of both worlds: as a paedomorph with larval characters, in a favorable aquatic environment. Yet it would have the ability to escape that environment by metamorphosing, should the environment deteriorate. Alternatively, as Newman (1988) has shown for Couch's spadefoot toad (*Scaphiopus couchii*), selection in ephemeral ponds may favor phenotypes that attain maturity quicker, whereas in slower drying ponds, selection will favor more slowly developing forms. These differences have a genetic basis (Newman, 1988).

Working on the salamander *Notophthalmus viridescens dorsalis*, Harris (1987) has shown that population density is the principal factor which determines whether or not juveniles undergo developmental conversion into paedomorphic forms. Harris arrived at this conclusion by experimentally raising larvae in tanks at different population densities. In one tank he kept 10 individuals, in the other he kept 40. What he found was that most larvae living in the low-density tanks became sexually mature while still in the larval stage, and remained as paedomorphic forms inhabiting the aquatic environment (Fig. 4-5). In the high-density tanks, many more larvae metamorphosed into efts (immature terrestrial forms). In the natural environment, drying up of ephemeral ponds would result, all things being equal, in increased population densities. Selection under such conditions favors individuals that metamorphose and do not remain as paedomorphs (facultative paedomorphs). However, should the terrestrial environment deteriorate, and the aquatic persist, then paedomorphosis may become obligate, and genetically fixed.

3.3.2. Sex as a Target of Developmental Conversion

We have demonstrated how developmental conversion can affect certain species, and may confer an adaptive advantage by nature of the increased extent of phenotypic variability. In many species, intraspecific heterochrony affects only certain sexes. In-

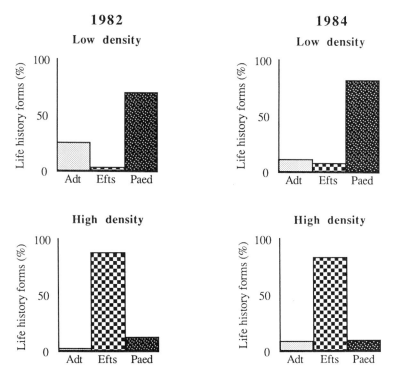

Figure 4-5. Histograms showing results of data obtained by Harris (1987) on relative frequencies of adults (Adt), efts, and paedomorphs (Paed) under experimentally controlled high and low population densities, revealing higher incidence of paedomorphosis under low-density regimes.

deed, the phenotypic expression of sexual dimorphism, per se, arises through the action of heterochrony, by differences in timing of onset of maturity (including humans, Chapter 7). Sexual dimorphism can vary from slight size and shape differences, brought about by minor differences between sexes in the timing of onset of maturity, to extreme size and shape differences, where one morph (almost invariably the male of the species) is so progenetically reduced, by virtue of its very early onset of maturity, that it can only exist by being parasitic on the female. As in developmental conversion the progenesis may be either reversible or irreversible. Both reversible and irreversible progenesis may be induced by extrinsic factors, or by intrinsic perturbations of the developmental system, which we shall discuss separately below.

Of the many examples of heterochronically induced sexual dimorphism documented in the Mollusca (see below), most appear to have evolved by intrinsic factors dominating. However, there are cases where extrinsic perturbations dictate the nature of sexual dimorphism. In the oyster *Ostrea puelchana* (d'Orbigny), for instance, one set of individuals lives for several years and develops a rhythmic cycle of hermaphrodism, with male, followed by hermaphroditic, then finally female phases. Another group occurs as progenetic males, living only for about 1 year (Table 4-1), attaining a smaller size and functioning only as males (Fernandez Castro and Lucas, 1987). These progenetic males

Table 4.1. Comparison of Sexual Development in Progenetic and Nonprogenetic Individuals[a]

	Nonprogenetic individuals			Progenetic individuals	
Age	Shell height (mm)	Development		Shell height (mm)	Development
1 mo	<5	Settlement anywhere		5–17	Settlement on bearer oysters
3 mo	5–15	Abortive juvenile sexuality; facultative phase for about 36% of nonprogenetic oysters		5–21	Gradual male sexual development that becomes functional before end of summer
1 yr	24–58	Functional males		20–26	Functional males, then detachment and death
2 yr	52–80	≈70% of individuals male followed by female phase; ≈30% female phase, brooding and larval release, then male followed in autumn by female phase		—	—
>3 yr	70–110	Successive rhythmic sexuality, female phase becoming increasingly frequent with increasing age		—	—

[a] From Fernandez Castro and Lucas (1987, Table 4).

attach to an enlarged region, located near the labial palps, at the anterior edge of the left valve, known as the "inhalant cavity." Morriconi and Calvo (1983) noted differences in proportions of progenetic individuals between populations, depending on population density. They argued that lower proportions of males develop as progenetic individuals in populations with high population density. This is in direct contrast to the situation in salamanders, where high population density leads to higher incidence of the production of paedomorphs. The presence of females may be a factor that controls the incidence of male progenesis in some invertebrates. When a larva of the parasitic eulimid gastropod *Stilifer linckiae* settles onto a starfish that it is to parasitize, it passes through a male phase to become a female (Lützen, 1972). However, if a larva settles where a female is already present, then it will be progenetically arrested in the male stage.

Similar sexual polymorphism induced by extrinsic factors has been described in copepods (Haq, 1965, 1972), fish (Briggs, 1953; Shapovalov and Taft, 1954), and decapods (Turvey, 1981). In this latter case, two male morphs occur in the decapod *Eustacus spinifer* (Heller): (1) the "normal" males, which are generally immature at weights of less than 50 g and a length of 45 mm, and which grow up to 130 g in length; (2) progenetic males that attain maturity at weights of between 1 and 5 g and lengths of between 12 and 20 mm, which rarely grow to a length exceeding 40 mm. These forms have highly inflated genital papillae and mature testes with an extremely high gonosomatic index. Turvey (1981) has suggested that the "normal" males arise from K-selection, whereas the progenetic males are classic r-strategists. Thus, the production of two such male morphs may be of particular adaptive significance in allowing the species to "hedge their bets" against long-term changes in environmental conditions. During periods of environmental instability, progenetic forms may develop preferentially, whereas during more stable times the K-strategist "normal" males develop.

3.3.3. Fungal Morphological Plasticity

The importance of extrinsic factors in affecting developmental pathways is not restricted to the animal kingdom. There is mounting evidence (Rayner *et al.*, 1987; Rayner, 1988) that fungi are similarly susceptible to environmental cues triggering alternative developmental pathways. The target of the extrinsic stimuli is the mycelium, a network of hyphae (branching, protoplasm-filled filaments) that extract water and nutrients from the environment and which periodically produce "fruiting bodies," such as toadstools and puffballs. Indeed, the pattern of mycelial development is a fundamental aspect of the evolutionary biology of fungi.

The possession by fungi of a hierarchical series of "superimposable switch mechanisms" (Rayner *et al.*, 1987) produces wide developmental versatility, conferring different functional properties that allow the occupation of a wide range of, often discontinuous, niches. These switches are triggered by a wide variety of extrinsic and intrinsic stimuli (Rayner *et al.*, 1987). Growth rates may be accelerated or retarded, while the timing of onset and offset of growth can be influenced by a host of extrinsic factors. Patterns of mycelial development will vary under different environmental conditions. By switching developmental modes, very distinctive forms can be generated. As Rayner *et al.* (1987) have shown, cultured fungi may grow either as unicellular individuals or as mycelia. These can adjust angles and frequencies of hypha branching. Moreover, they are able to slow down their growth to create a dense mass (by neoteny), or accelerate development (acceleration) to form a diffuse growth pattern.

Rayner *et al.* (1987) liken these changing growth rates to gear shifts. When in a higher gear, rapid mycelial growth occurs which facilitates the coverage of a wide domain and rapid extraction of nutrients. Such a situation they equate with selection in an r-selected regime. Intraspecific neoteny, in the form of lower growth rates in a "lower gear," aids in the establishment of an inoculum base (cells introduced into a medium for cultures), consolidation of territorial gains and stress tolerance, all features of a K-selected regime. The mycelia even have the ability to arrest their own development (progenesis) by inducing senescence, the "breaking system" of Rayner *et al.* (1987). This leads to the death or inactivation of cells (Rayner, 1988). Such a breaking system that produces progenesis of local growth fields may in fact be a prerequisite for the redirection of growth resources prior to the initiation of a novel morphogenetic pathway. It has even been suggested that "fairy-ring" fungi may be conditioned by an intrinsic switching system of this type. These switches may be triggered by external stimuli, such as damage or variations in light levels. However, they also seem to have the ability to switch without any obvious extrinsic input. Biochemically, it would appear that this localized progenesis is caused by the activation of phenol oxidase systems and melanization.

3.4. Evolutionary Significance of Extrinsic Perturbations to the Developmental Program

While many genotypes exhibit a range of phenotypic variation under the influence of extrinsic factors, what is their capacity to be the springboard for future evolutionary events? After all, it may be that the lability of the organism's developmental program that allows it to react to external stimuli in such a way as to generate a wide range of adaptively "useful" morphotypes by heterochronic change is the adaptively significant aspect of the species. With such species, environmental perturbations will not prove deleterious but will merely open up new phenotypic pathways. Such is the situation in many fungi and in animal and plant species that display a high degree of developmentally induced plasticity. Without recourse to a series of sequential developmental switches that allow different developmental options to be followed under different environmental conditions, species would only be able to occupy much narrower niches. As it is, the developmental flexibility allows much broader niches to be occupied. This is the adaptation that has led to the success of these species.

Examination of the fossil record enables us to assess the extent to which extrinsic factors have played a role in inducing intraspecific heterochrony. Ammonoids have long been recognized as showing dimorphism, akin to that seen in some living molluskan groups (Landman, 1988). There has been much debate as to whether or not this dimorphism reflects the expression of different sexes (Callomon, 1963, 1969, 1981, Makowski, 1963, 1971). The arguments that have been presented in support of the idea that this dimorphism has a sexual basis include:

1. Both small and large forms must have identical initial (? juvenile) stages.
2. In the adult stage there must be no intermediates between the two morphs.
3. The numerical ratio of the two forms needs to be comparable with that documented in living cephalopods.

In some instances the dimorphism may not be in size, but in other characters, such as degree of shell coiling (the involuti- and evoluticonchs of Zakharov, 1978). There is, however, no consensus among those who believe that the dimorphism is sexual dimorphism as to which morph is female and which male. Makowski (1963) suggested that small forms were males, and large forms, females. Cope and Smith (in Callomon, 1969), however, suggested that in some genera the macroconchs may have been males, in others, females. Surprisingly, until Matyja's (1986) analysis of developmental polymorphism in some Oxfordian ammonites, there had really been little attempt to relate dimorphism in ammonites to heterochrony, apart from a recent suggestion (Kennedy and Wright, 1985) that micromorphs might best be regarded as "paedomorphic dwarfs." This is perhaps even more surprising, given the pivotal role of ammonites in the evolution of the concept of heterochrony over the last 100 years or so (Landman, 1988; see Chapter 1). Ivanov (1975) has discussed how variations in shell size and ornamentation related either to differences in growth rates, longevity, or maturation; in other words, rates, onset, and cessation of growth, the three prime variants in heterochrony. Smaller morphs, which also possess fewer septa, fewer whorls, and more simplified ornament, might be considered to be progenetic relative to the larger morphs (which are therefore relatively hypermorphic to the smaller morphs), having attained sexual maturity earlier (Makowski, 1963; Callomon, 1963). This, of course, is based on the presumption that small size equates with earlier onset of maturity, and not simply a reduced growth rate (Chapter 2) (see Dommergues, 1988, for a discussion of the use of septa in ammonites for assessing the true age of ammonite shells). However, it is well recorded in living cephalopods that growth is strongly retarded at the onset of maturity (Wells and

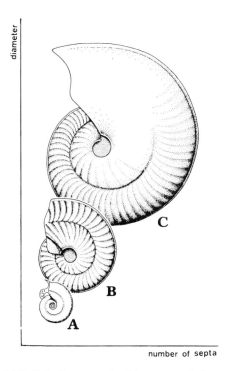

Figure 4-6. Trimorphs of the Middle Oxfordian ammonite *Ochetoceras caniculatum*. Reproduced from Matyja (1986) with permission of the author.

Wells, 1977). It is implicit in this interpretation that for the morphs to be sexual dimorphs, there should always be two types: a micromorph and a macromorph. Yet in his work on Oxfordian ammonites, Matyja (1986) has demonstrated that rather than just two morphs, three discrete morphs are frequently present, similar to the situation found in a number of living cephalopods (see Section 3.2.2). These he terms miniconchs (the smallest), microconchs, and macroconchs (the largest) and has recognized them in a range of genera, such as *Quenstedtoceras, Cardioceras,* and *Ochetoceras* (see Fig. 4-6). While the three may occur together, frequently a single morph can exist over a wide geographic area for an appreciable period of time. In Matyja's opinion the presence of trimorphs undermines the whole concept of ammonite polymorphism being related simply to sexual dimorphism. It is his belief that heterochronically induced differences in shell size and morphology may equally well have been due to variations in the time needed for sexual maturation of individuals hatched in particular spawns, under the impact of changing temperature regimes. Thus, ammonite polymorphism may well have its foundations in extrinsic perturbations to the developmental system.

Role of Temperature in Trilobite Evolution

Next we demonstrate that the fossil record provides evidence that extrinsic perturbations of an organism's developmental system have played an important role in the

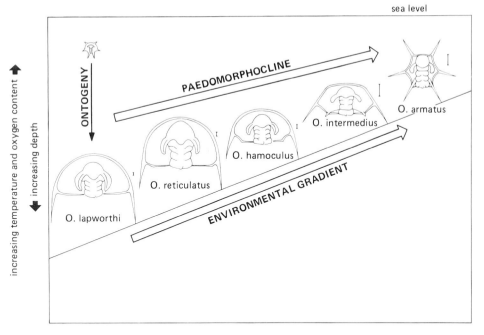

Figure 4-7. Paedomorphocline in the Scottish early Cambrian trilobite *Olenellus*. Reproduced from McNamara (1982b).

generation of new species. One example that adequately demonstrates this is a suite of five species of the Early Cambrian trilobite genus *Olenellus* from northwest Scotland (Cowie and McNamara, 1978; McNamara, 1978). Morphological differences between adults of these species are related to differences in degree of ontogenetic development. Although the five species do not occur at different stratigraphic levels, outgroup comparison indicates that four of the species arose sequentially by paedomorphosis. The five species lie at various points along a heterochronic morphological continuum (a paedomorphocline—see Section 3 in Chapter 5). By outgroup comparison with other olenellids, *O. lapworthi* can be shown to represent the ancestral morphology and *O. armatus* the most derived. This latter species possesses the most overt larval characteristics of any of the species. Not only does it retain cephalic spines, which otherwise only occur on olenellid protaspids, but it also has fewer thoracic segments (9, compared with 14); it is also the smallest species. *O. reticulatus*, *O. hamoculus*, and *O. intermedius* represent intermediate morphotypes along the paedomorphocline between *O. lapworthi* and *O. armatus* (Fig. 4-7).

While showing a high degree of morphological variability at the generic level, the five species form discrete units, with no intermediate morphotypes. Such morphological separation is thought to have been a principal feature in allowing the species to evolve. It led to ecological and reproductive isolation, leading to subsequent allopatric speciation of the descendant morphotypes. Because of niche constraints, there would have been a limit to the number of species that could have evolved from the ancestral apaedomorph, *O. lapworthi*. It has been suggested (McNamara, 1978) that the paedomorphic process that operated in these species of *Olenellus* was global progenesis: successively more paedomorphic species being smaller and, in the case of *O. armatus,* having fewer thoracic segments. If onset of sexual maturity was precociously initiated during ontogeny, then fewer thoracic segments would have been generated.

It has also been argued (McNamara, 1983a, 1986b) that poor developmental regulation in early Cambrian trilobites was a major factor influencing the extent of morphological variability. But what of the selection pressures that favored particular heterochronic morphologies? In the case of the Scottish olenellids it is considered that the five species developed along an environmental gradient from deep to shallow water. The variation in timing of onset of maturity, and the ensuing morphological differences, have been considered to have been the result of extrinsic factors (McNamara, 1978, 1982b). There is abundant evidence that apaedomorphic forms, such as *O. lapworthi,* inhabited relatively deep water as adults. On the other hand, it is generally accepted (Fortey and Chatterton, 1988) that trilobite protaspids were pelagic. Consequently, extreme paedomorphs such as *O. armatus* are likely to have become mature and remained in the pelagic "juvenile" environment for their entire life cycles. Higher water temperatures in shallow water might have caused premature maturation of the pelagic larvae of the ancestral species of *Olenellus.*

There is further evidence that water depth may have been involved in the niche partitioning of the five species of *Olenellus.* This is deduced indirectly by analysis of the extent of genal ceca (raised, ramifying markings on the outer cheeks of the trilobite cephala). These ceca are considered to have functioned as a secondary respiratory base in trilobites (Jell, 1978), and to represent the surface expression of the canals that carried the hemocoel around the trilobite cephala. In *O. lapworthi* they cover the extensive extraocular area possessed by this trilobite (Cowie and McNamara, 1978). The progressive reduction in extraocular area between *O. lapworthi* and *O. armatus* resulted in a reduction in this secondary respiratory base. Consequently, by virtue of earlier maturation induced by higher water temperatures, successive paedomorphs may have been restricted to progressively more oxygenated, shallower waters. Thus, the effect of temperature was not only to induce variable timing of maturation, but also, as a by-product, effective niche partitioning. In the Early Cambrian seas, vacant niche space is likely to have been common, enabling effective ecological isolation, restriction in gene flow, and subsequent allopatric speciation. Extrinsic perturbing factors, particularly temperature, therefore were particularly crucial in initiating heterochronic changes early in the Phanerozoic.

4. SELECTION AND INTRINSICALLY PRODUCED HETEROCHRONY

We have shown how extrinsic factors may directly induce heterochrony both by generating phenotypic variation (phenotypic modulation) and by producing polymorphs (developmental conversion). Similar phenotypic effects may also arise from heterochrony that is generated by intrinsic factors. General intrapopulational genotypic variation may involve variation predominantly in genes controlling developmental regulation. Heterochrony is not just a phenomenon for the generation of saltations and "monstrous hopefuls," but is the prime factor in most cases of intrapopulational genotypic and phenotypic variation (Chapter 3). Phenotypic variation attributable to genetic perturbations of the developmental system that may be unrelated to externally generated environmental factors has also been discussed in Chapter 3. In this section we focus on its interplay with the external environment.

4.1. Intrinsically Produced Heterochrony

An elegant demonstration of the effect of intrinsic heterochrony is the morphological variation in cranial features of domestic dogs, compared with cats (Wayne, 1986). It exemplifies the importance of slight intraspecific differences in allometries producing the "normal" range of phenotypic variation seen in vertebrates and invertebrates. Individuals can be relatively compared; an individual vertebrate possessing a bone that has a slightly higher positive allometry (due either to its faster relative growth rate of the local growth field—the bone—or a slightly lower global growth rate) than another member of the population is relatively peramorphic in this feature. Alternatively, the mean value of the allometric coefficient of the bone can be compared between populations to assess relative degrees of intrapopulational heterochrony. In the case of domestic animals, artificial tinkering with growth allometries allows comparison of heterochronies to be made between different breeds. It is also instructive to compare the range of allometric differences between different species of domesticated animals. The range of morphological types in different dog breeds is far greater than in cats (Wayne, 1986). This difference has its basis in allometric growth of the skull. Wayne has demonstrated that skulls of domestic dogs grow with pronounced positive allometry in certain directions, whereas in cats skull growth in all directions is almost isometric (Fig. 4-8). Difference in ontogenetic scaling, by extending or contracting the juvenile growth period, produces dog breeds not only of very different sizes, but also of very different skull shapes. Small dogs, retaining juvenile skull shape, are therefore paedomorphs, by progenesis. Larger dogs are relative peramorphs, produced by hypermorphosis. However, in domestic cats, extension or contraction of the growth period, or onset of maturity, will have little morphological effect, and morphological variability is thus minimal. Pigs also show strong changes in skull proportions during growth, and like dogs, hundreds of breeds are recognized (Wayne, 1986). Thus, without the pronounced ontogenetic morphological change in dogs and pigs, there would be little variation.

Where growth of traits is allometric, any factors that influence growth rates or timing are likely to have a phenotypic effect. Common influential factors in mammals are litter size, age and condition of the mother, differential mortality of smaller young, and, in experimental studies, laboratory conditions (Creighton and Strauss, 1986). Basically, any differences in adult size, which may affect shape, can be accounted for in the standard three parameters of size at birth, growth rate, and length of growth period, or any combination of these. In assessing the influence of these factors in intraspecific differences in size and shape of traits in cricetine rodents, Creighton and Strauss (1986) studied variations in intraspecific scaling of brain size with body size caused by heterochronic changes in growth patterns. They compared two named subspecies of *Pero-*

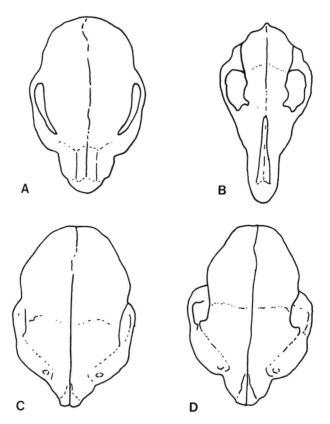

Figure 4-8. Dorsal view of skull of (A) a domestic dog neonate and (B) an adult dog, contrasted to that of (C) a domestic cat neonate and (D) an adult cat. Note the much greater degree of morphological change during ontogeny in the dog skull compared with the cat. Redrawn from Wayne (1986, Fig. 9).

myscus maniculatus and found that growth trajectories for body weight and brain weight were the same from birth to 8 days, but after that the allometric relationships between the two subspecies varied from day 10 through to day 40, due, they argue, to differences in the timing of offset in cranial growth. One subspecies, *P. m. bairdii*, is therefore paedomorphic with respect to the other, *P. m. gracilis* (Fig. 4-9), by progenesis of a local growth field (the brain). While falling within the range of intraspecific variation of allometric coefficients for brain size in vertebrates of 0.2 to 0.4 (Jerison, 1973), their values lie close to the two extremes (0.21 for *P. m. bairdii* and 0.37 for *P. m. gracilis*). This is even though the differences in offset growth of brain size between the two subspecies varied by a factor of only about 4 days. Therefore, although such phenotypic effects of heterochrony induced by progenesis or hypermorphosis are often lightly dismissed as being merely the effects of allometric scaling, having no genetic basis, the potential adaptive significance of such changes can be quite profound phylogenetically.

While such examples demonstrate the importance of heterochrony in generating phenotypic variation, unless specific heterochronic morphotypes are preferentially selected, there will be no evolution. However, we must once again stress that the heter-

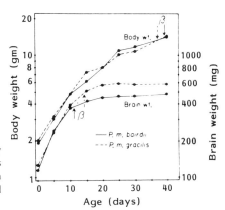

Figure 4-9. Postnatal growth in brain weight and body weight for *Peromyscus maniculatus bairdii* and *P. m. gracilis* showing differences in offset time of brain growth between the two subspecies. Reproduced from Creighton and Strauss (1986, Fig. 10), with permission of the authors.

ochronic variability, such as that in domestic dogs, is by virtue of the pronounced allometric growth, the very factor responsible for the success and adaptability of the species. Such variability, arising from pronounced allometry of key structures, has the potential to be an important precursor to evolutionary change.

Intraspecific heterochrony that operates on local growth fields can result in quite distinctive adult morphotypes. This can be demonstrated by the existence of such forms in the fossil record. An example of extreme postdisplacement of a specific trait has been documented in the echinoid *Eupatagus antillarum* from the Eocene of the southeast United States (McKinney *et al.*, 1990). In this echinoid, as in most spatangoid echinoids, the ambulacral columns on the upper surface of the test are developed as two sets of paired petals, and an anterior, axial ambulacrum. The paired petals each comprise rows of pore pairs through which, during life, passed tube feet adapted solely for respiration. The pore pairs and tube feet in the anterior ambulacrum are quite different, being simpler in design, and being adapted for sensory or funnel building activities. While these two different sorts of ambulacral structures are markedly different in adults, both morphologically and functionally, in the early, postlarval echinoid they show little morphological differentiation. The standard growth model in spatangoids is for relative peramorphic growth of the pore pairs and tube feet in the paired petals, while those in the anterior ambulacrum retain the morphological characteristics of ancestral regular echinoids (McNamara, 1988a). In some individuals of *E. antillarum*, however, the anterior ambulacrum develops all the morphological features of the petals. Presumably in most postlarval spatangoids, accelerated development is restricted to the paired petals, the gene sequence responsible for switching on this accelerated development not activating the anterior ambulacral structures. However, in the "deviant" *E. antillarum* the genes responsible for accelerated growth are activated. While it could be argued that such heterochronic "monsters" are just ecophenotypes, there are a number of arguments against this viewpoint, favoring, on the contrary, the view that there is a regulatory genetic basis to the evolution of this trait:

1. The "deviant" forms are not restricted to just one time horizon and they occur with normal forms.
2. In all other respects, the "deviants" resemble "normal" morphs.

3. Only this species is affected; contemporaneous species of *Eupatagus* do not develop this morphotype.
4. No known ecophenotypic traits in living spatangoids approach this degree of extension of an ontogenetic pathway.
5. Perhaps most significantly, two spatangoid genera are recognized (*Gibbaster* and *Plesiaster*) that are characterized almost solely by the possession of this "deviant" trait.

4.2. Selection and Heterochrony: A Case Study

Translation of intrapopulational heterochronic change to interspecific change involves the coming together, at the right place, and the right time, of heterochronic morphotypes that are either higher up the adaptive peak than their immediate ancestors and can more effectively occupy the species' preoccupied ecological niche, or have developed a novel morphotype that occupies a completely different adaptive peak in a completely different ecological niche. This might either be a vacant niche (in which case we are likely to be dealing with a major evolutionary breakthrough), or a niche already occupied, but by a species lower on the adaptive peak for that niche. We discuss the significance of this to rates and directions of evolutionary change in Chapter 5. Here we merely show how intraspecific heterochrony has the capacity to be translated into interspecific change.

For intraspecific heterochrony to be successfully translated into speciation, not only must there be large-scale morphological change during ontogeny, but at each ontogenetic stage a high degree of variation. We are dealing with the cascading effects of heterochrony: at the first level, differentiative and growth changes during ontogeny; these lead to a higher level of intraspecific variation by heterochrony causing a number of ontogenetic changes. Then there is an even higher level of interspecific heterochrony, whereby features of the intraspecific variation become genetically fixed as new adaptive peaks are conquered and new ecological niches attained.

This can best be demonstrated in a fossil lineage where there are many forms that undergo pronounced morphogenetic changes during ontogeny, and where there is pronounced intraspecific variation. Our example is a heart urchin called *Lovenia* which occurs in Miocene deposits of southern Australia. The lineage comprises three species: *L. forbesi*, which ranges from the Late Oligocene to the Early Miocene; an undescribed species from the Middle Miocene; and *L. woodsi*, from the Late Miocene. These three species are thought (McNamara, 1989) to constitute a single lineage. The details of the trend and the selection pressures inducing particular morphological trends are discussed in more detail in Chapter 5. Here we wish to show that the morphological changes that occur during ontogeny are those that produce morphological variation, i.e., intraspecific heterochrony, and that it is these self-same characters that form the targets of selection.

Perhaps the most distinctive feature of the test of this genus of echinoids is its tuberculate nature (Fig. 4-10). [Although not unusual in echinoids *per se*, in heart urchins large tubercles are relatively uncommon. Their reappearance represents the reawakening of a previous developmental program controlling tubercle growth (McNamara,

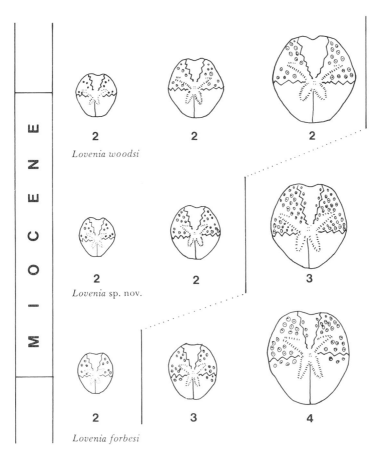

Figure 4-10. Paedomorphic evolution of tuberculate interambulacral areas in three species of the heart urchin *Lovenia* from the Tertiary of southern Australia. Numbers refer to number of tuberculate columns (half test).

1988a).] During life these tubercles would have supported long spines, which in living species of this genus are known to serve a defensive function (Ferber and Lawrence, 1976). In the oldest species, *L. forbesi*, the tubercles (and hence spines) show a progressive increase during ontogeny, not only in number, but also in areas of the test on which they occur. Thus, we are not looking here at allometric changes during growth induced by changing levels of mitotic activity, but differentiative changes—an increase in the numbers of tubercles and spines generated over a period of growth, as the test increases in size. However, the principles are the same for either differentiative or allometric changes during ontogeny. In large adults, tubercles are present in four interambulacral columns on each side of the test. They first appear, however, in the central pair of columns (Fig. 4-10) at a test length of about 15 mm. They then start appearing in the posteriormost column, generally between 20 and 30 mm. Finally, at test sizes generally between 23 and 32 mm, tubercles appear in the anteriormost column (although they can appear at test lengths as short as 20 mm and as long as 40 mm in a few instances). Variations in the timing of appearance of tubercles in any column represent either predisplacement, if they start appearing earlier than is the norm for the population, or postdisplacement, if they appear later.

In studies such as this, we are confronted by differentiative changes (or allometric changes) and size change of the entire organism. When dealing with fossil material (and frequently neontological material), actual timing of events is unknown. So it is necessary to resort to size of the whole organism as a gauge of time. This is not entirely unreasonable in this case, because there is no doubt that the size of the organism is providing us with some measure of time. They did not appear fully formed at particular sizes, but took time to grow. Our problem is to establish whether the actual growth rates of two individuals were the same or not. As discussed in Chapter 2, most studies in recent years have taken size as an implicit proxy for time. Thus, if there are two specimens, both 23 mm in test length, and one has no tubercles in the posteriormost column, whereas the other has two, it has been assumed that the specimen with two shows relative predisplacement in tubercle generation in that column (or alternatively, the other shows relative postdisplacement). However, without information on actual growth rates one could equally well argue that the specimen with no tubercles just had a faster growth rate than the other, which took longer to attain 23 mm in length and thus had a longer period in which to produce the tubercles. In this scenario, had the specimens without tubercles continued to grow past 23 mm it would have initiated tubercle growth at the same actual time as the other, but at a larger size! But in the real world, would rate of growth of specific structures be totally divorced from the size increase of the whole organism? How can we resolve this dilemma? Is it the case that without any data on time it is impossible to establish which heterochronic processes have operated?

If only one character is looked at, then it is probably not possible to resolve the dilemma. However, in most heterochronic studies many different characters can be examined, and not all are just peramorphic, or just paedomorphic. Different structures on the same organism may be affected differently, either by post- or predisplacement, or neoteny or acceleration of any differentiative or mitotic characters. By analyzing changes through ontogeny in different characters, the likely heterochronic process can be ascertained. However, to be entirely certain, absolute markers are needed (such as growth increments) preserved.

In the example just discussed (let us call the *Lovenia* without tubercles in the posterior column "A," the one with two tubercles "B") we want to investigate whether A shows relative postdisplacement. If the difference in tubercle number was merely a function of a faster overall growth rate in A, then all morphological features should appear relatively paedomorphic, because there had been less time in which to generate the various morphological features during ontogeny. In this case we could argue that the apparent paedomorphic features in A are just a result of a faster overall growth rate of the test, while growth rate of specific structures was not any faster. Therefore, a dissociation of growth rates would be necessary. In contrast, if other characters do not show this seeming paedomorphosis, then it could be argued that there is no evidence for increased global growth rate. Thus, while A may have no tubercles in the posterior column at 23 mm, it might have generated ten sets of pore pairs in the ambulacra, compared with only eight in B. Thus, the individual structures must be the ones whose timing of onset of development, or rate, have been changed.

Not only can this method be used to look at intrapopulational heterochrony, but it can equally well be used to examine interspecific changes. In *Lovenia,* the number of tubercle-bearing columns decreases along the lineage. Large adults (say 35 mm long) of the oldest species, *L. forbesi,* have tubercles in four columns on each side of the test. In the next (the undescribed) species, they occur in only three, whereas in the youngest species, *L. woodsi,* they are found in only two (Fig. 4-10). Small adults of all three species have them in only two. The natural argument would be that the postdisplacement has occurred in descendant species. Such postdisplacement is so extreme that the structures are not just delayed in their timing of appearance, but they are not generated at all.

Assessing rates of tubercle growth, either intra- or interspecifically, is done using the same technique. Because some interspecific tubercle number changes, compared with increase in test size, appear to be faster, while in other columns they are slower, then we can be sure that we are dealing with true growth rate changes between species. The extent of intraspecific variation in tubercle number is great in these species. For instance, in *L. forbesi,* at a medium size test length of 25 mm, the number of tubercles in the central columns may vary between 6 and 12. This reflects a range of rates of tubercle production or of timing of onset or offset of tubercle number between individuals. Potentially, therefore, if structures, such as tubercles, are targets of selection, a gradual change in the mean tubercle number over time, could produce a gradual phyletic transformation from one morph into another. In the descendant species, the range of tubercle number is only 5 to 8, while in the youngest species, *L. woodsi,* it is further reduced to 4 to 6. If the argument holds that such variation on the intraspecific level is purely ecophenotypic, as might be argued, how then can it be translated into a selective character between species if it has no genetic basis? Alternatively, depending on the most prominent agents of selection, gradual selection for, in this instance, fewer tubercles, might not have been the order of the day. Extreme variants might have been preferentially selected, at the expense of the phenotypic norm. This would engender rapid change, and produce a punctuated event in the fossil record. A combination, therefore, of both pronounced ontogenetic morphological change, and the consequent substantial morphological plasticity caused by intrapopulational heterochrony, provided a huge resource on which natural selection could act. Heterochrony generated the raw material; extrinsic agents dictated the timing and course of speciation. Why this particular evolutionary trend went in the direction that it did will be discussed in Chapter 5.

4.3. Selection, Heterochrony, and Sexual Dimorphism

As we have already stressed, the evolutionary significance of polymorphisms arising from perturbations to the developmental system may not always lie with the evolutionary potential of such polymorphisms, but from the fact that the polymorphisms themselves are very often the traits that have been selected for. As a consequence, they provide improved fitness for the species. In such cases they are not so much generators of evolutionary change, but the products of evolution. This is particularly so in the many cases of sexual dimorphism, which are often so extreme that the fitness of the species is appreciably increased by the existence of these "monsters." This gross disparity in morphologies (and consequently the life habits) between the sexes frequently arises from major perturbations to the developmental program of only one sex. This is not a case, however, of natural selection acting only on one sex, because the survival of the species, both dimorphs, is likely to be contingent upon such developmental polymorphisms conferring some sort of selective advantage as we discuss below.

Differences of timing of onset of maturity between sexes may be permanent or transient, as with developmental conversion. This clear dimorphism or polymorphism within one or both sexes is the feature that allows conquest of the adaptive peak. The more transient progenetic state of one sex can be demonstrated in the galeommatacean bivalve *Pseudopythina rugifera*. In this species, females (which attain a valve length of up to 15 mm) are ectocommensal on either a species of mudshrimp, or one of two species of polychaete worms. They often house within their mantle cavity a minute, sexually mature male, reaching no more than 1.25 mm in length (O Foighil, 1985). However, these males do not remain in this state for their entire life span. After fertilizing the host female, these progenetic males continue their development outside of the female, becoming first hermaphrodites, then females. They are able to move out of the female by nature of their possession in the progenetic male stage of an enlarged foot that cannot be withdrawn into the shell. Similar progenetic males (generally referred to in the literature as "dwarf" males) have been described in other members of the Galeommatacea. Progenetic males have occasionally been found inside the female mantle cavity in *Pseudopythina subsinuata* (Morton, 1972). These appear to follow a similar life history to *P. rugifera*. In two other galeommataceans, *Chlamydoconcha orcutti* (Morton, 1981) and *Montacuta phascolionis* (Deroux, 1960), it is still not clear whether or not progenetic males found associated with large, hermaphroditic conspecifics remain as progenetic males for their entire life, or switch to a hermaphrodotic stage, as Jenner and McCrary (1967) have suggested. In species of *Montacuta,* progenetic males occur either in interbrachial space (Deroux, 1960), mantle tissue, or the mantle cavity (Jenner and McCrary, 1968).

Mollusks do not only show reversible progenetic male stages and male dimorphism. In many groups, varying degrees of more straightforward sexual dimorphism can be seen, induced by heterochrony. This is probably best exemplified in the prosobranch gastropod family Eulimidae, recently reviewed by Waren (1983). All species within this family that have separate sexes show a range of sexual dimorphism, with males varying between 0.1 and 0.7 times the size of the females (Fig. 4-11). For example, *Monogamus entopodia*, a species that parasitizes the echinoid *Echinometra mathaei*, is monogamous. The female and relatively progenetic male, which attains only half the size of the female, inhabit an enlarged tube foot of the echinoid (Lützen, 1976), the snail's shell partially protruding from the infected tube foot. In other eulimids the smaller male will occur parasitically on the female. In *Thyca crystallina,* the progenetic male, only about one-tenth the size of the female, clings to the right side of the foot of the female (Fig. 4-12). Whereas the female has a strongly ornamented shell, the male shell is smooth and very thin. The female itself is parasitic, being found on species of the starfish *Linckia.*

However, this degree of sexual dimorphism is as nothing when compared with the extraordinary degree of dimorphism found in one group of prosobranch gastropods that parasitize holothurians: *Enteroxenos, Gasterosiphon, Entocolax, Entoconcha, Thyonicola,* and *Diacolax* (Lützen, 1968) (Fig. 4-13). For instance, in *Enteroxenos oestergreni* the female is endoparasitic in the intestine of the synaptid holothurian *Stichopus tremulus.* The gastropod male is so precociously developed that for a long time the species was

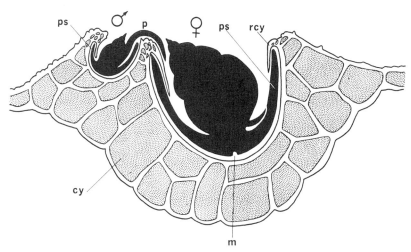

Figure 4-11. Sexual dimorphism in the parasitic gastropod *Paramegadenus arrynchus*. cy, cyst on the dorsal side of the host echinoderm; m, mouth; p, penis; ps, pseudopallium; rcy, rim of cyst. Reproduced from Humphreys and Lützen (1972), with permission of Dr. W. F. Humphreys.

considered to be hermaphroditic. However, Lützen's work has shown that the so-called testis of *E. oestergreni* is not just a male gonad, but also an extremely progenetic male. Having lost its larval shell, the male enters the female through the ciliated tubule that connects the female to its host's esophagus. The minute male attaches itself to a special male receptacle. It then expands into little more than a testis. A similar example is found in the teredinid bivalve *Zachsia zenkewitschi*. The female form of this ship-worm bores into the rhizomes of sea grass (Turner and Yakovlev, 1983). The mature male is little more than a "tailed larva." As a larva the male crawls into a mantle pouch present in the

Figure 4-12. Dorsal and ventral aspects of *Thyca crystallina*, showing progenetic male attached to female. Redrawn from Weaver (1963, Figs. 1 and 2).

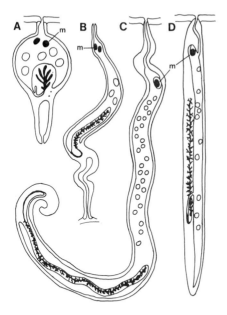

Figure 4-13. Extreme sexual dimorphism in eulimids endoparasitic in holothurians. The progenetic male (m) lives *within* the female, which itself lives parasitically within the holothurian. *A, Entocolax; B, Entoconcha; C, Thyonicola; D, Enteroxenos.* Redrawn from Lützen (1968).

female at the base of the siphons. There, the male metamorphoses. The presence of the male within the female again ensures a high percentage of fertilization of the eggs in a relatively unstable habitat. In these forms, the strongest selection pressure that ensures optimum survival of the species is one whereby extremely small male size enables localized fertilization. The target of selection is therefore optimum fecundity. When considered in an evolutionary context, a hypothetical ancestral species, where the sexes are of similar size and the male does not live within the female, would have lower reproductive success. Such a form would occupy a lower slope on the adaptive peak. So, it is not the morphological characteristics of the male that have been the target of selection, merely its small size, which enables optimum fertilization: virility selection (Travis, 1988). It is worth considering that the hypothetical ancestral species and the highly derived species are likely to occupy the same fundamental ecological niches. However, the males' realized niches are quite different. Heterochrony, therefore, has the capacity to act on only one sex, but can result in much greater species fitness as a result.

In groups other than mollusks, progenetic, parasitic males do not occur within the bodies of the females, but on them. In ophiuroid echinoderms, examples are known (Mortensen, 1936), such as *Astrochlamys bruneus,* where a number of progenetic males are carried on the upper surface of the larger females (Clark, 1976). In three other species of ophiuroids, the progenetic males are positioned with their mouth directly in contact with the female's mouth; the female, as Mortensen (1933) so aptly described it, "so far from being an affectionate mother nursing its young in a self-sacrificing manner, being a passionate mistress living in continuous close embrace with its male lover." The females of the three species that have this association are themselves epizoic, living on echinoids: *Amphilycus scripta* on *Echinodiscus; Ophiodaphne materna* on *Clypeaster;* and *Ophiosphaera insignis* on *Tripneustes* (Clark, 1976). Even in some vertebrates, progenetic male parasites are known. The deep-sea fish family Ceratiidae (seadevils) are character-

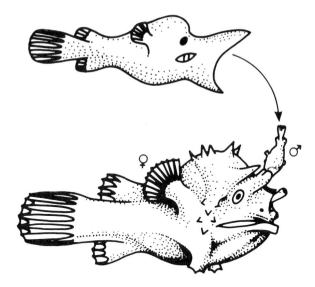

Figure 4-14. Progenetic male of the deep-sea angler fish *Photocorynus* that lives parasitically attached to the head of the female. The male is about 15 cm in length, the female about 1 m.

ized by large females up to 145 cm in length, but extremely small males, never exceeding 16 cm in length. In species such as *Ceratias holboelli,* the progenetic male attaches to the underside of the female (Pietsch, *in* Smith and Heemstra, 1986). In *Photocorynus,* however, the 15-cm-long male attaches to a process on the head of the 1-m-long female (Fig. 4-14).

5. CONCLUSIONS

Biological evolution arises from the complex interplay between extrinsic factors, such as natural selection, and intrinsic factors, such as the nature of an organism's developmental system. It is this "tug-of-war" between these two opposing forces that lies at the very heart of evolutionary change (McKinney, 1988b): "the dialectic of 'inside' vs. 'outside' influences (structural and developmental 'constraints' vs. environmental selection)" (Gould, 1988a, p. 9). Extrinsic factors may exert a direct perturbing influence on an organism's developmental system and contribute toward producing novel phenotypes within populations. On the other hand, intrinsic perturbations to the developmental system may occur in a relatively passive extrinsic setting and generate a wide variety of heterochronic morphotypes as part of "normal" phenotypic variation. Whether or not speciation ensues is dependent on a wide range of *subsequent* factors, one of the most important being the availability of niche space for the novel morphotype(s). At the very heart of the mechanisms of evolution lies the activity of heterochronic processes *within* species, for the vast majority of intraspecific morphogenetic variation arises from heterochrony, even including phenomena such as polymorphism and dimorphism.

Phenotypic plasticity, reflecting wide variation from the phenotypic "norm," may be caused by any of the heterochronic processes: changes of onset, offset, or rates of growth. Such changes to developmental rates that produce a continuum of morphotypes may not necessarily be induced by environmental perturbations, but may arise from intrinsic alterations to the developmental program, by changes to the genetic regulation of development. This we describe as "phenotypic variation." Thus, the distinction between **polymorphism** and **phenotypic variation** is merely that the former produces discontinuous structural traits, while the latter forms a continuous array of traits within a population. Localized populations within species with relatively high degrees of phenotypic plasticity induced by intraspecific heterochrony, are those which have the potential for being likely targets of selection at the species level, as the novel morphologies that they possess may be better suited to the outer zones of the organism's fundamental niche than the "normal" phenotype of the ancestral population. The evolution of high levels of phenotypic plasticity need not necessarily be a precursor of speciation, but may in itself be of particular adaptive significance and have been the target of selection. It is this very lability, the power of the organism to be externally "manipulated," that may be the important target of selection. In terms of the duality between extrinsic and intrinsic factors, extrinsic factors in some instances may induce intrinsic changes, but in others intrinsic factors may operate initially, and may then subsequently be selected, by extrinsic factors.

Extrinsic factors can directly trigger developmental change in two ways:

1. By **phenotypic modulation**, where generalized phenotypic variation results from environmental factors affecting rates or degrees of expression of the developmental program with no alteration of the organism's genetic program;
2. By **developmental conversion**, in which specific environmental cues activate alternative genetic programs controlling development. This may cause the development of alternative morphotypes (polymorphs), or lead to premature cessation of the developmental program.

Extrinsic perturbations of hormones can induce major phenotypic changes, which may induce differential niche utilization if trophic apparatuses are involved. This applies to both intra- and intersexual polymorphs. Among abiotic factors that can induce heterochrony by acting on the organism's hormonal system, probably the most important in ectotherms is temperature.

Only certain stages of development tend to be affected by developmental conversion. This occurs either by the activation of an "on–off" switch or by the expression of quite different morphotypes. Either way, such changes induce morphological saltations. The role of extrinsic factors is not so much to cause changes, but to trigger nascent developmental pathways that are already genetically programmed. Factors such as the influence of pheromones and changes in population density are agents that have been shown to cause developmental conversion.

Where there has been an intrinsic "push," rather than an extrinsic "pull" inducing changes in intraspecific developmental regulation, heterochrony can occur in two ways: as **growth heterochronies**, where the heterochronic variants occur as a phenotypic continuum within the population, and some growth and some **novel differentiative heterochronies**, where the variants are discontinuous (Chapter 3).

The phenotypic expression of sexual dimorphism arises through the action of heterochrony affecting the timing of onset of maturity. Differences in the timing of onset of maturity between male and female members of the same species are a common phenom-

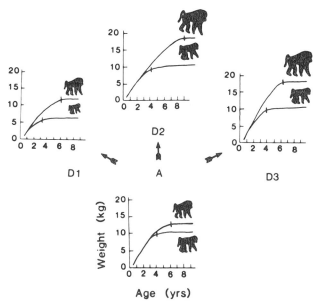

Figure 4-15. A schematic representation of how body weight dimorphism can occur by various rate and timing processes. In D1, increased weight dimorphism occurs by progenesis in the female. In D2, the male attains a larger size than the ancestor (A) by hypermorphosis, so increasing the degree of bimaturism. In D3, the size and degree of dimorphism are as in D2, but timing of maturation does not differ from the ancestral condition. Reproduced from Shea (1988), with permission of the author.

enon in animals. This **bimaturism** will result in dimorphism where growth is allometric (see Shea, 1988, Fig. 7). Increasing the degree of bimaturism will result in increased dimorphism (Fig. 4-15). Extreme progenesis in one sex can be of great adaptive significance as it can lead to miniaturization of the male which is able to live parasitically on or within the female. This ensures higher rates of fecundity. Such progenesis may be either reversible or irreversible and be the product of both extrinsic and intrinsic agents.

The translation of intrapopulational heterochronic change to interspecific change involves the coming together, at the right place, and the right time, of heterochronic morphotypes that are either higher up the adaptive peak than their immediate ancestors and can more effectively occupy the species' preoccupied ecological niche, or have developed a novel morphotype that occupies a completely different adaptive peak in a completely different ecological niche. This might either be a vacant niche (in which case we are likely to be dealing with a major evolutionary breakthrough), or a niche already occupied, but by a species lower on the adaptive peak for that niche.

The effect of pronounced ontogenetic morphological change, followed by the consequent substantial morphological plasticity, caused by intrapopulational heterochrony, provides a huge resource on which natural selection can act. Heterochrony generates the raw material which, in concert with extrinsic agents, directs the timing and course of speciation.

Chapter 5

Heterochrony in Evolution
Direction, Rates, and Agents

*Time present and time past. Are both perhaps
present in time future. And time future contained
in time past.*
T. S. Elliot, *Four Quartets*

1. INTRODUCTION

In Utopia the creation of traits in response to changing environmental conditions could occur *ex nihilo,* with no restrictions based on what had occurred in the past. There would be no discernible patterns in the appearance of new characters. But life is not like that. The paleontological and neontological records teem with evolutionary patterns, many repeated in disparate groups. Biological evolution is strongly constrained by what structures exist at each step of the evolutionary journey. Rather than new structures being created from scratch by genetic mutations, evolution involves "tinkering" with preexisting parts (Jacob, 1977). Furthermore, this tinkering needs to be carried out while the "machine is running" (Frazzetta, 1975), for the organism must remain viable even while these changes are occurring. Consequently, structural changes need to be made as rapidly as possible and to as few working components as possible. When complete, the organism must work as a fully functioning, integrated unit.

This leads to a dilemma, illustrated by that much-used analogy for natural selection, the watchmaker, who has at his call a wide range of preexisting parts from which he can manufacture new "watches." In the neo-Darwinian perspective the component parts with which the watchmaker is supplied are unlimited in size and shape, fashioned by him into an optimum, finely tuned form. Such a process is of necessity slow (particularly if the watchmaker is blind—see Dawkins, 1986) and far from efficient, since alteration of any part will have a cascading effect on others, so inducing change elsewhere.

A much quicker process is to avoid such a "piecemeal" approach, and to change parts of the system as integrated wholes. The watchmaker would be able to call upon a limited number of larger and already integrated parts. The problem is that each part, while well integrated within these larger units, would not always be as well adapted to the external environment as if it were created piecemeal. The principal constraints here will thus be the suboptimality of component parts, while in the piecemeal approach it would be the longer time and difficulty of creating integrated components.

These mechanics have major implications for evolutionary theory. Where the component parts are "finely tuned" to a small degree by gradual environmental changes, the influence of the watchmaker, the extrinsic agent of selection, will be paramount. But our watch is not a passive structure. It is not just clay from which parts can be created *ex nihilo* by the hands of the all-creating watchmaker. Our watch self-assembles (Chapter 3), a key factor introduced by the watchmaker early in the piece. All he need do in order to fashion a new watch is to tinker with the self-assembly, though as noted, he must not interfere with its "running engine." Being an expert in systems management, he can perhaps cause this vitalistic watch to cease growing a little earlier than it should. But sometimes his control is less strongly defined and major internal perturbations in the watch can result in rapid, global changes over which he has little control. Many of the "unintended" covariations so produced may result in suboptimal or adaptively neutral traits, yet the new form may, by dint of its new size or unique organization ("shape"), be adaptive in a new niche of its own "creation." Of course, too many of the unintended traits might be mechanically (intrinsically) maladaptive, in which case the entire form becomes unviable and grinds to a halt (see Chapter 8).

Our main argument in this chapter is that the heterochronic hierarchy, by modulating breadth and depth (number of parts, and degree of change per part) of its effect on the developmental system (Chapter 3), makes for a very labile process. In terms of rate changes, it allows both rapid, large-scale changes, enabling new adaptive peaks to be created in unoccupied ecological niches, as well as minor adjustments to the adaptive peak to accommodate minor changes to the ecological niche. In addition to the role that it plays in determining rates of interspecific evolution, heterochrony can affect the direction of evolution. It operates on a restricted number of parts and it deals with a preexisting directional component—the organism's own ontogeny. Thus, its role in controlling the direction of evolution within lineages is paramount.

In this chapter our aim, therefore, will be to show the impact of internal, developmental change on organismic evolution. Any such change will always have both a direction and a rate at which it is changing. We shall explore the degree to which heterochrony can influence the development of evolutionary trends, and the part that it plays in the current debate on the relative importance of rapid (punctuated equilibria) and gradual speciation (phyletic gradualism). Lastly we shall examine the extrinsic factors that sort these heterochronic changes, and so drive them at particular rates and in particular directions—these are the agents of selection.

As many of our examples are drawn from the fossil record, we shall firstly address the vexing question of polarity—identifying ancestor/descendant relationships. While it may be relatively simple, when comparing the relative degrees of development of two forms, to say that one has traits that are comparatively paedomorphic (or peramorphic) compared with the other, we shall examine the question of the reliability of both paleontological and cladistic methods in revealing ancestor/descendant relationships.

2. ASSESSING POLARITY IN HETEROCHRONY

2.1. Introduction

We have shown how heterochrony deals with changes in developmental timing between ancestors and descendants. This may occur between two generations, with the progeny of an individual organism showing either marginally different developmental rates from its parent, or perhaps changes in the timing of onset or cessation of development of particular organs or structures. Determining whether there has been peramorphosis or paedomorphosis is therefore dependent on being able to assess ancestor/descendant relationships in order to establish whether the trait is becoming relatively more "juvenile" or relatively more "adult" in appearance. Without this ability, polarity cannot be established, and any hypotheses on the significance of heterochrony in evolution become purely conjectural. In studies of intraspecific heterochrony it is relatively easy to observe and measure directly the direction of heterochronic change—its polarity. For instance, Harris (1987) undertook his experiments on the effect of varying population densities on the generation of paedomorphosis in descendant populations of salamanders (see Chapter 4) and by doing so was able to study generation-by-generation heterochrony, and identify the direction of heterochronic change.

However, it is only under experimental conditions dealing with intraspecific heterochrony that polarity can be determined with absolute certainty. To establish polarity in cases of interspecific heterochrony, ancestor/descendant relationships can only be recognized indirectly. This has proved to be one of the major stumbling blocks in studies of heterochrony for over a century. Indeed Hyatt's neglect in determining polarity with any degree of rigor (see Chapter 1) was largely instrumental in initiating the demise of the Biogenetic Law.

In order to determine character polarity in heterochronic studies, three methods have been used in recent times. These are: (1) the *ontogenetic method*, (2) the *outgroup* (or cladistic) *method*, and (3) the *paleontological* (or stratigraphical) *method*. Often these methods have been used independently of one another, but in a number of studies attempts have been made to combine the different methods, with the hope of arriving at as close an approximation to the phylogenetic truth as possible. Bear in mind that none of the methods will provide unequivocal evidence of true ancestor/descendant relationships. What they will provide, however, are adequate phylogenetic hypotheses which are both refutable, in the best Popperian tradition, and testable.

2.2. The Ontogenetic Method

It has been argued that the ontogenetic precedence of character states can be used to detect character transformations. There has been a great deal of debate in recent years over how ontogenetic criteria should best be used. While some systematists (Nelson, 1978) have claimed that the ontogenetic method is independent of hypotheses of relationships, others have insisted that it can only be used in conjunction with outgroup comparison, or even that it is itself a form of outgroup comparison (Kluge and Strauss, 1985). Nelson (1978) redefined the biogenetic law in terms of ontogeny as:

> given an ontogenetic character transformation, from a character observed to be more general to a character observed to be less general, the more general character is primitive and the less general advanced.

Fink (1988) has expressed it another way. When an observable ontogenetic transformation of character state **a** to state **a'** occurs, since **a** precedes **a'** ontogenetically, we consider **a'** to be the derived state of the character. Thus, state **a** is more primitive than state **a'**. Thus, state **a** must be viewed as being ancestral to state **a'**. However, this relationship is only valid so long as the ontogeny of the character has been one of terminal addition.

It has been argued (Rieppel, 1985; de Queiroz, 1985) that ontogenetic transformations are themselves characters; in other words the whole ontogeny evolves. It is therefore not possible to say that any particular part of an organism is more conservative and contains more evolutionary history than any other. Furthermore, they argue, one should not expect to obtain a direct insight into ancestry just by a close examination of ontogeny. If, as Wake (1989) maintains, we accept that the diversity of organisms has occurred as a result of evolutionary processes, then it becomes appropriate to accept de Queiroz's view that ontogenetic transformations are themselves characters. Consequently, ontogenetic criteria alone cannot be used for assessing character polarities, because the ontogenies themselves are the foci of interest.

2.3. The Outgroup (or Cladistic) Method

Of methods using extant organisms, this is the most widely used method of determining polarity (for cladistics in heterochronic analysis, see Stevens, 1980; Maddison *et al.*, 1984; Fink, 1988). This is achieved by comparing a character in the group being analyzed with a similar character in related taxa. If the character occurs in all or some of the taxa being studied, it is said to be derived and indicates a common ancestry of forms within the lineage. See Krauss (1988) for a major review of ontogeny polarization and heterochrony. A character which occurs in the ingroup as well as in some or all of the outgroup taxa is said to be primitive. However, a major problem with the method of outgroup comparison is the high degree of subjectivity arising from inadequately established relationships.

Some workers (O'Grady, 1985) consider that organisms pass through discrete ontogenetic stages, each of which is comparable with similar stages in related organisms. By using outgroup analysis the discrete characters can be polarized. Thus, changes in ancestral ontogenies can be assessed. However, when studying heterochrony, ontogenetic trajectories, not just discrete stages, are analyzed. The difference between these two methodologies is that one does not need to identify comparable stages when comparing ontogenetic trajectories. While it has long been argued that ontogenetic sequences are conserved during phylogeny, this, as Alberch (1985) has pointed out, need not necessarily be so. Thus, for example, significant changes early in ontogeny result in magnified changes in later ontogenetic stages. Wake (1989) calls this "repatterning." It is the

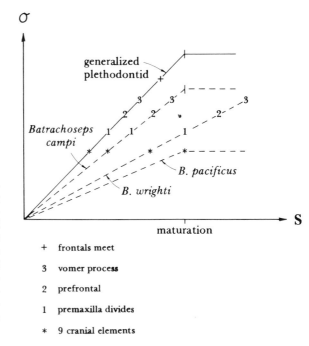

Figure 5-1. A generalized representation of four ontogenetic trajectories for the cranium of plethodontid salamanders of the genus *Batrachoseps* and a postulated ancestor. Vertical axis is shape, horizontal axis is size. Size is normalized relative to size at sexual maturity. Compared with the outgroup, the sequence *B. campi–pacificus–wrighti* shows increasing paedomorphosis. Redrawn from Wake (1989).

+   frontals meet

3   vomer process

2   prefrontal

1   premaxilla divides

*   9 cranial elements

whole ontogeny that is subject to evolutionary change; so that a true phylogeny will be based on evolutionary transformations of the entire ontogeny.

Heterochrony in the Salamander *Batrachoseps*

As an example of the use of outgroup comparisons in an attempt to formulate polarities and thus derive heterochronic phylogenies, we will discuss Wake's work on the salamander *Batrachoseps*. Here we have an example of a living genus where no discernible information on polarities is available from the fossil record. Only outgroup comparison offers the possibility of providing a reasonable means of trying to establish species interrelationships and evolutionary polarities. So by analyzing the differences in the entire ontogenies, by comparison with a suitable outgroup, it is hoped that character polarities can be established and the nature of any heterochronic change assessed.

Batrachoseps is a plethodontid salamander, which includes eight species found in North America. Like other members of the tribes Bolitoglossini and Plethodontini, *Batrachoseps* is fully terrestrial. Eggs are laid on land; development is direct, the aquatic larval stage having been abandoned. In order to assess the role of heterochrony in the evolution of the eight species, Wake plotted the first appearance of a number of cranial elements during ontogeny in three of the species: *B. campi*, *B. wrighti*, and *B. pacificus* (this species representing a number of similar species). These are compared with a postulated plethodontid ancestor, which provides the outgroup for comparison.

Four important morphogenetic events occur during the ontogeny of outgroups of *Batrachoseps*. These need to be characterized before the polarity of the species under consideration can be assessed. The four events, represented by symbols on Fig. 5-1 are:

1. Number of premaxilla in the larva. In some adults a single premaxilla is present, while in others the bone divides during hatching or metamorphosis. Outgroup analysis suggests that the divided bone is plesiomorphic (primitive or generalized) for adults. Division is roughly coincident with metamorphosis in those species with larvae, and with hatching in those with direct development. While the presence of a single bone may indicate paedomorphosis, in some species the bone fails to divide because it has undergone secondary fusion, resulting in increased jaw strength. This is, by contrast, a peramorphic event.

2. Late in ontogeny a prefrontal bone forms in some species. Rarely the appearance of minute spots of bone in species generally lacking the prefrontal indicates atavistic reversal (Alberch, 1983).

3. The larval vomer (a bone in the nasal region) becomes a small, tooth-bearing process at metamorphosis. Species in which it is absent are considered to be paedomorphic (Wake, 1966), although some otherwise peramorphic species may also lack this process (Wake, 1963).

4. During early ontogeny the frontal and parietal bones are widely separated from their bilateral counterparts. As ontogeny progresses, the bones join along the midline in many plethodontids. However, in all species of *Batrachoseps* a gap is always present.

By comparison of the species data with the outgroup, there is evidence that paedomorphosis has occurred in *Batrachoseps*. In addition to features such as the failure of the frontal and prefrontal bones to join, all species have four digits in each hind limb. This is known to be a paedomorphic trait in small amphibians (Alberch and Gale, 1985).

Wake has argued that presented with the ontogenetic data alone, phylogenetic analysis would be extremely difficult. Two possibilities exist:

1. Paedomorphosis increases with cladogenesis.
2. Extreme paedomorphosis is a synapomorphy (shared derived character) of the species, and is reversed to varying degrees.

If the ancestral stock was extremely paedomorphic, *B. pacificus* would be the ancestral form. This clade contains many species and is widespread. *B. wrighti* and *B. campi* would thus be considered to be derived species which have undergone peramorphic change from the extreme paedomorphic ancestral stock. If, however, paedomorphosis had been progressively increased during phylogenesis, *B. campi,* the "least" paedomorphic species, would represent the earliest branch, as its ontogeny most closely resembles that of the ancestral outgroup. In this scenario, *B. wrighti* and *B. pacificus* would be derived, more paedomorphic species.

As the entire ontogeny of *Batrachoseps* has evolved, we cannot use ontogenetic criteria alone to reject either of these two contradictory hypotheses. Further evidence can be obtained from a phenetic analysis of the species involved. Comparative analysis of allozymes indicates that *B. campi* and *B. wrighti* are sister taxa. While genetic distances between species in this genus are generally high, between these two species they are small. This might indicate support for the hypothesis that *B. pacificus,* the extreme paedomorph, is ancestral to other species. A third hypothesis, however, is that *B. wrighti* is ancestral, *B. pacificus* is progenetic (because of its truncated ontogeny), and *B. campi* is neotenic, passing more slowly through the ontogenetic stages of its ancestor.

The use of such ontogenetic data in isolation highlights the difficulties in assessing polarities, and hence formulating concise phylogenetic hypotheses. As Wake has argued, the concept that ontogenies are conservative may be misguided. Evolution can affect every point along the ontogenetic trajectory.

2.4. The Paleontological Method

The use of the fossil record in phylogenetic studies has long been criticized (Eldredge and Cracraft, 1980; see Fink, 1982, 1988). This criticism has focused on the typical paleontological assumption that ancestor/descendant relationships can be inferred from a study of the fossil record. Critics have pointed to (1) the poor state of the fossil record; (2) the paucity of data, when compared with the data which can be obtained from recent organisms; and (3) problems in assuming that the stratigraphic ranges of fossils reflect their true age ranges. In a recent attempt to assess the adequacy of the fossil record for phylogenetic studies to answer these points, Paul (1985) has demonstrated that far from being secondary to neontological data, paleontology deserves equal consideration.

2.4.1. Adequacy of the Fossil Record

No one denies that the fossil record is an incomplete representation of the history of life. However, what is generally ignored by neontologists is the high degree of variability in the quality of the fossil record between different groups of organisms. To take an extreme case, there is little doubting that the fossil record of butterflies is very poor compared with the fossil record of mollusks. Yet no paleontologist in his right mind undertaking a phylogenetic study based on fossilized organic remains is likely to use butterflies in preference to mollusks.

The original habitats of the organisms, as well as their structural components, also greatly affect the degree of completeness of their fossil record. Organisms living in or around environments of deposition are more likely to be fossilized than those inhabiting regimes of high erosion. Thus, plants favoring growth sites near streams and lakes will have a better fossil record than alpine floras. Even within classes, the fossil record can vary enormously. Regular echinoids are epifaunal and often inhabit environments of high hydrodynamic activity. Consequently, after death, the echinoid tests are generally rapidly broken up and become part of the sediment, rather than being fossilized in their entirety. Most irregular echinoids, on the other hand, occur in areas of high fossilization potential: they are infaunal, living, feeding, and dying within the sediment. They are thus much more likely to be fossilized than their regular counterparts.

Irrespective of the overall incompleteness of the fossil record, the stratigraphic sequence of fossils must be in the correct order with respect to their evolution. It's all a question of correct sampling and interpretation of the fossil record.

2.4.2. Information from Neontological and Paleontological Sources

Paul (1985) has pointed out that while the chances of an organism being preserved in the fossil record are astronomically small, they are really little different from the chances of a living organism ending up in a museum or in a research collection. Biologists do not need to collect every specimen of a species to understand its biology. Like paleontologists, biologists deal with very small samples. Thus, in the two disciplines, sample sizes tend to be similar and comparable.

Although biology potentially offers more information in the form of soft-part anatomy, genetics, and embryology, many living organisms are known from incomplete specimens. Paul (1985) cites the example of living Jamaican nonmarine mollusks, of which about 90% are known only from their shells. To all intents and purposes they might as well be fossils. However, this lack of knowledge of the soft anatomy does not invalidate what we do know about the nonmarine mollusks. To imply that because we cannot gather all the information from fossils, they are useless, is quite misleading. In fact, there are many invertebrates in which the morphological characters used to differentiate living species are features which are preserved in the fossil record.

While the fossil record is incomplete in having a bias toward organisms with hard parts, the neontological record is incomplete in only having information on species living today. If the average life of a species is 6 million years, and 600 million years is the length of the Phanerozoic, then living species represent only 1–5% of species that have ever lived. The neontological record, therefore, is a very poor sample of Phanerozoic life. Its strength lies in the biological information which it can offer. The strength of the paleontological record lies in the sequence of events it preserves over a very long period of time. Neither the neontological nor the paleontological records are any better than the other. Each provides different, but valuable, information for phylogenetic analysis. Unless information from both sources, if available, is used, a distorted and incomplete history of life will be perceived.

2.4.3. Validity of Stratigraphic Ranges

A major criticism of the fossil record is that stratigraphic sequences may not represent the true period of existence of organisms. However, the only way in which fossils can be preserved in the wrong stratigraphic order is if they coexisted in time. Thus, Cambrian trilobites and Cretaceous ammonites will never occur in the wrong stratigraphic order. Using the same average history of a species as being 6 million years. Paul (1985) has calculated that over the approximately 600 million years of the Phanerozoic, overlapping ranges will occur in only 1–5% of possible comparisons of species pairs. This percentage will not alter with variations in the degree of completeness of the fossil record. Thus, in 90–99% of possible comparisons between species pairs, there is no chance that they could be preserved in the wrong stratigraphic order. Indeed, if this were not so, correlation by the use of fossils would be well nigh impossible.

In evolutionary studies we are concerned with comparing species that are morphologically similar and have similar ontogenetic trajectories. They will also be stratigraphically proximal. Even in such situations the chances of species being preserved in the wrong stratigraphic order are far less than their chances of being correctly preserved. Take for example a period of time from T1 to T4 in which two species, A and B, one ancestral to the other, have occurred. There are two alternative time relationships be-

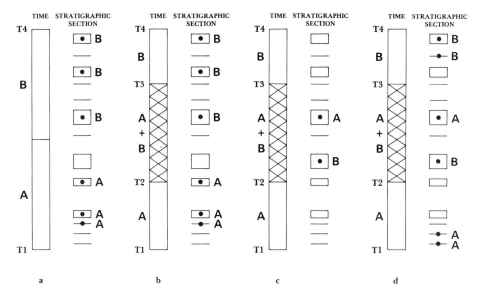

Figure 5-2. The validity of stratigraphic ranges: (a) between time T1 and T4, species A is replaced by species B: sampling a section would reveal A occurring lower than B; (b) between T2 and T3, A and B have temporal overlap; sampling through a section corresponding to T1 to T4 still shows A occurring before B; (c) where sampling is incomplete and only carried out in the region of temporal overlap, inverted polarity could be inferred, as B could be found lower than A; (d) however, collecting the entire section and entire time range, the true polarity will be deduced, even though specimens can be located at only six horizons.

tween the species: either they did not overlap in time (Fig. 5-2a) or they did overlap (Fig. 5-2b–d). A stratigraphic sequence will contain blocks of sediment representing parcels of time. If the species did not coexist, there is no way that one could appear in the fossil record before the other, even with incomplete sampling. Many examples of anagenetic speciation, where ancestral species became extinct and were replaced by a descendant form, have been documented (see below). In such situations the order that the fossils are found in a stratigraphic sequence represents the order in which they appeared in time.

In the situation where the two species coexisted for a limited period, three parcels of time can be considered: T1–T2, when species A alone existed; T2–T3, when species A overlapped in time with species B; and T3–T4, when species B existed alone, species A having become extinct. Collecting in a stratigraphic section in such a situation could yield spurious results if sampling was insufficiently rigorous. Thus, it is possible that if only two horizons were collected within the time range T2–T3, species A could be found stratigraphically higher than species B (Fig. 5-2c). However, if T2–T3 is adequately collected, or if the full stratigraphic sequence representing T1–T4 is sampled (Fig. 5-2d), it will become clear that species A predates species B. Therefore, contrary to the opinion of a number of neontologists, the chances of species being misplaced in stratigraphic sequence are remote.

However, neither paleontologists nor neontologists can unequivocally say that any one species is ancestral to another. All that can be achieved is a working hypothesis based on a combination of morphological, ontogenetic, and stratigraphical criteria.

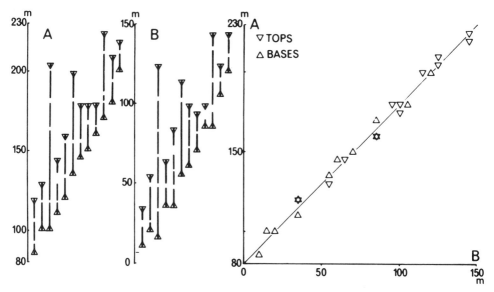

Figure 5-3. A and B represent logs of two sections on which the known ranges of 12 species common to both sections are plotted. One section is selected as a reference section. Bases and tops of ranges are plotted as a scatter diagram (right) and a regression line is fitted to the scatter. The regression equation is used to calculate which of the two sections has the lower first (and the higher last) occurrence for each species. These values are plotted on the reference section. Repetition for all available sections produces a composite standard section. Reproduced from Paul (1985, Fig. 2) with the permission of the Palaeontological Association.

2.4.4. Determination of a Correct Stratigraphic Sequence

In practical terms, what is the best method for determining whether or not a sequence is correct? Perhaps the most suitable is that defined by Shaw (1964). In his system a single stratigraphic section is chosen as a reference section. First and last occurrences of all species common to this section and a second one are plotted as a scatter (Fig. 5-3). It is possible to establish in which section the first occurrence is lower and in which the last is higher. By repeating this procedure with other sections a composite standard section, in which total ranges of all species are accurately recorded, is obtained. The composite standard section enables the sequence of first appearance of species to be established, and therefore the order in which they appeared.

The existence of an extinct species allows three time components to be established: E1, the period before its existence; E2, the period of its existence; and E3, the period after its existence. Only E2 can be identified with accuracy, because the actual boundary of E2 with E1 or E3 can never be known with absolute certainty. The co-occurrence of two extinct species allows five time components to be established: E1–5. Only the period of overlap can be identified with certainty. The best known ranges are usually those of the most common species. Extending the amount of collecting is likely to extend the overlap.

Thus, while the nature of sedimentary deposition is such that stratigraphic sections are too incomplete to allow generation-to-generation changes to be traced (Sadler, 1981), patterns of change can still be recognized. This is particularly so if the morphological characteristics of a species undergo relatively little change over prolonged periods. In such situations it is unnecessary to have complete sequences.

2.5. Application of Both Outgroup and Paleontological Methods

Rather surprisingly, the application of a combination of both outgroup and paleontological methods is unacceptable to many cladists (Eldredge and Cracraft, 1980). But as McNamara (1986a) and Fortey and Chatterton (1988) have pointed out, it is rather a shortsighted attitude to neglect stratigraphic information. There has only been one detailed attempt to interpret the impact of heterochrony on the phylogenetic history of one particular monophyletic group specifically using cladistic analysis in combination with stratigraphic data. This work, carried out by Miyazaki and Mickevich (1982) on the Mio-Pliocene bivalve *Chesapecten,* showed peramorphosis. Despite claims to the contrary (Nelson, 1978; Fink, 1982, 1988), undoubted problems can be encountered when just cladistic or outgroup methods are used in sequences showing paedomorphosis. Without stratigraphic information, spurious results can be obtained. The importance of stratigraphic data in formulating an accurate polarity is illustrated using an example of two species of the spatangoid echinoid *Breynia* from western Australia.

2.5.1. The Bivalve *Chesapecten*

Chesapecten is a common genus in Mio-Pliocene deposits on the eastern coast of North America, particularly around Chesapeake Bay, and south to North Carolina. Over a period of 11 million years (from 23 to 13 million years BP), seven species evolved (Ward and Blackwelder, 1975). In their study, Miyazaki and Mickevich used both cladistic (outgroup) and stratigraphic (paleontological) data to construct a phylogeny and assess whether heterochrony had played a significant role in the evolution of this monophyletic lineage.

Miyazaki and Mickevich collected 17 samples of *Chesapecten* from 15 localities, ranging in age from the middle Miocene to the early Pliocene. Most localities yielded a wide size range of shells, thus providing detailed ontogenetic information. Patterns of character differentiation were also inferred from individual development by analyzing changes in growth lines. Comparison of character state differences between species provided information for phylogenetic analysis. Samples ranged in size from larval prodissoconchs to the largest individuals at each locality. Up to 30 mm shell height a byssal notch is present in all species. While remaining in some earlier species at larger sizes, in many later species it became occluded. Miyazaki and Mickevich used this size to designate onset of the adult stage. It also transpired that above this size there was a reduction in degree of morphological change.

In their analysis 11 characters were used: (1) length/height ratio; (2) umbonal angle; (3) dorsal to anterior half-diameter ratio (posterior plane); (4) anterior to posterior half-diameter ratio (for valve symmetry); (5) convexity to height ratio; (6) diameter of muscle scar to height ratio; (7) anterior to total auricle length ratio; (8) height to length of anterior auricle; (9) depth of byssal notch to anterior auricle length ratio; (10) rib number; (11) height of central rib to valve height ratio.

The mean value for each character in each sample was divided by average standard deviation for all samples to remove bias introduced by incomplete variability represented in some samples. The equation used was:

$$K_{ij} = \frac{u_{ij}}{(\sum^n =_1 SD_{ij})/n}$$

where i = sample, j = character, u = mean character value for each sample, n = number of samples, K = character condition of each sample, $\sum SD_{ij}/n$ = mean standard deviation for a character across samples.

Miyazaki and Mickevich's aim was to use their data to construct a phylogeny, without requiring any assumptions or preconditions about the nature of the evolutionary process. Once formulated, the phylogenetic hypothesis would be compared with the stratigraphic information to see the degree of congruence between the two methodologies. A procedure was used to obtain a best-fit cladogram, called transformation series analysis (Mickevich, 1981, 1982). An individual character transformation on the best-fit cladogram would show the sequence of character states as linked on the cladogram. According to best-fit criteria, the number of shaped character states ought to be maximized on the cladogram.

The shared state can transform on the cladogram independent of the numerical value of the character states. So while character states 1, 2, 3, 4, 5 may be considered a transformation ordering, transformation series analysis does not use this ordering. A cladistic transformation for a character of 1, 4, 2, 5, 3 is still compatible with the best-fit cladogram. Thus, according to Miyazaki and Mickevich, transformation series analysis allows for heterochronic character state changes.

The cladogram of the relationships of different forms of *Chesapecten* produced by transition series analysis on the characters shows a high degree of congruence with stratigraphic data (Fig. 5-4). Even a transitional form, 2–3, recognized by Ward and Blackwelder (1975), falls neatly between these two groups on the cladogram. Even though stratigraphic gaps are present, the overall relationship of the taxa on the cladogram and the stratigraphic sequence agree with one another. There is, however, one notable inconsistency, regarding the stratigraphic species *C. middlesexensis*. This highly variable species occurs in three places on the cladogram.

Character state changes in phylogeny were shown to be strongly correlated with ontogenetic changes. In many characters, such as overall shell shape, umbonal angle, auricle length, and depth of byssal notch, phylogenetic changes parallel ontogenetic changes, indicating peramorphosis in these features. In addition to an increasing degree of development in later species, they also underwent a size increase along the lineage. Early species are small and light, later species are large and heavy. While this implies hypermorphosis as a possible heterochronic mechanism, some of the species differences are due to changes in the timing of appearance of new characters. For example, byssal notch occlusion occurs progressively earlier in successive species—this is predisplacement.

2.5.2. The Echinoid *Breynia*

Two species of *Breynia* are known from western Australia: *B.* aff. *carinata,* from the Middle Miocene in the Cape Range district; and the living species *B. desorii.* This species

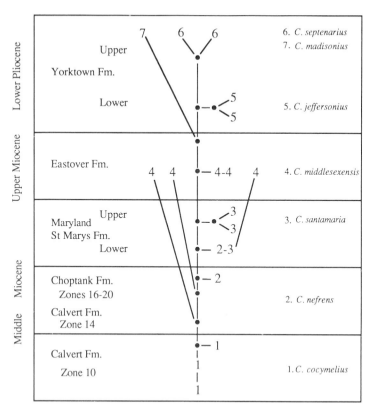

Figure 5-4. Cladogram of *Chesapecten* sample relationships using transformation series analysis, aligned with the stratigraphic sequence. Numbers refer to stratigraphic positions of separate samples. Redrawn from Miyazaki and Mickevich (1982, Fig. 4).

also has a fossil record back into the late Pleistocene (McNamara, 1982a). These two species differ in one important respect. On the adoral surface, the juxtaposition of the labrum, a long thin plate positioned immediately behind the peristome (mouth), with other plates occurs in two ways. In the living species, *B. desorii*, this labrum is in contact with the large paired plates of the plastron in both juveniles and adults of almost every specimen studied (Fig. 5-5). This pattern is characteristic of the vast majority of spatangoids, and is a primitive character. The Middle Miocene *B.* aff. *carinata,* on the other hand, differs in that while the juveniles and small adults have the labrum and plastronal plates in contact, in large adults adjacent ambulacral plates have translocated between the labrum and plastronal plates, so separating them (McNamara, 1987a). Ontogenetically this character is derived. Moreover, outgroup comparison shows it to be a phylogenetically derived character. Thus, if no stratigraphic information was available, cladistic analysis would suggest that *B. desorii* was closer to the ancestral morphology than *B.* aff. *carinata.* Thus, the polarity would indicate a peramorphic transformation from a condition in which plates were in juxtaposition to one in which they became separated.

Stratigraphic data, on the other hand, suggest an alternative explanation because *B.* aff. *carinata* predates *B. desorii.* The condition found in the living *B. desorii* formed by

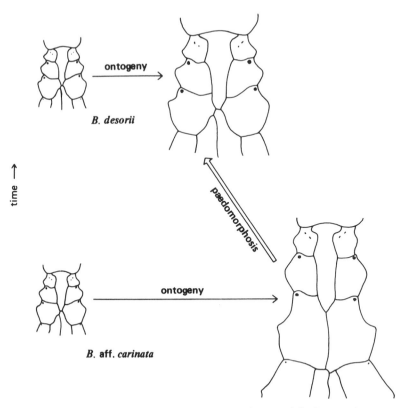

Figure 5-5. Paedomorphic plate translocation in the labrum/plastron of the heart urchin *Breynia*. *B.* aff. *carinata* is a Miocene species: *B. desorii* is a Pleistocene to Recent species. Reproduced from McNamara (1988a), with permission of Liverpool Geological Society.

paedomorphosis from the Miocene form. The separation of the labrum and plastronal plates occurs in all known Miocene species of *Breynia*. Among living species it is only known to occur in about 1% of *B. desorii* specimens (McNamara, 1982a). The rare presence of this character lends support for the contention that these two species of *Breynia* are genetically very closely related. In all other adult living *Breynia* species the plates are in contact.

Thus, an outgroup comparison alone would have provided a false polarity, and hence incorrect assignment of the type of heterochrony. Addition of stratigraphic information, in this instance, allowed for a more plausible explanation of the polarity and hence the type of heterochrony.

3. HETEROCHRONY AND THE DIRECTION OF EVOLUTION

3.1. Introduction

For a long time there has been a tendency to see directionality in evolution as purely the product of extrinsic factors, largely because of the great preoccupation in evolutionary studies with natural selection (Maderson *et al.*, 1982). In this section we demonstrate how heterochrony, the most abundant, and perhaps most innovative, intrinsic factor in evolution, plays a crucial role in the generation of evolutionary trends. We examine the complex interplay between changing timing of development and changing environmental factors. We demonstrate just how much impact this tug-of-war between extrinsic and intrinsic factors, each under different conditions exerting different forces, has had in generating directionality in evolution. Where possible, we assess the degree to which heterochrony and evolutionary trends are tied to the production of adaptations. After all, adaptations (the functional expression of morphological trends) arise, as Maderson *et al.* (1982) note, "from a directional principle intrinsic to the relationship between organisms and their environment." To what extent are the structures themselves the targets of selection? Are they often merely by-products of selection for other factors, such as reproductive strategies or life history strategies? It is our contention that much of the directionality in evolution arises from the inherent parallelism between an organism's ontogeny, and the directionality present in so many environmental gradients. The evolutionary trends with which we shall be dealing are usually morphological gradients. However, behavioral or life habit trends are also encompassed within this concept.

Evolutionary trends may range from small- to large-scale. Where phyletic changes can be documented within single species lineages, there is some evidence that heterochrony is involved. Many of the examples with which we shall be dealing will be at the species or generic level. But many large-scale evolutionary trends also have their basis in heterochrony (McNamara, 1990). Since Eldredge and Gould (1972) proposed their concept of "punctuated equilibria," there has been a tendency to regard many of the so-called evolutionary trends described from the fossil record as little more than arbitrary elements chosen by investigators because they appear to conform to a morphologically graded series. In other words, they are little more than the fanciful wanderings of paleontologists' trend-oriented imaginations. Nevertheless, it is clear that statistical tendencies do indeed occur, whether it be in a single clade or branching clade. What we shall be largely talking about in this section is directionality at the species level. While we will be addressing the question of the significance of heterochrony in slow versus rapid evolutionary change in the section on rates, this is only part of the story of the role of heterochrony in directionality in evolution.

There have been a number of alternative models proposed to explain the generation of evolutionary trends. They have been interpreted as: (1) the product of "directional speciation" (Grant, 1963; Stanley, 1979), induced by "mutation pressure"; (2) "phylogenetic drift" (Raup and Gould, 1974), induced by "genetic drift"; (3) "species selection" (Stanley, 1975, 1979), induced by "natural selection" pressures; (4) the result of evolution toward increased specialization in species-specific characters—the "effect hypothesis" of Vrba (1980); (5) "environmental orthoselection," for example, a consistent decrease in temperature (Futuyama, 1986b); (6) "co-evolutionary interactions" (Futuyama, 1986a), including the "arms race" concept of Vermeij (1987); (7) changes in "variance" [Gould, 1988b, based on Stanley's (1973) notion]. The problem with these ideas is that most deal only with either nebulous intrinsic (1 and 2) or extrinsic factors (3 to 6), rather than with a combination of the two, while one (7) tries to examine large-scale trends outside of the context of intrinsic versus extrinsic effects. However, even here, where Gould (1988b) reduces the emphasis on natural selection, the very variance that he shows is changing (by virtue of having nowhere else to go) is more often than not variance produced by heterochrony.

Recently, a view has emerged that heterochrony has been an important agent in directing morphological variation along particular, constrained, pathways (Ede, 1978; Gould, 1980; Alberch, 1980; Levinton and Simon, 1980; McNamara, 1982b; Maderson *et al.*, 1982; McKinney, 1988b). Jablonski *et al.* (1986) have stressed the importance of intrinsic factors in trend generation. They have noted how there is a tendency for size changes in evolutionary trends. The implication of this is that heterochronic changes which involve size change play a significant role in trend generation. However, intrinsic factors alone are unlikely to generate trends. The many close morphological–environmental associations that have been recorded (McNamara, 1988b), which we discuss below, argue that migrations up and down ontogenetic trajectories are tracking environmental trajectories with which they are coadapted.

As we argue throughout this book, heterochrony can be seen as the most important factor in the generation of intrinsic phenotypic change. It is such phenotypic change which is the target for selection, whether it be focused on shape or size, or on life history or behavioral strategies. Whatever the agents of selection, they do not operate on a haphazard array of traits. They are targeted on a limited series of factors, restricted in diversity by the nature of organisms' ontogenies (an aspect that we discuss at length in the section on targets of selection). To return to the watchmaker analogy, heterochrony supplies the parts. The big question is, how restricted is the supply?

Developmental programs are series of constrained structural organizations, as discussed in Chapter 3. For instance, limb morphology in living vertebrates has changed very little from that in Devonian vertebrates [Darwin's "unity of type" (Shubin and Alberch, 1986)]. This invariance in limb development probably arose from historically conserved mechanisms of morphogenesis (Oster *et al.*, 1988). All limbs form under a single set of basic "construction rules." This restricts the extent and degree of variability that can be generated. And these rules impose a developmental order independent of adaptive pressures and other extrinsic factors (Oster *et al.*, 1988). Thus, developmental constraint arising from these construction rules imposes a preexisting directional component, should perturbations occur. Such perturbations can only act within the limit (preexisting foundation or "Bauplan" in a global sense) of the inherent construction rules and so result in heterochronic change with a preexisting directional component (part suppliers to the watchmaker with a limited inventory). Combine this with extrinsic directionality (the watchmaker's predilections) and the directions in which evolution can proceed are heavily constrained.

If heterochrony has played a major role in evolutionary trends, then the fossil record should be rife with examples of lineages showing patterns of both inter- and intraspecific heterochronic changes in particular directions, either paedomorphic or peramorphic. Therefore, we should be able to pose a number of specific questions concerning the role of heterochrony in evolutionary trends: (1) what is the nature of heterochronic trends; (2) do different heterochronic processes produce different heterochronic trends; (3) can heterochronic trends only be recognized in the fossil record, or can the neontological record also provide evidence for them; (4) can heterochronic trends operate on different structures in different directions within the same lineage; (5) what is the nature of the selection pressures that induce the changes; (6) what features are being selected for: size, shape, behavior, or life history strategy; (7) why do the trends go in any one particular direction? Such questions form the major bridges between ecology, ontogeny, and evolution. While it has become common to "decouple" paleontology from neontology, the only truly unified study of evolution will result from such connections. This does not preclude emergent processes at coarse scales, but a complete understanding will come only from determining connections among processes interacting at different temporal and spatial scales.

3.2. Heterochrony and the Generation of Evolutionary Trends

3.2.1. Paedomorphoclines and Peramorphoclines

Many sequences of fossils that have been considered to represent single evolving lineages reveal a pattern wherein structural features (either their shape, number, or size) become either "more" paedomorphic or "more" peramorphic. Rarely does the same structure in a lineage show reversals wherein the character is, say, paedomorphic in the first descendant, then peramorphic in the next, followed by paedomorphic in the next.

Consider first a series of ontogenies that change through time by paedomorphosis. The ancestral, nonpaedomorphic form (termed the apaedomorph) passes through a number of morphological stages, say A to M, during the course of its ontogenetic development from embryo to adult (Fig. 5-6). If the apaedomorph gives rise to a descendant form in which all or some of the characters are paedomorphic, then the descendant will, for the character, or series of characters, pass through only, let us say, stages A to K (bearing in mind that for organisms with continuous growth, the defined stages are merely arbitrary, although in organisms that grow discontinuously, or whose phenotypic expression is episodic, such as arthropods, discrete stages are directly applicable). The postulates of the model are that subsequent to the evolution of this first paedomorph, another form evolves from the first paedomorph. This second paedomorph may pass only through stages A to I and may itself subsequently give rise to a further paedomorph, that passes only through stages A to G. This pattern continues to the last species to evolve in the lineage, the most paedomorphic, that only passes through stages A to C. The six adult morphologies in the lineage constitute a morphological gradient through time, M–K–I–G–E–C. These temporal adult morphological stages follow the opposite morphological pathway to the ontogenetic development of the earliest species, the ancestral apaedomorph. This sequence of adult morphologies displaying a temporal morphological gradient of increasingly more juvenile characters has been called a **paedomorphocline** (McNamara, 1982b).

The operation of peramorphic processes will produce a similar pattern, but one that is a mirror image of the paedomorphocline. The lineage will consist of a sequence of increasingly more peramorphic species. This has been termed a **peramorphocline** (McNamara, 1982b). Collectively, paedomorphoclines and peramorphoclines may be termed **heterochronoclines**. In a peramorphocline, the ancestral, aperamorphic, species passes through ontogenetic stages A to C during ontogeny. By the operation of one of the peramorphic processes, a descendant form will evolve that passes through stages A to E during its ontogeny. Subsequently, a descendant arises that passes through yet more

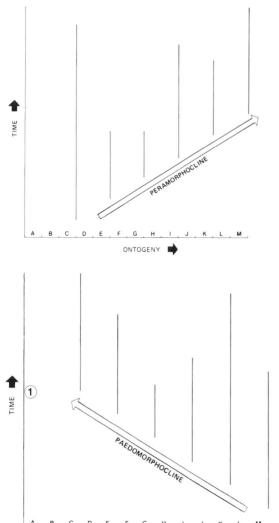

Figure 5-6. Paedomorphocline and pera-morphocline: morphological gradients of progressively more paedomorphic and more peramorphic species through time. A to M represent arbitrary ontogenetic stages. Species time range is for the adult morphology. Time 1 represents a situation whereby all paedomorphs and the apaedomorph could coexist temporally: ecological separation of morphotypes will ensure that they do not coexist spatially. Reproduced from McNamara (1982b).

stages, A to G, and so on. The six forms in Fig. 5-6 have adults that constitute a morphological gradient through time of C–E–G–I–K–M. This morphological trend is the mirror image of that developed in the paedomorphocline.

Heterochronoclines can be of two basic forms: **anagenetic**, or **cladogenetic**. In the former, there is, on the geologic time scale, no apparent overlap in temporal ranges between the species and there is no overall increase in numbers of species: as a new species arises, so its ancestor becomes extinct. Where there is perceptible overlap in stratigraphic (time) ranges between species in a heterochronocline, then the speciation event is cladogenetic, the ancestral species surviving as one of the branches. In this scenario, species numbers will increase. The selection pressures that determine whether the heterochronoclines are anagenetic or cladogenetic are discussed below.

In the original model of paedomorphoclines and peramorphoclines (McNamara, 1982b, Figs. 1 and 2), the morphological gradients were considered to be discontinuous, on the basis of evidence from empirical examples. However, it could be argued that if the heterochronic changes are induced by, say, gradual changes in allometries, then these changes could be continuous. While this can be demonstrated in some intraspecific lineages (see below), it has yet to be determined for interspecific cases. It may be that a species' genetic homeostasis or gene pool integration is so interlocked that species changes occur only in discrete packages even in a continuously graded environment, consistent with the "punctuated equilibria" model.

Originally, the concept of heterochronoclines was proposed as a model to explain directional *interspecific* evolutionary trends, where the morphological change was induced by heterochrony. Examples to support this have been described principally in marine invertebrates, including brachiopods, echinoids, trilobites, ammonites, and other mollusks (McNamara, 1988b). In all of these examples the transitions are between what are interpreted, on the basis of morphological criteria, as species. Levinton (1983, 1988) has criticized this approach to explaining evolutionary trends at the species level on the grounds that the temporal changes occur without cladogenesis. He has argued that unless cladogenesis can be demonstrated it is not possible to be certain that the morphotypes that are being dealt with represent different species. Unfortunately, Levinton was mistaken in interpreting the lineages that McNamara described as being anagenetic patterns, where there was no temporal overlap between species. On the contrary, in those examples of temporal heterochronoclines described in the initial paper, all show considerable overlap in the time ranges of the "morphospecies." This, we contend, therefore demonstrates cladogenetic events and is perhaps best demonstrated in the *Tegulorhynchia–Notosaria* paedomorphocline (McNamara, 1982b, 1983b), a lineage of Cenozoic rhynchonellid brachiopods.

Fossil and living species of these two genera are confined to the Indo-West Pacific region. In this lineage (Fig. 5-7) the ancestral rhynchonellid brachiopod species, the Paleocene *Tegulorhynchia boongeroodaensis,* underwent appreciable morphogenesis during ontogeny. Its shell became relatively broader; the individual valves increased appreciably in depth, reflecting a great increase in internal volume; the commissure (line joining the two valves) developed a strong median plication; the number of ribs increased from about 25, at a shell length of 2 mm, to about 80 at a maximum shell length of 18 mm; the beak reduced in height resulting in an increase in umbonal angle; and the foramen became relatively smaller, as the deltidial plates joined, almost closing the foramen, through which the pedicle would have passed. This reflects a relative reduction in pedicle thickness through ontogeny.

The second species in the paedomorphocline is the New Zealand Late Eocene to Early Miocene *T. squamosa.* This long-ranging species may in fact be synonymous with the living species *T. doederleini,* according to Lee (1980). If so, then all subsequent forms to evolve along the paedomorphocline did so by cladogenesis. *T. squamosa-doederleini* has fewer ribs than *T. boongeroodaensis* (only 60), slightly narrower shell, less strongly developed plication, larger foramen, and slightly larger shell size. The presumed descen-

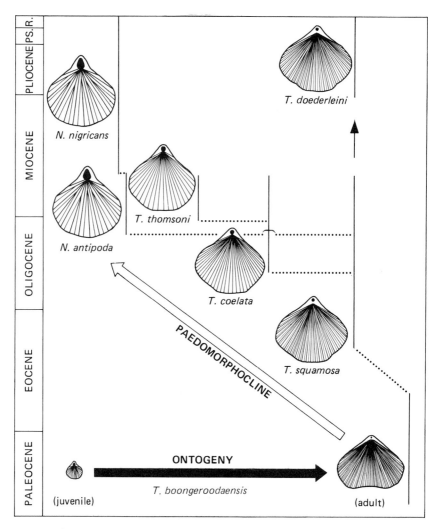

Figure 5-7. Development of a paedomorphocline in the evolution of the Australasian Tertiary rhynchonellid brachiopods *Tegulorhynchia* and *Notosaria*. The paedomorphocline follows an environmental gradient of deep to shallow water. Reproduced from McNamara (1983b).

dants, *T. coelata,* which first appeared in the Middle Oligocene, and *T. thomsoni,* from the Early Miocene, continue the trends apparent between the older species: fewer ribs, narrower shell, weaker plication, and larger foramen. The end members of the paedomorphocline are fossil and living species of *Notosaria, N. antipoda* and *N. nigricans,* respectively. The adults of these species are morphologically closest to the earliest ontogenetic stages of the apaedomorph, *T. boongeroodaensis,* showing few ribs (less than 25), relatively very narrow shells, most disjunct deltidial plates, and so largest foramen (see Fig. 6-20). There can be little doubt that most speciation events (for so they are) were cladogenetic. Likewise, all spatial heterochronoclines (where the species coexist temporally) must have arisen by cladogenesis [e.g., the *Olenellus* paedomorphocline—see

Chapter 4; the spatangoid echinoid *Breynia* paedomorphocline—see McNamara (1982a); and the portulasiid echinoid peramorphoclines described by David (1989)]. Levinton's argument that heterochrony has had little impact on the generation of interspecific evolutionary trends, therefore, does not stand up to scrutiny.

3.2.2. Heterochrony and Trend Generation

Using fossils to assess which heterochronic processes have operated in the generation of particular heterochronoclines poses the usual problem discussed in detail in Chapter 2: general absence of age data. However, use of such data when available, in combination with size criteria and information derived from the operation of dissociated heterochrony, allows some degree of assessment of heterochronic processes to be made. In the case of the *Tegulorhynchia–Notosaria* paedomorphocline, the paedomorphosis appears to have been global in its effects. It has been proposed (McNamara, 1983b) that neoteny was the principal heterochronic process. While the presence of very conspicuous growth lines in these genera provides the potential for testing this hypothesis, this has not been performed to date. A very small paedomorph, *T. sublaevis* (Thomson), that was contemporary with *T. squamosa,* has been interpreted as being progenetic, on account of its small adult size and paedomorphic characters (McNamara, 1983b), but this is really allometric progenesis (see Chapter 2). While it is conceivable that interspecific changes in the *Tegulorhynchia–Notosaria* paedomorphocline could have been gradual, particularly where changing rates of mitotic growth and differentiation are involved, there is indication that the pattern of evolution within this lineage was one like that depicted in Fig. 5-6: a cladogenetic paedomorphocline, with long periods of morphological stasis within species, and rapid transitions between them. Evidence for this comes from *T. squamosa-doederleini.* This "species"-pair appears to have undergone an extremely long period of stasis, extending for up to 45 million years (Lee, 1980).

The primary effect of differing modes of heterochrony on the generation of evolutionary trends is on the patterns of heterochronoclines that are developed. As we shall discuss more fully in the section dealing with rates of evolution, the type of heterochronic process can determine whether interspecific changes are gradual or episodic. Where either global progenesis or hypermorphosis were the heterochronic processes, the likelihood of interspecific changes being episodic is much greater. This is because where allometric changes during ontogeny are substantial, or where rates of differentiation are high, then abrupt radical changes in timing of maturation can produce "punctuated" events. Evidence from polymorphisms (see Chapter 4) demonstrates that such substantial morphological jumps can occur between generations. Thus, in some trilobite paedomorphoclines, where progenesis caused size reduction combined with thoracic segment reduction, progression from one species to another is likely to have been episodic. In the *Olenellus* paedomorphocline (Chapter 4), the apaedomorph and first paedomorph have 14 thoracic segments, but the fourth paedomorph, the terminal member of the paedomorphocline, only 9. In this character at least, change was episodic. The operation of global progenesis or hypermorphosis in organisms, such as arthropods, whose phenotypic expression itself is episodically induced by molting, combines to produce abrupt changes between species, and thus a pronounced **stepped heterochronocline**. Two sorts

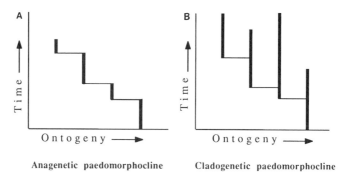

Anagenetic paedomorphocline   Cladogenetic paedomorphocline

Figure 5-8. (A) In the anagenetic paedomorphocline there is no temporal overlap of species, and thus number of species is not increased through time. (B) In the cladogenetic paedomorphocline there is temporal overlap and increase in numbers of species through time.

may exist: *stepped paedomorphoclines* or *stepped peramorphoclines*. Where there is no overlap of species' ranges, the pattern would be a *stepped anagenetic paedomorphocline* (or *peramorphocline*) (Fig. 5-8a). If the evolution can be demonstrated to be intraspecific (and we shall document just such a case below), then the pattern of the evolutionary trend can be described as just an *anagenetic paedomorphocline* (or *peramorphocline*). In those cases where there is overlap between species' ranges, the pattern can be described as a *stepped cladogenetic paedomorphocline* (or *peramorphocline*) (Fig. 5-8b).

What is clearly being implied is that the nature of the organism's growth pattern will play a crucial role in determining patterns of evolutionary change, in particular whether evolutionary changes will be episodic or gradual. While we shall examine this point more closely below, it is worth pointing out here the different effects that are created by the same heterochronic process affecting organisms that grow continuously (e.g., vertebrates) and those whose growth (or at least the external expression of that growth) is episodic (e.g., arthropods).

In Fig. 5-9 we demonstrate this effect. In the stepped ontogenetic pattern, typical of arthropods, an individual passes through molt stages A to C, and in the process generates 5 spines: 1 at molt A, 3 at molt B, and 5 at molt C, which we can take as representing maturity, attained at time 4. In the case of the organism that grows continuously, it too generates 1 spine at A, 3 at B, and 5 at C; but it will also pass through a stage intermediate between A and B when a second spine is generated, and then between B and C when a

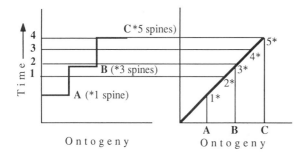

Figure 5-9. Effect of different growth patterns in determining patterns of evolution. In the stepped pattern, 1 spine is produced at molt A, 3 at molt B, 5 at molt C. In continuously growing forms, intermediate numbers of spines are produced throughout growth. Thus, pattern of growth will affect the heterochronic result (see Fig. 5-10).

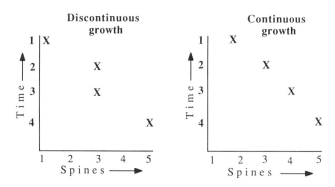

Figure 5-10. Differences in paedomorphoclines in organisms that undergo discontinuous growth compared with those that undergo continuous growth. Pattern of spine production through ontogeny is as shown in Fig. 5-9.

fourth spine is generated. Should progenesis occur at time 3, the specimen that molts will only be at morphological stage B with 3 spines, whereas the continuously growing organism will be between B and C and have produced 4 spines. Should a paedomorphocline develop by subsequent progenesis occurring at time 2, then the molted specimen will produce a form identical to that at time 3, in having reached only morphological stage B with 3 spines. In the continuously growing form, however, progenesis at time 2 results in a form that has only 3 spines. Progenesis at time 1 will produce a form in which 2 spines have been developed. In the molting species only 1 spine will have been produced at time 1.

The resultant paedomorphocline of the molting form will have a more pronounced stepped appearance than that developed in the continuously growing form (Fig. 5-10). In the former, adults will have 5 spines in the ancestral form, 3 in the descendant, and 1 in the youngest. In the latter, the adult paedomorphs along the paedomorphocline will show progressively 4, 3, then 2 spines. Not only will the paedomorphocline patterns vary, but the "extent" of effective paedomorphosis will vary between discontinuous and continuous growth forms; for progenesis at time 1 in the discontinuous form will produce an adult with only 1 spine, whereas at the corresponding time the continuous growth form will have 2 spines.

Although much study of heterochronoclines has centered on examples at the specific and generic levels, there is every reason to believe that where phyletic gradualism can be demonstrated from the fossil record, paedomorphoclines and peramorphoclines can be recognized intraspecifically. While there has been little attempt to link intraspecific trends with heterochrony, we can demonstrate gradual heterochronic changes along an intraspecific heterochronocline, on the basis of some as yet unpublished data on a lineage of the sand dollar *Echinocyamus planissimus* from western Australia (K. J. McNamara, in preparation).

E. planissimus is first recorded in Pliocene sediments from core samples from the Perth Basin of western Australia. A fairly continuous sequence of core deposits shows that the species persisted through the Pleistocene to the present day off the western Australian coast. While most of the characters showed morphological stasis over this period, a few dissociated growth fields show regular heterochronic trends. These include a relative decrease in production of pore pairs; and a decrease in number of tubercles, but a relative increase in size of the tubercles (see McNamara, 1990, Fig. 9.4). Evidence suggests that these changes occurred progressively within the lineage over a period of about 3 million years ago to the present day.

Pore pairs in the five dorsal petals increase in number during ontogeny. For example, the oldest, Pliocene, members of the species have on average two pore pairs at a test length of 3 mm, increasing to six pore pairs at a test length of 5.5 mm. In Early Pleistocene forms, four pore pairs are present at 5.5 mm length, and increase to six at 6.5 mm. In living specimens only three are present at 5.5 mm, and a maximum of six at 8.5 mm. As the species shows a progressive increase in size along the lineage (maximum lengths: Pliocene, 5.5 mm; Pleistocene, 7 mm; Recent, 10 mm), and as many other characters show no apparent morphological change along the lineage, the relative decrease in number of pore pairs along the lineage is most likely due to a reduction in rate of pore pair production (neoteny). Such paedomorphic reduction in the lineage would therefore be an example of an intraspecific paedomorphocline. Likewise, the decline in tubercle number could also be construed as being a neotenically generated paedomorphocline.

As true growth rates are unknown, we are faced with another example whereby an assumption has been made that rate of size increase of the entire test is the same in all three morphs. However, the increase in size along the lineage might in fact reflect an increase in rate of overall size increase of the test. The relative decline in number of pore pairs might then simply be interpreted as a consequence of the shorter time the younger forms had in which to generate pore pairs. Thus, if the Pleistocene form attained a test length of 5.5 mm quicker than the Pliocene form, and the Recent form attained the same size even quicker, then this might explain the reduction in pore pair number along the lineage. If this was the case, then all local growth fields ought to show an effective reduction in growth rates. However, the tubercles show an increase in size and form an apparent peramorphocline (as tubercle size increases ontogenetically). If the apparent changes in local growth fields were purely a result of the increase in overall test growth rate, then the tubercles, like the pore pairs, would not only have decreased in relative number, but also in size. This is not the case. Their increase in relative size may indicate real intraspecific peramorphosis, either by acceleration or by predisplacement. Alternatively, it is possible that the tubercles underwent local hypermorphosis, offset of tubercle growth having been delayed. The significance of the occurrence of both paedomorphoclines and peramorphoclines in the same lineage is discussed below. The potential inherent in the recognition of intraspecific evolutionary trends engendered by changes in internal developmental regulation is that by assuming the trends will continue at the same rate into the future as they have in the past, it is possible to predict what the species might look like in, say, 1 million years. In the case of this species of *Echinocyamus*, it would, as an adult, be larger, perhaps up to 12 mm in length; have even fewer pore pairs, perhaps only four in the largest adults; and fewer, but larger, tubercles. Likewise, with a heterochronocline identified from the fossil record, predictions can be made of the likely morphology of an undiscovered extension to the heterochronocline.

Changes within a lineage brought about by the operation of heterochronic processes primarily affecting subtle changes in rates of development can be considered to be the real bread-and-butter evolutionary changes: the fine tuning of more significant evolutionary innovations produced by more radical heterochronic activity, generally at early stages of development. *Echinocyamus,* like the other fibulariid genus *Fibularia,* probably arose as a result of heterochrony. Rather than being ancestral, primitive sand dollars, these tiny fibulariids are derived, specialized forms that probably arose by progenesis. This is indicated by their very small size. As has been argued by a number of research workers, progenesis may have played a significant role on a number of occasions in the evolution of higher taxa (see below). Subsequent changes within the evolving lineages are more subtle. This holds true for the many lineages of echinoids that arose during the Tertiary by selection of heterochronic morphotypes. Selection favored morphotypes generated by slight changes in growth allometries and rates of differentiation of morphological structures [principally pore pairs and tubercles, but also of coronal plates, on and within which these structures are carried (McNamara, 1988a)]. For instance, most spatangoid lineages in the Australian Tertiary show abundant examples of paedomorphoclines and peramorphoclines [e.g., *Schizaster* (McNamara and Philip, 1980a); *Pericosmus* (McNamara and Philip, 1984); *Breynia* (McNamara, 1982a); *Protenaster* (McNamara, 1985a); *Hemiaster* (McNamara, 1987b) (Fig. 5-11); *Psephoaster* (McNamara, 1987b); *Lovenia* (McNamara, 1989)].

In all of these cases, with the exception of the first, and to some degree the second example, there is little size change along the heterochronoclines, because it is specific structures that are undergoing heterochronic change, not the entire organism. In *Schizaster,* however, there is a size increase, suggesting that hypermorphosis might have operated. Moreover, some structures also show acceleration, such as the increase in rows of pore pairs in the anterior ambulacrum of *Schizaster* (*Ova*) (McNamara and Philip, 1980a,b). *Psephoaster* parallels the sand dollar *Echinocyamus* in that it probably arose as a genus by global progenesis. Subsequent speciation involved the operation of peramorphic and paedomorphic processes on local growth fields.

3.3. Dissociated and Mosaic Heterochronoclines

Many lineages, such as those of the brachiopods *Tegulorhynchia–Notosaria* and the trilobite *Olenellus,* can be classified as global paedomorphoclines. In other words, all heterochronic changes are paedomorphic. However, in many lineages dissociated heterochrony occurs. Here only specific morphological traits are affected, and where some might be paedomorphic, others are peramorphic (Chapters 2 and 3). Dissociated heterochronoclines occur when some structures consistently show acceleration, hypermorphosis, or predisplacement of local organs or growth fields between successive species, to produce a peramorphocline, while other structures always show neoteny, progenesis, or postdisplacement between species to produce a paedomorphocline. If the traits are all peramorphic (or all paedomorphic), this is a **dissociated heterochronocline**. Different traits might be affected by different processes. In other words, some traits might be affected by acceleration, others by predisplacement or hypermorphosis, all being peramorphic processes. If some traits are affected by paedomorphic processes, while others are affected by peramorphic ones, a **mosaic heterochronocline** can be produced.

The implication of this in terms of targets of selection is quite significant. It suggests that local growth fields, or suites of growth fields, may be under specific selection

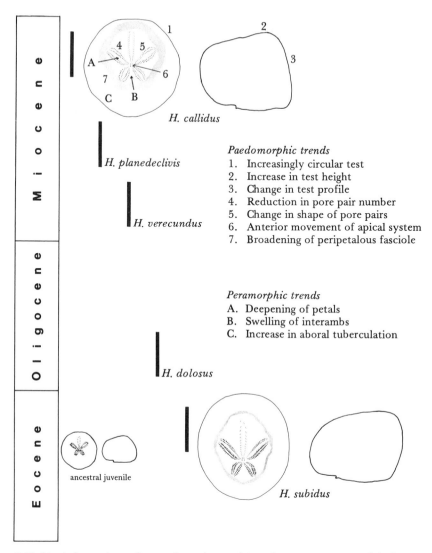

Figure 5-11. Mosaic heterochronoclines in the evolution of Australian Tertiary species of the heart urchin *Hemiaster*. The heterochronoclines follow an environmental gradient of coarse to fine-grained sediments, thought to correspond to an increase in water depth (McNamara, 1987b).

pressure. This "constructional reductionism" is somewhat at variance with some recent arguments (Gould and Lewontin, 1979) that selection often does not favor specific adaptive traits but focuses on aspects such as life history strategies. Any consequent morphological changes are seen as being merely by-products. While the number of dissociated and mosaic heterochronoclines in the literature is quite small, we believe that this reflects a preoccupation with heterochrony involving large-scale shape and body size changes late in ontogeny (late-stage global growth heterochrony—see Chapter 3). Thus, examples of body size increases or reductions, and concomitant global morphological changes are more easily recognizable than those that involve just subtle shifts of allometric coefficients or rates of differentiation of only a few structures. It is likely that the incidence of dissociated and mosaic heterochronoclines in evolutionary trends will prove to be quite commonplace.

The frequency of dissociated and mosaic heterochronoclines will depend on the nature of the ontogenetic development of the organisms. Most work on this has been carried out in echinoids (McKinney, 1984; McNamara, 1988a). In some groups, such as spatangoids and holasteroids, that undergo pronounced allometric changes compared with most other groups of echinoids, the reductionist nature of the targets of selection is such that different coronal plates grow with different allometries (often some being positive, while others are negative). Even different axes within a single plate can show different allometries. Through time these opposing allometries can become even more polarized as different heterochronic processes act on the different axes (McNamara, 1988a). The result can be the evolution of quite extreme structures. For instance, in the evolution of portalesiid echinoids (David, 1989), many features demonstrate the occurrence of peramorphoclines within lineages, such as the great increase in longitudinal growth of the coronal plates, at the expense of the transverse growth, which grows with negative allometry and forms a paedomorphocline. The extreme expression of these dissociated heterochronoclines is *Echinosigra*, which attains an elongate, flask shape (Fig. 5-12). Such dissociated heterochrony shows not that every structure is under individual selection pressures, but that suites of characters may be changing under the same pressure: trait covariation (Chapters 3 and 8). As we discuss below, there is the possibility that where heterochrony acts on a two-tiered level in some colonial organisms, the style of heterochrony at one tier might directly affect the nature of the heterochronic

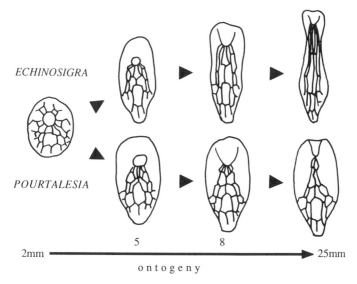

Figure 5-12. Ontogenetic change in two pourtalesiid echinoids. *Echinosigra* and *Pourtalesia*, illustrating how differential plate allometries can result in the growth of markedly different test shapes. Redrawn from David (1989).

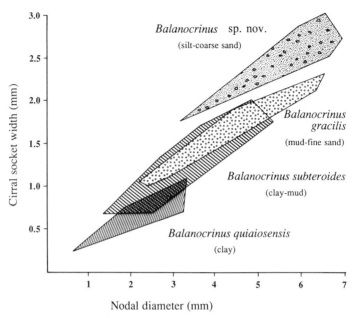

Figure 5-13. Paedomorphic development of cirral socket width in early Jurassic species of the crinoid *Balano-crinus*. In juveniles, cirral socket width is relatively large. As the stem increases in diameter, the socket width shows negative allometry. Combined with peramorphic increase in stem diameter, coarser-grained sediments were able to be colonized. Redrawn from Simms (1988).

processes operating at the other tier. In the case of differing axial plate allometries in echinoids, the reduction in plate allometry in one direction must also affect that in another direction. If it were not so, then the net effect would be for no phylogenetic change. Developmentally, it is probably also energetically more efficient if there is a net reduction in allometric growth in one direction while another increases. Thus, the areal extent of the echinoid plate might not change, but its shape can undergo appreciable change and its contribution to the functions of the organism can, as a consequence, be altered.

Often, mosaic heterochronoclines may show occupation of a sequence of niches along an environmental gradient. For instance, the four species of an early Jurassic crinoid *Balanocrinus* lineage show a paedomorphocline in cirral socket size, but a peramorphocline in stem diameter, as it doubled its width (Simms, 1988). The combination of more robust stems and cirri allowed occupation of coarser-grained sediments by descendants (Fig. 5-13).

The importance of dissociated and mosaic heterochrony in evolution is that they provide the potential for the development of a huge range of shapes. For instance, although work on the significance of heterochrony in the evolution of plants has received relatively little attention, recent work by Guerrant (1982, 1988) has demonstrated the power of heterochrony in generating novel adaptations. Guerrant has shown how the flowers of the hummingbird-pollinated larkspur *Delphinium nudicaule* resemble the buds of more generalized bumblebee-pollinated species. He argued that the external appearance of *D. nudicaule* was brought about by a reduction in rate of development of the floral parts, producing a paedomorphic descendant. "Juvenile" characters are typified by the budlike forward-pointing orientation of the sepals, which produces a tubular floral form characteristic of hummingbird-pollinated plants. However, not all characters of the plant are paedomorphic. The nectariferous petals show an acceleration in rate of development, producing a shape "beyond" that of the generalized larkspur. Selection acted on the larkspur flower in different ways, retarding some features, accelerating others. The result was a flower shape capable of attracting and rewarding a new pollinator—for a larkspur a major adaptive breakthrough.

Perhaps a more startling example of mosaic heterochrony in plants is the gigantic parasitic genus *Rafflesia* from Sumatra and Borneo. By what might be interpreted as extreme postdisplacement, the genus has effectively lost all leaves, stem, or roots. This extreme paedomorphosis, however, has not affected the flower, which reaches a diameter in excess of 1 m (Ismail, 1988). This large size is probably a consequence of the long generation time causing hypermorphosis (peramorphosis) in the flower. Smaller species that have flowers reaching only 20 cm in diameter probably have shorter generation times, and may therefore be considered as being "less" hypermorphic. The more derived nature of the larger flowering species is indicative of a peramorphocline in this feature between the various species.

Such disparate shape changes represent a shuffling and reordering of existing characters. No dramatic genetic mutations are necessary to evolve vastly different shapes, just critical changes in times or rates of growth, controlled, perhaps, by small perturbations in the developmental program. The adaptive and evolutionary consequences, as we have shown, can be quite profound.

Another form of extreme mosaic heterochrony has been documented by Dommergues and Meister (1989) in which they perceive mosaic paedomorphoclines and peramorphoclines operating on the same structure within the growth of individual ammonites. Rather than different structures showing different modes of heterochrony, the same structure shows different modes at different stages of its ontogeny. In some harpoceratine ammonites from the Early Jurassic of northwest Europe, the growth paths of the shell may follow, say, a paedomorphocline in the early growth stages, but then in later growth stages follow a peramorphocline. Such a change occurs because of changing allometries during growth. After all, in most organisms, structures that grow allometrically do not do so at a constant rate during the entire course of ontogeny (Chapters 2 and 3). For example, let us suppose that growth of a particular structure in a hypothetical ammonite shell is initially negative, then isometric for a short period, then finally positive. If only the early negative allometry becomes extended between successive species in a lineage, then the shell will show a paedomorphocline. However, if the late stage shows positive allometry and is also extended, then the later part of shell growth will form a peramorphocline between species. Thus, in Fig. 5-14, **A–A'–A"** will form an allometric paedomorphocline, shape being less developed at comparable sizes along the lineage, and **B–B'–B"** a peramorphocline, as shape is more advanced at comparable sizes. ("Allometric heterochrony" is discussed in Chapter 2.)

3.4. Heterochronoclines and Environmental Gradients

Having discussed heterochronoclines, largely from an "internal" perspective, we now concentrate on how they form and why they go in the directions that they do.

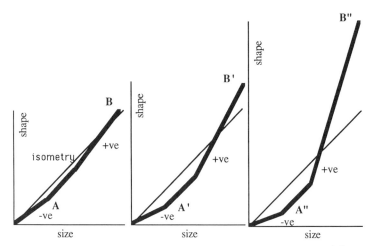

Figure 5-14. Effect of changing allometries during ammonite growth and evolutionary shifts in timing of their changes on development of pera- and paedomorphoclines. Thus, **A–A′–A″** will form a paedomorphocline, while at **B–B′–B″** a peramorphocline is developed.

Perhaps a useful analogy to help in understanding the role of heterochrony in directing evolution is to consider the lineage of species as a train. It has internal fuel, such as coal, diesel, or electricity, that corresponds to heterochrony; an external driving force, the driver, corresponding to agents of selection; and lastly, and this is the topic with which we shall be dealing with here, the rails along which the train travels, corresponding to the environmental gradient along which the heterochronocline is developing. The significant point in this analogy is that the train can only begin and finish its journey if all these factors are operating in concert. If a harmonious integration of these factors is not accomplished, then the heterochronocline will not develop. Without the fuel (heterochrony) the trend will not move; without the driver and rails it has nowhere to go!

In earlier examples we have shown how paedomorphoclines developed in some trilobite lineages along environmental gradients of shallow to deep water; in brachiopods from deep to shallow water. Many echinoid lineages (McNamara, 1988a,b, 1990) show heterochronic evolution along environmental gradients of coarse to fine-grained sediments, which may also correlate to evolution along a gradient from shallow to deep water (McKinney, 1984, 1986). Water depth and sediment grain size are among the most common environmental gradients in the marine environment that affect invertebrates (McNamara, 1988b). In the terrestrial environment, temperature and elevation are common environmental gradients along which lineages evolve. For instance, Grant (1963) recognized the importance of environmental gradients in controlling speciation in five species of the herbaceous plant *Polemonium*. Here the species evolved along a gradient toward lower temperatures at higher elevations.

3.4.1. The Adaptive Landscape

To understand how heterochronoclines develop, it is useful to consider each species as occupying an adaptive peak (Wright, 1932) along the gradient. Wright envisaged a landscape of adaptive peaks and valleys. The summits of peaks are occupied by the genetic "elite," whose genotypes produce morphotypes most fitted to a particular environment. There has been a tendency (e.g., Simpson, 1944, 1953) to view adaptive peaks as a combination of morphological/functional characteristics of a taxon in combination with the environmental regime that it occupies, the adaptive zone. This combining of morphological with ecological features has led to some confusion, and produced the introduction of terms such as "biospace" and "ecospace" (Valentine, 1969), in place of the evolutionary niche concept. We prefer to view the adaptive peaks purely in terms of the phenotypic character, being products of intrinsic change, while ecological niches we regard as the product of extrinsic factors (biotic and abiotic environment). While adaptive peaks are only as effective as the ecological niches into which they fit, the recognition by many workers of *exaptations** (formerly known as *preadaptations*) argues that adaptive peaks are restricted by the phenotype. The very existence of an exaptation equates with an unoccupied niche. The niche becomes available when an exaptation, or series of exaptations open up or create the niche. Niches "preexist" as potential roles in this sense, though one may also say that organisms "create" their own niches from a potential pool. The unoccupied niche then becomes a realized niche, while the exaptations create a new adaptive peak. In the scenario that we propose, attainment of adaptive peaks is carried out by heterochrony [or genotypically, heterochronic genes (Slatkin, 1987)].

Because of the nature of evolution, heterochronoclines develop along a single axis in the overall adaptive landscape. The gradient can be considered, in simplistic terms, to consist of a series of ecological niches into which the adaptive peaks slot. Each species will consist of a normal range of heterochronic phenotypes, only some of which will reach the adaptive peak. For instance, in a paedomorphocline some phenotypes within each species will show extreme development of paedomorphic structures. If these do not lie along the axis of the environmental gradient, then they will barely scrape the base of the adaptive peak and not contribute toward the evolution of the paedomorphocline. Other paedomorphs will attain varying heights up the adaptive peak, the "fittest" occupying the top of the peak. Where only the ancestral species of an incipient paedomorphocline occupies the initial adaptive peak that corresponds to the first ecological niche along a prospective environmental gradient, then extreme paedomorphs in that species which develop along the environmental gradient have the potential to overcome the adaptive threshold that lies between the existing adaptive peak and the next potential peak along the niche series. To do this these paedomorphs must be capable of slotting into the next vacant niche along the niche axis. Should this occur, occupation of the next niche and development of a new adaptive peak will result in ecological, and thus genetic, isolation. However, it must be emphasized that the separation need not necessarily be geographical, and thus the speciation event need not be considered allopatric in the generally accepted sense of the term (Mayr, 1970). It may be geographically parapatric, or even sympatric (Chapter 6).

The adaptive threshold between two peaks need not remain static, principally

---

* *Exaptations* (Gould and Vrba, 1982) are features of an organism which evolved either for a particular function, then were coopted for a new use, or which evolved, but had no particular function, but became coopted for a use at a later stage.

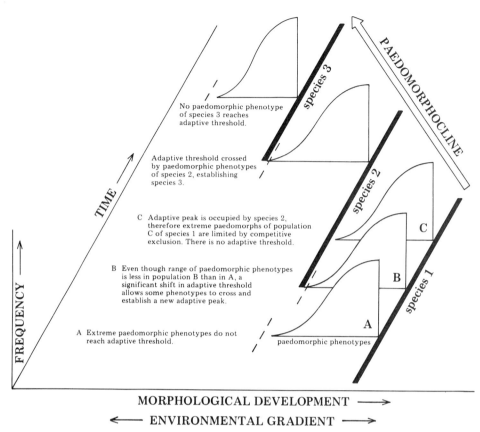

Figure 5-15. Suggested mechanism for the development of a cladogenetic paedomorphocline (or peramorphocline). It is not until a population of species 1, such as population B, has paedomorphic phenotypes capable of crossing the adaptive threshold and occupying a new adaptive peak, that a paedomorphic species may arise. Shifting adaptive thresholds and variation in range of paedomorphic phenotypes affects the timing of evolution of species 2. For instance, even though the range of paedomorphic phenotypes is greater in population A than in B, the position of the adaptive threshold is such that it is not crossed by phenotypes of population A. Extreme paedomorphs of population C of species 1 are limited by competitive exclusion. There is no adaptive threshold. The persistence of such populations, however, effectively blocks reverse speciation from species 2 to species 1. Selection of extreme paedomorphic phenotypes of species 2 which can cross an adaptive threshold results in the establishment of a further paedomorphic species, species 3. The paedomorphocline is therefore established and unidirectional speciation ensues. Broken lines represent adaptive thresholds. Reproduced from McNamara (1982b).

because the ecological niches may vary in their breadth along the axis by niche contraction and expansion. Thus, it need not be mandatory for an extension of the range of paedomorphic phenotypes to occur for the adaptive threshold to be crossed (Fig. 5-15). If the adaptive threshold moves closer to the earlier adaptive peak (e.g., via environmental change), then existing extreme paedomorphs will move across to the new adaptive peak. In reality, both factors will be changing over any period of time such that a slight shift in the adaptive threshold and extension of the phenotypic range will ensure successful occupation of the next adaptive peak along the environmental gradient. A heterochronocline may therefore be considered as a series of adaptive peaks that have been established sequentially through time and which fit into a parallel series of ecological niches that lie along an environmental gradient. The position of each peak will be determined by a number of factors, but principally by the position of the preceding peak as well as the distance between ecological niches. These factors themselves will be determined by the morphological or ecological separation between ancestors and descendants, sufficient for them to avoid significant competition for resources.

Variation in rates of morphological development in the ancestral apaedomorph, or changes in timings of onset or offset of growth, might be expected to produce a sufficient number of morphotypes to occupy all of the potential adaptive peaks on the heterochronic gradient. In such a scenario the last species in a paedomorphocline could therefore potentially develop directly from the apaedomorph by a large variation in rate or timing of development. Other, intermediate, forms on the heterochronic gradient could subsequently develop at random. Evidence from paedo- and peramorphoclines described from the fossil record shows that the adaptive peaks are not filled in this random manner. They are filled sequentially. This implies that there are constraints within the system imposed on the development or the selection of the heterochronic phenotypes. Each species is restricted in the extent to which it can undergo perturbations to its developmental system and still remain viable. It can only produce occasional phenotypes capable of crossing the adaptive threshold and occupying the next adaptive peak. Furthermore, ecological constraints, or the availability of niche space may restrict the extent to which the descendant phenotype may go along the environmental gradient.

In many heterochronoclines a number of traits covary along environmental gradients. In the case of peramorphoclines, such ontogenetic–environmental extensions probably form coadapted suites. That is, while some (e.g., Jablonski *et al.,* 1986) consider that ontogenetic channeling itself causes internal directional determination, much of the directionality is in fact already "preaimed" or "predetermined" by earlier external selection (see Chapters 6 and 8).

3.4.2. Factors Limiting Heterochronoclines

The extent of the heterochronocline will be determined by intrinsic factors, or extrinsic factors, or a combination of both. The limiting intrinsic factor will be the degree to which the species are constrained by developmental and structural considerations. For instance, if the principal target of selection is the number of appendages, and a paedomorphocline has been established wherein appendage number is decreasing, then a minimum figure may be reached below which the organism will not be able to function: the adaptive peak can never be created. Intuitively, it would seem that paedomorphoclines would likely be subjected to such constraint more than peramorphoclines. Where structures are being added the same constraint may not apply. However, constraining factors would still occur. For instance, if the same appendages are increasing in number, but the organism is not increasing in size, then there will be size constraints that will impose an upper limit on appendage number, and hence extent of the peramorphocline. Energy and behavioral considerations will also affect the viability of such forms. The principal extrinsic factor that constrains heterochronocline development is the availability of niche space along the environmental gradient.

While environmental variables, such as temperature, sediment grain size, water depth, or elevation, are usually gradational, adaptations are not so gradational. Many adaptations also covary in suites. These are often size related (see Chapter 6).

Cladogenetic heterochronoclines, where ancestral species persist following the evolution of the descendant, may, to some degree, be considered to be **autocatakinetic** (Hutchinson, 1959), i.e., self-generating in a particular direction. In a recent article, Gould (1988b) has pointed out that many evolutionary trends may be no more than an increase in variance, and thus by inference autocatakinetic. Cladogenetic heterochronoclines are a subset of this, since they increase variance as they branch. Once established, such a heterochronocline can be thought of as self-generating in two ways. First, the development of a heterochronocline along a sequence of niches is necessarily a directional process. More importantly, what drives the heterochronocline in one direction is the persistence of (and the competition from) the ancestral species. If one considers a modal distribution of heterochronic phenotypes, then those phenotypes of the second species in a paedomorphocline that are on the peramorphic side of the modal phenotypic expression would be outcompeted by the paedomorphic phenotypes in the ancestral species. These phenotypes are "fitter" in their own particular ecological niche than the descendant's peramorphic phenotypes would be (see Fig. 5-16a). This blockage in one direction means that evolution can only occur in the other direction and is autocatakinetic.

In contrast to cladogenetic heterochronoclines, the driving force behind anagenetic heterochronoclines lies outside of the heterochronocline system, for example in the form of predation pressure. Therefore, we shall discuss this in the section dealing with the agents of selection. One aspect of trend generation that does need considering here, however, is the driving force behind heterochronoclines that develop intraspecifically. We have demonstrated, using the example of the echinoid *Echinocyamus,* that heterochronoclines can be established within individual species lineages. Further research along these lines is likely to show that many documented examples of anagenetic phyletic gradualism involve changes in intraspecific heterochrony (Fig. 5-16b).

The most likely explanation for the "movement" of such anagenetic heterochronoclines is that the ancestral morphotype occupied only part of the niche axis under consideration. Once established, the same *modus operandi* will apply as in cladogenetic heterochronoclines: autocatakinesis by ancestral blockage will drive the heterochronocline unidirectionally while there is temporal overlap of both morphs. Extinction of the ancestral morph may ensue either because of intraspecific competition or because of differential survival controlled by extrinsic factors, such as predation pressure. However, if it can be argued that the morphological changes have no adaptive significance, then one must resort to explanations involving a form of "morphogenetic drift," somewhat akin to "genetic drift." Such phyletic gradualism that has an underlying morphological basis in heterochrony might also develop if the niche itself is gradually changing unidirectionally (causing orthoselection). In the case of *Echinocyamus,* there are no data to indicate what adaptive significance the morphological changes might have had. While the increase in size may have been a target of selection, the dissociated nature of the heterochronoclines argues for selection on more than size.

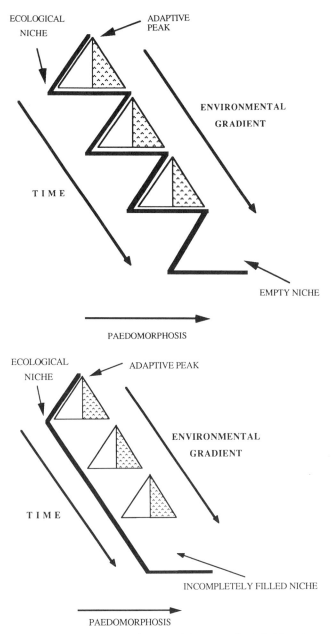

Figure 5-16. (a) Diagrammatic representation of a series of disjunct ecological niches along an environmental gradient into which adaptive peaks, attained by progressively more paedomorphic forms, are episodically fitted. (b) Diagrammatic representation of a broad ecological niche that lies along an environmental gradient and which is gradually filled through time by progressively more paedomorphic forms.

4. HETEROCHRONY AND RATES OF EVOLUTION

4.1. Introduction

In recent times there has sometimes been a tendency to see heterochrony merely as an instigator of rapid evolutionary change, internally steered and leaping like an evolutionary Tarzan from adaptive peak to adaptive peak. The neo-Darwinian (or modern synthetic) view, on the other hand, sees evolution as gradual and externally steered: the slow, laborious conquest of successive adaptive peaks driven by the continuous grind of natural selection. But does heterochrony only produce morphological saltations? In this book we argue that it is a more fundamental aspect of evolution, underpinning most morphological change, be it gradual or rapid.

The polarized view that evolution, if gradual, is just under the control of natural selection (i.e., extrinsic factors), and if fast is generated by heterochrony (i.e., intrinsic factors), is unrealistic. There is a continuum between gradual changes and rapid changes, both intrinsic and extrinsic. As we have stated throughout this and the preceding chapter, evolution cannot work by extrinsic agents alone; neither can it work just by intrinsic generation of shapes, sizes, or behaviors. Any consideration of the role of heterochrony in rates of evolution in any given lineage must be viewed from the perspective of both intrinsic (1 and 2) and extrinsic (3 and 4) factors:

1. An organism's growth pattern (i.e., is it continuous or episodic, as in arthropods?).
2. The heterochronic process occurring (i.e., will it more likely engender rapid morphological change by the generation of morphological saltations than gradual phyletic change?).
3. The structure of the population (i.e., is there polymorphism, how much and how extreme is it?), plus the changing nature of the extrinsic agents of selection (i.e., are predators, for instance, selecting two disjunct size classes or just one?).
4. Ecological niche boundaries (are they continuous or discontinuous?).

The recent preoccupation in the paleontological and biological literature with rates of evolution has focused on whether evolution at the species level occurs rapidly or gradually [termed "punctuated equilibria" and "phyletic gradualism," respectively, by Eldredge and Gould (1972); see Levinton (1988) for a recent critique]. Heterochrony has been cited by many authors (e.g., Bonner, 1982; McNamara, 1982b) as a likely mechanism for inducing rapid morphological change. In our view, heterochrony can equally well be the progenitor of gradual change through incremental effects. Whether it is rapid or slow will be dependent upon the complex interplay between the four interacting factors, of growth pattern, heterochrony, population structure, and niche characteristics. The dominance of one or more of the four parameters over the others will strongly influence whether the evolution will be gradual or rapid. We will show that if there are saltations in any of the four, the likelihood is for rapid speciation. Only if there are no saltations in any of the four factors is evolution likely to be gradual. If rapid, the extent of the morphological change (in other words, the breadth of the saltation—Chapter 3) will determine not only the taxonomic level at which it is expressed, but also its importance in the generation of major morphological novelties.

We look first at the more rapid, drastic heterochronic changes. Garstang (1928), for

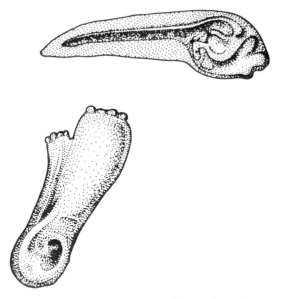

Figure 5-17. Ascidian larva and adult. de Beer (1928) suggested that paedomorphosis in such a larva may have led to the evolution of vertebrates.

example, suggested the importance of rapid paedomorphosis in evolution. He concentrated on its role in the evolution of vertebrates from invertebrates, believing that they may have arisen from the pelagic larva of a tunicate-like deuterostome invertebrate (discussion in Raff and Kaufman, 1983). This would have occurred by progenesis at an early larval stage. The free-swimming tunicate larva (Fig. 5-17) possesses all the fundamental chordate characters: a notochord; dorsal hollow nerve cord; gill slits; and postanal propulsive tail. Precocious sexual maturation would have caused the retention of such ancestral larval characters into the adult phase and a consequent major adaptive breakthrough. Providing firm evidence to substantiate claims that heterochrony has been instrumental in morphological changes of such a magnitude is well nigh impossible. But this example highlights some of the factors that need to occur in combination to produce such a major evolutionary change. There would be a drastic heterochronic event, such as global, or at least broad-scale progenesis, at a very early stage in development. The morphological change that the ancestral organism goes through during ontogeny would have to be large and often combined with an appreciable habitat shift (in this case benthic to pelagic).

In the rest of this section we discuss the major factors operating on the intrinsic and extrinsic parameters that influence the rates of evolution, firstly to produce rapid evolutionary change (Section 4.2), then gradual change (Section 4.3). Finally, we discuss the factors that might induce macroevolutionary changes.

4.2. Morphological Saltations

To account for the absence of intermediates in many fossil lineages, Eldredge and Gould (1972) proposed the punctuated equilibria model in which species arise very rapidly, not by the transformation of the entire species, but in small, isolated populations. Following their expansion of range, these new species remain morphologically stable until they become extinct. In the original model, such speciation is cladogenetic, the ancestral species persisting temporally with the descendant. In recent years the role of intrinsic, in particular developmental, aspects has come under increased attention as a major factor that produces long periods of morphological stasis followed by a rapid burst of morphological change. Levinton (1988) has been concerned with whether or not these sudden morphological changes are speciation or just an indication of variable phyletic evolution, i.e., a sudden acceleration in rate of change with a single species lineage. He cites the study of changes in shape in the *Globorotalia plesiotumida–G. tumida* lineage (Malmgren *et al.*, 1983). Here, following a long period of stasis the lineage undergoes a relatively rapid change at the Miocene–Pliocene boundary, then in its new morphology persists relatively unchanged to the present day. Arguments such as these over whether this is a case of "phyletic" evolution (Levinton, 1988) because there is no cladogenesis, or whether (as we believe) it is sometimes more appropriately termed "punctuated anagenesis" (Gould, 1985) or "punctuated gradualism" (Malmgren *et al.*, 1983) do little more than obfuscate the real meaning behind such patterns. In our view, of more significance to evolutionary theory is the interplay between the underlying intrinsic and extrinsic mechanisms that caused the changes.

A potent intrinsic factor maintaining morphological stasis (which none of the protagonists in the punctuated equilibria/phyletic gradualism debate seem to deny exists) is **developmental constraint** or **canalization** (Chapter 4), which prohibits or severely restricts change in character space (Williamson, 1987). Developmental constraint is sometimes associated with the concept of **genetic homeostasis.** Genetic homeostasis is defined by Lerner (1954) as "the property of the population to equilibrate its genetic composition and to resist sudden changes." As Mayr (1970) points out, genetic homeostasis determines to what extent a gene pool can respond to selection. Coadapted genotypes are generally separated by "nonadaptive valleys" (Mayr, 1963). Thus, change from one coadapted state to another is difficult. The consequence of this is that the species is "buffered" to resist evolutionary change (Williamson, 1987). When the gene pool is finally pushed out of equilibrium, a **genetic revolution** occurs, rapidly leading to a new equilibrium. "Stabilizing selection" has been advocated as the most important extrinsic factor in preserving the morphological status quo of a species (Lande, 1980; Hoffman, 1982b; Charlesworth *et al.*, 1982).

If we look at these possible causes of stasis from a heterochronic perspective, then developmental canalization means that there is little if any heterochrony occurring in the population. Stabilizing selection implies that any heterochronic phenotypes that do develop are selected out. In such a scenario it may be changes in extrinsic factors, such as breakdown in the intensity of stabilizing selection, that result in the rapid evolution of a new species (Fig. 5-18). Stabilizing selection, which may be induced by high levels of predation on "phenodeviants" or occur by virtue of a narrow ecological niche, acts to dampen phenotypic plasticity (an important source of potential new species—see Chapter 4). Once this extrinsic constraint is released, there can be a rapid explosion of morphological change that overflows into an adjacent niche with the resultant evolution of a new species. Similarly, release of developmental constraint, perhaps by the effect of extrinsic agents (Williamson, 1981), can allow the rapid proliferation of a much wider

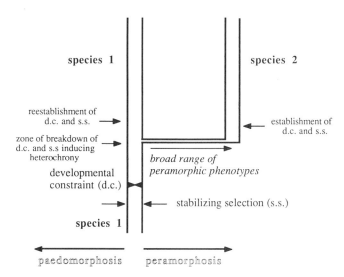

Figure 5-18. Illustration of the way in which a combination of developmental constraint and stabilizing selection can maintain morphological stasis through time. Subsequent breakdown of these constraining factors can lead to rapid selection of extreme paedomorphic or (in this case) peramorphic phenotypes. With the reestablishment of developmental constraint and stabilizing selection, a new species is developed.

range of phenotypes, generated by intrapopulational heterochrony. As Williamson (1987) has pointed out, and as we have argued consistently throughout this book, the dominance of either internal "constraint" or "selection" in our thinking represents an artificial polarity. *Both* extrinsic and intrinsic operate in concert to maintain species' geographic and temporal integrity; likewise they may act in concert in the rapid evolution of new species, although changes in relative dominance of the two are crucial.

A **saltation** (= "macromutation") is a rapid morphological transformation. Whether one occurs will depend upon whether the heterochronic processes are acting globally or regionally. Global changes will by definition produce broader overall morphological change (more "parts" are affected—Chapters 3 and 8) than regional or local changes. Of the three styles of heterochrony, i.e., onset, offset, and rate of development, any may produce rapid morphological transformations. The globality ("breadth") of the saltation, and thus its contribution to rapid speciation, will depend upon the relative difference in onset or offset times between ancestor and descendant. Thus, the greater the difference in timing, the greater are the morphological and size changes, assuming that there are no changes in allometric coefficients. The degree of saltation also depends upon the ontogenetic phase during which the heterochrony occurs. If the change occurs during crucial periods of early differentiation, then the consequent morphological change can be appreciable. If growth is extended (hypermorphosis) and there is also acceleration, then large morphological saltations can occur (such combinations of heterochronies may be common—Chapter 3).

Changes in timing of onset of growth can have large cascading effects later in ontogeny (Chapter 3) perhaps causing "rapid evolution." Such cascades in zebras have been shown by Bard (1977). Differences in striping patterns between the two zebra species *Equus burchelli* and *E. grevyi* can arise from slight variations in the timing of

inception of stripe development in the embryos. Minor differences in the onset of striping early in embryonic development produced quite different adult patterns. Bard noted
that timing of onset of stripes occurs during the third week of development of E. burchelli, but in the fifth week in E. grevyi. Because the stripes are deformed to varying
degrees, depending on the timing of initiation, the adult patterns of the two species look
quite different. It should also be noted that similar effects can also be produced by
changing the embryonic growth rate, as this will change the body areas over which
diffusion can occur.

The significance of pre- and postdisplacement in generating saltations is particularly important in vertebrates, where much of the differentiative growth occurs in the
early phase of embryogenesis. Thus, a delay in cell differentiation of a limb bud can result
in substantial morphological saltation, even to the extent of an "either/or" situation—
limb or no limb. Whether or not a limbless condition is maladaptive (and thus strongly
selected against), nonadaptive (in which case such phenotypes can be carried by the
population), or adaptive (because of changed environmental conditions or exploitation
of a new environment) depends upon the environmental opportunities that exist. On the
other hand, subsequent growth heterochronies of differentiated traits (Chapter 3) can
contribute to rapid speciation by change in rate or offset of that growth, in particular by
peramorphic increase in rate or delay in offset time (see Chapter 7 for discussion of this
with regard to the human brain).

With global or regional progenesis or hypermorphosis, not only will greater differences in timing of offset between ancestor and descendant produce a greater phenotypic
effect but so will higher allometric coefficients of growth of particular traits. The more
positive or negative the trait growth allometry, the shorter the timing difference between
ancestor and descendant has to be in order to achieve the same phenotypic effect.
Nevertheless, even though greater allometry and greater differentiation in offset timing
can induce more pronounced morphological saltations as with rate and onset heterochronies, this has to be counterbalanced by the viability of such a "monstrous" phenotype. Not only must it be physiologically viable, but to form the basis of a new, rapidly
evolved species, there must be a suitable niche in which its adaptive peak can be manifested. Moreover, there is the problem of who the monster mates with. Arthur (1984) has
suggested that mutations that produce such monsters could be either sex-linked or
recessive, so hiding gene expression until it has been transferred through a number of
individuals. Also possible is the nonrandom occurrence of such mutations, causing similar monsters to appear independently at the same time (Alberch, 1982).

In the case of organisms, such as many arthropods and amphibians, which undergo
metamorphosis from the juvenile to adult phase, there is a "built-in" ontogenetic morphological saltation. In such cases, global progenesis can be a particularly potent force,
inducing major morphological saltations and rapid speciation, simply because no intermediate morphology exists. Because such organisms already occupy two quite different
niches during their ontogeny, the adult and juvenile will possess different morphological
and behavioral exaptations. For instance, paedomorphic amphibians are morphologically and behaviorally quite different from apaedomorphs because of their retention of
larval characteristics that allow the organism to reproduce and spend the entire life cycle
in an aquatic environment. That such a major shift in life history strategy was achieved
with relative ease is illustrated by the ubiquity of paedomorphic amphibians in the fossil

record. Many temnospondyl amphibians are paedomorphic, such as members of the Brachyiopidae, Chigutisauridae, and Plagiosauridae (Fig. 6-8) (Milner, 1989). Where there are such ontogenetic morphological saltations at metamorphosis, there is almost always a jump from one niche to another quite different niche. Such niche changes may be aquatic to terrestrial, as with amphibians, or aquatic, terrestrial, or fossorial to aerial. The implications of this in inducing major evolutionary breakthroughs are discussed below in Section 4.3.

Intrapopulational heterochrony can result in morphological saltations in the form of polymorphs (see Chapter 4). While, as we have argued, the existence of polymorphs contributes to a species' fitness by virtue of extending the phenotypic variability of the species, the differentiation of polymorphs into different subniches can have the potential to provoke rapid speciation if the subniches become genetically isolated. This can be said to be occurring in the case of the apple maggot fly *Rhagoletis pomonella* (see Chapter 6).

In the case of discontinuous polymorphisms, selection will promote rapid evolutionary change, without the involvement of intermediate morphotypes. Jaanusson (1973), for instance, believed that many of the sudden transitions seen in graptolites arose through polymorphism. How this can work from a developmental perspective is shown by Sage and Selander's (1975) study on polymorphism in cichlid fishes. The polymorphisms involve quite abrupt differences in the types of pharyngeal teeth: those fish with heavy molars feed on snails, others that have no molars, but many more, small teeth, feed on algae. From a developmental perspective the possession of few large molars can be interpreted as a relative paedomorphic trait in the differentiative component of tooth number, but a peramorphic trait in the attainment of large tooth size. The algae feeders have many more teeth (peramorphosis) that are much smaller (paedomorphosis). Both polymorphs can coexist because there is no effective niche overlap—each feeds quite differently.

Persistence of both polymorphs, or even the evolution of others with different tooth morphologies will be contingent upon the abundance of the relative food sources. The question is whether the evolution of two such disparate polymorphs would have been by a rapid change in the developmental program of the ancestral polymorph, or a gradual evolution of one from the other through intermediate morphotypes. Extrinsic factors are likely to have been of paramount importance. The fitness of a polymorph with intermediate pharyngeal morphology is liable to be less than that of the two extreme polymorphs that are able to more effectively deal with such very different food sources. Developmentally, there is no reason why intermediate morphotypes could not have developed. But any such individuals are likely to have been outcompeted by the more efficient snail or algae eaters, on the assumption that no suitable, intermediate food source was available. Thus, the adaptive peak had the potential and the ability to be generated, but there was no effective ecological niche available. Here the internal, heterochronic changes may only be gradual; it is discontinuous extrinsic factors that have dictated the discontinuous nature of the polymorphisms and their likely rapid appearance.

To establish the rate at which polymorphs evolve, or the rate at which they might yield potential new species, it is imperative to look at both intrinsic and extrinsic factors. These will vary with circumstances. Which factor is relatively more important overall is an empirical question awaiting research.

4.3. Phyletic Gradualism

Theoretically, gradual evolutionary changes within or between species, will arise only when intrinsic heterochronic change is gradual, as are the changing extrinsic factors. For example, small, directional neotenic and accelerated shifts between successive generations can lead to an apparent gradual morphological shift. If the traits are adaptations to a particular niche that lies along an environmental continuum, such as shallow to deep water or coarse to fine-grained sediments, then a gradual shift in morphological structures can theoretically ensue. Rather than quantum leaps from one adaptive peak to another occurring, the adaptive peak shifts in a certain direction along an environmental gradient—in other words, characters form heterochronoclines paralleling environmental gradients. Whether these are driven by genetic drift, or extrinsic factors such as competition or predation is discussed below.

There have been few attempts to relate morphological changes along heterochronoclines to rates of evolution. Many of the spatangoid heterochronoclines described from southern Australia (McNamara, 1989) have all the criteria that we believe are necessary for interspecific change to be gradual. However, stratigraphic detail is insufficient to test this. The morphological changes largely involve shifts in allometries and rates of differentiation of secondary structures, providing adaptations to new niches that lie along an environmental continuum, such as coarse to fine sediments, and shallow to deep water (see Figs. 5-11, 6-18, 6-22). If the agent of selection, in these cases gastropod predation (see below), was gradually increasing in intensity, then shift from one habitat and one morphological state to another could have occurred gradually. However, even in this apparently gradual scenario, all it would take would be one parameter, such as fluctuating predation pressure, to result in speciation rates being rapid. Thus, if the population structure of the species established an equilibrium with a particular level of predation, and if this level was maintained for an extended period of time, then the species could show morphological stasis. If the level of predation increased rapidly, due to a population explosion in the predator species, then only extreme heterochronic morphotypes that existed in areas of slightly reduced predation pressure would survive. Maintenance of this higher level of predation pressure could result in selection of extreme heterochronic morphotypes, and the appearance in the fossil record of a rapid speciation event.

If just one of the four components involved in rates of speciation (the nature of the organism's growth pattern, the heterochronic process, the population structure, or the nature of the ecological niche) has a saltatory aspect, then speciation is likely to be rapid. Given the nonlinearity of most natural processes, the likelihood of all four factors being gradual or continuous in nature is much less than one of them being episodic. It would be expected that interspecific changes are often rapid.

Thus, while anagenetic heterochronoclines may, in the classic phyletic gradualism model, involve a gradual change from one species to another, they can also be rapid, resulting in a pattern of "punctuated anagenesis." Following a long period of morphological stasis, where the species' developmental system may be constrained by stabilizing selection agents, such as predators, a rapid change can induce a rapid shift in the morphological structure of the population, and consequent rapid speciation. Another scenario can also be visualized (Fig. 5-19) where, following a long period of stasis, there

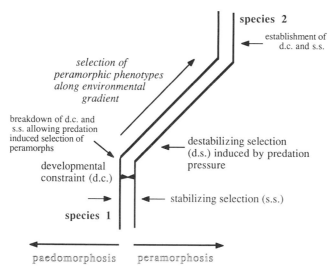

species 2

selection of
peramorphic phenotypes
along environmental
gradient

establishment of
d.c. and s.s.

breakdown of d.c. and
s.s. allowing predation
induced selection of
peramorphs

developmental
constraint (d.c.)

destabilizing selection
(d.s.) induced by predation
pressure

stabilizing selection (s.s.)

species 1

paedomorphosis   peramorphosis

Figure 5-19. Breakdown of developmental constraint and stabilizing selection can be followed by a protracted period of destabilizing selection along an environmental gradient. Selection of gradually more paedomorphic or (as in this case) peramorphic phenotypes can then lead to the establishment of renewed developmental constraint and stabilizing selection, once the new phenotype maintains a state of equilibrium with the destabilizing force.

is a gradual shift from one morphotype to another. This could occur if the intensity of stabilizing selection and developmental constraint is reduced. This will lead to a breakdown of genetic homeostasis until it is reestablished in a different adaptive peak. Such a scenario has been described by Cisne *et al.* (1982) in the Ordovician trilobite *Flexicalymene senaria*, through a 60-m section that represents sedimentation over a period of about 4 million years. By analyzing the number of pygidial rings through the sequence, Cisne *et al.* (1982) have been able to show how, following 1 million years of stasis, there was a change in pygidial ring number over roughly 800,000 years. This was followed by another 1 million years of stasis at the new level. Phenotypically, what occurred was a gradual change in the mean number of pygidial rings from about 5.5 to about 7, over this extended time period (Fig. 5-20). They attribute this change in ring number to heterochony and consider that it reflects a paedomorphic retention of higher juvenile pygidial ring number. The gradual change may be evidence of parapatric speciation over an extended time period.

Peramorphosis can equally well be an important mechanism in gradual phyletic evolution. Olempska's (1989) study of the heterochronic change in a lineage of the ostracod *Mojczella* has shown that by comparing equivalent molt stages, it is possible to document both size and shape changes through a stratigraphic sequence. Not only does the relative increase in size at each molt stage result in a change in overall carapace shape, but other characters covary. The most distinctive is the increase in growth of the crest on the side of the carapace (Fig. 5-21). In early forms two separate crests are present. In later forms they are separate in early growth stages, but merge in later ones to form a single crest that gets progressively more sinuous.

The polarity in opinions that exists between rapid and gradual rates of evolution is

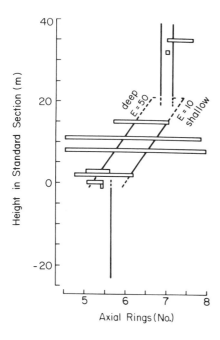

Figure 5-20. Spatiotemporal change in the segmentation of pygidial segments in the trilobite *Flexicalymene senaria*. Ordination score contours (E = 10, E = 50, signifying shallow and deep, respectively) show the average number of axial rings expected in a pygidium of average size in the intervals before, during, and after the evolutionary episode that occurred in a 20-m time interval. The clinal gradient represents a nearly 1 million-year-long parapatric speciation event. Reproduced from Cisne *et al.* (1982) with permission of Dr. J. L. Cisne.

reflected in the widely held view that these changes affect the whole organism. A species is said to show either stasis or gradual morphological change over time. In reality, species can show a mixture of both stasis in some traits, and gradual change in others ["mosaic evolution"—see Stanley (1979) for examples]. We have already described just such an example in the case of intraspecific morphological change in a species of the echinoid *Echinocyamus* (see Section 3.2.2). Here many traits show no apparent change from the Pliocene to the present day, whereas others show a gradual change, involving rates of differentiation of pore pairs and tubercles, and changes in mitotic rates of growth of tubercles, and spines. Mosaic (and dissociated) heterochronoclines may be viewed as mosaic evolution with a spatial vector in addition to the temporal one.

Among the few detailed examples from the fossil record that appear to show examples of phyletic gradualism, evolution in tarsier-like omomyid primates from the early Eocene of Wyoming also demonstrates this pattern of mosaic evolution of traits by the operation of different heterochronic processes. Rose and Bown (1984) have shown that in an area of excellent stratigraphic control through a 700-m sequence that spans about 4 million years, the nature of the omomyids' dentition changed appreciably. They provide evidence for a gradual transformation between *Tetonius homunculus* and *Pseudotetonius ambiguus*. In this lineage the dentition shows both dissociation and covariation. The anterior dentition evolved dramatically, but the molars and last premolar (P_4) underwent little change (Fig. 5-22). Early in the lineage, P_2 was lost; in other words, there was extreme postdisplacement to the extent that the tooth failed to erupt. Subsequently, there was a trend toward gradual reduction in size of P_3, the canine (C), and incisor (I_2) by paedomorphosis. The large anterior incisor (I_1), on the other hand, was dissociated from this set of covarying traits, and shows a gradual peramorphic increase in size. Along with these size changes were concomitant changes in morphological complexity. The adaptive significance of these changes was for a shorter-jawed omomyid with improved anterior grasping teeth, reflecting a change in diet.

Such gradual changes involving heterochrony may not simply involve temporal changes, but may be more complex in also involving spatial changes. In an elegant study, Cisne *et al.* (1980) have shown how changes in head proportions, caused by changing the offset "time" of development of allometric traits in the Ordovician trilobite

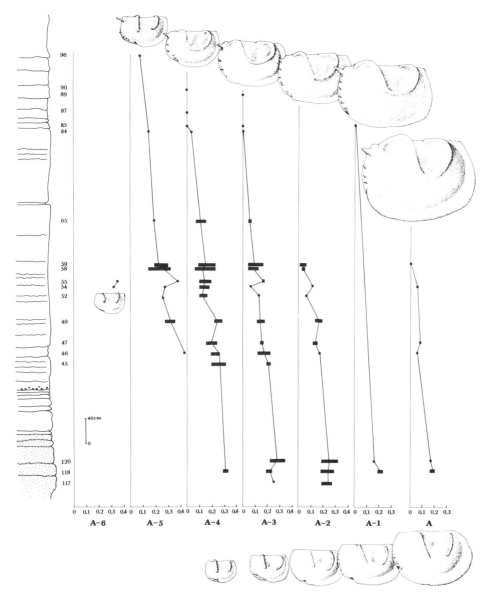

Figure 5-21. Gradual peramorphic evolution in the ostracod *Mojczella*. Reproduced from Olempska (1989), with permission of the Lethaia Foundation.

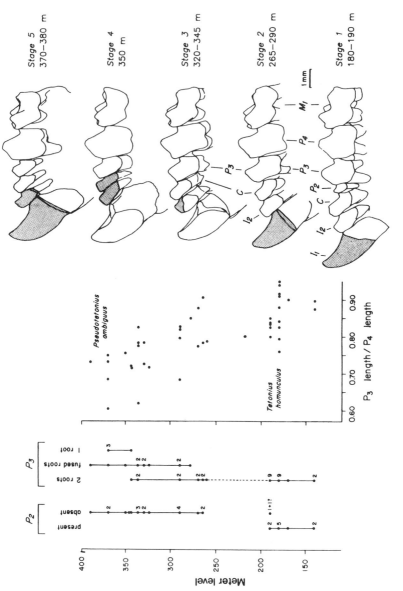

Figure 5-22. Gradual evolution of the anterior lower dentition in early Eocene omomyid primates of the *Tetonius homunculus–Pseudotetonius ambiguus* lineage showing paedomorphic reduction in premolars (P), canine (C), and incisor (I₂) and peramorphic increase in anterior incisor (I₁). Reproduced from Rose and Bown (1984), with permission of Macmillan Magazines Ltd.

Triarthrus, occurred gradually both spatially along an environmental gradient, as well as temporally. The two putative species *Triarthrus becki* and *T. eatoni* represent extreme end member morphotypes along a graded spatiotemporal cline. Anatomical changes over time reflect selection for morphs that predominantly inhabited one end of the cline. With increasing size, specimens pass from the *eatoni* morph to the *becki* morph because of allometric growth of the cranidium. Thus, *T. eatoni* is a relatively paedomorphic form. Along the environmental gradient of increasing water depth, the morphology changed from the larger, relatively peramorphic *becki* morphotype to the smaller, relatively more paedomorphic *eatoni* morphotype. The gradual change from *becki* to *eatoni* up through the stratigraphic sequence is considered by Cisne *et al.* (1980) to reflect not so much an evolutionary change, but a transgressive phase. As the water deepened during regional transgressions during the Late Middle to early Late Ordovician, the structure of the population changed to reflect a greater number of paedomorphic *eatoni* morphotypes. Apparently gradual evolutionary changes may often, as in this case, be changing spatio-temporal clines, reflecting preferential selection of particular heterochronic morphotypes. This also shows how heterochronoclines can be spatial as well as temporal (see McNamara, 1982b).

4.4. Macroevolution

4.4.1. Heterochrony and Major Morphological Discontinuities

While we have concentrated primarily on the role of heterochrony at the specific and subspecific levels, one of the major "claims to fame" of heterochrony has been that it is an ideal mechanism for the sudden appearance of major new morphological novelties, resulting in the evolution of new families, orders, even phyla. (Here we use "macroevolution" to mean the origin of higher taxa.) Maderson *et al.* (1982, p. 307) have argued that the "developmental constraints are the basis of any discontinuities and clumping in organic morphospace. The sparse and clumped distribution of morphology does not represent a set of optima constructed by natural selection from a set of unbounded possibilities." They go on to argue that heterochrony provides an appropriate mechanism for evolutionary change that does not conform with the traditional patterns of the modern synthesis of evolution. It does this in a number of ways (Maderson *et al.* 1982):

1. Large-scale changes at the phenotypic level, with little genetic change.
2. Macroevolutionary changes in a geologically relatively short space of time, following long periods of morphological stasis.
3. Appearance of traits that are of no selective value and were not selected for *per se,* but happened to be the traits that were available when development ceased.
4. Possibility of "rejuvenation" of a phyletic lineage due to gross changes in body size and morphology. Reappearance of these, perhaps under a different environmental regime, may lead to the establishment of a major new clade.

The fossil record is replete with examples of such taxonomic groupings appearing *geologically* almost instantaneously, with no apparent or obvious ancestor. The most important is the so-called "Cambrian explosion" about 570 million years ago, when major groups, such as brachiopods, echinoderms, arthropods, and mollusks, appeared and diversified at an extremely high rate. One way to explain this is to invoke heterochrony. For instance, we have discussed in Chapter 4 the importance of heterochrony during early trilobite development at this time. Furthermore, we discuss in Chapter 6 how the sudden appearance of many new families of trilobites at the end of the Cambrian can be attributed to a change in heterochronic styles brought about by alterations in developmental regulation, along with the rise in dominance of many new predators. Campbell and Marshall (1987) have similarly highlighted the relatively rapid appearance of many new classes of echinoderms during the early Paleozoic, with little evidence of convergence of morphologies between the classes at an early phylogenetic stage. Among the explanations that they propose for these "cryptogenetic" appearances include the possibility that low levels of competition enabled a wide variety of structural designs to be more easily selected and the possibility of large morphological change being induced with minimum genetic change. From the perspective of intrinsic change, the rapid appearance of evolution novelties such as those that occurred during the first 50 million years of the Phanerozoic, has been attributed to perturbations of the regulatory and developmental systems (Valentine and Erwin, 1987).

The discovery of factors such as interrupted coding sequences, transposable elements, multigene families, pseudogenes, and other complex aspects of the genome allows for the possibility of extensive morphological divergence arising from small changes in the genome. Among living organisms this can best be seen by analyzing and comparing factors such as mitochondrial DNA between genera and families, so as to highlight any close genetic similarities between morphologically disparate groups. This has recently been done by Bruns *et al.* (1989), working on false-truffles. While they are strikingly dissimilar morphologically (Fig. 5-23), the structure of the mitochondrial genome is identical in the false-truffle *Rhizopogon* and the mushroom *Suillus,* on the basis of the 15 regions of the genome studied. Furthermore, there is a high level of similarity between the genera in nucleotide sequences from a portion of the mitochondrial large subunit ribosomal RNA gene. Bruns *et al.* (1989) attribute the close genetic similarity but different morphology to small genetic changes and intense selection. Their model for the evolution of *Rhizopogon* involves changes in the genes regulating development. As they point out, the evolution of the underground false-truffle *Rhizopogon* from a suilloid ancestor, such as *Suillus,* may have been via an intermediate secotioid form, in which the normal pileus expansion to form the characteristic mushroom shape, was arrested. Further paedomorphic loss of ancestral adult features, including loss of the stipe, size reduction of the basidiocarp, loss of cystidia, loss of sterigmata, and completion of development underground, resulted from selection for reduced water loss and animal (specifically rodent) dispersal. The importance of developing an enclosed spore-producing tissue in the secotioid form is that it would lead to increased inbreeding and a rapid increase in frequency of recessive alleles in the offspring. Any intermediate morphologies between suilloid, secotioid, and *Rhizopogon* forms would be maladapted to either air or animal dispersal and would be eliminated.

Trying to apply such models involving change in the regulation of expression of the genome directly to marine organisms that existed 550 million years ago is obviously difficult. However, indirect evidence from the frequency of heterochrony in trilobites

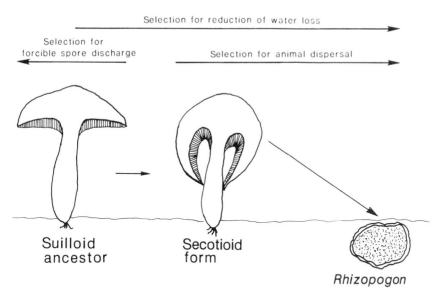

Figure 5-23. Paedomorphic origin of the false-truffle *Rhizopogon* from a suilloid ancestor via a secotioid intermediate form. Reproduced from Bruns *et al.* (1989), with permission of Macmillan Magazines Ltd.

(McNamara, 1986b) suggests that regulatory control of the developmental system was poor early in the evolution of this group. Thus, the potential existed for major saltations. More problematic is the viability of organisms when they undergo major perturbations of their developmental systems (a point we discuss below). Certainly there would have been a lot of "unoccupied niches" at this time. If we consider niches in terms of "resource utilization function" (Arthur, 1987), at times such as the early Cambrian there were many points on the resource axis that were available for utilization, but were not being utilized. Heterochrony had the ability to generate morphological novelties, with minor genetic change, in a regime where resources were particularly underutilized, both in breadth and in depth. If genomes were less tightly canalized during these periods, heterochrony would have been a common phenomenon, as the fossil record for trilobites seems to indicate. Minor genetic changes would induce major shifts in the developmental program, causing major morphological jumps. If the situation during the early Cambrian had been one of only "empty niches" and no heterochrony, then the Cambrian explosion would not have occurred. Likewise, heterochrony alone, with no available niches, would have precluded the explosion. Again it is the combination of the intrinsic and extrinsic factors that led to such major adaptive radiations. To argue that the extent of "adaptive space" was *the* primary factor in the rise of major evolutionary novelties in metazoans (Erwin *et al.*, 1987) is misleading. (Further discussion of the Cambrian explosion is in Chapter 8.)

Heterochrony has often been invoked for the evolution of many other major groups at times other than during the early Paleozoic, such as vertebrates (Garstang, 1928), hexapods (Manton and Anderson, 1979), birds (Thulborn, 1985), flightless birds (de Beer, 1958), parasitic worms (Conway Morris and Crompton, 1982), conifers (Rothwell, 1987), monotremes (Gregory, 1947), amphibians (Long, 1990), many brachiopod orders and families, amphibian families, and irregular echinoids (McNamara, 1988a). In many of these groups, paedomorphosis has been seen as the process responsible for the evolutionary novelty. In other words, morphological traits already in existence at another ontogenetic stage are restructured to operate in the adult phase. This may allow new structural adaptations to be selected (such as flightlessness in birds, or a pelagic mode of existence in tunicates). Alternatively, size might have been the major target of selection, as in some parasitic worms (see Chapter 6). However, utilization only of existing structures, even if they are early juvenile features, means that a restricted set of structures are being used. While they may open up a new niche, this is often a deadend (e.g., blindness and lack of pigmentation in troglodytes). Even so, de Beer (1958) argued that paedomorphosis had a much greater evolutionary potential than peramorphosis. He cited the case of the evolution of hexapods from multisegmented arthropods. He believed that peramorphosis produced structures that were specialized, while paedomorphic ones were simpler, and therefore held greater evolutionary potential. In our branching tree model, peramorphosis would be seen as creating too many branches, whereas paedomorphosis "unburdens" the tree. However, we consider that peramorphic froms have the same evolutionary potential as paedomorphs. For instance, the peramorphic event might occur early in ontogeny (novel differentiative heterochrony) and lead to a new adaptation so opening up new niches. Or it might be only a single peramorphic event in a mosaic of paedomorphic and peramorphic events that is the critical one in opening up a new adaptive pathway. This is exemplified by the hypermorphic development of our brain (see Chapter 7). Another example is the evolution of ratites. de Beer (1958) and others have argued that here is a prime example of paedomorphosis leading to the evolution of a major new group. Attention has always focused on the reduced wings, brought about by paedomorphosis. Yet, as we argue in Chapter 6, flightlessness confers no adaptive advantage. Indeed, from the point of view of susceptibility to predation, it could be construed as being maladaptive. More likely it is two other features that were of adaptive significance, notably the development of larger, more powerful legs, and, in many cases such as emus, rheas, moas, and ostriches, large body size. Both of these features were the product of *peramorphosis,* not paedomorphosis. (Further refutation of the greater evolutionary potential of paedomorphosis is in Chapter 8.)

Another important factor in the role of heterochrony in macroevolution is the dissociability among covariant sets of characters. As Gould (1982) has pointed out, very often where progenesis has opened up new adaptive potentialities, not all morphological features are reduced to the same degree. Certain sets of structures may covary to differing degrees, so producing a mix of ancestral juvenile and adult features. Many characters are bound in covariant sets, and so dissociate in blocks (McKinney, 1988b). While some may be tied very closely to maturation, and therefore "dragged along" by progenesis or hypermorphosis, others are less strongly bound (Chapter 8). An example quoted by Gould is that of the veneracean mollusk *Turtonia minuta.* While being only 1 to 2 mm long, this progenetic species possesses a mixture of ancestral adult and juvenile features, such as an adult ligament, and larval gills and siphons (Ockelman, 1964). As a consequence of dissociation of covariant sets, morphological change can be substantial and contribute toward macroevolution. (More examples are in Chapters 2 and 3.)

We have discussed Garstang's idea that progenesis resulted in a tunicate larva reproducing and inhabiting quite a different niche from its adult phase (pelagic as opposed to benthic) thus leading to the evolution of vertebrates. Largely because of the inability to test this proposal effectively, and problems associated with the linear concept of ontogenetic development, this idea has tended to flounder in recent years under the impact of an opposing view that vertebrates originated from calcichordates (Jeffries, 1986). However, there is support (Anderson, 1987) for another old view, that hexapods arose from a myriapodous ancestor by paedomorphosis. This centers around the proposal that there must have been some hypothetical early Devonian myriapods that hatched as hexapodous juveniles before completing development by the addition of a further 11 leg-bearing segments, so reaching sexual maturity with 14 pairs of trunk limbs. The occurrence of progenesis very soon after hatching would result in an adult form with just three pairs of limbs. As Anderson (1987, p. 148) has observed:

> Since growth, moulting and sexual maturation in hexapods are under hormonal control, it is possible that our hypothetical myriapod species could suddenly spawn a sprinkling of sexy little hexapods through a small change in the genetic control of the hormonal control of post-hatching development.

Even more fundamental than the evolution of new phyla is the possibility that the evolution of meiosis from mitosis was facilitated by heterochrony. Recently, Margulis and Sagan (1986) have suggested that the evolution of meiosis occurred by a delay (a postdisplacement) in the division of centromeres in the cell. A consequence of this was that the number of chromosomes was halved, instead of staying constant, during cell replication. A delay in division by the centromeres would have resulted in only halves of each chromosome being pulled to the centrioles. When the centromeres eventually divided, paired chromosomes would have split and the cells divided, with each cell containing only one set of chromosomes. Such heterochronic change may have been the very basis of the evolution of sexual reproduction and thus the evolution of the plant and animal kingdoms. Whereas in this example it was the onset of development at an extremely early phase that was involved, the heterochronic process was still a paedomorphic one.

The evolution of parasitism may sometimes have been instigated by progenesis. Conway Morris and Crompton (1982) have proposed that the evolution of the phylum Acanthocephala from a priapulid ancestor was facilitated by progenesis. The Acanthocephala are small endoparasites that have an indirect life cycle that alternates between an arthropod intermediate host and a vertebrate definitive host. This phylum possibly evolved from priapulids that burrow in the sediment and whose larvae inhabit interstices between sediment grains. The result of early progenesis in this case was that the gut failed to develop. The subsequent parasitic life-style was associated with the development of internal fertilization that had the selective advantage of maximizing fecundity (a point on which we elaborate in Chapter 6).

The importance of exaptations in allowing the evolution of taxa at the class level has been illustrated by Thulborn's (1985) suggestion that birds may have arisen from theropod dinosaurs by paedomorphosis. Thulborn has suggested that juvenile theropods possessed feathers, their role being insulation. Retention of such structures into a larger adult that retained other juvenile theropod characters, such as very large orbits, inflated braincase and retarded dental development, reduction in fibula width, reduction in

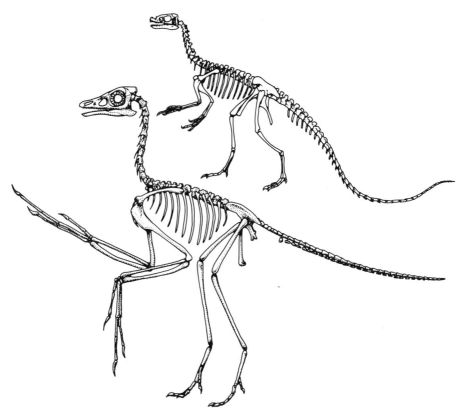

Figure 5-24. Skeletal reconstructions of juvenile theropod dinosaurs (upper) and adult *Archaeopteryx* to illustrate that apart from retention of paedomorphic characters, such as feathers, by *Archaeopteryx,* some important characters, notably digit length in the forearm, underwent peramorphosis, allowing development of a wing sufficiently large to be used in flight.

number of tarsals and metatarsals, contributed toward the attainment of flight. However, as with "paedomorphic" flightless birds, a major characteristic of *Archaeopteryx* is a peramorphic feature, namely its greatly enlarged fingers, that allowed support of the wing (Fig. 5-24). Perhaps most critical in such innovative breakthroughs is the dissociation of certain characters to produce adaptive zone-opening specializations (Chapter 8). Dissociation allowed digit size increase, leading to wings in this case.

4.4.2. Atavisms and Polymely

Another way for the developmental system to produce rapid morphological change is for structures to be expressed that have long been suppressed. These are **atavisms**. The role of atavisms in macroevolution has been examined recently by Lande (1978), Riedl (1978), Gould (1982), Alberch *et al.* (1979), and Raff and Kaufman (1983). Atavisms involve the reappearance of major traits, apparently lost, by reexpression or rearrangement of tissue interactions. This appears to be consistent with the argument that changing early development by such atavisms can result in a reorganization of morphology and major discontinuous macroevolutionary jumps.

Atavisms point out the relative simplicity with which development can change from one pathway to another and lead to substantial modifications of the phenotype. Experimental work on animals such as guinea pigs shows that different developmental mechanisms can be triggered concurrently in the same organ. An atavistic digit V appears by a prolongation of growth of metatarsal V (hypermorphosis in a local growth field), whereas atavistic digit I is produced by development of a new metatarsal I (Hall, 1984b). What many such atavistic traits show are reappearance by local peramorphosis, these traits having been lost by paedomorphosis during the phylogenetic history of the group. Such is the case in guinea pigs and in horses, where extra digits can form. Like horses, guinea pigs experienced a phylogenetic trend of digit reduction through the Tertiary (Hall, 1984b). Similarly, some limbless vertebrates (e.g., snakes) are known to produce atavistic limbs. A number of atavisms are known in whales. Nemoto (*in* Berzin, 1972) records about 1 in every 5000 sperm whales producing atavistic limbs, while a specimen of the humpbacked whale *Megaptera novaeangliae* has been described which had a hind limb more than 1 m in length (Andrews, 1921). Embryos of all whales possess limb buds. In all the above examples, development of the embryonic bud was not terminated, but continued, reflecting (atavistic) hypermorphosis in a local growth field.

While nobody would argue that such atavistic phenotypes are likely to be the progenitors of a major new clade, they do provide good evidence for the ease with which major morphological saltations can be produced by minor changes in the timing of development of particular structures. As Hall (1984b, p. 119) states, "Atavisms should not be an embarrassment to the evolutionary biologist. They are the outward and visible sign of a hidden potential for morphology change possessed by all organisms."

There is another way that teratologies may theoretically contribute to macroevolution, and that is through **polymely.** This is the appearance of supernumerary legs, digits, or both, at relatively high frequencies in certain populations. These may or may not be functional. Polymely has mainly been recorded in amphibians. Van Valen (1974) has proposed a theory whereby populations with such largely nonadaptive structures have the potential to give rise to higher taxa. He has argued that the gradation from "normal" limb or digit number through to the teratological abnormalities (which might also include forms with *reduced* number of limbs or digits) suggests that disruption of developmental canalization has occurred, rather than a simple switching from one developmental pathway to another. Where these polymelous polymorphisms have been studied there is no evidence of genetic control. However, there is direct evidence of environmental control, principally temperature, and viruses. For instance, polymelous polymorphs of *Rana esculenta ridibunda* mostly occur in cold spring water. Voitkevitch (1958) has shown that retardation (neoteny) of development under these conditions can cause polymely.

Van Valen (1974) has argued that if most of these polymorphisms are environmentally induced, the range of environments in which they can occur, and their expression, will be genetically controlled, and thus alterable by selection. He considers that these polymorphs provide a useful macroevolutionary model. Selection will focus on those genotypes that are most sensitive to the environmental factor influencing the variation.

Less environmental disturbance will be required as the population genome becomes more sensitive, resulting in the character appearing in most, or even all, environments. Thus, a phenotype that is normally maladaptive can be established and maintained, perhaps because it is merely nonadaptive under certain environmental conditions, and therefore not selected against. Or it may even in certain cases actually be of adaptive significance. As the environment changes, or the population enters a new environment, the trait might become adaptive. It will have appeared by a saltation, but was able to be carried through by virtue of not being maladaptive. It then has the potential to be the precursor of a new higher taxon.

4.4.3. Effect of Patterns of Ontogeny on Macroevolution

In most studies of heterochrony, but perhaps more so in those that have focused on its role in the evolution of higher taxonomic units, there has always been the tacit assumption that development was a linear, progressive process (Garstang, 1928; de Beer, 1958; Gould, 1977) and that onset of maturity could slide up and down this developmental pathway, operating at any level. However, Cohen and Massey (1983) have questioned this premise. As they point out, if the developmental program is an integrated unit, then the juvenile phase will anticipate the future structure of the adult, and the adult will be constrained by the structure of the juvenile. So to change developmental timing and produce a new adult, the genetic program must be changed in order to have an influence throughout ontogeny. This, Cohen and Massey believe, can be both radical and dangerous. As a counterproposal, two other developmental models have been formulated, termed the *parallel development model* and the *successive development model* by Cohen and Massey.

The parallel development model has been particularly applied to organisms that undergo metamorphosis, such as insects (Sander, 1982), amphibians, and lower chordates (Barrington, 1968). This model sees juvenile and adult developmental pathways as diverging during the course of phylogeny and being separable during ontogeny. As Cohen and Massey (1983, p. 322) describe it, "the how-to-make-a-maggot-kit" has few parts in common with its complementary "let's-build-a-blowfly-kit." The transfer of information only occurs at metamorphosis, when there is destruction of larval tissue, morphogenesis of adult tissue and gonads. For a larval form to perform as an adult by progenesis, it would be necessary for the sexual function to be transferred from one program (adult, quiescent in the larva) to the other (larval, endopolyploid, cued to self-destruct by the same hormones which cause maturation of the sex organs). Thus, although such progenesis has been invoked for the evolution of major morphological novelties, in particular some phyla, such instances are likely to be rare, because of these innate developmental problems. As Cohen and Massey point out, many animals almost achieve such progenesis, but fail at the crucial point, a striking case in point being mayflies (see Section 5.2), silk moths, the frog *Pseudis paradoxa,* and Pacific salmon, all of which have to metamorphose to the adult, albeit for an exceedingly short period, just to reproduce. If reproduction occurred just 2 hours earlier in some mayflies, for instance (i.e., in the larval phase), then the morphological saltation between the ancestral type and this progenetic descendant would be immense, probably sufficiently large to provoke taxonomists into erecting a new higher taxon, perhaps at the ordinal or class level.

However, there are some examples, particularly among the Crustacea, such as in some copepods and ostracods, where progenesis has crossed the juvenile/adult boundary, and reproduction does take place in the juvenile phase.

The successive development model of Cohen and Massey (1983) applies to those many organisms where the larval form is not so much constrained by physiological adaptation, but by developmental necessity. The early morphogenesis of most organisms is independent of their own zygote genomes. Morphogenesis determined by zygote genomes only occurs after normal nuclear–cytoplasmic ratios have been established by cleavage when the organism's body plan is forming. This formation, where the maternal morphogenetic program is controlling, is known as the *phylotypic* stage (Sander, 1982). In chordates it is when the neurula appears; in annelids, the trochophore. The phylotypic stage is a period when very disparate eggs converge to produce comparable morphologies, prior to divergence to different adult body plans. Thus, the simple morphology of phylotypic larvae is a function of the organism's developmental stage, rather than any innate physiological requirement. Cohen and Massey (1983) argue that if progenesis were to occur at this stage, then the effect of breeding by the phylotypic larva would be to make the zygote genome morphologically redundant. The zygote program would, however, still play a part in constructing the new organism's ovaries, and thus eggs. The consequence would be the evolution of a quite different form of postlarval development because of the formation of a novel zygotic program after the phylotypic phase.

Consequently, the idea that larvae may have been progenitors of new phyla can still be accommodated, even if there are difficulties with the linear and parallel models. New structures may not only arise from adaptations from preexisting adaptive larval structures, but also from nonadaptive, phylotypic larvae. While progenesis involving adaptive larval structures may be a source of macroevolution, progenesis at an even earlier stage has the added advantage of allowing the descendant to escape from the evolutionary dictates of the ancestors. This can be achieved (Cohen and Massey, 1983) either, as in mammals, by reducing the maternal prescription to zero, so allowing experimentation with many different early embryonic pathways; or by reducing zygotic morphogenesis to almost nothing, as in nematodes, and utilizing a uniform, but very versatile morphology.

4.4.4. In Conclusion: Macroevolution and Heterochrony

When we consider the role of heterochrony in macroevolution, rapid major morphological change can occur by the operation of any heterochronic process, and paedomorphosis and peramorphosis can both equally well be involved. Furthermore, the particular phase of early development that is affected can have a great impact on the evolutionary potential of the heterochrony. If the developmental perturbation affects the postphylotypic phase, then the effect is most likely to be pronounced when the ancestral organism involved undergoes a sudden, large-scale morphological transition (metamorphosis) during ontogeny, wherein the two phases occupy quite different ecological niches. For such a transition to be really effective, ancestral juvenile characters must be exapted (preadapted) to function in a different way in the descendant adult stage. Such traits may be adaptive in a different way in the descendant juvenile, or they may even be nonadaptive. Yet when transposed to the adult stage, and in a different setting, they may function in a novel way that has adaptive advantage.

Perhaps the one facet of heterochrony that will prove to be the most challenging to workers over the next decade will be the assessment of the contribution that it has made to the origin of higher taxa. Any number of stories can be written on the basis of superficial similarities between juvenile traits in one group and adult traits in another. What needs determining is just how important the small, relatively unspecialized paedomorphs have been in establishing major new groups of organisms, compared with the role of peramorphs. If cladistic studies are to play their part in this program, they must encompass comparisons of entire ontogenies, not just adult traits. Detailed embryological studies must be undertaken in an intellectual framework that sees heterochrony as an integral part of evolution. If not, we shall continue to stagger on reporting ". . . arguments about putative phylum-level transitions [that] have too often been promulgated in the speculative mode that once gave heterochrony such a bad name among thoughtful scientists concerned with testability. . ." (Gould, 1988a).

5. AGENTS OF SELECTION

Most of the literature on heterochrony has dealt not only with the direction and rate of evolution, but also with changing heterochronic frequencies and targets of selection (see Chapter 6). However, one rather fundamental question that has been lacking in the vast majority of these studies is that most basic of questions—why? Why, for instance, should a particular paedomorphocline develop in the direction that it does? What is the force that drives it? As we have noted in our train analogy earlier, the fuel (heterochrony) is there, as are the rails (environmental gradient) passing through a number of stations (ecological niches). But who, or what, is the driver?

5.1. Competition

We have shown how cladogenetic heterochronoclines may be interpreted as autocatakinetic, i.e., self-driving. The persistence of the ancestral species may provide enough impetus to drive the heterochronocline. Competition will be the prime motivating force, the agent that drives the course of evolution along pathways highly constrained by these intrinsic (developmental) and extrinsic (ecological) factors. However, in contrast to the beliefs of some cladists, the fossil record shows the common occurrence of anagenesis, whereby one species *replaces* another along heterochronoclines (Hallam, 1989). Since the introduction of the concept of punctuated equilibria (Eldredge and Gould, 1972), anagenesis and cladogenesis have become linked with rates of evolution: anagenesis with the gradual intraspecific transformation of one species into another; cladogenesis with rapid, punctuated speciation, following periods of morphological stasis. How heterochrony fits into these models we shall discuss in the next section. Here we wish to clarify our use of the terms "anagenesis" and "cladogenesis" and demonstrate that the agents responsible for driving anagenetic trends are different from those that drive cladogenetic trends. *Both* are used here in an interspecific sense, with no

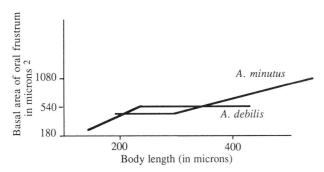

Figure 5-25. Reduction of competition between closely related species by heterochrony. In the two copepods *Amphiascus minutus* and *Amphiascoides debilis,* resource differentiation occurs by variations in growth allometries during ontogeny. Redrawn from Marcotte (1983).

connotation of rate adhering to either term. Cladogenesis involves the evolution of one species from another, with persistence of the ancestral species. Anagenesis involves the wholesale evolution of one species from another (be it gradual or rapid) which results in the replacement of the ancestral species by the descendant. One might well argue that such an anagenetic scenario could arise from progressive competitive exclusion of the ancestor by the "fitter" descendant. While this, on occasion, may occur (*vide* the cotton rats—Chapter 6), the evidence from a number of groups of fossil invertebrates is that ancestor and descendant occupy different niches, and thus are unlikely to have been involved in direct competition for resources.

Competition between closely related species can be minimized by heterochrony. The meiobenthic copepods *Amphiascus minutus* and *Amphiascoides debilis* are sympatric in marshes in Nova Scotia, Canada, and compete for the same resources: bacteria and diatoms growing on spheres of organic detritus and clay (Marcotte, 1983). However, the two species appear to differentiate the resource by size. The size of the food that the copepod can handle is determined by the size of the oral frustrum, which is allometrically related to body size. Marcotte (1983) found that when the size of this structure in the two species converges below a morphometric ratio (larger/smaller) of 1.4, the species' areal distributions are different. When above this competitive threshold, the species' distributions overlapped spatially.

In *A. minutus* the oral frustrum is small in the juvenile phase, but enlarges rapidly with body length in adults. In *Amphiascoides debilis,* on the other hand, the structure increases rapidly during juvenile growth, but does not increase in size with adult growth (Fig. 5-25). Marcotte (1983) found that when both species are present, *Amphiascus minutus* avoids competition with *Amphiascoides debilis* by demographically reducing the numbers of individuals in the size range near the overlap in oral frustrum size between the two species. This is achieved either by paedomorphosis, and thus retention of smaller oral frustra allowing feeding on smaller food particles, or by a developmental saltation to the larger adult size, bypassing the body size at which oral frustrum size was the same in both species. Marcotte (1983) has observed that the common occurrence of heterochrony in infaunal harpacticoid copepods may have arisen in just such a way so as to reduce levels of interspecific competition.

5.2. Predation

Recent research on a variety of groups of organisms, from fossil echinoids and fish, to living fish, mayflies, and fleas (see below), is providing evidence to suggest that a prime motivating force (and one that has largely gone unrecognized) in the generation of anagenetic heterochronoclines and major heterochronic evolutionary trends in general, is predation. In contrast to cladogenetic heterochronoclines where the driving force comes from within the system by means of interspecific competition, the driving force in anagenetic heterochronoclines always comes from a perturbing force outside of the system. While the importance of predation in affecting life history strategies has been the subject of much research (Kerfoot and Sih, 1987; Vermeij, 1987), there has been little attempt to link these life history changes to heterochrony. However, not only may heterochronically induced changes in life history strategies be driven by predation; but there is also evidence from the fossil record that both within specific lineages and also in the higher-level trends within major taxonomic groups, morphological changes induced by heterochrony, leading to consequent occupation of new ecological niches, were driven by predation.

5.2.1. Predation-Directed Heterochronoclines

There are a number of documented examples in the fossil record of the direct effects of predation. These examples rely upon the evidence provided by damage to the prey inflicted by the predator, which may or may not be lethal (Vermeij, 1987). Either way this provides a strong indication of the sorts of selection pressure to which many animals were subjected. Among the clearest, and most common, examples particularly in Mesozoic and Cenozoic strata are boreholes made by predatory gastropods, generally in mollusks and echinoderms. A recent study (McNamara, 1990) on predation levels in the spatangoid echinoid *Lovenia* has shown that variations in intensity of predation, as measured by changing frequencies of lethal boreholes, can be correlated with heterochronic changes in defensive, antipredatory strategies. The combination of changing levels of predation, habitat shifts, and heterochronoclines of structures that reduce predation pressure all point to the direct influence that predation has had in directing the evolution of the lineage.

The evolutionary changes that occurred both intra- and interspecifically between the three species in the *Lovenia* lineage have been discussed in Chapter 4. Within the lineage, heterochrony led to the evolution of morphotypes with improved defensive structures (primary spines) and with adaptations to progressively finer-grained sediments along the lineage. These were habitats with reduced levels of predation. This morphological and ecological channeling appears to have been directed by strong predation pressure selection. The frequent evolution of heterochronic morphologies adapted to fine-grained sediments in many Mesozoic and Cenozoic irregular echinoid lineages may also have been driven by predation pressures (McNamara, 1990).

The most notable morphological structures that show heterochronic change are: (1) the number of primary spine-bearing dorsal interambulacral columns; (2) the number of primary spines within these columns (as shown by the number of tubercles); and (3) the number of tubercles on the ventral lateral interambulacra, which in life supported primary spines adapted for burrowing. Primary dorsal spines in living species of *Lovenia* are known to serve a defensive function (Ferber and Lawrence, 1976). Both dorsal and ventral spine number increased during ontogeny, but at different rates in each of the three species. Compared with the ancestral *L. forbesi*, the Middle Miocene *L.* sp. shows an increase in dorsal tubercle number, by peramorphosis. Its Late Miocene descendant *L. woodsi* shows a paedomorphic reduction in tubercle number. Such a reversal is unusual in spatangoid lineages, where trends are generally unidirectional, resulting in the formation of paedo- or peramorphoclines. As we have discussed (Chapter 4), the lineage also shows the formation of a paedomorphocline in the numbers of interambulacral columns which possess primary tubercles and spines, from eight in adults of the oldest species, *L. forbesi,* to only four in adults of the youngest species, *L. woodsi* (Fig. 4-10).

Ventral tubercles, which supported burrowing spines, show a peramorphic increase in descendant species, probably by predisplacement. The sediments in which the species of *Lovenia* are preserved show a progressive decrease in grain size through the Miocene. Increase in number of these burrowing ventral spines probably reflects adaptation to burrowing in finer-grained sediments. However, changes in dorsal spine density are more likely to reflect adaptations to predator pressure, as discussed below. Many other spatangoids that lack primary defensive spines show heterochronic adaptations to occupying deep burrows in finer-grained sediments (McNamara, 1990), but *Lovenia* is not a deep burrower.

Echinoids are attacked by a wide variety of predators, including other echinoids, asteroids, fish, gastropods, crustaceans, birds, sea otters, and even arctic foxes (Fell and Pawson, 1966). Of these, only gastropods, which cut or drill holes in the test, leave an unambiguous trace in the fossil record. Living echinoids, including irregular burrowing echinoids, suffer quite high levels of predation from the family Cassidae (Hughes and Hughes, 1981). It is possible to identify cassids as the predators of fossil echinoids by their distinctive method of penetration of the echinoid test. Whereas other predatory gastropods, such as naticids, bore into their prey, cassids cut a disk from the prey's test (Hughes and Hughes, 1981). Failed predation attempts on some *Lovenia* tests show evidence of this method of attack. Many Miocene echinoid populations from the Australian Cenozoic show evidence of high levels of cassid predation. Furthermore, there is a close correlation between the time of arrival of cassids in Australia, in the Late Oligocene (Darragh, 1985), and the first evidence of gastropod predation.

Gastropod-induced mortality was highest in the Early Miocene *L. forbesi* (28%), reducing to 20% in the Middle Miocene *L.* sp., then to only 8% in the Late Miocene *L. woodsi*. It is also perhaps notable that *L. woodsi* shows a lower percentage of successful borings, only 70%, compared with 84% in *L. forbesi* and 95% in *L.* sp. The increase in density of protective spines in the second species, *L.* sp., may have occurred in response to improved predation techniques by the gastropod in Middle Miocene times.

The influence of the primary spines in protecting the echinoid against gastropod predation is shown by the low frequency of predation holes either in the region of the test covered by tubercles, or immediately posterior, in the area covered by the canopy of extended spines (Fig. 5-26). The high concentration of holes around the periproct in *L. forbesi* suggests predator preference for this region, akin to that shown by the living cassid *Cypraecassis testiculus* on *Diadema antillarum* (Hughes and Hughes, 1981).

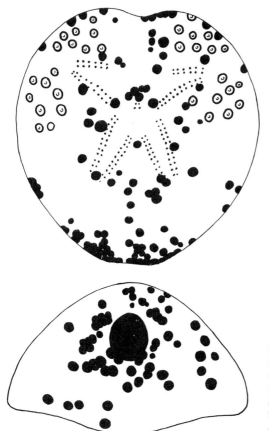

Figure 5-26. Composite reconstruction of distribution of gastropod boreholes in the test of the Early Miocene heart urchin *Lovenia forbesi*. Regions most free of predation are the areas covered by, and immediately posterior to, the tubercles. These would have been protected by an array of primary spines.

Predation is recognized as being a major agent of selection (Vermeij, 1982), capable of inducing in the prey a wide variety of antipredation responses. Reduction in the level of gastropod predation in the *Lovenia* lineage, combined with movement of successive species into regimes of finer-grained sediments, suggests that those morphotypes capable of occupying finer-grained sediments, by the possession of more burrowing spines, preferentially survived. Data on the bathymetric distribution of living cassid gastropods (Abbott, 1968) show that species richness declines into deeper water.

The highest concentration of defensive dorsal spines in the intermediate species of *Lovenia* suggests that selection pressure initially favored morphotypes with greater concentrations of these spines in response to the high predation pressures. The combination of this higher density of dorsal spines with ventral spine increase (to allow habitation of finer-grained sediments in deeper water) proved to be insufficient to cope with the high predation levels. Consequently, subsequent selection favored the *L. woodsi* morphotype, which was capable of inhabiting a region of lower predation by further peramorphic increase in ventral burrowing spine number. Because it was able to colonize an area of much lower predation, there was less selective pressure to maintain a high density of aboral defensive spines. So by the evolution of antipredation strategies and structures produced by heterochrony, Cenozoic spatangoid lineages were able to counter the effects of strong gastropod predation pressure by crypsis (in this case occupying a burrow-ing niche in finer-grained sediments in deeper water, in a region of low predation pressure) and/or by the peramorphic evolution of a dense cover of defensive spines. Thus, the directionality of paedomorphic and peramorphic trends and the establishment of an anagenetic pattern may have been induced by predation, acting as the main agent of selection. Strong predation pressure caused the extinction of ancestral species incapa-ble of withstanding it. Descendant species persisted for only as long as predation was below a critical level. Above this level, only morphotypes which had evolved improved antipredation devices persisted, i.e., survived the "Arms Race" (see Vermeij, 1987).

It has been suggested (Stanley, 1979) that the disappearance of Paleozoic echinoids with flexible tests, and their replacement by forms with rigid tests, may have been partly influenced by predation pressure. Likewise, the great increase in diversity of infaunal echinoids, gastropods, and bivalves since post-Paleozoic times has been shown to corre-spond to an increase in diversity of predatory gastropods (Stanley, 1977). The great morphological diversity developed by irregular echinoids during the Tertiary has been attributed to the widespread occurrence of heterochrony (McNamara, 1988a). It is possi-ble that the development of such a wide range of morphologies was strongly influenced by predation pressures. Heterochrony provided the fuel for morphological change; the direction in which it was driven was largely controlled by predatory selection pressures.

The importance of predation in influencing the selection of particular heterochronic morphotypes in different habitats can likewise be demonstrated in vertebrates. Bell (1988) has shown how pelvic reduction in the three-spined stickleback fish *Gasterosteus* occurred by heterochrony, as a direct response to predation pressure. Generally less than 75 mm in length, and bearing three dorsal spines, a row of lateral bony plates, and a stout pelvic girdle, species of *Gasterosteus* have invaded freshwater habitats, where pro-nounced phenotypic differentiation has occurred, in contrast to the homogeneous ma-rine populations. According to Bell, the primitive fully formed pelvic girdle in marine populations is a most effective defensive structure against vertebrate predators because it bears a prominent spine, combined with a posterior process and fin ray. Pelvic reduction by paedomorphosis, involving loss of some of these elements, occurs in some *Gasteros-teus*, and in other genera of sticklebacks (Fig. 5-27). This has only been documented in some freshwater habitats and is interpreted by Bell as occurring in response to a change in selection pressure. In the marine environment the main agents of selection are verte-brate predators. The presence of pelvic spines and processes forms a most effective defense against other fishes that try to swallow the sticklebacks. Vertebrate predation therefore ensures strong selection pressure for phenotypes with robust pelvic girdles, i.e., phenotypes that have *not* undergone paedomorphic reduction in this structure.

However, in a number of freshwater habitats, there has been a shift in predation pressure away from vertebrates toward invertebrates, predominantly insects. Whereas the pelvic spines and processes were effective armor against vertebrate predators and are thus useful adaptations, in the freshwater environment the same structures become maladaptive because they appear to be useful to the predatory insects, who use them to improve their grip on the small fish. In these instances where predatory fishes are absent, but predatory insects present, selection pressure has favored paedomorphic phenotypes that show a *reduced* degree of pelvic development (see Fig. 5-27). Bell has shown that the sequential loss of: (1) pelvic spine and separation of the girdle into anterior and posterior

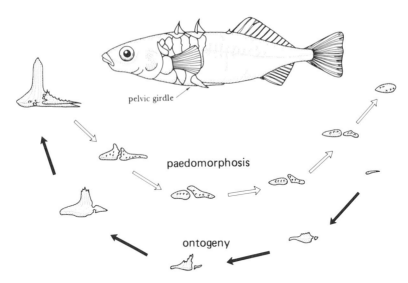

pelvic girdle

paedomorphosis

ontogeny

Figure 5-27. Paedomorphic reduction of the pelvic girdle in the three-spined stickleback *Gasterosteus*. Re-drawn from figures in Bell (1988).

vestiges; (2) progressive reduction in size of posterior vestiges until they disappear; and (3) reduction and loss of anterior vestiges follows the opposite course to the ontogenetic development of these structures in the marine precursor. There is also evidence that such paedomorphic pelvic reduction is moderately heritable.

In addition to observing this phenomenon in living species, Bell has also recorded it among fossil populations. In a 110,000-year sequence of Miocene deposits, fossil specimens of *Gasterosteus doryssus* show paedomorphic pelvic reduction. The sequence forms an intraspecific paedomorphocline where, as fully formed pelvic phenotypes declined, intermediate vestiges appeared, followed by anterior vestiges, then specimens in which the pelvis was lost altogether. The total absence of other fishes from these deposits is used by Bell as an argument for increased selection pressure against phenotypes with a fully formed pelvis brought about by high insect predation pressure. Although there has been repeated occurrence of pelvic reduction in different stickleback lineages, a persistent clade of sticklebacks entirely lacking a pelvis has not occurred.

5.2.2. Effect of Predation on Life History Strategies

Predation can have an effect on life history strategies by selecting for behavioral and habitat patterns (which have been affected by heterochronic processes) that act to reduce the impact of the predation pressure. One way that this can be achieved is by alterations in the timing of maturation, for many organisms will be under different levels of predation pressure at different stages in their life histories (Werner and Gilliam, 1984). For instance, let us assume that there are two populations of an organism and each takes ten units of time to pass through its entire life cycle from birth to death. In one population (A1) the juvenile phase occupies six of these time units, while in the other

(A2) it occupies eight (Fig. 5-28). If the juvenile phase is spent in a different habitat from that of the adult, and one where predation pressure is much less than during the adult phase, then selection will favor the population that spends a shorter time in the more vulnerable adult phase. This change in timing of maturation relative to the entire life span could be viewed as occurring in two possible ways: either by delaying the onset of maturity, i.e., hypermorphosis (thus, the juvenile phase is extended at the expense of the adult phase), or by a contraction of the adult phase, i.e., early offset (progenesis) of the adult phase. Evidence from morphological characteristics of the short-lived adults of the mayfly, *Dolania americana,* suggests that postdisplacement has played an important role. Adults lack functional mouthparts and the digestive system has largely failed to develop (Soldán, 1979). Thus, while the relative extension of the juvenile phase would appear to indicate hypermorphosis, the paedomorphic nature of the adults suggests that the delay in onset of maturity could be considered a postdisplacement phenomenon. Hypermorphosis extends the entire life cycle, including the adult phase. In the mayflies there has been no concomitant increase in adult phase; on the contrary, there has been a foreshortening. Therefore, in terms of the heterochronic model this is best viewed as an example of postdisplacement of a number of features, including the onset of maturity. After all, the underlying molecular mechanism will be the same for delay in producing morphological traits and for the onset of sexual maturity: delay in production of the relevant hormones to stimulate the growth or the metamorphosis.

Such a severely foreshortened adult phase in response to predation pressure is found in a number of families of Ephemeroptera (mayflies—Fig. 5-29), such as the Oligoneuriidae, Leptophlebiidae, Behningiidae, Euthyplociidae, Polymitarctidae, Palingeniidae, and Tricorythidae (Edmunds and Edmunds, 1980). In these mayflies the adult phase has been effectively reduced to a minute proportion of the entire life span, often for less than 2 hours out of a total life span of 2 years. The adult stage of the male *Dolania americana* lasts for about 1 hour, the female for only about half an hour, out of a life span of 14 months (Sweeney and Vannote, 1982). Thus, the adult phase occupies only about one ten-thousandth of the entire life span of the organism. Peters and Peters (1988) have observed that in the period since mayflies evolved during the Devonian, the major trend in mayfly evolution has been for this relative elongation of the juvenile phase, at the

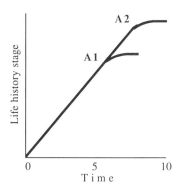

Figure 5-28. Effect of changing maturation times on levels of predation. Under a regime of higher adult predation pressure, forms in population A1 that attain maturity earlier will be at a selective advantage over those in population A2 that spend longer as juveniles under higher predation levels.

Figure 5-29. The American sand-burrowing mayfly. *Dolania americana,* which, under predation pressure, has reduced its adult phase to a mere 2 hours out of a total life span of 2 years. Reproduced from Peters and Peters (1988), with permission of the American Museum of Natural History.

expense of the adult phase. In its juvenile phase the mayfly is aquatic and a predator. During the mass mating of mayflies (such as *D. americana*) during the dawn hours, both the males and females (which incidentally are relatively paedomorphic to the males in reproducing in the subimago stage) are subjected to extremely high levels of predation by birds, bats, insects, and spiders (Sweeney and Vannote, 1982). Predation occurs both in the air and on the water surface, from beetles. Thus, the very intense predation pressure which is focused only on the adult phase of these mayflies has led to heterochronic reduction in the adult phase.

5.2.3. Effect of Predation on Selection of Size

One of the more important adaptive features of body size is that it strongly affects the risk of predation. Optimum foraging strategy by predators (Sih, 1987) generally reduces predation pressure on small and large sizes: small, because the energy expended by predators in attacking small prey does not produce a sufficiently high return on the investment of energy; large, because too much energy needs to be expended for an adequate return, as well as basic physical constraints of tackling a large item of prey. Such selection pressure therefore favors the survival of forms that reproduce earlier at a small size (progenesis) or later at a much larger size (hypermorphosis). However, there are also obvious survival advantages in rapidly attaining the optimum adult body size by size acceleration. This applies to both predator and prey. For prey species, it takes them out of range of the optimum prey size quicker, while for predators it allows them to function within this predatory range earlier.

The adaptive significance of a prey species attaining a smaller adult size by progenesis is exemplified by the bivalve *Nucula turgida*. This species inhabits shallow areas of Dublin Bay and rarely exceeds 10 mm in shell length. Wilson (1988) showed that the boring gastropods that preyed upon the bivalve rarely took the smaller individuals. As the bivalve reproduces most efficiently when 5 to 7 mm in length at 3–5 years of age, predation of larger size classes reduced competition among the prey, favoring the smaller forms that have greatest reproductive potential. Obviously, there needs to be a sufficient range of variation in timing of onset of maturity so that progenetic forms can survive the predation pressure, i.e., multiple alleles of "timing" genes that can be sorted in the gene pool (Slatkin, 1987).

Because some organisms' abilities to either avoid predators or harvest resources from different types of habitats change during ontogeny, the timing of shift from one strategy or habitat type to another is often directly related to size and to growth rates. Werner and Gilliam (1984) have reviewed the importance of these crucial **ontogenetic niche** changes and the degree to which this affects population structures. Importantly, as we discuss in Chapter 6, increase in body size during ontogeny and the accompanying change in niches means that there is often avoidance of direct overlap in resource use. During ontogeny, food or habitat use may change continuously as individuals grow, or there may be more abrupt changes, as we have discussed earlier with reference to amphibians. The timing of ontogenetic habitat shifts is therefore a critical factor in terms of both the organisms' life history strategies, and interspecific interactions.

The bluegill sunfish (*Lepomis macrochirus*) undergoes such habitat shifts, from littoral vegetation to open-water or pelagic habitats in lakes, several times during its life history (Werner and Hall, 1988). After hatching in the littoral zone the fish migrate to the pelagic zone to feed on zooplankton, then return to the littoral zone for some years to feed on invertebrates. They may then, at larger body sizes, once again migrate to the pelagic zone. Intensity of predation on the bluegill by largemouth bass affected the sizes at which habitat shifts occurred. In five lakes that Werner and Hall (1988) sampled, the mean body sizes at which the switch from feeding on vegetation to zooplankton occurred varied from 52 mm to 83 mm. The density of bass in different lakes was positively correlated with the size at which habitat shifts occurred: the higher the density, the larger the size at which the switch occurred (Fig. 5-30).

Experiments have shown that the predation risk for bluegills is much higher in open water. Conversely, high density of vegetation in the littoral zone resulted in very low predation rates. Even though the pelagic zone is an inherently more hostile habitat, it is energetically imperative for the bluegills to migrate to that habitat at some stage in their life history in order to attain sufficient food for their larger size. There is therefore a strong element of trade-off between higher predation risk and attaining a more profitable food source. As the risk diminishes with increasing prey body size, the larger the body size at which the switch takes place, the greater is the selective advantage.

Thus, the habitat switches that are body size related depend on the presence of their major predator, the largemouth bass. These fish effectively confine small bluegills to the vegetative littoral zone. The higher bass densities that result in this confinement of the bluegills until they attain a larger size class also have other consequences. Because of the

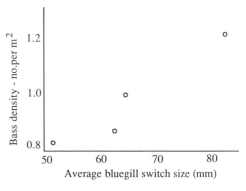

Figure 5-30. Effect of intensity of predation pressure by bass on timing of habitat shifts in bluegills: the higher the predator density, the larger the size at which the shift occurs. Data from Werner and Hall (1988).

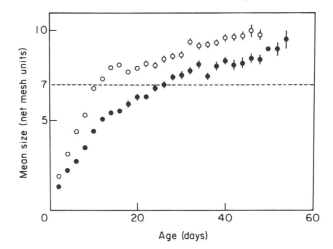

Figure 5-31. Growth of *Daphnia* sampled at the end of culling experiment after large sizes had been removed, showing change in growth rates and timing of onset of reproduction. Each point gives the mean size of *Daphnia* as a function of age in populations of small individuals (open circles) and large individuals (closed circles). Reproduced from Edley and Law (1988), with permission of the Linnean Society.

direct correlation of size increase with time, the effect of higher predator density is to cause the bluegills to spend a longer time in the littoral zone. In the two extreme lakes, the bluegills that underwent the habitat shift at 52 mm spent only 2 years there, whereas those in another lake where predator densities were higher and did not shift until an average body size of 83 mm had been attained, had spent 4 years there. This had a direct impact on growth rates, for as more age classes are confined to the vegetation and population density increases, so growth rates are lowered. This has the effect of further extending the time that it takes for the bluegill to attain an optimum size for habitat shift.

5.2.4. Induction of Heterochrony by Predation

In all the previous examples that we have discussed, the role of predation has been as an agent of selection. It has been an extrinsic factor that operates only after the intrinsic production of particular heterochronic morphotypes. However, some recent experimental evidence indicates that, counterintuitively, predation pressure may itself actually *induce* heterochrony. In a study of the waterflea *Daphnia magna,* Edley and Law (1988) used clones from seven sites in eastern England. The populations were maintained by parthenogenesis and two size-specific culling experiments were carried out on three replicate populations. Having divided the entire size range into 11 classes, they repeatedly removed size classes 0 to 6 from one group and sizes 7 to 11 from another, mimicking the effects of a predator. While the numbers culled from the 0 to 6 group showed no significant change over the time period of the experiment, culling of the large sizes resulted in a steady decline in numbers in this category.

At the end of the culling period, the structures of the populations had changed, because growth rates had undergone a change, as had the timing of onset of reproduction. After culling large sizes from one population continually, clones had been selected that grew more slowly (Fig. 5-31). Furthermore, they reproduced and died at a smaller size than clones from other treatments (Fig. 5-32). There was also a reduction in age at first reproduction. Thus, after culling large sizes, clones that were selected grew more slowly through growth stages 0 to 6, whereas in treatments where small sizes had been preferentially culled, the clones grew rapidly through this early growth phase, by a factor of about twice the rate of the large culled treatment.

Artificial predation resulted in the selection of clones that tended to be in vulnerable size ranges for relatively short periods; i.e., where culling was on small size classes, selected clones grew through this phase quicker, whereas where large sizes had been culled, clones took much longer to reach this more vulnerable stage. Size-specific culling would seem to cause genetic differentiation in growth rates and timing of maturation. Escape from vulnerable states of predation is therefore achieved by heterochrony. Compared with the "ancestral" population, the large-size-culled clones show a combination of neoteny and progenesis to produce paedomorphic descendants, whereas the small-size-culled clones show relative acceleration and hypermorphosis to produce peramorphic descendants. Thus, extrinsic predation pressure has the ability to cause prey to evolve lower reproductive values during life history stages when they are most vulnerable, with a shift in reproductive value toward sizes not cropped.

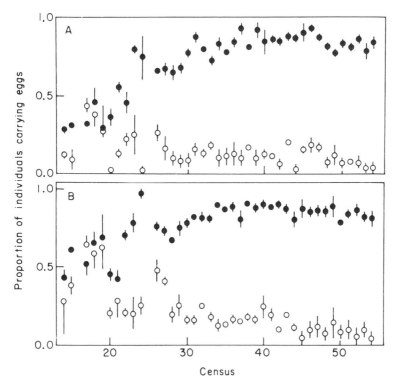

Figure 5-32. Proportion of individual *Daphnia* carrying eggs at two different sizes, during culling. Open circles represent small size classes, closed circles largest size classes. Reproduced from Edley and Law (1988), with permission of the Linnean Society.

5.2.5. Predation and Heterochrony in Major Evolutionary Trends

While we have stressed the importance of predation at the level of single lineages, major evolutionary trends may also have been propelled by predation pressure. For instance, as we note in Chapter 6, the increase in frequency of peramorphosis in early Ordovician trilobites may have been strongly affected by extrinsic factors, such as increased predation pressure. The evolution of a much wider range of morphological features than occurs in Cambrian trilobites from, at one extreme, increased spinosity to, at the other extreme, effacement, conferred a selective advantage against predation by adoption of a cryptic habit. Spinosity increased camouflage, while effacement evolved in parallel with features that facilitated burrowing. Increased predation in post-Cambrian times is likely to have come from the adaptive radiations of both cephalopods and fish. As we discuss below, preferential selection of such morphologies generated by peramorphic processes may have been a major contributing factor in the cryptogenetic appearance of so many new trilobite families in post-Cambrian times.

Similarly, the impetus for the evolution of burrowing irregular echinoids as a whole may have been the evolution of a wide range of predatory gastropods during the Mesozoic (Vermeij, 1987). This was permitted by the heterochronic changes in growth patterns in Jurassic echinoids (McNamara, 1987a). The evolution of increased septal complexity in ammonoids and the corresponding evolution of these forms into deeper-water habitats may also have been predator-driven (Hewitt and Westermann, 1987), predation pressure being higher in nearer-shore environments. Thus, the ability of organisms to evolve morphological traits by heterochrony allowed the occupation of new niches where predation pressure was less. Had morphological traits that permitted occupation of new niches not evolved, or alternatively, if the niches were not present, then extinction of the lineage would have ensued. So long as the prey species kept one step (or one niche) ahead of the predator, the lineage survived. Once this advantage was lost, the prey lineage was likely to have become extinct.

Chapter *6*

Heterochrony and Targets
of Selection

*. . . the search for explicative laws in natural facts
proceeds in a tortuous fashion. In the face of some
inexplicable facts you must try to imagine many
general laws, whose connection with your facts
escapes you. Then suddenly, in the unexpected
connection of a result, a specific situation, and one
of those laws, you perceive a line of reasoning that
seems more convincing than the others. You try
applying it to all similar cases, to use it for making
predictions, and you will discover that your
intuition was right.*
Umberto Eco, *The Name of the Rose*

1. INTRODUCTION

To understand the importance of heterochrony in evolution, one of the main aspects that needs to be considered is the target of selection. By this we mean intrinsically generated structures are being preferentially selected under extrinsic regimes. We have demonstrated how the raw material for evolution can be produced at the intrapopulation level by the generation of heterochronic morphotypes as part of normal phenotypic variation. Extrapolating this to the interspecific level and above, we need to examine what is being selected, e.g., shape, size, life habits, behavior, or all of these.

Historically, specific morphological structures alone have been considered as the targets of selection. An ammonite that increased the convolutions on its suture line through successive ontogenies did so because this new design conferred a functional advantage over the ancestral type. However, since Gould (1977) explored other avenues in his attempt to understand more fully the targets of selection, and looked at size and ecological strategies, the area of examination has spread widely (see Werner and Gilliam, 1984, and Ebenman and Persson, 1988, for reviews). In this section we provide a broad overview of targets of selection. We shall of course look at the evidence for specific morphological structures being the prime targets, but we also explore recent arguments for size, ecological strategies (in particular those relating to reproductive strategies), and behavior as important targets of selection.

But first there is another category of "targets" that requires assessment: the nature of the heterochrony that has been selected, either the style or the process. We have outlined in our introductory chapter how, in the late 19th century, the only targets of selection that were perceived as being of evolutionary significance were those produced by "recapitulation." With a shift in fashion in the early part of this century, the prime targets were perceived to be paedomorphic structures. Gould (1977) cogently argued that neither one heterochronic style nor the other should prevail, among or within particular groups. Thus, any group should contain approximately equal numbers of paedomorphs and peramorphs. Yet some of the detailed studies undertaken in the last 10 years have shown often quite substantial differences *among and within particular groups,* in the proportions of heterochronic styles (McNamara, 1988b). (Note that this does not mean that one style, e.g., a paedomorphic one, is *overall* more important when all groups are considered.) So in this section we shall also look at these differences in proportions of heterochronic styles and try to assess why they should have arisen, focusing on differences between groups and within groups over time. In particular, we shall investigate whether the apparent differences might have arisen from intrinsic or extrinsic factors.

In this chapter we firstly examine the effect of different growth strategies on the types of heterochronic processes that occur. These include differentiative and growth heterochronies, hierarchical growth strategies in colonial organisms, and modular growth strategies in plants. We then show how heterochronic targets in some groups of organisms have changed through time and discuss what intrinsic and extrinsic factors have influenced these targets. We then explain why in other groups the targets have not changed through time and to what extent intrinsic factors, such as genome size, have contributed to the dominance of paedomorphosis in some groups. The next major sections deal with shape and size as targets of selection. Specifically, we discuss the influence of heterchrony on size and shape, focusing on adaptations such as feeding, habitat, and locomotion, and how it affects size both globally and in local growth fields. Finally, we examine the impact of heterochrony on life history strategies.

2. INFLUENCE OF GROWTH STRATEGIES

2.1. Differentiative versus Growth Heterochronies

Extrinsic factors that contribute to selection are limited to the raw material on which they operate. Thus, intrinsic growth processes play a paramount role in determining the targets of selection. The changing frequencies of different processes of heterochrony between related groups of organisms may often reflect biases of growth processes in the different groups. We have discussed in Chapter 3 the differences between differentiative versus growth heterochronies. The former arises from changes in the timing of tissue induction and cytodifferentiation; the latter from later mitotic growth of the differentiated parts. In different groups of organisms the relative frequencies of differentiative versus growth heterochronies may vary (e.g., depending on when cell fates are determined). For instance, in vertebrates differentiation predominates during the early phase of embryogenesis; thereafter, mitotic growth is dominant. In many invertebrates, however, there can often be a greater proportion of differentiation later in ontogeny than occurs in vertebrates. For example, in echinoids, crinoids, blastoids, edrioasteroids, carpoids, asteroids, and cystoids the morphological targets of selection are constrained by the relative proportions of differentiative versus mitotic growth (McNamara, 1988b) (see below).

In echinoids there is also a disparity between growth strategies in regulars and irregulars (McNamara, 1988a) which might have ultimately led to differences in the range of morphotypes generated by heterochrony, and thus differences in the range of niches occupied. The test of regular echinoids grows largely by the differential addition of new plates. These then mitotically increase in size close to isometry. Differences among plate rates of growth are small. Acceleration or retardation in rate of plate differentiation will induce local peramorphosis or paedomorphosis. These rates are often higher in regulars than irregulars. Consequently, global progenesis or hypermorphosis can have a structurally profound effect on the number of plates in the test. Irregular echinoids, particularly groups such as spatangoids and holasteroids, produce far fewer plates, but different plates grow with markedly different allometries from their neighbors, as mitotic rates vary substantially. They may even have different allometric coefficients along different plate axes (in the same manner that cranial elements change shape during ontogeny in vertebrates). Thus, the same extent of progenesis or hypermorphosis in a regular and irregular echinoid, will have little effect on plate number in the irregular. However, because of pronounced plate dissociations in spatangoids and holasteroids that develop with these pronounced allometries (McNamara, 1987a), local neoteny and acceleration can have substantial morphological effects. For the structure concerned, such as a single plate, both its shape and size will be changed. As its function will be a consequence of both, it is the two parameters that are important. Take, for example, a ventral plate that carries spines that are used for burrowing. Increasing the allometry of this plate by acceleration will change not only its shape, but also its size. The size increase will allow a relatively greater number of burrowing spines to be accommodated. This, of course, is contingent upon the rate of differentiation of these spines remaining constant between ancestor and descendant. If the target of selection is an increase in the number of these spines, both the allometric increase in plate size by acceleration in mitosis *plus* an acceleration in rate of spine differentiation could occur. Here then, a local size increase confers a clear functional advantage on the organism. And it has been well documented (McNamara, 1988a) that a greater number of these spines allows the descendant to occupy a niche different from that occupied by the ancestor (a finer-grained sediment).

Acceleration of local growth fields has been very important in other areas of echinoid development. As we discuss below, in the spatangoid echinoid *Protenaster*, acceleration in the growth of support structures for food-gathering tube feet has allowed occupation of a different feeding niche by the descendants. In a number of unrelated spatangoid lineages, the targets of selection have often been the same, with the same functional advantages accruing. The parallel evolution of these lineages, involving evolution of a test morphology that allowed occupation of finer-grained sediments, with no global size increases along the lineages, argues for selection of suites of morphological characters that confer a distinct functional advantage in an adjacent, vacant ecological niche.

In carpoids and blastoids, similar large plate allometries occur, and again structural changes and their consequent functional changes can be attributed to localized allometric neoteny and acceleration. This is exemplified by the Devonian blastoid *Eleutherocrinus*, where acceleration occurred in growth in some plate axes, while others, such as the deltoid axes of the D-ray radial and side plates on the D-ray ambulacrum, show a neotenic retardation. If body size proxies age, then local mitosis occurred at a slower rate in the descendant than in the ancestor. As we discuss, the adaptive significance of acceleration of growth of zygous basal plates in *Eleutherocrinus* was the development of low-level suspension feeding (Millendorf, 1979). Likewise, the action of (allometric) neoteny and acceleration on different plates in *Pentremites* led to the evolution of a range of thecal shapes that were able to adopt new feeding strategies (Waters *et al.*, 1985).

In other echinoderm groups, however, such as crinoids, edrioasteroids, and, to some degree, asteroids, the dominance of differentiation of many new plates during ontogeny greatly influenced the nature of the descendant morphotype following the operation of heterochronic processes. Thus, (allometric) acceleration in rate of generation of brachial plates and arm length in *Promelocrinus* (Brower, 1976), and paedomorphic reduction in pinnule length and plate production in disparids (Frest *et al.*, 1979) are typical of the heterochronic changes that occur in crinoids. Within groups such as the microcrinoids, progenesis not only resulted in the evolution of crinoids with very small cups, but also led to a reduction in the number of plates produced (Lane and Sevastopulo, 1982; Lane *et al.*, 1985).

One of the more striking effects of local paedomorphosis induced by reduction in rate of differentiation, either by neoteny or by postdisplacement, occurred in the evolution of comatulids from stalked isocrinids (Simms, 1988). "Primitive" isocrinids such as *Pentacrinites* possessed long, highly cirriferous stems and were probably pseudopelagic (Simms, 1986) (Fig. 6-1). *Eocomatula*, probably the earliest member of the lineage that gave rise to the comatulids, shows a paedomorphic reduction in number of stem plates and was probably benthic. The derived *Paracomatula* with fewer stem plates that taper distally, led ultimately to the comatulids that lost stem plates altogether, as differentiation of these plates ceased altogether.

2.2. Hierarchical Growth Strategies in Colonial Animals

Evaluating the role of heterochrony in colonial animals, such as corals, bryozoans, and graptolites, entails the analysis of developmental shifts at two levels (Pandolfi, 1988). Heterochrony can involve either or both of the two levels of development within the colony (Lidgard, 1986). First, there is the **ontogeny** of each individual zooid or theca that makes up the colony. These individuals can be regarded as modules that together construct the whole colony. Second, the developmental history, or **astogeny**, of the colony as a whole can be influenced by heterochrony. The colony can be considered as

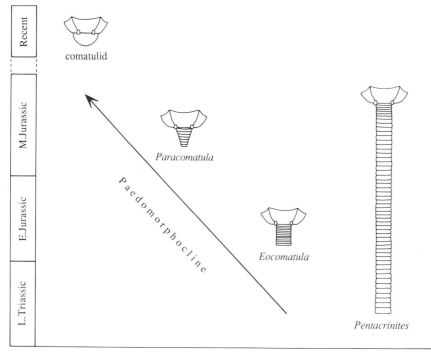

Figure 6-1. Suggested paedomorphic evolution of comatulid crinoids from stalked isocrinoids, based on Simms (1988).

an individual organism, composed of generations of modules. Growth of the colony will not be constrained by the increase in size of modules. Yet the potential for the colony to reproduce is dependent upon at least one of the modules developing this ability. The colony can be treated as an individual organism for the following reasons: the colony recruits as an individual; it does not undergo sexual reproduction until the colony has reached a specific age or size; different species have characteristic colonial growth patterns and can repair damage in order to restore the original shape of the colony; material can be translocated within the colony; and mortality rates that are ecologically significant are those of the colony, not individual modules.

One fundamental difference between many colonial and most acolonial animals is that astogenetic growth may be indeterminate. Some, however, such as graptolites (Finney, 1986), may have determinate growth. Because the onset of sexual maturity plays such an important role in some processes of heterochrony, identification of this in colonial organisms is imperative. The link between onset of sexual maturity and growth rates can be determined for colonial organisms. Alternating periods of high asexual growth rates and low growth rates during periods of sexual reproduction have been documented in ascidians and hydroids. Early astogenetic growth is often very fast to enable colony establishment. Sexual reproduction may be delayed until the colony is established. Where early growth rates are low, onset of sexual maturity will be earlier.

While the detection of the onset of sexual maturity is relatively straightforward in

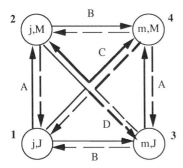

Figure 6-2. Model showing morphological results of heterochrony in clonal colonial animals. Redrawn from Pandolfi (1988, Fig. 3). Unbroken lines represent paedomorphosis, broken lines peramorphosis. Bold lines represent astogenetic heterochrony, normal lines ontogenetic heterochrony. J, juvenile colony; j, juvenile module; M, mature colony; m, mature module. At A, heterochrony is only at the astogenetic development level; at B, only at the ontogenetic level; at C, at both developmental levels, producing either both paedomorphosis or both peramorphosis; while at D, it operates at both developmental levels, but if paedomorphosis is at one level the other is peramorphosis, and vice versa.

acolonial animals, by the use of factors such as age of onset of first reproduction, which is usually correlated with a reduction in growth rate, its identification in heterochronic colonies is more complicated. The timing of onset of sexual maturity of a module may be affected in the same direction as the colony as a whole, thus *enhancing* the heterochronic result of either earlier or later onset of sexual maturity in the colony as a whole. Conversely, if the heterochronic processes are acting in different directions at the ontogenetic and astogenetic levels, the heterochronic result of earlier or later onset of sexual maturity will be *impeded* (Pandolfi, 1988). Thus, it is possible for hypermorphosis to occur at the ontogenetic level in individual zooids, but astogenetic acceleration would not result in earlier onset of maturity. Astogenetic development can directly influence ontogenetic development by its effect on the timing of sexual maturity. Zooids may therefore be inhibited from reaching sexual maturity by virtue of their position in the astogenetic development of the colony as a whole.

Astogenetic or ontogenetic growth rate variations may have an effect on the overall growth form of the colony (Jackson, 1979). Extrinsic factors, such as variations in sedimentation rate, can have a direct influence on growth rates. This morphological flexibility of colonial organisms provides the basis for developmental changes to occur and become heritable.

When dealing with development at two hierarchical levels, four possible combinations of heterochronic effects can occur (Fig. 6-2):

1. Heterochrony only at the astogenetic level inducing either astogenetic paedomorphosis or astogenetic peramorphosis.
2. Heterochrony only at the ontogenetic level inducing either ontogenetic paedomorphosis or ontogenetic peramorphosis.
3. Heterochrony at both developmental levels in the same direction, producing both ontogenetic and astogenetic paedomorphosis or ontogenetic and astogenetic peramorphosis.
4. Heterochrony at both developmental levels, but in different directions, producing astogenetic paedomorphosis and ontogenetic peramorphosis, or astogenetic peramorphosis and ontogenetic paedomorphosis.

Astogenetic heterochrony is much more common than ontogenetic heterochrony (Anstey, 1987; Pandolfi, 1988). Astogenetic neoteny can occur in a number of ways (Pandolfi, 1988): when astogenetic units of a given stage asexually bud at a slower rate than in the ancestor; when the modules of these units at a given stage take longer to develop; or when astogenetic morphological characteristics develop at a slower rate in the descendant. Astogenetic progenesis arises from precocious onset of maturity in astogeny, or because astogenetic characters have a shorter growth period. Size changes between ancestor and descendant may not be observable within astogenetic stages.

Astogenetic acceleration will arise when the units at a given astogenetic stage asexually bud at a faster rate in the descendant; when the units just develop at a quicker rate; or when astogenetic morphological characteristics develop at a faster rate in the descendant. Astogenetic hypermorphosis occurs by delay in onset of maturity of the whole colony, allowing prolonged development of astogenetic units of early astogenetic stages.

The level of integration between modules, involving aspects such as the amount of skeletal and soft-tissue fusion, will affect the link between ontogeny and astogeny, and affect the degree of cohesion in heterochronic processes at the different developmental levels. Highly integrated colonies would be expected to have a strong link between ontogeny and astogeny. In poorly integrated colonies, such as rugose corals, cohesion will be reduced or entirely decoupled (Coates and Oliver, 1973). Pandolfi (1988) considers that the degree of interrelationship between ontogeny and astogeny might have a strong impact on heterochronic changes, particularly the timing of colony maturation. Colonies with strong astogenetic–ontogenetic ties are thought to undergo astogenetic heterochrony regardless of the ontogenetic stage, while ontogenetic heterochrony will have profound effects on colony development. For instance, the evolution of a complex meandroid growth form in the living coral *Trachyphyllia bilobata* is thought to have arisen (Foster *et al.,* 1988) by dissociated peramorphic changes in ontogeny resulting in a paedomorphic modification of astogeny. On the contrary, where colony integration is weak, astogenetic heterochrony is likely to have little effect on individual modules or on ontogenetic changes during astogeny, as in favositid corals (Noble and Lee, 1990).

These interrelationships can be seen to operate to varying degrees in different groups of colonial organisms that show varying degrees of colony integration. In graptolites, it is high and the morphology of the module is strongly affected by its position in astogeny. The result of this high level of colony integration in gratolites is for astogenetic to ontogenetic heterochrony to show a ratio of greater than 3:1 (Pandolfi, 1988). Similarly high levels of integration in bryozoans are reflected in a dominance of astogenetic heterochrony. Where ontogenetic heterochrony did occur, it was inexorably linked with astogenetic changes (Anstey, 1987). While few examples of heterochrony have been described in corals, the low levels of integration of rugose corals indicate that ontogenetic heterochrony would dominate. Pandolfi (1988) has postulated that changing levels of colony integration might result in changes in proportions of astogenetic to ontogenetic heterochrony. For example, tabulate and scleractinian corals show periods of high integration during reef development, and low integration when reefs were weakly developed or absent. Thus, periods of reef growth may be times when astogenetic heterochrony dominates.

2.3. Modular Growth Strategies in Plants

Although having been the subject of much less research than in animals, there is much evidence to suggest that heterochrony has had a similar profound effect on the evolution of plants (Takhtajan, 1972; Rothwell, 1987; Guerrant, 1988). Because plants have fundamental differences in growth strategies from animals (although they do show some similarities to colonial animals, particularly with regard to their modular construction and indeterminate growth), the effects of heterochrony are expressed in a different manner in plants. This arises not only from the indeterminate growth habit and modular construction of plants, but also because of the two genomes that operate in plants—the phenomenon of alteration of generations. Thus, both the haploid gametophytic and diploid sporophytic generations are influenced by heterochrony. It is following meiosis that the most striking differences between the effect of heterochrony in plants and animals are seen. While the gametes that are produced in most higher animals by meiosis are only haploid cells, in plants the haploid products of meiosis, meiospores, divide mitotically to form the gametophyte generation.

Like colonial animals, plants differ from the unitary body plan of acolonial animals in being of modular construction. The plant may be considered as a group of three basic types of organs: leaves, stems, and roots (Guerrant, 1988). While many growth parameters are indeterminate in plants, some higher plants, such as the angiosperms, have determinate growth that produces a genetically predetermined number and arrangement of cells. For instance, leaves and flowers typically exhibit determinate growth. At the cellular level, plants differ fundamentally from animals in producing new cells only at the apex. Each cell has only a limited ability to increase mitotically thereafter, the majority of organ size in plants arising from hydrostatically driven cell enlargement. As in colonial animals, two levels of development can be distinguished in modular plants (Tomlinson, 1982), and heterochrony can operate at the two levels. The two levels are *ontogenetic development,* which refers to the sequence of addition of new modules; and *primordial development,* where particular modules are *elaborated* from some meristematic precursor (Guerrant, 1988). Changes of timing that affect these two growth modules may be considered to equate with our concept of differentiative and growth heterochronies, respectively, although whereas in animals, growth will be mitotic, in plants it will not be variations in rates or timings of mitosis that affect growth heterochronies, but in onset, offset, and rate of cell *expansion.*

Heterochrony has played an important role at the gametophyte level in flowering plants. Takhtajan (1972) has argued that paedomorphosis led to the evolution of the pollen and embryo sac in angiosperms. Production of gametes occurs at such a young and early stage in flowering plants that the ancestral gametangia do not form at all. The truncated range and altered sequence of events inhibit complex structures from subsequently developing. Similarly, heterochrony has greatly influenced evolution in mosses (Mishler, 1986). The size and form of leaves change along the moss stalk from small "juvenile" leaves to larger "mature" forms. Each is ecologically advantageous. From "juvenile" leaves new individuals (ramets) can develop, while in some "mature" leaves peramorphic features occur that aid capillary uptake of water by the stalk and in gas exchange in leaves covered by water. Specialized brood leaves are likely to be paedomorphic in origin.

Problems of homology abound in plants, particularly in structures that have indeterminate growth. However, structures such as leaves and flowers generally have determinate growth and allow the effect of heterochronic processes to be assessed. Consequently, as with growth structures in animals, allometric changes can be determined. For in-

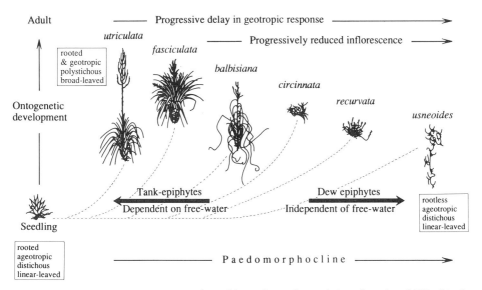

Adult ——————— Progressive delay in geotropic response ——————→

utriculata

rooted & geotropic polystichous broad-leaved

fasciculata

——————— Progressively reduced inflorescence ——————→

balbisiana

circinnata

Ontogenetic development

recurvata

usneoides

Tank-epiphytes
Dependent on free-water

Dew epiphytes
Independent of free-water

rootless ageotropic distichous linear-leaved

Seedling

rooted ageotropic distichous linear-leaved

——————— Paedomorphocline ——————→

Figure 6-3. Diagrammatic representation of possible paedomorphic evolution of species of *Tillandsia*. Redrawn from Tomlinson (1970).

stance, differences in mature fruit shape in squash (*Cucurbita*) and peppers (*Capsicum*) arise from differences in allometric coefficients. It was found in *Cucurbita* that these differences arise from the action of a single gene. The developmental determinants of size and shape in squash fruits are independently inherited (Sinnott, 1963). Shape is under one-gene, two-allele Mendelian control; size, however, is under polygenic control.

Changes in leaf form along a single axis (heteroblasty) can be viewed in terms of heterochrony (Guerrant, 1988). Reduction in size and simplification of "juvenile" leaves is associated with earlier maturation of these structures. However, the initial size of the meristem can influence ultimate leaf size. Furthermore, growth of some modules can influence the development of other modules. For instance, seed size in the legume *Desmodium paniculatum* has been shown (Wulff, 1985) to affect the timing of transition from "juvenile" to "adult" leaves.

While there are problems with homology in plant studies, if aspects of developmental relationships that produce individual structures are compared, rather than the structures themselves, then plant structure can be compared among individuals at the whole organism level (Guerrant, 1988). This has been demonstrated from a study of the phylogenetic relationships between species of the genus *Tillandsia*, a member of the pineapple family. Tomlinson (1970) has documented a case whereby changes in phyllotaxy and heteroblasty form part of the development of a paedomorphocline from, at one extreme, the ancestral morphotype exemplified by the epiphyte *T. utriculata,* to the small, rootless, linear-leaved *T. usneoides*. The large apaedomorph species is dependent upon free water, and has leaves that change from being initially negatively geotropic, to positively geotropic later in ontogeny. A sequence of species (Guerrant, 1988, Fig. 5) develop that retain the characteristics of progressively more juvenile ancestral species (Fig. 6-3), and which become dew-epiphytes, independent of free water. The principal bases of comparison are developmental homologues, in this case phyllotaxy and geotropic response, rather than specific structural elements, which are the *products* of the developmental processes. This allows the confounding effects of indeterminate growth to be removed from a study of comparative development in plants.

Like animals, plant growth can be affected by changes in the timing of onset of sexual maturity. Work on the genetic regulation of changes from the juvenile to adult phase in maize (Poethig, 1988) has resulted in the identification of a series of mutations that prolong the expression of juvenile development and are by definition hypermorphic. The allele, which is unusual in showing non-cell-autonomous behavior, is thought to control the production and distribution of a diffusible factor that regulates development. Whether or not this is the hormone gibberellic acid is not clear (Poethig, 1988).

Because of the contingencies of alternating haploid and diploid generations in plants, heterochronic processes that operate at one level are likely to influence development at the other level, thus producing a cascading effect. For instance, as Guerrant (1988) has shown, if small shoot apex size is preferentially selected in an angiosperm, it could have the incidental effect of reducing the size of floral primordia. This could lead to less sporogenous tissue in the anthers of individual flowers undergoing meiosis; thus, fewer pollen grains would be produced. As the pollen produces the growth hormone gibberellin, floral opening may not ensue because of levels of the hormone being too low (Lord, 1984). The point is that targeting of selection on a nonreproductive trait may have the incidental effect of influencing the plant's reproductive strategy.

3. CHANGING HETEROCHRONIC TARGETS THROUGH TIME

3.1. Intrinsic Factors Influencing Targets

Of all groups of organisms in which heterochrony has been reasonably well documented, one stands out as showing appreciable changes in frequencies of styles of heterochrony through their phylogenetic history—the Trilobita. The early members of this group that lived during the Cambrian do not show equal frequencies of paedomorphosis and peramorphosis over this time period: they show a dominance of paedomorphosis (McNamara, 1983a, 1986b, 1988b). Of the 17 described examples, 14 show paedomorphosis (Fig. 6-4). Moreover, there is a predominance of one particular process, with 11 of the 14 examples being due to allometric progenesis. (As in all fossil studies, we can identify pattern: paedo- or peramorphosis, and allometric heterochronies, but not age-based heterochronies—Chapter 2.) This highly biased selection of one particular mode of heterochronic expression raises the question of whether this reflects true relative incidence of heterochronic processes at this time (i.e., the intrinsic aspect was the controlling factor), or whether extrinsic factors were favoring the selection of particular styles of ("equally produced") heterochrony. It has been argued that high developmental flexibility of early Cambrian trilobites, induced by poor developmental regulation, contributed toward the high incidence of allometric progenesis (McNamara, 1983a). In other words, hormonal control on timing of development was poorly regulated. This

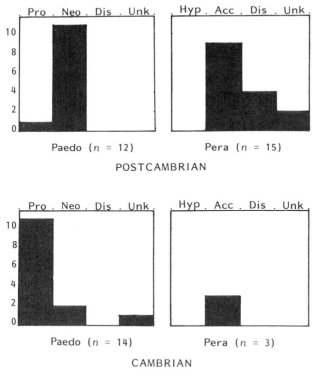

Figure 6-4. Histograms of frequencies of described examples of heterochronic processes in trilobites. Pro, progenesis; Neo, neoteny; Dis, displacement; Unk, unknown; Hyp, hypermorphosis; Acc, acceleration. Note that these are all examples of allometric heterochrony, whereby size has been taken as a proxy for time. Reproduced from McNamara (1988b).

would favor a higher incidence of either allometric progenesis or hypermorphosis. Yet, surprisingly perhaps, to our knowledge no examples of allometric hypermorphosis have been recorded in Cambrian trilobites. Thus, while an intrinsic factor has partially determined the heterochronic style, the ultimate determination of whether it was allometric progenesis or hypermorphosis seems to have rested with extrinsic factors. The most likely possibility at this time is vacant niche space, in particular the nektonic environment.

A number of workers (e.g., Vermeij, 1974; Runnegar and Bentley, 1983; Conway Morris and Fritz, 1984) have also expressed the opinion that imprecise genetic control due to poorly regulated developmental systems may have resulted in large degrees of morphological variation in other groups of Early Cambrian organisms. Pronounced intraspecific morphological variation in the enigmatic lapworthellids has been attributed (Conway Morris and Fritz, 1984) to "diffuse genetic control in the absence of adequate regulatory mechanism." Similar strong variation has been recorded in other Early Cambrian taxa, such as the bivalve *Pojetaia runnegari* (Runnegar and Bentley, 1983); and to a number of tommotiids (see Conway Morris and Fritz, 1984, for references). Likewise, early land plants appear to show marked morphological variation (Edwards, 1979; Edwards *et al.*, 1983) inferring that heterochrony may be particularly common early in the phylogenetic history of many groups. This may, in turn, have played a significant role in the propagation of many early adaptive radiations. Conway Morris and Fritz (1984) have, however, cautioned against a general acceptance of the idea that poorly regulated developmental systems alone were the cause of this variability. They suggest that an alternative aspect might have been that the prevailing ecological pressures in shallow-water communities played a more dominant role. As Whitney (1961) has demonstrated, where species of fish have been introduced into a competition-free (and presumably predation-free) environment (somewhat akin to the ecological situation prevailing in the early Cambrian), phenotypic latitude is high. As the population numbers increased, so the range of phenotypic variation declined. Unraveling which was the major cause of this apparent developmental flexibility, intrinsic or extrinsic factors, is perhaps one of the major challenges with which we are faced in evolutionary studies today.

As we have indicated, one reason for the frequent occurrence of paedomorphosis in Early Cambrian trilobites may well have been that control of hormonal mechanisms responsible for developmental regulation was poorly established early in the phylogenetic history of the group (McNamara, 1983a). It has been suggested (McNamara, 1986b) that as the Cambrian progressed, so trilobite hormonal systems became more efficient, resulting in more controlled regulation of developmental programs. This would have led to a decrease in the incidence of allometric progenesis, and hence paedomorphosis. A consequence of this perceived improvement in developmental regulation has been that classification at progressively higher taxonomic levels is relatively easier in younger trilobites. This can be shown by the variability in thoracic segment number. It varies intraspecifically in Early Cambrian trilobites when developmental control was at its weakest; between genera in Middle–Late Cambrian forms as it improved; and only between suborders in some post-Cambrian forms when it was much more rigorously controlled. For instance, four important suborders that arose in the Ordovician, the Trinucleina, Illaenina, Calymenina, and Phacopin, have 5–7, 6–10, 13, and 11 thoracic segments, respectively. In contrast to this high degree of developmental control in thoracic segment generation in many late Cambrian and post-Cambrian trilobites, most Cambrian orders show high segment number and high variability. For instance, segment number in the order Redlichiida varied between 9 and 50+ and in many Cambrian superfamilies of the Ptychopariida, between 5 and 40+. Moreover, Fortey and Chatterton (1988) have demonstrated how there was not only a trend for a reduction in degree of thoracic segment number variability within the mid-Cambrian to late Ordovician suborder Asaphina, but also a trend for paedomorphic reduction in segment number. This can be traced within asaphine families. For example, the earliest ceratopygid, *Proceratopyge*, has nine segments, while the younger *Dichelopyge* has only six. Some families are more stable, genera within the Asaphidae, for example, always having eight segments.

This concept of "improved" control of development regulation is an example of evolutionary "hardening" of integration and contingencies, resulting in development being harder to modify. At the cellular level (Chapter 3), this means that cellular aggregation and assembly has produced more integration and specialization. Levinton (1988) has termed this effect the Evolutionary Ratchet model. This he describes (Levinton, 1988, p. 217) as "the evolution of timing, rates, and localization (which) leads to a

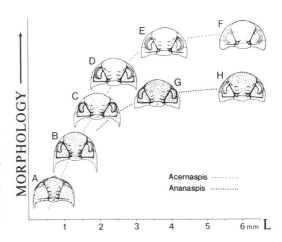

Figure 6-5. Schematic ontogenetic trajectories for the trilobites *Acernaspis* (broken line) and *Ananaspis* (dotted line) demonstrating proposed paedomorphic relationship of *Ananaspis* to *Acernaspis*. All ontogenetic stages drawn at same scale. Reproduced from Ramsköld (1988), with permission of the Lethaia Foundation.

complex developmental process that can be disrupted less and less easily as time progresses" (see Chapter 8 for further discussion).

Trilobites provide, arguably, the clearest example of changing styles of heterochrony through the phylogenetic history of the group. While paedomorphosis played a significant role in Cambrian trilobite evolution, it played a subsidiary role to peramorphosis in the evolution of post-Cambrian trilobites. Furthermore, even the paedomorphic processes are different, with a much greater emphasis on allometric neoteny as the prime process. This can be demonstrated by the nature of heterochronic change in the Silurian phacopid trilobites *Acernaspis* and *Ananaspis*. Ramsköld (1988) considers that *Ananaspis* evolved from *Acernaspis* by neoteny (probably global), and that subsequent evolution within the *Ananaspis* lineage was also by allometric neoteny (Fig. 6-5). This involved the retention of juvenile *Acernaspis* characters in descendant adult *Ananaspis*, such as relatively short eyes; deep palpebral furrow; more deeply incised glabellar furrows; pronounced tuberculation; and retention of pronounced occipital lobe. Examples of allometric neoteny in the Cambrian are rare, the only documented example being a species of *Xystridura* (McNamara, 1981).

The change in the nature of trilobite faunas across the Cambrian/Ordovician boundary is quite remarkable. Some 41 families of trilobites became extinct at the end of the Ordovician, while 32 new families appeared. Stubblefield (1959) thought that this "cryptogenetic" appearance of so many new families might have been caused by paedomorphosis. Evidence suggests that, on the contrary, the change was more likely due to a much higher incidence of the selection of peramorphic forms in the post-Cambrian. These new forms generally show greater morphological diversity than their predecessors, having evolved spines, or massive fringes around the head, or becoming very smooth and swollen.

Compared with Cambrian trilobites, the ontogenetic development of many post-Cambrian trilobites began at a more advanced morphological stage. For example, protaspids and early meraspids of the lichids *Amphilichas* and *Acanthopyge* are morphologically more advanced than similar growth stages in Cambrian trilobites (Fortey and Chatterton, 1988) (Fig. 6-6). The same is true for the protopygidia of the remopleuridids

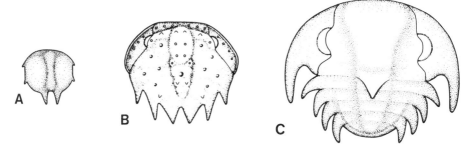

Figure 6-6. Protaspid (earliest postlarval) growth stages of (A) Middle Cambrian corynexochid *Bathyuriscus?*, (B) Early Devonian lichid *Acanthopyge bifide*, and (C) Early Devonian scutelluine *Dentaloscutellum hudsoni*, showing, at comparative growth stages, greater degree of morphological development in the post-Cambrian forms. Reproduced from Fortey and Chatterton (1988), with permission of the Palaeontological Association.

Menoparia and *Remopleurides,* which began postembryonic development at a later morphological stage than in asaphids. Similarly, the encrinurid *Balizoma* is morphologically very advanced in the protaspid stage, compared not only with the typical Cambrian protaspids, but also with other encrinurids. Edgecombe *et al.* (1988) have noted that these protaspids are conspicuously larger than comparable encrinurid protaspids. Whether they are showing predisplacement or the advanced morphological characteristics occurred as a result of a longer period of precalcified development, is not clear. This pattern is evident in many Silurian and Devonian trilobites. For example, Chatterton (1971) has commented on the "adult" appearance of the protaspis of *Dentaloscutellum* (Fig. 6-6). Morphologically, it is very advanced, with a broad, clavate glabella and posteriorly positioned eye lobes. Within the heterochronic framework, such earlier appearance of ancestrally advanced characters in descendants is predisplacement. Even if the subsequent rate of morphological development had been the same in the post-Cambrian forms as in their Cambrian precursors, the resultant adult will be more advanced morphologically, often strikingly so if certain structures grew with strongly positive or negative allometries.

A further factor that greatly influenced the morphological diversification of post-Cambrian trilobites was the increased frequency of allometric acceleration in morphological development, producing peramorphosis. Whittington (1981) had commented on the similarity between protaspids of many post-Cambrian forms, even though their adult morphologies show a range of variation, much greater than between Cambrian trilobites. This increased degree of morphological divergence during ontogeny may be explained in terms of increase in the frequency of occurrence of acceleration of allometries in many lineages. The effect of accelerating ancestral allometries to cause an increase in morphological divergence between adults can be demonstrated in phacopid trilobites. Not only are phacopid protaspids and meraspids in an advanced morphological stage due to predisplacement (Whittington, 1956, Plate 24, Figs. 1–5, 1957, Fig. 16; Chatterton, 1971, Fig. 19; Fortey and Chatterton, 1988), but increased positive allometry in lateral glabellar growth resulted in phacopids undergoing a much greater degree of morphological development than in most Cambrian trilobites and in their probable ancestors, the dalmanitids. Although both have similar meraspids, increased positive allometry in phacopids caused the great increase in glabellar growth by acceleration. This was accompanied by a general reduction in depth of glabellar furrows and in the size of the cheeks. Similar cephalic development occurred in the Asaphacea and Illaenacea, where extensive ontogenetic change in the shape of the glabella was accompanied by effacement. As in phacopids, illaenaceans underwent extensive morphological change during ontogeny (Ludvigsen and Chatterton, 1980).

There are many other similar examples of peramorphic change in the glabella in post-Cambrian trilobites. These changes are likely to reflect large-scale changes in the alimentary system and nature of the cephalic appendages. Thus, peramorphosis by both (allometric) predisplacement and acceleration in post-Cambrian trilobites resulted in great increases in morphological diversity and consequently exploitation of a much greater range of habitats. This is not to say that paedomorphosis ceased at the end of the Cambrian. Indeed, paedomorphosis has been proposed as a major factor in the evolution of the schizochroal eye in phacopid trilobites (Clarkson, 1971, 1975). But here, as in other examples, paedomorphosis involved local growth fields, rather than being global, as in Cambrian trilobites.

While it can be argued that internal changes in regulation of development were a major factor in the decline in dominance of paedomorphosis, the increased incidence of peramorphic forms as targets of selection in the early Ordovician may have been affected more strongly by extrinsic factors. The appearance of morphological characters, such as increased spinosity, and, at the other extreme, effacement, may both be considered to reflect the selection of cryptic morphologies (that is, morphologies that confer selective advantage against predation). Spinosity will have had the effect of increasing camouflage, while effacement combined with burrowing are both indicative of an increase in selection pressure induced by increased levels of predation. The early Ordovician saw an increase in dominance of cephalopods and fish, both of which are likely to have preyed upon trilobites. The preferential selection of the morphologies produced by peramorphic processes may therefore be the cause of the cryptogenetic appearance of so many new trilobite families early in the Ordovician.

A further development which became apparent in post-Cambrian trilobites was the increased frequency of occurrence of dissociated heterochrony, where some morphological features are peramorphic, but others paedomorphic. As we have stressed, such dissociated heterochrony has played an important role in the evolutionary diversification of many groups of organisms. In trilobites it has been well documented in some Silurian encrinurids by Edgecombe and Chatterton (1987). They have shown how rather than occurring in a haphazard manner, the effects of heterochrony in these encrinurids are ''regionally global.'' In other words, cephalic features are entirely paedomorphic, while pygidial features are peramorphic. They ascribe this phenomenon to dissociated hormonal regulation of development of each region. Little attention has been paid to this particular phenomenon in other organisms, but as we discuss in the section on hierarchy of heterochrony below, a somewhat similar pattern of heterochronic regional dissociation can be seen in some colonial organisms, notably bryozoans (Anstey, 1987).

A transformation from predominantly paedomorphic targets to peramorphic ones may not be unique to trilobites. Although the earlier part of the phylogenetic history of Mesozoic and Cenozoic echinoids has been less well studied than trilobites for the effects of heterochrony, small paedomorphs (possibly progenetic) similarly appear to have been

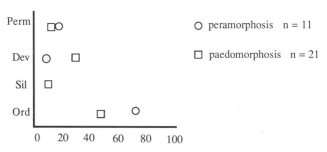

Figure 6-7. Relative frequencies of paedomorphosis and peramorphosis in Paleozoic bryozoans (data from Anstey, 1987, Tables 2 and 3).

primary targets of selection at this time. This was mainly centered on early Mesozoic "regular" forms, such as *Eodiadema, Mesodiadema, Serpianotaris,* and the families Saleniidae and Tiarechinidae. While in later "irregular" echinoids progenesis does occur (e.g., fibulariids and neolampadids), many of the changes in test architecture that characterize irregular echinoids arose from the selection of peramorphic traits. As in trilobites this resulted in the evolution of a much wider range of morphotypes that were adapted to a much wider range of habits. As we shall discuss below, the agent of selection that directed this evolution may well have been the same as that responsible for directing the evolution of post-Cambrian trilobites.

The ultimate reason for the different targets of selection between early Mesozoic and later forms arises from changes in the developmental system of later irregular echinoids. The outcome of this was a fundamental change in the patterns of growth of the echinoid test (McNamara, 1987a). While regular echinoids grow principally by the addition of new coronal plates, with only relatively minor subsequent size increase (allometries being low and of similar values among plates), irregular echinoids produce fewer plates, but these plates undergo greater allometric growth. Furthermore, growth allometries in different plates can vary appreciably (particularly in holasteroids and spatangoids), even between different axes within the same plate. Acceleration or hypermorphosis under these conditions produced a great diversity of morphologies that were adapted to a range of habits (principally infaunal) that were not open to occupation by their ancestors. Once again there is the production of suitable developmental changes allowing the attainment of adaptive peaks in a substantially new "volume" of ecological "hyperspace."

3.2. Extrinsic Factors Influencing Heterochrony

The Bryozoa are another group of invertebrates that provide evidence for changing frequencies of heterochronic styles over part of the phylogenetic history of the group. Anstey's (1987) analysis of heterochrony in Paleozoic bryozoans shows a predominance of peramorphosis in earlier forms (Fig. 6-7), in direct contrast to the situation in trilobites. Of 11 documented examples, 8 occur in the Ordovician, 1 in the Devonian, and 2 in the Permian (Anstey, 1987, Table 2). While 10 examples of paedomorphosis have been documented from the Ordovician, this represents only 47% of known examples. There are 2 examples in the Silurian, 6 in the Devonian, and 3 in the Permian (Anstey, 1987, Table 3).

The reason for the changes in heterochronic styles might be linked to changes in life habits of the bryozoans. Anstey has shown how peramorphosis is more common in offshore bryozoans, while nearer-shore forms show trends toward paedomorphosis. Recurrent trends toward paedomorphosis in later Paleozoic forms in trepostome bryozoans indicate the possibility of colonization of onshore habitats by descendants of offshore species. Upright branching and frondescent colonies, largely peramorphic in form, dominate Late Ordovician faunas in North America. A further aspect may be the particular morphological targets of selection, for as Anstey has revealed, different characters are affected by different styles of heterochrony. As we have discussed above, in the section (2.2) dealing with the hierarchical aspects of heterochrony, the frequencies of paedomorphosis and peramorphosis differ at the ontogenetic level of individual zooids and the (astogenetic level) colony as a whole. Paedomorphosis is more important at the colonial level, whereas peramorphosis is concentrated more on the individual zooids. Changing targets of selection through the Paleozoic may therefore have played a significant part in affecting the change in dominance of styles of observed ("successful") heterochronies.

4. SELECTION OF HETEROCHRONIC TARGETS THROUGH TIME

4.1. Equal Frequencies of Paedomorphosis and Peramorphosis

Those groups, just discussed, that show changing modes of heterochrony over time are, on the basis of available evidence, greatly outnumbered by those that show no obvious changes. In this latter category are two groups: (1) one that demonstrates roughly equal proportions of paedomorphosis and peramorphosis as "successful" targets of selection; and (2) another where one style consistently predominates. A critical factor in assessing into which of these two categories any group should fall is the available documentation of heterochrony within the group. This is highly variable and depends on whether paleontological or neontological material is being examined. A number of groups, in particular ammonites, and to a lesser extent fossil brachiopods and echinoids, have a long history of assessment for the frequency of heterochrony. Others, notably fossil gastropods and corals, have yielded few examples of heterochrony. Yet, as recent research on fossil gastropods by Geary (1988) and corals by Pandolfi (1988) has shown, this is more a function of the predilections of workers in that field rather than any innate developmental conservatism that inhibits heterochrony in these two groups of organisms.

Historically, as we have recounted, ammonites were thought to show only peramorphosis. Then they were thought to show only paedomorphosis. Work in recent years has shown, however, that both paedomorphosis and peramorphosis occur, and probably in similar proportions. However, even here there is every indication (Landman, 1988) that different sets of structures are targeted by different kinds of heterochrony. Global allometric progenesis has been documented in a number of Cretaceous ammonites (McNamara, 1988b, Table IV) and resulted in the evolution of many small taxa. Examples of allometric neoteny are less well documented, but do occur. While the specific designation of peramorphic evolutionary change has not often been made in ammonites in recent times, Landman (1988) has shown that many of the fundamental evolutionary changes affecting the architecture of the septal wall, producing the characteristic pattern of sutures on internal molds of the shells, can be attributed to peramorphic processes, often by the activity of more than one process simultaneously. The result is both a progressive increase in numbers of sutural lobes and saddles, plus an increase in "frilling." The recurrent theme in ammonoid evolution for increase in septal wall complexity is attributed to a strengthening of the outer shell by the production of stronger internal supports. This allowed the occupation of deeper-water habitats by improving resistance to higher hydrostatic pressures. Furthermore, it may also have served an antipredatory function (Hewitt and Westermann, 1987).

Other morphological features of the shell, notably shape and ornamentation, seem to show a tendency toward domination by paedomorphosis. For instance, the first appearance of fine ornamentation in *Hoploscaphites nicolleti* is progressively delayed in stratigraphically higher specimens (Landman, 1988). This therefore also shows the operation of these processes intraspecifically, as well as at higher taxonomic levels. While the dominance of peramorphosis in sutural development is likely to be real, the apparent dominance of paedomorphosis in these other features is likely to be an artifact. After all, if there had been evolution of shells with fewer ribs from ones with many, there is likely to have been, at a correspondingly earlier period in the phylogenetic history of the lineage, peramorphic evolution of multiribbed froms from ones with fewer ribs. Indeed, as many of the examples of paedomorphosis that have been documented in ammonites are from later Mesozoic specimens, it would be interesting to see if much earlier ammonoid lineages showed a predominance of peramorphosis in these characters.

4.2. Extrinsic Factors and the Dominance of Paedomorphosis

Recent research on heterochrony in gastropods has focused on both paleontological (Geary, 1988; Majima, 1985) and neontological examples (Lindberg, 1988). Geary has shown how in one lineage of the Miocene *Melanopsis* paedomorphosis occurred, while in another lineage of the same genus peramorphosis prevailed. Majima (1985), like Geary, has demonstrated both paedomorphosis and peramorphosis in the naticid *Glossaulax*. The impression from these two studies is that with further research, both peramorphs and paedomorphs will be shown to have often been targets of selection. In contrast to the few studies undertaken on fossil gastropods, an appreciable amount of work has been carried out on heterochrony in living gastropods. As Lindberg (1988) has observed, many workers have suggested that heterochrony has figured prominently in gastropod shell evolution, such as in the Zygobranchia (Garstang, 1929); pulmonates (Gould, 1969); and in the origin of the Scissurellidae, Eotomariidae, Temnotropidae (Batten, 1975), Fissurellacea (McLean, 1984), and mesogastropods (Robertson, 1985); Tissot

(1988) has demonstrated its activity both intra- and interspecifically in living Indo-Pacific cowries. Significantly, however, these examples demonstrate an overwhelming predominance of paedomorphosis. Lindberg's examination of heterochrony in the soft anatomy of gastropods likewise shows that paedomorphic characters were the primary targets of selection, notably the paedomorphic appearance of the alimentary system (where the stomach lost the gastric shield and style and the looped hindgut underwent shortening) and the trend for reduction in numbers of tooth fields in the radula.

Whereas the early dominance of paedomorphosis in trilobites has been argued to have been partly a function of intrinsic factors, in gastropods extrinsic factors may have been more significant. We have demonstrated in Chapter 4 the often quite extraordinary changes which occurred in some gastropods to achieve internal fertilization. Lindberg (1988) has likewise stressed the frequent incidence of trends toward a change from external fertilization and planktic, nonfeeding development to internal fertilization, passive or active brood protection, and feeding larvae. Such brooding species, as Lindberg points out, are often progenetic. Clearly the facility for progenesis is inherent in the gastropod's developmental system, as, no doubt, is hypermorphosis. But whereas progenesis induces a profound alteration in the organism's life-style that has been the target of selection in many lineages, hypermorphosis may produce no such fundamental phenotypic manifestation that is selectively advantageous to the organism.

The overwhelming dominance of paedomorphosis in brachiopods is demonstrated by the compilation made by McNamara (1988b, Table VI). Of 30 examples, 25 are of paedomorphosis, and of these 16 have been interpreted as being the result of allometric progenesis. Until data from growth lines are analyzed, we will not know if this represents "true" (age-based) progenesis. After all, small body size can also be produced by body neoteny and postdisplacement (Chapter 2; McKinney, 1988a). What is significant is that many of the examples do possess small body size. As we discuss below, this, or life history events associated with small size, might have been the main target of selection. This preponderance of paedomorphosis has resulted not only in the evolution of a range of genera and species, but also some suborders, such as the Thecideidinea. Paedomorphic traits will be important targets of selection when there is a distinct biphasic life history. In vertebrates, such as amphibians, this may be a juvenile aquatic phase and a terrestrial adult phase. In brachiopods it more likely reflects life history strategies, as we discuss below.

Among other major groups of invertebrates, the data are not sufficient to determine whether or not paedomorphs have been the dominant targets of selection. However, some recent detailed studies are beginning to pick up patterns in the frequencies of heterochronic styles at lower taxonomic levels. For instance, in a study of heterochrony in calceocrinid crinoids, Brower (1988) showed that the dorsal cup in most of the nine lineages that he studied showed paedomorphosis. However, many features of the arms showed what he termed "increases in number of branches" or "formation of elongate plates," both characters that are likely to have been the result of peramorphosis. Thus, a strong dissociation seems to have occurred in regional growth fields, suggesting that quite different selection pressures were operating on different parts of the organism. However, it is not unlikely that the viability of these lineages was in fact prerequisite upon just such a dissociation. A similar relationship between paedomorphic and peramorphic structures has been observed in bryozoans (Anstey, 1987).

4.3. Genome Size as an Intrinsic Factor in the Dominance of Paedomorphosis

If paedomorphic features dominate in some groups of organisms, it can be argued that this may be achieved not only by the dominant influence of extrinsic factors, such as we have demonstrated in gastropods and brachiopods, but by intrinsic factors. There is mounting evidence to suggest that intrinsic factors may have been crucial to the dominance of paedomorphosis in some groups of amphibians. In living amphibians, paedomorphosis has most often been recognized in plethodontid salamanders (e.g., Sprules, 1974; Larson, 1980; Alberch and Alberch, 1981). In many lineages, such as those of *Bolitoglossa, Batrachoseps, Aneides, Ambystoma,* and *Notophthalmus,* there appears to have been production of a number of paedomorphic targets, principally a short tail, fully webbed hands and feet, reduction and loss of some bones in the hands and feet, and general reduction in the degree of ossification in the skull. These paedomorphic changes occur not only by neoteny but also by progenesis. Progenesis is indicated by the small size of some amphibian genera combined with a shorter juvenile phase of development. Ecological factors selecting the paedomorphs include fluctuating environmental parameters, such as temperature, food, and habitat (Sprules, 1974). As ancestral plethodontids had a biphasic life cycle (an initial aquatic larval phase followed by metamorphosis to a semiterrestrial adult phase), paedomorphic evolution has produced genera that are restricted to the ancestral juvenile aquatic environment throughout their entire life history. Other genera have gone in the opposite direction and lost the juvenile aquatic phase entirely. As well as these global heterochronic changes, there are many cases of localized heterochronic changes, many of which (though by no means all) are paedomorphic (Wake and Larson, 1987).

The paleontological record is similarly rife with examples of paedomorphosis in amphibians. In a review of heterochrony in fossil amphibians, McNamara (1988b) found that 85% of examples showed paedomorphosis. Prominent among these were a number of families of Permian and Triassic temnospondyls (Boy, 1971, 1972; Milner, 1982; Warren and Hutchinson, 1988). If the small size of forms such as *Doleserpeton, Tersomius, Microbrachis,* and *Branchiosaurus* was a product of a truncated juvenile growth phase, then they are progenetic. *Microbrachis* never exceeded 300 mm in length, had lateral line sulci, gill ossicles, unossified carpals and tarsals, small limbs, and elongate trunk (Fig. 6-8). The shape of the skull was distinctly different in these forms from typical nonpaedomorphs, such as capitosaurids, which undergo pronounced allometric changes during ontogeny (Fig. 6-9). Other paedomorphic skull characteristics in temnospondyls include relatively larger eye orbits, positioned relatively anteriorly, and the rounded skull shape caused by an absence of elongation of the snout, characteristic of nonpaedomorphs. These are all features possessed by juvenile capitosaurids (Fig. 6-9).

If we are to accept that the overwhelming dominance of paedomorphosis over peramorphosis is a real phenomenon (and everything points to that being the case), then what are the relative extrinsic and intrinsic factors that have made paedomorphic structures the main target of selection in these groups for many hundreds of millions of years? Was de Beer (1958) correct in his belief that major evolutionary trends are more likely to have originated from small "less specialized" forms, such as paedomorphs? (See Chapters 5 and 8.) Ecologically, there was a great advantage to these amphibians in adopting a paedomorphic morphology. Many of the fossil amphibians were facultatively

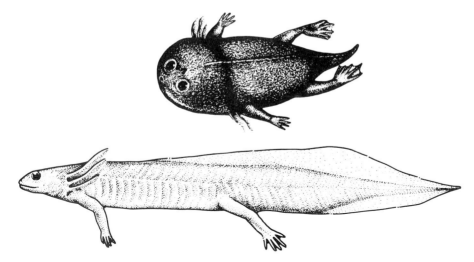

Figure 6-8. Examples of paedomorphic Paleozoic amphibians: *Gerrothorax* (upper) and *Branchiosaurus* (lower). *Branchiosaurus* redrawn from Milner, 1982 (Fig. 8).

or permanently paedomorphic relatives of small terrestrial temnospondyls, exploiting plankton-feeding niches in lakes (Milner, 1982). As we have shown in Chapter 4, extrinsic factors, such as variation in population pressures, are sufficient to generate paedomorphosis. Research in recent years has indicated that this "ease of going paedomorphic" has an underlying intrinsic cause. This research has focused on cell size, in particular the phenotypic effect of changes in the size of the eukaryote genome. It is known that much DNA in cells has no genic utility (Cavalier-Smith, 1985). But to what extent does the remaining "secondary DNA" (Hinegardner, 1976) contribute to the nature of the phenotype? In particular, what is the significance of variations in genome size to developmental rates, and thus to heterochrony?

The **genome size** is defined as the mass of DNA in an unreplicated genome (Hinegardner, 1976). It is also known as the **C-value**. There is a constancy in C-value within species, but huge interspecific variation. However, the variation in eukaryote genome

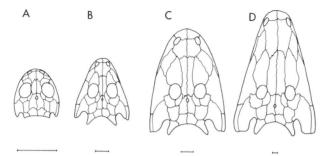

Figure 6-9. Diagrammatic drawings of changes in skull roof proportions in postmetamorphic growth of capitosaurid amphibians: A, at metamorphosis; B, juvenile; C, immature adult; D, mature adult. Bars = 10 mm. Reproduced from Warren and Hutchinson (1988), with permission of the Palaeontological Association.

size bears no relationship to organismic complexity; but there is a correlation between genome size and developmental rate (Cavalier-Smith, 1985). Four principal hypotheses have been proposed to explain evolution in genome size and the effect of these changes on rates of growth and development:

1. *Junk DNA*—it has been suggested (Ohno, 1980) that much of the genomic DNA is basically redundant, consisting of functionally inert sequences, reflecting ancient, nonfunctional relicts of gene duplication.

2. *Selfish DNA*—More recently it has been proposed that the sole function of secondary DNA is merely to increase the amount of DNA in the genome, so increasing its size (Orgel and Crick, 1980; Doolittle and Sapienza, 1980). It has been argued that there would be a progressive evolutionary increase in genome size, stopped only when the genome reaches such a size that it imposes no survival advantage to the organism, by restricting rates of growth and development.

3. *Nucleotypic DNA*—This implies that developmental rates may be determined by the size of the nuclear genome, in an inverse manner (Bennett, 1972).

4. *Skeletal DNA*—This regards genome size as being largely an adaptive feature, maintaining a metabolically favorable ratio between cell volume and nuclear volume. It proposes a generalized inverse relationship between evolutionary changes in size of the genome and developmental rates. As Cavalier-Smith (1978) describes it:

> . . . natural selection acts powerfully on organisms to determine their cell size and developmental rates (which are inversely related). The mean cell volume of an organism is the result of an evolutionary compromise between conflicting selection for large cell size and rapid developmental rates: the particular compromise reached for a particular species will depend on its ecological niche and organismic properties. Since larger cells require larger nuclei, selection for a particular cell volume will secondarily select for a corresponding nuclear volume, producing a close correlation between cell and nuclear volumes in different organisms. I suggest that the basic nucleotypic function of DNA is to act as a nucleoskeleton which determines the nuclear volume; small C-values are therefore required by small cells with small nuclei, and large C-values by large cells needing large nuclei. The DNA C-value of an organism is therefore simply the secondary result of selection for a given nuclear volume, which in turn is the secondary result of the evolutionary compromise between selection for cell size and for developmental rates.

Cavalier-Smith (1985) asserts that on the basis of available evidence, a positive correlation exists between eukaryote genome size and a number of quantitative characters. These include:

1. Total volume and mass of metaphase chromosomes
2. Cell volume and weight
3. Length of cell cycle
4. Duration of meiosis
5. Pollen maturation time
6. Minimum generation time in flowering plants
7. Time taken for embryogenesis from fertilization to the hatching tadpole stage in amphibians

However, most of these factors are not directly and mechanistically determined by genome size, but are mostly the result of indirect developmental and/or evolutionary correlations. The crucial factor is cell volume rather than genome size. Evolutionary variations in cell volume probably lead to the selection of correlated variations in these other characters.

While it has been argued that *large* C-values have often been the targets of selection, this may not always be the case. In fact, the three amniote classes—mammals, birds, and reptiles—are exceptional among eukaryotes in having small variation in their genome size (2- to 4-fold, rather than 10- to 100-fold as in other groups of multicellular animals and plants). Thus, if selection is on cell size, it favored small cell size predominantly in amniotes.

Frequently there is a strong (inverse) correlation between cell size and developmental rates. If so, cell size may be an important factor in determining modes of heterochrony, particularly strong biases toward any one style (such as paedomorphosis in amphibians). However, the relationship between genome size and rate of embryonic development is not always a simple correlation, as for instance in anurans (Oeldorf *et al.*, (1978). Sessions and Larson (1987) have suggested that large genomes appear to limit anuran embryos to slow development, but small genomes appear to allow a wide range of developmental rates.

Both lungfish and paedomorphic salamanders have exceptionally large genomes and a correspondingly high DNA content (Morescalchi and Serra, 1974; Morescalchi, 1979). Because of the positive correlation between genome size and cell volume, it has been possible to measure cell volume, and thus genome size, in some fossil material, allowing assessment to be made of long-term phylogenetic changes in cell volume. In amphibians and flowering plants there is an *inverse* relationship between developmental rates and cell volume. Horner and Macgregor (1983) measured the C-value in 18 species of amphibians and showed that species with the largest genomes take up to 24 times longer to reach a comparable state of development. They have argued that there will be a tendency for genome sizes to increase unless checked by natural selection. Likewise, Sessions and Larson (1987) compared the C-value of 27 species of plethodontid salamanders with developmental rates. These they established by analyzing rates of limb regeneration. C-values ranged between 13.7 and 76.2, almost a sixfold difference. Variation within species was low. Regeneration rates also showed about a sixfold difference between slowest and fastest. Plotting the two against one another (Fig. 6-10) shows an inverse relationship between C-value and regeneration rate. When the data were compared with a phylogenetic tree of the 27 species (Sessions and Larson, 1987, Fig. 2), 65% of lineages showed changes in genome size or growth rate. In 16 that showed large changes in genome size, 11 showed changes in growth rates; almost all showed an inverse relationship between the two. In terms of heterochrony, it could therefore be argued that paedomorphic organisms that have evolved by neoteny did so because of their extended developmental rates produced by longer cell cycles.

Animals that have large cells frequently have lower respiratory rates, while plants with large cells have greater shoot elongation rates. These may be important factors in favoring the evolution of larger cells and corresponding large genomes. Thus it could be argued that the paedomorphic nature of animals with large cells is in fact no more than a by-product of selection for lower respiratory rates, rather than direct selection for larger cell size. For sluggish animals that spend much of their time under starvation conditions, it is possible that the more economical energy metabolism made possible by larger cells

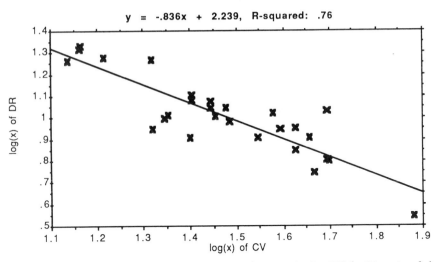

Figure 6-10. Log transformed plot of differentiation rate (DR) against C-value (CV) for 27 species of plethodontid amphibians, based on data from Sessions and Larson (1987, Table 2). Differentiation rate was computed from the number of developmental stages per day multiplied by 100.

may be enough to constitute a net selective force in favor of large cells. It has been observed (Szarski, 1983; Olmo, 1983) that amphibians with relatively large cells are generally torpid. Furthermore, they are paedomorphic and do not undergo metamorphosis to air-breathing adults. Cavalier-Smith has argued that plethodontid salamanders, which often have unusually high cell volumes, lack lungs and respire through their skins, may have lost their lungs in order to grow larger cells. However, the fish with the largest cell volume are lungfish, which undoubtedly do utilize lungs to respire. Like amphibians, though, lungfish are able to survive low levels of oxygen, supporting the idea that selection was for low metabolic levels, favoring large cell size. Thomson (1972) has presented data (Fig. 6-11) showing a steady increase in osteocyte volume during lungfish evolution.

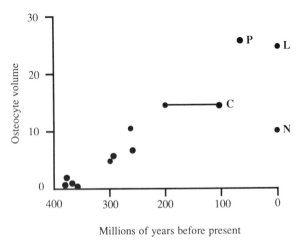

Figure 6-11. Changes in osteocyte volume during the evolutionary history of lungfish. Redrawn from Cavalier-Smith (1985, Fig. 4. 6). N, *Neoceratodus*; P, *Protopterus*; L, *Lepidosiren*. Data from Thomson (1972). The cell size of *Ceratodus* (C) remained constant for nearly 100 million years.

This increase may have arisen from selection of large cells that favored lower metabolic rates and corresponds in lepidosirenids with the evolution of estivation in the Permian. The modern lepidosirenids (*Lepidosiren,* 121 pg DNA per haploid genome, and *Protopterus,* 142 pg DNA per haploid genome) have a much larger cell size than other living lungfish that do not estivate (*Neoceratodus,* 80 pg DNA per haploid genome). Furthermore, *Neoceratodus* has shown no increase in cell size over the last 250 million years. Like plethodontid salamanders, the evolution of larger cell size in lungfish was accompanied by a host of paedomorphic changes. Bemis (1984) has shown that these include the loss of a heterocercal tail; fusion of median fins; reduction of fin rays; loss of cosmine; change in scale shape from rhombic to round; reduced ossification (Fig. 6-12). While there may only have been two periods of cell size increase in lungfish, several different groups of amphibians independently evolved larger cell volumes (Thomson, 1988).

And what of phylogenetic changes in cell size in other groups of organisms? Perhaps one obvious group to investigate are single-celled organisms, such as the Foraminifera, which have an extensive fossil record. Many lineages of Tertiary foraminifers show trends of size increase (Boltovsky, 1988). But as to correlation between this and developmental rate, little is known. If cell and genome size in foraminifers was also indirectly correlated with developmental rate, then there should be numerous examples of paedomorphosis and a preponderance of this in the fossil record. Unfortunately, the state of knowledge of heterochrony in this group is very poor. But what little work has been done suggests the possibility that this style of heterochrony may be common. For instance, paedomorphosis has been demonstrated in one lineage of *Globorotalia* (Scott, 1982), while it appears to have played a dominant role in evolution within the Textulariina and in the evolution of *Haplophragmoides* (Brasier, 1982).

Of the small number of examples of heterochrony documented in another group of microfossils, the Conodonta, all except one show paedomorphosis (McNamara, 1988b). However, on such a small sample size (four) little weight can be placed. A recent attempt to assess changes in genome size in conodonts has been made by Conway Morris and Harper (1988). Measuring epithelial cell sizes that ranged over a period of 270 million years, they found initial cell sizes to have been small, but static for the first 50 million years, even while phenotypic diversity was increasing. Later cell sizes are larger, but show a wide size range, providing no firm support for the premise of a general increase in genome size in multicellular organisms.

5. SHAPE AS A TARGET OF SELECTION

5.1. Introduction

Selection can act directly on size, shape, total developmental time, or timing of certain developmental events (Fig. 6-13). In this section we focus on shape as the principal target of selection. We have discussed in earlier chapters how developmental mechanisms place potential intrinsic constraints on morphological evolution. The result of this developmental canalization is the production of patterns of **morphologic covariation**. Shape changes often covary with size changes. However, selection for shape alone

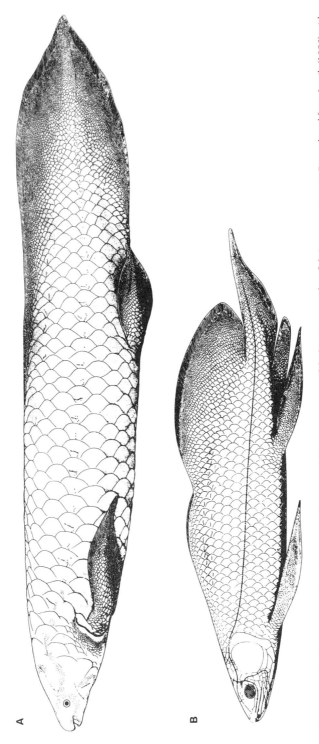

Figure 6-12. (A) The living lungfish, *Neoceratodus forsteri*, and (B) reconstruction of the late Devonian lungfish *Scaumenacia curta*. Reproduced from Jarvik (1980) with permission of Academic Press.

rarely produces a change in size (Bonner and Horn, 1982). Suites of traits often covary, with the result that while selection may be targeting just one trait, others are "dragged" along by covariation. The outcome is that these traits are "aimed" in a certain direction (Chapter 8; McKinney, 1988b). While such "dragged" traits may be initially nonadaptive, they may be exaptations and so subsequently become targets of selection.

Recently, a number of authors (Lande, 1979; Atchley *et al.*, 1981; Cheverud, 1982) have discussed the claim that patterns of phenotypic change may not reliably reflect underlying genetic patterns. However, these studies by and large focus on adult traits alone and not the developing organism as a whole. Research by Shea (1985b) on primates indicates that the similar patterns of phenotypic growth seen in African apes and their extension to larger sizes are likely to reflect underlying patterns of genetic allometry and selection for increased rates of body weight growth, larger global size, and extension of many ancestral ontogenetic allometries. Thus, during development, different morphological structures that arise share the same ontogenetic history. Genetic variation in a developmental precursor is likely to generate **genetic covariance** between traits (Riska,

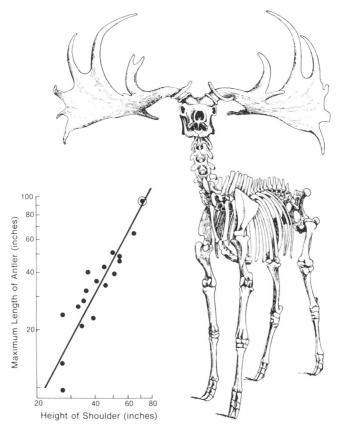

Figure 6-13. Relationship between body and antler size in cervine deer. Antlers generally become proportionately larger as they develop with positive allometry. The largest antlers were possessed by the Pleistocene Irish Elk, *Megaloceros* (circled dot on graph), and are a classic example of hypermorphosis. From *Evolution* by Theodosius Dobzhansky *et al.* ©1977 by W. H. Freeman and Co. Reprinted with permission.

1986). However, there is evidence (Atchley, 1984) to indicate that phenotypic and genetic covariation may change during postnatal growth, perhaps reflecting changes in the developmental processes responsible for regulating patterns of covariation (Zelditch, 1988). In a recent work, Cheverud (1988) has suggested that phenotypic correlations are in fact likely to be fair estimates of their genetic counterparts.

In any case, covariation can produce a suite of traits, some of which may well be suboptimal (Chapter 8). The time of development when selection acts can strongly affect evolutionary change of traits (Zelditch, 1988). Zelditch has pointed out that if selection acted upon rats as they were undergoing weaning, the jaw and cranium could be affected separately. However, if selection acted upon the adult, the jaw could not be modified without concomitant effects on the neurocranium.

In terms of heterochrony, the effects of the six fundamental processes on selection of shape changes can be considered in two groups. First are *local changes* where there are structural changes in particular traits, or suites of traits, but no global size changes. Second are *global changes* where there is integrated size and shape change (e.g., global progenesis and hypermorphosis). Where there is such global heterochronic change, the alteration of maximum adult body size will cause shape changes in local traits if these are following a nonisometric ontogenetic trajectory. Attempts to link life-history strategies with heterochrony (Gould, 1977; McKinney, 1984, 1986, 1988b) have concentrated on this latter group. Other studies, of the functional significance of structural heterochronic changes, have focused on the former group where only individual or suites of local growth fields are affected by heterochronic change. Although we shall consider these two groups in different sections, it must always be remembered that organisms are subject to a wide variety of selection pressures so there is some element of artificiality in such a distinction. Even where global body size is the dominant target of selection, covariation will produce trait changes that may also have secondary functional significance, i.e., they are "aimed" as we have said. Indeed, the importance of such secondary trait changes in such a situation places a strong element of constraint on selection of size alone.

Even the distinction between "size" and "shape" that featured so prominently in Alberch and co-workers' (1979) classification of heterochrony, is fundamentally artificial, as we have discussed in Chapter 2. In this chapter, *shape* change refers to any *local growth field* change (teeth, feathers, limbs, any field or organ) and *size* change refers to *global* (whole-body) change and its attendant allometric by-products. In our discussion of shape as a target of selection we include all changes in multidimensional parameters, e.g., changes in mitotic growth, which result in changing allometries.

Identifying heterochronic structures as adaptive and the sole targets of selection can only be done when there is no phylogenetic size change. Local growth fields alone are the targets of selection. It is also important to be able to look at the nature of the new adaptive peak, i.e., the suite of shape changes in the descendant form, and interpret the ecological niche into which the adaptive peak "fits." We use the rest of this section to illustrate a variety of examples from an ecological perspective, including the reinterpretation of some classic evolutionary studies in a heterochronic framework. We shall show how the attainment of new adaptive peaks is constrained by three fundamental factors: (1) the nature of the organism's developmental strategy, (2) the kind(s) of heterochrony, and (3) the nature and breadth of potential ecological niches. These will be examined in

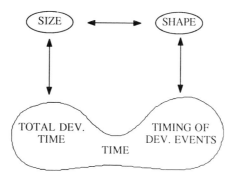

Figure 6-14. Diagram showing the relationship between size, shape, and time. Selection acts on all four corners of the diagram. Redrawn from Bonner and Horn (1982, Fig. 1).

terms of adaptations for feeding, for occupation of particular habitats, and for locomotion.

A prime factor that we highlight in this chapter is the importance of heterochrony in the interspecific evolution of new morphotypes able to colonize new habitats. Speciation occurs with the appearance of novel morphotypes that are adapted to a novel habitat niche that has the potential of providing an immediate obstacle to reproduction. Once this genetic isolation is accomplished, the establishment of a new species may ensue. Herein lies one of the most important, but largely ignored, aspects of heterochrony as an evolutionary *process*. By the production of novel morphotypes, or changes in reproductive timing, genetic barriers can be created by intrinsic, *not* extrinsic, factors. In the classic speciation models, such as that of allopatric speciation, extrinsic factors are seen as the prime agents in producing barriers to gene flow. However, the part played by heterochrony as the principal intrinsic factor is to create novel morphotypes, which function differently from their ancestors. The resulting ecological or behavioral separation may not result in subsequent geographic isolation in a new habitat. Thus, the restriction of the concept of allopatric speciation to physical, geographical separation is too simplistic in this view. Genetic isolation may be achieved sympatrically, in geographic terms, by the action of heterochrony. We shall elaborate on this when we deal with the importance of reproductive strategies as targets of selection (see below).

As with other evolutionary studies, interpretations of the adaptive significance of heterochronic structures have in the past generally focused upon the functional utility of specific morphological traits. Thus, the massive horns of the Irish Elk, *Megaloceros* (Fig. 6-14), would be seen as having evolved because they somehow improved the species' fitness, having been primary targets of selection. But as Bonner and Horn (1982) noted, the covariation of size and shape caused the large antler size and its distinctive shape, as body size increased. The peramorphic increase in sutural complexity in ammonoids that we discussed earlier is viewed as an adaptation that allowed descendants to inhabit deeper-water niches than their ancestors (Hewitt and Westermann, 1987), the extra convolutions increasing shell strength. The scientific literature abounds with a myriad of such examples (McNamara, 1988b), a few of which we shall elaborate on below. But can all morphological changes induced by heterochrony be viewed solely in this reductionist light? Does each part that is affected serve some functionally "useful" purpose in the descendant, molded by the pressure of natural selection?

In their "critique of the adaptionist programme," Gould and Lewontin (1979) criticized many such studies that viewed each morphological trait of an organism, whether or not it was viewed in heterochronic terms, as adaptively significant. More importantly, they stressed the need for an understanding of the underlying mechanism(s) for size and shape changes (for as they say, "male tyrannosaurs may have used their diminutive front legs to titillate female partners, but this will not explain *why* they got so small"). Organisms must be viewed as whole, functionally integrated entities. Many morphological structures, they argue, can be construed as nonadaptive, being merely by-products of changes that have affected other structures or the organism as a whole. Thus, the Irish Elk was interpreted by Gould (1974) as having evolved by hypermorphosis. Delayed juvenile offset resulted in prolongation of fast juvenile growth rates. Consequently, the strongly positive allometric growth of the antlers resulted in the evolution of a large animal with proportionately greater horns. Selection was acting on large body size, *not* on large antlers, *per se*. Of course, this is not to say that the horns did not come to serve a functionally useful purpose, as we shall elaborate below. In studies of heterochrony made during the last 10 years, this view that other aspects of an organism's biology, such as life history strategy or size, may also be targets of selection has become more widespread. The more classical adaptive interpretations are being viewed as unfashionable. Because of the central role that heterochrony plays in such changes, this issue must be addressed in any attempt to gain an understanding of targets of selection.

While a number of studies (Gould, 1977; McKinney, 1984, 1986, 1988b; Hafner and Hafner, 1988) have suggested that certain examples of heterochrony may be interpreted as selection for traits other than specific morphological ones (discussed below), we believe that it is still possible to make a case for different morphological traits being affected by different heterochronic processes. Some traits will confer a functional advantage to the descendant, allowing it to occupy a niche different from that of its ancestor, while other morphological changes may have conferred no adaptive advantage to the descendant, being merely by-products of other changes in life history strategies. In this section dealing with shapes as targets of selection, we shall be concentrating on the growing body of data that focus on the importance of dissociated heterochrony. Furthermore, we present guidelines for assisting in interpreting whether particular heterochronic traits are likely to have been the specific targets of selection. In order to make any meaningful statements about an organism's "adaptiveness" to its niche, it is important that we first discuss in brief the concept of adaptation and define "grades of adaptiveness."

5.2. The Concept of Adaptation

The concept of adaptation is one of the cornerstones of evolutionary theory (Goux and Roubaud, 1978; Dunbar, 1982; Krimbas, 1984). However, there is no clear agreement on just what adaptation actually is, many different interpretations having been proposed (Dobzhansky, 1968; Bock and Von Wahlert, 1965; Lewontin, 1957, 1977;

Stern, 1970). The term has been used in three fundamental ways (Medawar, 1951, 1960; Dobzhansky, 1968; Stern, 1970):

1. The property possessed—referring to the general or specific property, trait, feature, or characteristic possessed by an organism or population (known also as an *adaptive trait*)
2. The state of being adapted (*adaptiveness*)
3. The process of becoming adapted

Neo-Darwinists apply the term in two fundamental ways, referring either to a feature of an individual, or alternatively to a group. In this way it has been used from the molecular to the ecosystem level (Goux and Roubaud, 1978), and from this has arisen some confusion over the meaning of the term. Within the individual it has been used as a general property of the organism (Dobzhansky, 1942; Simpson, 1953; Lewontin, 1957; Mayr, 1963; Bock and Von Wahlert, 1965), or as an attribute of specific traits.

The concept of adaptation is frequently intermeshed with function. As Waddington (1960, p. 382) notes:

> Fishes are admirably designed for swimming, birds for flying, horses for running, snakes for creeping, and so on, ...

Roughgarden (1979, pp. 4–5) describes adaptation in functional terms this way:

> An adaptation is a trait that permits an organism to function well in its environment, a trait that endows an organism with capabilities especially appropriate in its particular environment.

To some authors, adaptation and function are even considered to be one and the same thing (Ruse, 1971, 1972).

Adaptation is often linked directly with the concept of "fitness," an adaptive trait facilitating the survival and reproductive capability of an organism. Dobzhansky, who in his writings consistently equated adaptation and fitness, also viewed adaptation in terms of development. He wrote (Dobzhansky, 1956, p. 347):

> An adaptive trait is . . . an aspect of the developmental pattern which facilitates the survival and/or reproduction of its carrier. . . .

Adaptation is generally seen as a measure of evolutionary success. A more "well-adapted" individual is ecophysiologically more efficient than most other members of the population (Mayr, 1982). Consequently, it will have greater survival potential and higher reproductive success. It has been argued that if traits have particular functions that in combination improve the organism's fitness, then a nexus must exist between adaptation and selection. In recent years there has been an attempt to break this nexus by those who consider that evolution can involve selection of new traits that might not be adaptive in performing specific functions that are advantageous to the survival of the new organism (Gould and Lewontin, 1979)—in other words, selection without adaptation. Conversely, they argue that adaptations can exist with no selection. Such adaptations are extrinsically induced; e.g., the different phenotypic colonial forms of corals that can develop under different flow regimes (Krimbas, 1984), although even here it could be argued (as we have in Chapters 4 and 5) that the very nascent plasticity that allows such adaptations to appear could itself have been the target of selection.

In neo-Darwinian writings an organism's attainment of the state of being adapted, of having achieved adaptation, has all the overtones of a quest for nirvana—the realization of a state of perfect being. Gould and Lewontin (1979) likened it to Voltaire's character in *Candide,* Dr. Pangloss, to whom this world is the best of all possible worlds and everything in it is of the best. They strongly criticized what they term the "adaptionist programme" in which organisms are "dissected" into parts, each of which is assigned a functional interpretation that is thought to confer improved fitness to the organism. As Dr. Pangloss so aptly put it (quoting from Gould and Lewontin, 1979, p. 149):

> Things cannot be other than they are. . . . Everything is made for the best purpose. Our noses were made to carry spectacles, so we have spectacles. Legs were clearly intended for breeches, and we wear them.

While this expresses a somewhat extreme viewpoint, the point is clearly made—a functional/adaptive interpretation should not be sought for every trait. Many features of an organism may have originated merely due to constraints imposed by the nature of the organism's architecture (spatial contingencies)—its **Bauplan**—and developmental (temporal) contingencies. While this view has been criticized (Mayr, 1983), the concerns expressed by the nonadaptionist school have highlighted the often nonscientific aspect of the adaptionist program and broadened our horizons in the quest for understanding the targets of selection. Mayr (1983) has expressed the view that the target of selection is always a whole individual, which is a developmentally integrated whole. Thus, he sees adaptation as a necessary "compromise between the selective advantages of different organs, different sexes, different portions of the life cycle, and different environments" (p. 331).

As we show below, when heterochronically induced size change is a target of selection, heterochrony may play a significant role in producing such new traits that are not themselves adaptations. Take, for example, the case of the human chin (Lewontin, 1978). The chin arises from the regression of two growth fields of the lower jaw. The alveolar (on which rest the teeth) regresses at a higher rate than the dentary (the lower part of the jaw), resulting in the development of the chin. The chin itself was neither a target of selection, nor an adaptation that existed for the sole purpose of allowing Mr. Darwin to demonstrate his hirsuteness. It was merely a by-product of this differential regression of parallel growth fields from selection for smaller teeth (as our diet became softer). A significant aspect of the nonadaptionist view then is the covariation of allometric fields. Because of the pivotal position of allometry in heterochrony, there is the prospect of viewing many heterochronic changes as nonadaptive. We shall present arguments here that heterochrony can be involved both in the selection of specific adaptive traits (i.e., shape as a target of selection) and also in the evolution of nonadaptive traits (i.e., when heterochronically induced body size change is a target of selection and shape changes occur as a by-product).

5.3. Grades of Adaptiveness

In terms of adaptiveness, a trait can be considered as: (1) *adaptive,* (2) *nonadaptive,* or (3) *maladaptive.*

1. An *adaptive* trait is one in which a morphological change produced by a heterochronic process resulted in the evolution of a functionally "useful" structure. That is, it contributes toward the occupation of the ecological niche, by conferring improved reproductive capability or survival potential. This might be achieved by change in a single trait, or of a suite of traits. Adaptive traits must be seen as being relative phenomena, in which degrees of adaptiveness can exist (Huxley, 1963; Clutton-Brock and Harvey, 1979).

2. A *nonadaptive* trait arises from a heterochronic process that confers no functional advantage to the organism. Neither, however, does it confer a disadvantage. This raises the knotty question of the extent to which a so-called adaptive trait is merely in the eyes of the beholder, or interpreter. In the case of the Irish Elk Gould (1974) has argued that the principal target of selection was body size. The large antler size he considered to be just an ancillary effect of its positive allometry; extend the growth period and end up with relatively much larger antlers. Thus, the antlers could be viewed as being nonadaptive structures. However, as we have pointed out there can be degrees of adaptiveness. Who is to say that the very large antlers did not also serve an adaptive "purpose"? Let us suppose that in this group of cervids, antler growth was not positively allometric with respect to body size, but isometric. While the Irish Elk would have had larger antlers than its presumed smaller ancestor, they would have been relatively appreciably smaller than in the realized Irish Elk, by virtue of the isometric growth. Would this putative Irish Elk still have been as viable as the realized form? Indeed, Geist (1986) has argued that antler size does not change as a simple allometric consequence of total body size. He notes that some moose have relatively smaller antlers than would otherwise be expected from just extensions to allometric growth. There is no doubting the behavioral significance of antler size in cervids. For instance, Clutton-Brock *et al.* (1982) have discussed the importance of antler size to sexual selection. Furthermore, Clutton-Brock and Harvey (1979) have suggested that positive allometry may have been the result of direct selection for changing antler proportions, in response to shifts in mating structure. While larger body size in the Irish Elk might have conferred some selective advantage, it may have been that it was the combination of body size *and* much larger antler size that tipped the scales in favor of the evolutionary success of this species. In the long term it could have been (in fact probably was) the large body size that was deleterious to the organism under rapidly changing (warming) environmental conditions, not its large antlers.

While some structures might be nonadaptive in one situation, they might become adaptive in another—this is a so-called *exaptation* ("preadaptation"). In heterochronic terms all structures can be considered to be exaptations when present only in an ancestral juvenile, if they become fixed in the adult stage of a descendant paedomorph and confer a functional advantage. In their ancestral juvenile forms they may be nonadaptive, only becoming adaptive in the descendant adult.

Extrinsic factors may also dictate whether or not a structure is an exaptation. For example, consider a situation where at time T1 a hypothetical organism has a small mouth adapted for feeding on fine-grained sediment, and four spines that function as effective antipredator devices (Fig. 6-15). By acceleration, two descendants, **A** and **B** evolve at time T2: **A** with a larger mouth and six larger spines; **B** also with a larger mouth, but spine size and number not having been accelerated. The larger mouth enables the new forms to occupy a vacant niche of larger sediment grain size. The extra spines that **A**

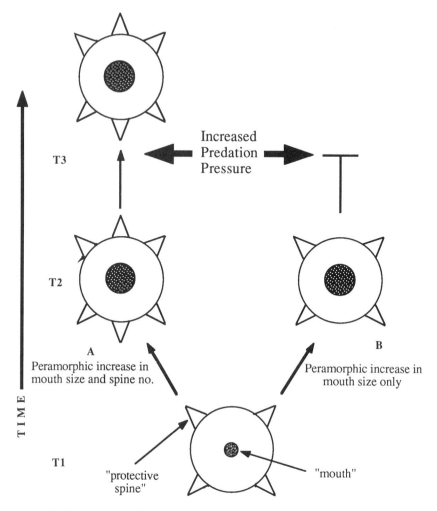

Figure 6-15. The effect of extrinsic factors on exaptations. At time T2, form A has evolved a larger mouth and more protective spines by peramorphosis. Form B evolved only the larger mouth. Increased predation pressure at time T3 meant that only form A survived, by virtue of its greater spine number. Thus, the possession of six spines was an exaptation.

has evolved confer no functional advantage as predation pressure is the same as that encountered by the ancestral form. However, with the introduction into the niche of a new, larger predator in time T3, form **B**, which retained the ancestral spine morphology, suffers much higher levels of predation from the new predator and becomes extinct. However, form **A** has sufficient spines of a large enough size to enable it to survive the new predator. Peramorphic increase in spine number at time T2 was therefore an exaptation and did not occur in response to increased predation pressure. Herein lies one of the major features of heterochrony, the fact that slight alterations in developmental rates or timing of maturation can affect a whole suite of structures, producing a large pool of potential exaptations. The role of extrinsic selective forces will therefore be only to sift out unsuitable morphologies that already exist: intrinsic change first, followed by extrinsic selection.

3. A *maladaptive* structure is one that is deleterious to the organism. Thus, in the scenario outlined above, should a form **C** have evolved at time T2 that had an accelerated mouth, but neotenic spine reduction, it would have suffered immediate extinction by predation pressure. An example of a maladaptive structure has been recorded in echinarachnid echinoids by Beadle (1989). He considers that retention of juvenile echinarachnid apical posterior eccentricity into the adult stage by paedomorphosis resulted in the evolution of *Dendraster*, probably from *Echinarachnius*, because it allowed this echinoid to feed in an upright position, rather than in the normal horizontal position. He cites the example of a specimen of *Echinarachnius parma*, originally recorded by Tower (1901), in which the normal ontogenetic pathway of posteriorly eccentric to centralized apical system had been accelerated such that the apical system had become anteriorly eccentric. While retention of a posteriorly eccentric apical system into the adult phase can be of adaptive significance, an anteriorly eccentric system is maladaptive.

Vestigial and rudimentary organs can also come into this category. Many of these are essentially paedomorphic, generally due to extreme postdisplacement. In humans, perhaps the most well known (because of its frequent maladaptive nature) is the appendix.

"Monstrous" teratologies can also be included within this category. As Alberch (1989) has observed, monsters are maladaptive and nonfunctional. However, they are still morphologically highly organized. Moreover, the development of maladaptive monsters does not occur in a haphazard way, but is also constrained by the organism's developmental system, as the *Echinarachnius* specimen illustrates. Thus, while two-headed monsters are not that uncommon in a wide range of vertebrates (e.g., snakes), three-headed monsters have not been documented. This is because trifurcations of growing cartilage are very difficult to achieve. Bifurcations, on the other hand, play a major role in vertebrate development as vertebrates have bilateral symmetry (Alberch and Gale, 1985).

5.4. Feeding Adaptations

Perhaps one of the most famous examples of the evolution of adaptations to different feeding strategies is that involving "Darwin's" Galápagos finches. Having been touted for so long as a classic example of the evolution of adaptations, there has been little attempt to explain the underlying mechanisms involved in generating the various beak shapes that allow the different species of finches to occupy different ecological niches. On his visit to the Galápagos Islands in 1835, Darwin wrote of 13 species of finches showing a great range of beak sizes and shapes, from species such as *Geospiza magnirostris*, which possesses a massive beak, to the slender-beaked *Camarhynchus parvulus*, with all grades in between. As Darwin (1845) wrote:

> Seeing this gradation and diversity of structure in one small, intimately related group
> of birds, one might really fancy that from an original paucity of birds in this archipe-
> lago, one species had been taken and modified for different ends.

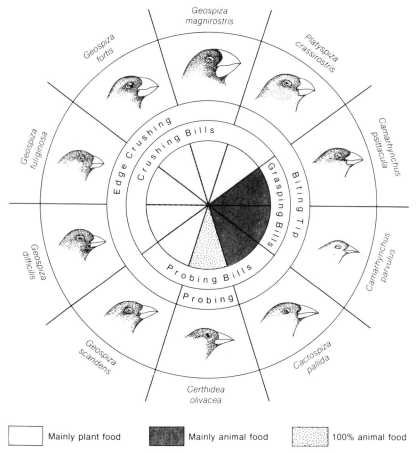

Figure 6-16. Variation in bill shape in ten species of Geospizinae from the Galápagos Islands. Variations in bill allometries allowed a wide range of food types to be utilized. From *Evolution* by Theodosius Dobzhansky *et al.* ©1977 by W. H. Freeman and Co. Reprinted with permission.

The general interpretation of such a diversity of beak structures is that (Dobzhansky, 1977):

> The size and conformation of the beak are adaptively adjusted to the kind of food a bird is dependent on.

Thus, in classic neo-Darwinian terms, extrinsic factors molded the shape of the bird's beak to allow it to feed most effectively from the kind of food it needed. This "cart-before-the-horse" approach presupposes that the birds needed to evolve a particular beak form in order to feed more effectively. This placement of emphasis solely on extrinsic factors results, we believe, in a false picture of the nature of the evolutionary processes that produced such a wide range of adaptive strategies.

Such changes in beak shapes and sizes involve changing allometries of beak length and beak height, relative to head size. The evolution of massive beaks, such as in *Geospiza magnirostris,* and to a lesser extent, *G. fortis* and *Platyspiza crassirostris,* arose from increased allometries of beak height (Fig. 6-16), probably as a result of a combination of acceleration and hypermorphosis in these larger birds. It was this intrinsic alteration of developmental pathways that produced a novel morphotype capable of establishing a new adaptive peak. Such beaks were admirably adapted to crushing. As suitable food sources were available, i.e., an appropriate feeding niche for this adaptive peak, then this became a viable morphotype. Developmentally it might have been just as feasible for a highly elongate beak to evolve; but if a suitable feeding niche was not present, then the structure was maladaptive. As we shall show, however, the evolution of the same structure in a different situation might have led to it being adaptive if a suitable feeding niche was available. Without the innate allometric capacity to alter beak growth in certain ways, the adaptive radiation in beak diversity would not have occurred in these birds, even given the availability of appropriate, unfilled niches.

Within species there is evidence to indicate that the dominant target of selection is beak size rather than beak shape, suggesting that at different hierarchical levels the targets of selection may show subtle changes: at a higher taxonomic level shape is favored; at a lower level, size. This has been elegantly demonstrated by studies of intraspecific variation in traits of *G. fortis*. During periods of drought, when food resources are low, selection favors larger birds that have correspondingly larger beaks. Larger beak size is a more important target than larger body size, because when food is scarce only birds with large beaks can crack open the remaining large, hard seeds, the smaller, softer seeds having already been depleted (Gibbs and Grant, 1987). Clearly the attainment of larger beak size at the intrapopulational level occurred by the attainment of increased body size. Body size increase therefore became a vehicle for the production of a larger-sized local trait that was the target of selection.

A similar scenario of allometric changes in beak shape and size by heterochrony can be proposed for the radiation of Hawaiian honeycreepers, the Psittirostrinae (Bock, 1970). Changing allometries involved:

1. Upper beak height
2. Upper beak length
3. Lower beak height
4. Lower beak length

Most trends involved peramorphosis by acceleration in growth of one or a combination of these four parameters. For instance, acceleration in upper beak length led to the evolution of *Hemignathus lucidus* and *H. wilsoni,* whereas acceleration in both upper and lower beak lengths resulted in the evolution of *H. obscurus* and *H. procerus.* Species with these curved, highly elongate beaks are adapted to feeding primarily from nectar, whereas their presumed progenitors, species of the genus *Loxops* that possessed shorter beaks, are insectivores (Fig. 6-17).

Acceleration in upper beak height alone produced the genus *Pseudonestor.* Acceleration in upper and lower beak heights produced species of *Psittirostra*, which possess a stout, parrotlike beak, convergent upon the Galápagos finch *Geospiza magnirostris*, because the same growth parameters were subject to change. A similar feeding niche accommodated this beak shape—crushing plant material. If Bock's (1970) interpretation of the phylogenetic relationships of the Psittirostrinae is correct, then paedomorphic trends can also be seen in the evolution of species such as *Loxops maculata,* where there has been a reduction in rate of beak growth.

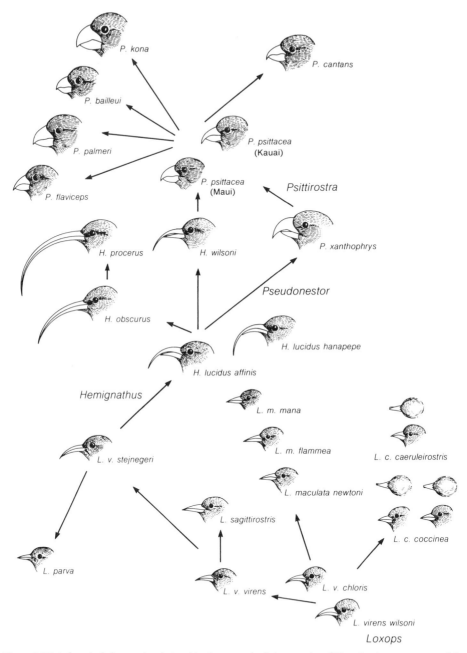

Figure 6-17. Inferred phylogenetic relationships between the living species of Hawaiian honeycreepers of the subfamily Psittirostrinae, showing the great diversity of beak shapes produced by variations in allometric growth. From *Evolution* by Theodosius Dobzhansky *et al.* ©1977 by W. H. Freeman and Co. Reprinted with permission.

In both the finches and honeycreepers, strongly dissociated heterochrony aimed at the same target (beak height) produced similar morphologies. However, the greater plasticity in growth of the honeycreeper's beak, particularly its length, provided a greater number of heterochronic products and so generated a wider range of morphotypes.

The parallel evolution of similar beak morphologies occurs because of the constraints imposed by the three dominant factors of the developmental strategies, heterochronic processes, and feeding niches. And because of the operation of these same factors in other geographical areas, further examples of convergent evolution can be seen. Thus, an adaptive radiation in Australian honeyeaters occurred in an analogous way to the Hawaiian honeycreepers, with allometric extensions in beak length producing long-billed honeyeaters. Allometric extensions in beak height produced Friar-birds. As with the Hawaiian honeycreepers, the attainment of these adaptive peaks is coincident with nectar-feeding niches for the long-billed forms and fruit for the Friar-birds (although both groups can also take insects). Honeyeaters that retain the ancestral small beaks, such as the Striped Honeyeater, are entirely restricted to a diet of insects. While little work has been carried out on beak growth rates in Australian honeyeaters, it is clear that hatchlings of all species have relatively small beaks (R. E. Johnstone, personal communication), and that speciation into different adaptive peaks has occurred by differential beak allometries. In all these three groups of birds, the dominant trend has been one of peramorphosis, which has allowed the exploitation of a wide range of feeding niches.

Not only do parallel patterns occur among families within classes, but the same three factors can be seen to operate across classes. Differential rostral (upper jaw) and mandibular (lower jaw) growth rates in Jurassic ichthyosaurs (McGowan, 1986) led to spectacular morphological and ecological differences among different genera. Whereas the study of bird beaks was restricted to living examples, and outgroup comparisons were necessary in order to establish polarity, stratigraphic information has allowed heterochronic polarity to be established for the ichthyosaurian lineage. Compared with the older *Ichthyosaurus*, the rostrum of the younger *Excalibosaurus* has a higher allometric coefficient compared with body size, presumably as a consequence of an increased rostral growth rate (i.e., rostral acceleration, based on the assumption that body size increases in the two genera were similar). The peramorphic rostrum was retained in the stratigraphically younger *Eurhinosaurus*; however, dissociation of the mandibular growth rate, leading to a reduction in allometric coefficient of this parameter, resulted in a form with very long upper jaw, but very short lower jaw (Fig. 6-18). As McGowan (1986) has noted, the biological consequences of this would have been profound and are likely to have led to the establishment of different feeding strategies. The principal result of the changes in jaw growth rates would have been the generation of different numbers of teeth. The longer the jaw, the greater the number of teeth that can be accommodated. Here we have an example whereby changes in growth allometries in one structure (the jaw) can induce complementary heterochronic changes in meristic structures (tooth production). If the increases in jaw length occurred by acceleration, then increase in tooth number would likewise have been as a result of acceleration in tooth production. Similarly, paedomorphic reduction in mandibular length would have produced a concomitant paedomorphic reduction in tooth number.

Such changes in tooth number have been documented in other extinct reptiles, and

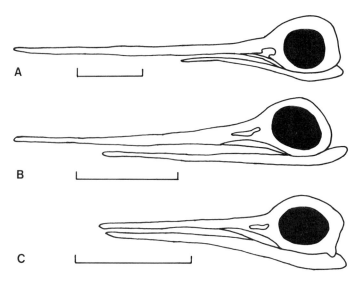

Figure 6-18. Allometric variation in rostral (upper jaw) and mandibular (lower jaw) growth in the Jurassic ichthyosaurs *Ichthyosaurus tenuirostris* (A). *Excalibosaurus costini* (B), and *Eurhinosaurus huenei.* Skulls drawn to the same skull lengths (A, B) and to same mandibular lengths (B, C). Redrawn from McGowan (1986).

are likely to have been of adaptive significance. In a detailed analysis of the phylogeny of Late Jurassic plesiosaurs, Brown (1981) documented many changes in allometries, mainly peramorphosis, of both cranial and postcranial elements. Tooth number increased during the Late Jurassic. In Triassic rhynchosaurs, on the other hand, the persistence of multiple row dentition in genera such as *Captorhinus, Labidosaurikos,* and *Moradisaurus* has been interpreted (Benton, 1984) as the paedomorphic retention of an ancestral embryonic type of dentition—a classic exaptation, as embryonic dentition can have no functional significance, but when persisting into the adult phase, can become functional, if the requisite feeding niche exists to allow the establishment of an adaptive peak.

The important effect of heterochrony on tooth development is not confined to the meristic change in tooth number, but also to changes in the shape of teeth themselves. This is perhaps best demonstrated in the evolution of elephant dentition, and the effect that this had on the establishment of new adaptive peaks. During ontogeny, elephant teeth grow by the addition of plates. In Middle Pleistocene forms only seven plates developed during the ontogeny of molar M3 (Maglio, 1972). Three descendant lineages each show a peramorphic trend of increased numbers of plates in the molars, with a maximum being reached in some species of *Mammuthus.* The increase in plate number was least in the *Loxodonta* lineage. In conjunction with this peramorphic increase in plate number, there was a commensurate paedomorphic reduction in thickness of the tooth enamel. The functional significance of these heterochronic changes was the evolution of "new adaptive zones" (= peaks) (Maglio, 1972) with improved shearing capabilities. The close developmental relationship between plate number and enamel thickness is shown by another lineage of *Elephas* where progenesis in *E. falconeri* resulted in the return to thicker, less-folded molars composed of fewer plates. Because the progenesis was global in its effects, this smaller elephant developed other paedomorphic traits, including a rounded cranium, with the absence of a frontal crest; small tusks and tusk sockets.

While the relationship of ontogeny to phylogeny in mammalian tooth development has received surprisingly little attention, evidence from the evolution of tooth morphology in horses attests to the importance of heterochrony in producing novel morphologies that allow the occupation of new feeding niches (e.g., Stanley, 1979). However, because of the important role of (global and local) hypermorphosis in horse evolution, separating the selective advantage of size increase from specific morphological changes is not easy. However, it is more than 100 years since Cope (1874) proposed that the complex multicuspate therian molar tooth had developed by addition during development of cusps peripheral to a single cusp. More complex tooth morphology would therefore be a peramorphic feature. Archer (1974) has noted how cusp development does not follow a strictly invariate sequence in marsupials. This he attributes to variations in rates of development of different cusps between different genera. Cusp initiation is the result of a cessation of mitosis at a point on the inner enamel epithelium, while active mitosis which continues to occur in areas between the cusps results in deposition of enamel in the valleys. Thus, the cusp destined to be tallest is the first to be initiated, and the lowest the last. However, tooth morphology is determined not only by the order of initiation of cusp growth, but also by growth rates of individual cusps. Archer (1974) has observed that these may vary. For example, a primary cusp could develop first, but if a secondary cusp grew at a faster rate, it could end up being larger. With such flexibility inherent in the onset, offset, and rate of cusp development, it is not surprising that such a wide diversity of tooth morphologies has developed in mammals by heterochrony. While there has been little attempt to interpret such dental morphology in heterochronic terms, it seems clear that the primary role of heterochrony in initiating changes in tooth morphology has allowed a wide diversity of feeding niches to be occupied by mammals.

The strong pressure on selection of heterochronic morphotypes that allow the occupation of new feeding niches is also reflected in many invertebrate groups. This has been best documented in a variety of echinoderm classes. Pronounced changes of plate allometries in some carpoids and blastoids by (allometric) neoteny and acceleration resulted in the attainment of novel morphotypes able to occupy different feeding niches. Because of allometric acceleration of growth of zygous basal plates in the Devonian blastoid *Eleutherocrinus,* a low-level suspension feeding strategy was attained (Millendorf, 1979). Similarly, allometric neoteny and acceleration in different plates in lineages of *Pentremites* resulted in the evolution of different thecal shapes that were able to adopt new feeding strategies (Waters *et al.,* 1985). Allometric acceleration in rate of generation of brachial plates and arm length in *Promelocrinus* (Brower, 1976), and paedomorphic reduction in pinnule length in plate production in disparids (Frest *et al.,* 1979) both resulted in the evolution of new feeding strategies.

The role of heterochrony was crucial in opening up a myriad of new feeding niches in echinoids. Among the most fundamental was the paedomorphic loss of the Aristotle's lantern (Smith, 1981). This dental apparatus possessed by regular echinoids is used to rasp algae from a variety of surfaces. With the reduction in development of this organ in most irregular echinoid groups, such as the clypeasteroids (sand dollars) and spatangoids (heart urchins), other feeding strategies were employed using modified tube feet around the mouth—the so-called phyllodal tube feet. These allowed the evolution of deposit feeding in echinoids. Secretion of mucus by the tube feet allowed sediment grains to be picked up and transferred to the mouth. Organic material upon and within the grains provided the source of nutrition. In a number of groups, such as cassiduloids, holasteroids, and spatangoids, there is an overall trend of paedomorphic reduction in the number of these feeding tube feet. This occurred either by postdisplacement or by neoteny, or by a combination of both processes.

Specific changes in the nature of these tube feet and associated supporting structures on the echinoid test can be followed within lineages and shown to be related to changing characteristics of the sediment upon which the animals feed. In the Australian heart urchin *Protenaster,* the earliest species, the Late Eocene *P. preaustralis,* fed upon coarse-grained calcarenite grains and adults possess a pair of pores through which each phyllodal tube foot passes (McNamara, 1985a). These pores are separated by a raised partition that forms a support for the tube foot musculature. As the lineage is traced through the Late Oligocene *P. philipi,* the Early Miocene *P. antiaustralis,* to the living species *P. australis,* the raised partition becomes breached in adult *P. philipi,* then regrows as a reniform ridge adjacent to a single pore in *P. antiaustralis* before finally developing a swollen platform in the living species (Fig. 6-19). This species, during its ontogeny, passes through the phylogenetic stages of its antecedents, demonstrating that these changes are peramorphic, having occurred by acceleration. In terms of surface area of this attachment site there was a general decrease in surface area, both ontogenetically and phylogenetically. Accompanying these morphological changes were changes in the sediment grain size that the echinoids fed upon. The reduction in muscle attachment area corresponds to a reduction in sediment grain size, allowing occupation and utilization of a finer-grained sediment niche by these burrowing echinoids. The strong selection pressure on such heterochronically produced feeding structures is further shown by the conservative nature of the rest of the echinoid's test. Over a period of at least 45 million years the overall character of the test remained very much the same, yet changes in the nature of the feeding apparatus were profound.

5.5. Nonfeeding Adaptations

Another dominant role that heterochrony has played in terms of adaptations is by physically facilitating the occupation of new habitats. In general, the more profound the morphological separation between ancestral and descendant morphologies, the greater the niche difference, and the greater the taxonomic separation. As a result of heterochrony facilitating the attainment of a markedly novel habitat, the macroevolutionary potential is profound (see Chapter 5).

We have already described (see Chapter 5) how paedomorphosis in the rhynchonellid brachiopod lineage *Tegulorhynchia–Notosaria* resulted in the evolution of a paedomorphocline from deep- to shallow-water species (McNamara, 1983b). This was attained by translation of ancestral juvenile morphological adaptations to the adult stage of descendants. The morphological characteristics of juvenile ancestral species were adaptations that functioned only in a certain size range in a deep-water environment and include a pedicle for attachment that was relatively large (undergoing negative allometry

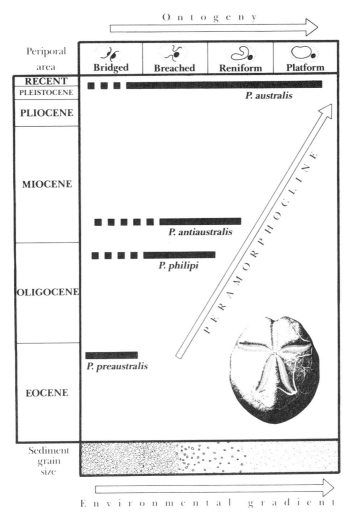

Figure 6-19. Peramorphocline in development of structures around the mouth associated with tube feet involved in feeding in the heart urchin *Protenaster* from the Cenozoic of Australia. Associated with changes in the feeding structure was a change from inhabiting coarse sediments to inhabiting finer sediments, in later forms. Reproduced from McNamara (1985a).

during ontogeny). Being relatively more unstable than the adult, a relatively larger pedicle by which it attached to a hard substrate was a functional adaptation. There was virtually no real increase in pedicle thickness (as determined by the size of the foramen in the umbonal region of the shell) during ontogeny of the ancestral species, *Tegulorhynchia boongeroodaensis*, following the early juvenile growth. However, extension of the early juvenile growth trajectory, seemingly by local hypermorphosis, produced a paedomorphic trait in later species. This apparent contradiction that hypermorphosis produced a paedomorphic character occurs because the growth of this trait was in fact controlled by another trait that underwent neoteny along the paedomorphocline! The

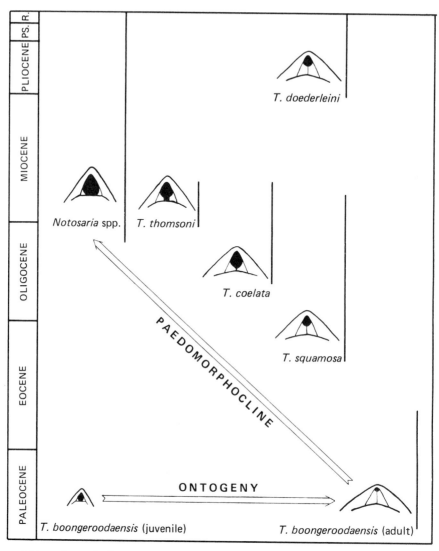

Figure 6-20. Paedomorphocline in size of foramen (and thus pedicle) in the Cenozoic rhynchonellids *Tegulorhynchia* and *Notosaria* from Australasia. This allowed an environment of higher hydrodynamic activity to be occupied by later species. Reproduced from McNamara (1983b).

foramen size (and thus pedicle size) was controlled by the rate of growth of the deltidial plates (Fig. 6-20). In the earliest species these grew rapidly during ontogeny, and caused the restriction in foramen size, and hence pedicle size. Neotenic reduction in growth rate of these plates through successive species along the paedomorphocline resulted in stratigraphically younger species attaining a progressively larger adult pedicle. This phenomenon, whereby the activity of one heterochronic process can instigate a different heterochronic process, has also been noted in the evolution of some encrinurid trilobites (Edgecombe and Chatterton, 1987) and in some bryozoans (Anstey, 1987).

From an adaptive point of view, the juvenile ancestral pedicle size was an exaptation, for when translated into adult shells it allowed the occupation of higher hydrodynamic regimes, in shallower water. A further exaptation was the low convexity of ancestral juvenile shells, reflecting the possession of a relatively small lophophore. This contrasts with the highly convex shell developed in ancestral adults due to positive allometric growth of the lophophore (a combined feeding and respiratory structure), a prerequisite for inhabiting a low hydrodynamic regime in deep water. Global allometric neoteny in the paedomorphocline resulted not only in the evolution of effective attachment structures in shallow water (species of *Notosaria* live in New Zealand today in the intertidal zone) but a smaller lophophore. Where current flow is much greater, a smaller lophophore is energetically more effective than a larger structure. Thus, while selection might have primarily focused on pedicle size (without which attachment would not have been possible), the incidental reduction in lophophore size was a useful added adaptive feature that allowed living species of *Notosaria* to occupy a higher level on the adaptive peak than they might otherwise have done with a larger lophophore.

Heterochrony has also been of importance in allowing some groups of other invertebrates to undergo major habitat transitions. Some bivalves have gone from an infaunal to epifaunal habitat and some echinoids have followed the opposite path from epifaunal to infaunal. All these invasions into new habitats were only possible *following* intrinsic changes in developmental pathways by heterochrony. While the role of extrinsic abiotic factors may have been passive, there is evidence that extrinsic biotic factors played a secondary role to heterochrony by affecting the timing and direction of these invasions.

The occupation of an epifaunal habitat by many bivalves was achieved by paedomorphic retention of a byssus (Stanley, 1972). Primitive bivalves all seem to have been infaunal (Pojeta, 1971) and used a byssus (a mass of threads for attaching to the substrate) in the postlarval stages to attain attachment to the sediment. By paedomorphosis this byssus appeared in later adults and allowed larger adult shells (particularly arcoid bivalves) to be stabilized on the sediment surface (Fig. 6-21). In species that are byssally attached as adults, a large foot is present in the juveniles. As the byssus develops, so the foot reduces in size (Stanley, 1972). However, in some later forms even this large foot has been paedomorphically transferred into the adult stage. This allowed these adults to revert not only to an infaunal existence, but to effect a burrowing mode of life in the sediment. Thus, not only was the initial change from infaunal to epifaunal habitat brought about by paedomorphosis, but so was the subsequent reconquest of the infaunal habitat.

One of the major keys to interpreting whether heterochronic changes were of primary adaptive significance is the degree of heterochronic dissociation of traits. As noted above, if heterochrony is global and also involves body size changes, then assessments of the relative degrees of importance of heterochronic changes in size or shape may often be subjective or unable to be substantiated. However, where there is only change in shape, and this can be functionally correlated with extrinsic changes, such as changing sediment characteristics, then the role of heterochrony in inducing the adaptive changes can be assessed with some confidence. In such instances dissociation occurs, and only certain traits are affected by heterochrony (see Chapter 5), and a mosaic of forms can potentially develop. If the organism is still able to function effectively even when different traits might be subjected to opposing heterochronic processes, then there

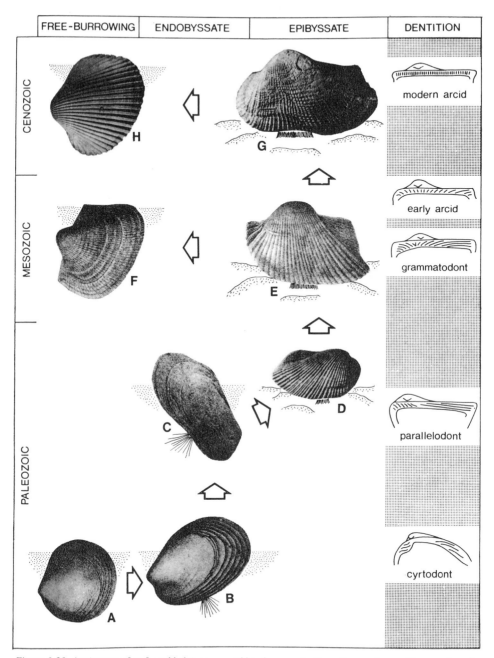

Figure 6-21. Attainment of epifaunal habitat in arcoid bivalves was facilitated by paedomorphic retention of a byssus. In species byssally attached as adults, a relatively large foot is present in juveniles. In later forms this was transferred by paedomorphosis into the adult stage, allowing recolonization of the infaunal habitat and efficient burrowing ability. Reproduced from Stanley (1972), with permission of the Paleontological Society.

is an immense potential for the evolution of a diverse suite of morphotypes. This will most readily be found in organisms which show high degrees of phenotypic plasticity, reflecting high developmental plasticity.

We have described how in particular lineages of spatangoid echinoids (such as the *Lovenia* lineage—see Chapter 4) such high intraspecific phenotypic plasticity has been instrumental in speciation by generating a diverse range of morphotypes. By a "shuffling" of developmental rates (i.e., some show a peramorphic increase in development, others a paedomorphic decrease), optimum forms can be produced that are able to reach the top of the adaptive peak. Within the *Lovenia* lineage, and a number of other Australian spatangoid lineages, dissociated heterochronic changes have resulted in the evolution of similar morphologies that allowed the occupation of the same habitat. Put simply, the early Tertiary part of lineages of *Lovenia, Hemiaster, Psephoaster, Pericosmus,* and *Schizaster* were characterized by species with morphological characteristics suitable for burrowing in a coarse-grained sediment (McNamara, 1989; McNamara and Philip, 1980a), including shallow petals (that carry respiratory tube feet) and narrow peripetalous fascioles (structures that secrete mucus that envelope the upper surface of the test in its burrow). By a variety of heterochronic processes within and among lineages, similar morphological changes occurred that allowed much finer-grained sediments to be inhabited by descendants. Because of the relatively more impervious nature of these sediments, morphological changes were favored that enhanced water flow over the test in its enclosed burrow (Fig. 6-22). The strong selection pressure toward migrating into finer-grained sediment habitats resulted in convergent evolution in many traits. This was facilitated by the constraining effects of the echinoid's developmental programs and the heterochronic processes. Similar morphologies were attained in *Schizaster* and *Pericosmus* by a combination of hypermorphosis and acceleration (all peramorphic features); in *Psephoaster* and *Hemiaster* by a dominance of peramorphic processes, but also some paedomorphic ones; in *Lovenia* by similar proportions of peramorphic and paedomorphic features. Only in the *Schizaster* lineage was there appreciable size change. The evidence therefore strongly supports the view that selection for new habitats was the dominant trend in the evolution of these lineages and involved selection for particular shape morphotypes. This is not to say that there are not cases of habitat selection being targeted on size, as we shall discuss below. But without the intrinsic ability of the organism's developmental system to show this dissociation, the suitable adaptive peaks would not have been attained, even though the vacant finer-grained sediment niches were in existence. The ultimate cause of the selection pressure that directed evolution into these particular niches is likely to have been predation pressure (see Chapter 5).

The migration of echinoids from an epifaunal environment to an infaunal one was probably also predator-induced. Epifaunal echinoids have developed a range of antipredatory devices, not the least of which is a dense covering of large spines. An alternative strategy, facilitated by heterochrony, was an infaunal habit. This behavior obviated the need for large spines. The paedomorphic reduction in spine size, but peramorphic increase in spine number to produce a dense coating of fine spines, had an ancillary adaptive role in that the peramorphic generation of many small spines facilitated the echinoid's ability (McNamara, 1988a) to burrow into sediment. The importance of heterochrony in the realization of refugial habitats has been greatly underestimated. Not only did it allow for the great adaptive radiation in irregular echinoids during the Meso-

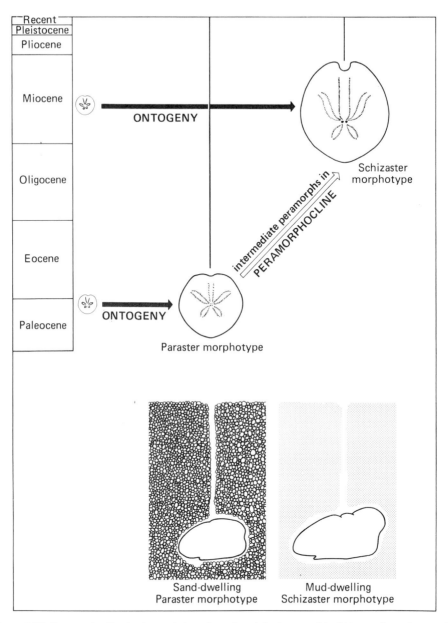

Figure 6-22. Peramorphocline in the evolution of species of the heart urchin *Schizaster* from *Paraster*. In addition to increased degree of development of many traits, there was also a progressive increase in size as later species evolved into finer-grained sediments in deeper water. Reproduced from McNamara (1982b).

zoic and Cenozoic, but the adaptive radiation in the diverse trilobite morphologies in the post-Cambrian can in part be explained by the evolution of a wider range of refugial and cryptic habitats in the face of increased predation pressure. As with echinoids, escape behavior in these arthropods was achieved by both the evolution of spines (which probably functioned more in a camouflage role), as in the Odontopleuridae, and burrowing. As discussed in Chapter 5, the increased frequency of acceleration of morphological development in some trilobite families, particularly those in the Illaenidae, led to effacement of the cephalon. The functional advantage of this was one of cryptic behavior, for the likely life orientation of genera such as *Bumastus* was with the body buried in the sediment, apart from the dome-shaped head projecting above the sediment/water interface. Whether this cryptic habit was to facilitate its role as a predator or reduce its role as a prey species is not clear. Either way, the attainment of such an adaptive peak in this niche occurred by the activity of heterochronic processes.

The wide range of ornamentation developed in many groups of ammonoids, and which was subjected to heterochronic changes in intensity and distribution on the shell, likewise probably served a major cryptic adaptive role. Swan (1988) has shown how in some Namurian Gastriocerataceae lineages a distinctive ornamentation evolved that may have served such a role. These ammonoids further show how many morphological features that are changed by heterochrony did so in response to prolongation or contraction of particular ecological phases in their life histories. Ancestral juvenile morphotypes are characterized by features such as high drag coefficient, low aperture orientation, and potentially cryptic ornament, and are considered (Swan, 1988) to be adaptations for benthic conditions. During subsequent ontogenetic development of the shell, streamlining occurred that reduced the drag coefficient and a high hyponome was developed, indicating that adults were nekto-benthic and had improved swimming ability. In the *Homoceratoides–Cancelloceras* lineage the paedomorphic retention of the ancestral juvenile characters by descendants, such as species of *Cancelloceras* (Fig. 6-23), suggests that benthic conditions became a suitable niche for the entire life span of the organism. The selection pressure for this adult adaptive peak may therefore have been an extension of the benthic niche, such that the entire ontogeny of the ammonoid could be supported by the benthic environment. Swan considers that this lateral expansion of the niche was important for ensuring genetic variability within the population.

The opposite pattern occurred in the *Reticuloceras–Bilinguites* lineage. Here the ancestral ontogenetic change in morphology, from one adapted to a benthic habitat early in ontogeny to a later nekto-benthic one, underwent allometric acceleration, perhaps accompanied by a benthic phase that was shortened. The adult nekto-benthic morphology of compressed, involute, smooth shell and other features that improved hydrodynamic efficiency occurred at smaller shell sizes (presumed to equate with earlier in ontogeny) in later species. Determination of the style of heterochrony was, in Swan's view, strongly controlled by the extent of the "ontogenetic niche" (this crucial concept is discussed elsewhere). If an ancestor was adapted to exploiting a benthic habitat early in ontogeny, and a pelagic one later in ontogeny, and if the benthic habitat contracted, either by abiotic factors (such as the effect of regressions) or by biotic factors (such as the appearance of a new predator), then peramorphic descendants were more likely to be selected. Conversely, if the pelagic habitat contracted, descendants that prolonged the juvenile morphology would be preferentially selected.

Such a situation, where the juvenile/adult transition involves a major habitat shift, is often accompanied by a major morphological shift. In other words, the organism's adaptive landscape has two peaks to accommodate two ontogenetic niches. The nature of the ontogenetic shift between these peaks can have an important bearing on rates of evolution. For instance, in the case of the ammonoids the peak shift will not be instantaneous, but will gradually shift as the shell changes shape. This shape change allows the

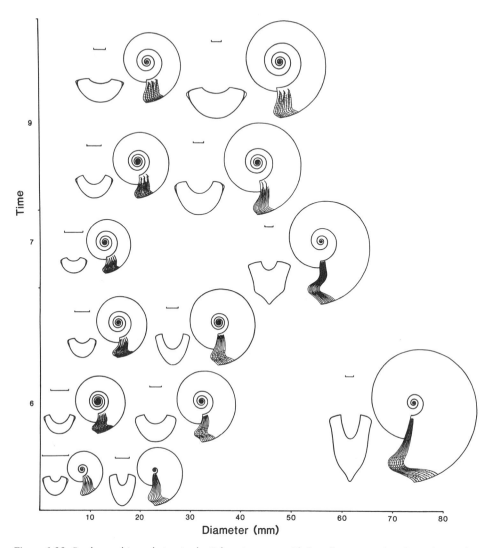

Figure 6-23. Paedomorphic evolution in the Paleozoic ammonoid *Cancelloceras*, attributed to selection for prolonged ancestral juvenile life history traits. Ancestral juveniles, with their broad shells, are thought to have been benthic, the more streamlined adults nekto-benthic. Selection for ancestral juvenile shell shape in descendant adults indicates a prolongation of life in the benthic habitat. Reproduced from Swan (1988), with permission of the Palaeontological Association.

transition into other niches to occur. But see how once again it is this change in shape, the intrinsic factor, that precedes the niche change. A progressive delay in adaptive peak transition will delay change in niche occupation. If the juvenile niche width contracts, then individuals that have accelerated adaptive peak shift will be at a selective advantage. If adaptive peak and ontogenetic niche transitions are more rapid (as in amphibians), then the effect of niche contraction on evolution may be much more rapid.

In the case of the Namurian ammonoids, Swan (1988) has suggested that the principal factor that was responsible for expansion of the benthic niche, which allowed the selection of more neotenic morphotypes, was eustatic transgressions that led to the expansion of the carbonate shelf environment. The contraction of this niche, that favored the allometrically accelerated morphotypes, may have been prompted by decreasing benthic oxygenation. Contraction of the neotenic morphotypes occurred, but with subsequent improvement in benthic conditions the paedomorphic forms reradiated, and the allometrically accelerated forms became extinct.

5.6. Locomotory Adaptations

The final group of adaptive strategies which we shall discuss in order to illustrate the adaptive importance of local shape heterochronies are those involving locomotion. This encompasses both the evolution of novel means of locomotion in particular groups and improved locomotory capacity within lineages.

There has been a long-standing view (e.g., de Beer, 1956) that many of the major groups of flightless birds, such as the ostriches, emus, rheas, and moas (the ratites), arose as the result of paedomorphosis from typical flying birds (carinates), their reduced wing size and downy, juvenile-like feathers being typical paedomorphic structures. However, if flightless birds did evolve from a flying ancestor, the truth, as with the case for humans being "neotenic" apes (see Chapter 7), is likely to be far more complicated (e.g., the large body size, including the large legs, would be best interpreted as being peramorphic structures produced by hypermorphosis, any paedomorphic traits, such as the wings and feathers, just attesting to the ubiquity of dissociated heterochrony). While the argument that the ratites evolved from the carinates seems persuasive, there are some (McGowan, 1984) who have suggested that evidence from the ontogenetic development of the tarsus in the two groups argues against a direct relationship between the two groups, favoring more the idea that they evolved from a common, probably theropod ancestor. Either way, heterochrony seems to have played a significant role in bird evolution for, as we discuss elsewhere, it has even been suggested that flight itself in birds arose as the result of paedomorphosis. If so, it is a trifle ironic that the very process which may have contributed to the evolution of flight in birds also appears at more recent stages in their history to have played a part in the loss of flight in at least one group of birds.

In 1500-year-old cave deposits on Hawaii, the remains of an extinct genus of goose called *Thambetochen* have been found (James and Olsen, 1983). Compared with a typical flying goose, *Thambetochen* possessed stout hind limbs, a large pelvis, small wing bones, and a small sternum which lacked a keel. In these and other respects, this large, extinct, flightless goose resembled the hatchling of a "normal" flying goose (Fig. 6-24). While James and Olsen claim that the paedomorphic nature of *Thambetochen* arose by neoteny (with the implication being that it was global in its effects), a detailed study of relative growths of individual skeletal elements would need to be undertaken in order to substantiate this. The reduced sternum and wings, which are the main contributors to the bird's flightlessness, would appear to be paedomorphic, but other structures, such as the stout limbs, may be peramorphic. It is interesting that, as with other flightless birds, there is this association of large, stout legs, a necessary ancillary adaptation to provide an

Figure 6-24. *Thambetochen* (upper left), an extinct Hawaiian flightless goose that shows paedomorphosis in loss of wings and reduced sternum, like the juvenile of a flying goose (lower figure), but peramorphic development of stronger, stouter legs than a typical flying goose (upper right). Redrawn from James and Olsen (1983).

alternative, effective means of locomotion. If, as seems likely, the large limbs are peramorphic, then it could be argued that the attainment of the flightless adaptive peak in these birds could only arise from a combination of paedomorphic (wings and sternum) and peramorphic features (legs), illustrating once again how suitable adaptations can only produce a functionally viable organism if there is dissociated heterochrony. Although paedomorphic reduction in wings and sternum would produce flightlessness, viability as a species and attainment of the adaptive peak would be prerequisite upon attainment of an alternative means of locomotion.

In organisms that use multiple appendages for locomotion, such as trilobites and echinoids, heterochrony can influence the ability to move in and upon different media. The high variability in thoracic segment numbers in early Cambrian trilobites (discussed elsewhere) would have meant commensurate variability in numbers of appendages. The general trend for overall reduction in thoracic segment, and thus appendage number, and the progressive evolution of fixed appendage number in post-Cambrian forms could be interpreted as strong selection pressure on reduced appendage number. However, intrinsic stabilization of the regulation of development in these later trilobites also had an important effect.

Within the spatangoid echinoid lineages that we have described with adaptations that allow occupation of finer-grained sediments, locomotory traits were also subjected to heterochronic change. Density of ventral tuberculation, reflecting the concentration of burrowing and locomotory spines, increases in all the lineages that evolve into habitation of finer-grained sediments. The denser the concentration of these spines, particularly the burrowing spines, the more efficiently the echinoids could burrow in the finer-grained sediment. Once again, it could be argued that the combined effect of all these dissociated heterochronic changes to feeding strategies, habitats, and locomotory efficiency, produced the optimum morphology that best suited the adaptive peak.

6. BODY SIZE AS A TARGET OF SELECTION

6.1. Introduction

The range in sizes of organisms is huge and spans more than 21 orders of magnitude, the blue whale being 10^{21} times heavier than the smallest microbe (McMahon and Bonner, 1983). Such a profound variation in sizes of organisms will have an immense impact on population dynamics and evolution. While classically many aspects of life history, such as mortality and fecundity, have been considered as functions of age, in many organisms size is more important than age in determining the fate of individual organisms. It has been cogently argued that size is perhaps the *most important characteristic* of an organism (Ebenman and Persson, 1988). Size is crucial in predator/prey relationships; size and morphology determine the ecological niche, and thus influence intra- and interspecific competition; and size can be a determinant of sex. Natural selection can therefore act directly on size, as well as on morphology (shape) and behavior. Stanley and Yang (1987) convincingly demonstrate that the *large majority* of published studies of evolution and evolutionary rates in fossils have documented body size changes (or some surrogate of body size). Size, they argue, is more labile than shape and we should therefore not be surprised at this predominance. This is highly relevant to heterochrony which is the proximal mechanism for the lability. As discussed in Chapter 3, rate and timing genes affecting body growth can be sorted individually. Here body size is a polygenic trait [e.g., mouse size is controlled by about 100 loci (LaBarbera, 1986)]. In other cases, mutations in hormonal activity (rate, offset) can involve only one or a very few genes (Prothero and Sereno, 1982). While purely environmentally induced plasticity can play a role, size in general is over 50% heritable in many animals (Atchley, 1983).

However, we must also consider the second half of the observed evolutionary lability: some of these readily produced, genetically motivated size changes must be selected for. We should not be surprised that it is often under strong changing selection because body size is "manifestly one of the most important attributes of an organism from an ecological and evolutionary point of view. Size has a predominant influence on an animal's energetic requirements, its potential for resource exploitation, and its susceptibility to natural enemies. It scales with physiological, life history, and population parameters" (Werner and Gilliam, 1984). In brief, body size is sensitive to a large number of selective influences; it reflects a compromise between these many competing forces. Given the changing nature of these forces and the sensitivity of body size to so many of them, it is clear that any compromise is usually temporary over long time spans.

Many organisms undergo shifts in ecological niches as they increase in size during ontogeny. Each of these niches will exhibit different selection pressures. However, as Ebenman and Persson (1988) argue, genetic correlations between traits expressed during a life cycle will constrain evolutionary adaptation to the different niches. This should induce strong selection pressure to reduce this genetic correlation between homologous characters which are expressed at different stages of the life cycle. Complex life cycles and metamorphosis may be ways of achieving relatively independent evolution at different stages in the life cycle.

Changes in niches during ontogeny will also greatly affect intraspecific competition. The degree of competition between individuals of different ages will more often than not depend on their relative sizes and thus their growth patterns (Ebenman and Persson, 1988). The effect of such competition within and among age classes may be to retard individual growth rates. This has been documented in a number of invertebrates and vertebrates (e.g., G. P. Jones, 1987; Wilbur and Collins, 1983).

In this section we focus on both overall body size and body size-related traits as targets of selection and assess the role that heterochrony plays in this. We have discussed the problems of isolating size from shape in Chapter 2. As we have noted, size is a general "vector" that describes overall covariant growth in all traits. Because shape change usually accompanies body size increase, selection on size will often also cause shape changes. Thus, trying to determine whether size is the prime target of selection rather than shape can often be difficult.

Given the situation where an organism gave rise to a descendant that was larger in maximum adult size, but in shape resembled (i.e., was isometric to) the ancestor, this would be termed **giantism** in heterochronic terminology. This implies that size has been decoupled completely from shape changes. Conversely, a size reduction would be termed **dwarfism** (Gould, 1977; Alberch *et al.,* 1979). Selection in these cases must therefore have been entirely on size or on factors of the organism's biology that affect overall body size. While many so-called examples of dwarfism have been documented, such as "dwarf hippopotamus" (Prothero and Sereno, 1982), "dwarf elephants," "dwarf deer" (Sondaar, 1977), and "dwarf molluskan" faunas (Mancini, 1978), detailed analysis of the morphological changes between these "dwarfs" and "normal"-sized forms indicates that they are not merely scaled-down versions of their ancestors. For instance, their small size may have occurred as a consequence of earlier onset of sexual maturity (progenesis). The consequence of this is that the scaling down of allometric trends within the organism's development produced descendants that were not only smaller, but morphologically dissimilar from their larger antecedants. This is because virtually no animal grows isometrically in all traits. Some shape change is bound to occur with body size scaling. For instance, the limbs of many island "dwarf" taxa, such as "dwarf hippos," were relatively much stouter than would be expected in other animals of similar size. The adaptive explanation proposed for this (Sondaar, 1986) is that unlike other small mammals with slender limbs that allowed rapid locomotion (usually away from an oncoming predator), "dwarf" mammals occurred on islands where there were no predators. Therefore, speed was not a target of selection, and ancestral leg allometries were adequate.

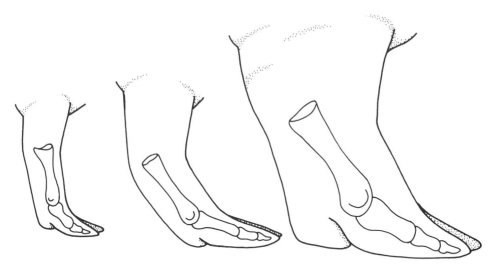

Figure 6-25. Differences in foot structure of "normal" and progenetic hippopotamus. The relatively much smaller toe bones in the progenetic hippopotamus produced a quite different articulation surface, with the animals walking on "tiptoe". This is thought (Sondaar, 1986) to have allowed the smaller forms to inhabit mountainous regions. Redrawn from Sondaar (1986).

Other examples of size-related shape change are seen in many herbivorous animals (such as hystricomorph rodents, suine artiodactyls, cervoid artiodactyls, and primates) where postcanine teeth show positive allometry relative to body size. That is to say, species with larger bodies have proportionately larger teeth. Thus, scaling down a hippo, if such a growth trajectory was being followed, should produce an animal with proportionately (but not isometrically) smaller teeth. However, Gould (1975) found that the postcanine tooth in "dwarf hippos" was relatively larger than would be predicted from a simple exercise of scaling down from the "normal"-sized version, demonstrating that tooth growth can be dissociated from body growth.

The interpretation of body size changes in evolutionary studies is a controversial subject. A major reason is the basic problem of trying to unravel the selective pressures on size *per se*, compared with selection on other aspects of the organism's life history strategies that make size itself an incidental outcome of some other change. Compounding this is the problem of allometry, and the need to separate the selective pressures on size from those on shape changes that occur as a consequence of body size change. As we have discussed in the example of intraspecific changes in the Galápagos finch *Geospiza fortis* (see above), body size changes were the vehicle that allowed another trait to change and be the primary target of selection. Similarly, the "dwarf" hippos on Crete and Cyprus possessed relatively much smaller toe bones than larger hippos. This produced quite different articulation surfaces. In large hippos the toes support a large foot cushion (Sondaar, 1986) (Fig. 6-25), but in "dwarf" hippos foot pads were not developed and the animals walked on tiptoe, rather like goats. This, Sondaar believes, produced a major adaptational shift, allowing these small hippos to inhabit mountainous regions.

While we concentrated on examples of shape change in the preceding section in this chapter, we focus here on the selective advantages of size, or of factors that induce size and ancillary shape changes, and attempt to unravel some of the problems associated with the complex interplay between shape and size in evolution. In the simplest example of size change alone, that of dwarfism and giantism, selection, as we have noted, is entirely on size and there is total dissociation of body size increase from shape changes. Whether such complete dissociation can actually occur is debatable and seems rare. For a given extension or contraction in timing of maturity, the *greater the number of traits* (breadth) that grow allometrically or of meristic characters which show an increase during ontogeny, the greater the number of shape changes for the same change in size. The more complex, then, is the unraveling of size or shape as targets of selection. Furthermore, the *greater the allometric coefficients for those traits* (depth), the greater the shape change between ancestor and descendant for the same size change. Intuitively, therefore, it would seem likely that the greater the influence of allometry when progenesis or hypermorphosis operates, the greater the influence of shape rather than size in terms of selective advantage. Most of the next five sections will concentrate on this aspect of the relative selective advantages of size and shape changes during phylogeny, as well as other factors that might be targets of selection that produce both size and shape changes.

6.2. Size, Selection, and Trends: Distinguishing Anagenesis from Cladogenesis

The very phrase "body size" is sure to connote *increasing* body size to virtually any evolutionary biologist. This trend, codified as **"Cope's rule,"** is found in many fossil lineages and clades (LaBarbera, 1986). It is often explained by saying that large body size has some intrinsic advantage and that, through natural selection, individuals of large size have more offspring resulting in larger species. This is not an accurate view. Small organisms are as well adapted to their environments, roles, and niches as are large ones. Indeed, they are more numerous and probably last longer as species since they are less sensitive to perturbations (discussed below). The problem, and solution, is found in the distinction between *anagenetic* (single lineage) and *cladogenetic* (branching) size change (McKinney, 1990).

Beginning with cladogenetic change, Stanley (1973) showed that many clades start off at small sizes and become larger as descendant species diffuse (and branch off) into environments (or niches in the same environment) that select for larger size. This has been elegantly demonstrated by MacFadden (1986) (Fig. 6-26) in the evolution of horses. MacFadden showed how, on the basis of 40 species of fossil horses, there was little change in body size from about 57 to 25 million years ago, combined with slow evolutionary rates. Thereafter, particularly between about 25 and 10 million years ago there was a major diversification, with some species representing up to an eightfold size increase. This occurred at higher evolutionary rates. This size increase has continued from the late Miocene, reaching a maximum in the middle Pleistocene. In our discussion of cladogenetic heterochronoclines in Chapter 5, we discussed the autocatakinetic aspect of such trends. In other words, persistence of ancestral species, combined with evolution along an environmental gradient, constrained evolution to such an extent that there was only one way to go. The same mechanism applies to body size as it does to

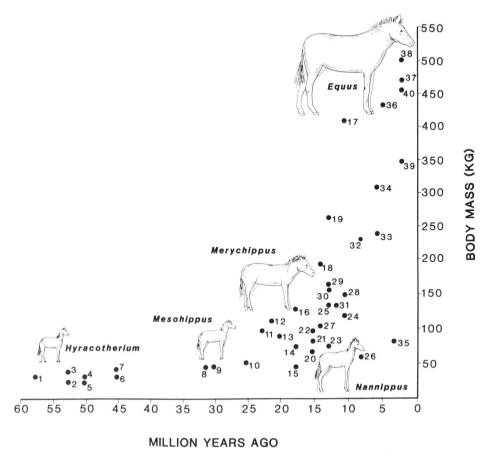

Figure 6-26. Distribution of estimated body mass calculated for 40 species of fossil horses through the Cenozoic, illustrating a progressive increase in size through time. Reproduced from MacFadden (1986), with permission of the Paleontological Society.

morphology. Thus, for horses there was nowhere else to go but "up," a trend of **clado-genetic asymmetry** (Gould, 1988b; also McKinney, 1990).

The reason for the small initial size, which is a major "cause" of the trend, was originally thought (Stanley, 1973) to be that larger organisms are so allometrically constrained that they are essentially too "specialized" to be potential raw materials for clade proliferation. However, this is a very fuzzy issue for at least three reasons: (1) there is no evidence that large animals are poor raw material. On a "bacterium to elephant" basis it is true that larger animals have more developmental "hardening" (see Chapter 8), but such gross comparisons are misleading distortions of scale because large versus small mice (for example) are vastly more similar than mice and bacteria (i.e., it is very doubtful that small mice are less "specialized" than large ones); (2) as noted by LaBarbera (1986) smaller species are much more abundant in most clades so that by sheer chance, even if all species have an equiprobable chance of being ancestral, smaller species will generally be progenitors; (3) a point never made to our knowledge is that many, maybe most, of the classic size trends began from ancestors who survived a mass extinction (e.g., early Cenozoic mammals). Given the greater susceptibility of larger animals to perturbations (Pimm *et al.*, 1988), smaller animals would be the only ancestral stock available even if they were not more abundant or less constrained.

In contrast to these clade-level branching trends, there are the largely up-and-down fluctuations of a single lineage. This has been best documented for the more recent, conspicuous animals, i.e., large mammals of the Pleistocene and Recent. Kurtén (1959, 1968) and Heintz and Garutt (1965) have shown how, in brown bears and mammoths, respectively, size in a single lineage (anagenetically) fluctuates with intimate correlation to climatic temperature. In the case of the brown bears (Fig. 6-27), cooler temperature promotes larger size (Bergmann's rule), for reasons of surface–volume regulation of homeothermy. The mammoths are an exception, showing the reverse relationship (Fig. 6-28). Perhaps the most striking aspect of the correlation between body size and temperature is that a single variable (temperature) can exert such a strong control when so many other variables are also present. It must be that these other external, selecting variables covary with temperature (e.g., predators and prey get larger too) or have enough "slack" that they are not too limiting. Clearly, where this covariation or slack does not exist, the conflicting size selection will lead to extinction or perhaps branching into two species if enough polymorphism is present and the diverging selection is slow enough.

6.3. Cope's Rule—Selection for Large Body Size

The underlying mechanism for size change is, as we shall show, heterochrony. This brings us to the key topic of what forces promote increases in the target of size, whether causing anagenetic fluctuations or, in new environments or niches, causing a small population to eventually branch off into a new species. The ecological literature is rife with many papers on this topic, many of which emphasize the need for controlled experiments to isolate the key causes. Among the factors that have been suggested as influencing size are predator escape, sexual selection, temperature effects (see above), and resource use. We elaborate on some of these below.

We have shown above how equids increased in size through the Tertiary. A similar pattern is discernible in three other groups of vertebrates: proboscideans, cetaceans, and pelycosaurs (see McKinney, 1990, for further discussion). The advantages of comparing these groups are that the former two contain within them representatives of the largest animals known from the land and the water, while the latter, along with equids, provides a useful "control" to see if the pattern is consistent in groups of smaller animals, including nonmammals.

In proboscideans a plot of third inferior molar length cubed (to more closely correlate to the volumetric measure of body weight) against time for the Gomphotheridae, Mammutidae, and Elephantidae reveals that proboscidean size increased at an exponential rate until recently (Fig. 6-29). All three families show the same pattern. In cetaceans a plot of body length cubed versus time produces a similar monotonic increase for maximum adult size, as do plots for pelycosaurs and horses. While exponential increases occur in all groups, adaptive differences among and within the groups appear to have

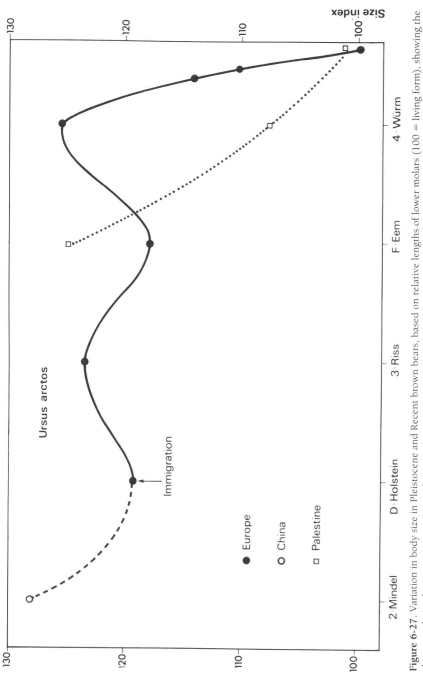

Figure 6-27. Variation in body size in Pleistocene and Recent brown bears, based on relative lengths of lower molars (100 = living form), showing the close correlation between temperature and size, in this case, larger body size and lower temperature (Bermann's rule). Reproduced from Kurtén (1968).

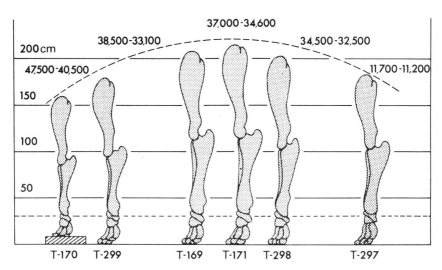

Figure 6-28. Size of foreleg in the woolly mammoth, *Mammuthus primigenius* from Siberia, showing an increase in size during warmer interstadials during the Late Pleistocene. Reproduced from Kurtén (1968).

little effect, such that filter-feeding mysticete whales increase in the same way as carnivorous odontocetes.

The pattern of cladogenetic size increase shows similarity to size change in individual ontogenies. In both there is initial, unconstrained exponential growth, which then encounters negative extrinsic feedback to stop growth. In ontogeny, multiplicative cell growth is slowed by "programmed" feedback controls, while multiplicative cladogenesis is slowed via reduction ("filling") of available new habitats (compare with discussion of Hayami, 1978).

When groups are considered at high taxonomic levels (e.g., ordinal level just discussed), size increase for the whole group (clade) is much less constrained by extrinsic factors. This is because the radiating lineages are not restricted to local environmental limits. The pattern, then, is one of cladogenetic size increase (Stanley, 1973; Gould, 1988b). At the generic level, however, many lineages show an anagenetic pattern (Chapter 5; McNamara, 1990), probably because the larger descendant outcompetes its smaller ancestor or is more predation-resistant (Chapter 5). For example, Martin (1986) has shown that one of the principal advantages of larger size in cotton rats is simply that it results in increased aggression in the larger forms, with the result that two species of different size cannot coexist in the same microhabitat. Large cotton rats will kill smaller species that transgress into their habitat (Martin, 1986). While this might result in the smaller form being driven into another habitat, it may ultimately lead to the extinction of the smaller form. The dominant long-term trend in cotton rats has been for an increase in body size, similar to that documented in other lineages, such as elephants and horses. Like these, there are also a small number of "dwarf" lineages.

The heterochronic mechanisms behind Cope's rule can be either an acceleration of body growth rates; delay in onset of maturity, so extending the juvenile period of exponential growth (hypermorphosis); or growth predisplacement. A common dictum is that large animals are large simply because they grow for a longer period (i.e., hypermorphosis) (Bonner and Horn, 1982; Gould, 1974). On a "bacterium to elephant" type of comparison this is no doubt true (McMahon and Bonner, 1983). However, within lineages, it can be seen that this is not the only way that size can increase, for the larger of the two closely related species often just grows at a faster rate (acceleration) for the same period of time (Chapter 2). There are clear survival advantages in the rapid attainment of a large body size quicker by acceleration. It takes prey species out of range of the optimum prey size sooner and it allows predators to function within this range earlier. It allows all individuals to reproduce sooner. Thus, there is strong selection to maximize growth rate (Werner, in Ebenman and Persson, 1988). However, for some organisms, more adaptive advantages can be gained by extending the growth period. For instance, in humans (the second largest primate), prolongation of the growth period allows an extended learning period, a major factor in the evolutionary success of the species (see Chapter 7). In such cases the more significant target of selection may have been the extended growth period, rather than on the larger size. The consequent size increase may have been of secondary consequence, although was probably also adaptive, in hunting and intraspecific aggression.

Large size, while being of adaptive significance in some environments, can be disadvantageous if there is a major environmental perturbation. Striking examples of this are the extinctions of the megafauna in the Late Pleistocene in a number of continents, induced by major climatic changes and the influence of a major new predator in the form

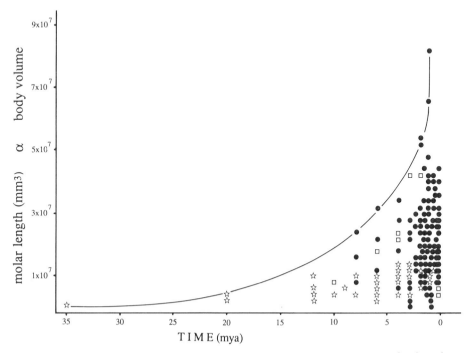

Figure 6-29. Plot of body volume (derived from molar length) against time for three groups of proboscideans, the Gomphotheridae (stars), Mammutidae (squares), and Elephantidae (circles), showing an exponential increase in size.

of man (Martin and Klein, 1984). While it might be considered that in such situations the adaptive peak itself becomes inappropriate for the altered ecological niche, a number of studies (e.g., Diamond, 1984) have suggested that larger animals are more prone to extinction than smaller ones because of their relatively lower population sizes. However, a recent study of British island birds by Pimm *et al.* (1988) has shown that a more complex situation exists when other factors are considered.

Even though heterochronic processes allow larger sizes to be attained, they do not explain why this trend of increasing size should be so common (i.e., what *pushes* the cladogenetic tree "up"—just because there is "nowhere but up" to go does not guarantee it will go up unless a force directs it). One of the most important is that it is the easiest way for new adaptive peaks to be attained, for body size change is relatively "easy" to do physiologically (Chapter 3; Anderson, 1987), with a simple genetic basis (Prothero and Sereno, 1982; Chapters 2 and 3).

Two general types of situations can foster size increase. One is where resources are rich and the environment is stable enough to support large body sizes, so large individuals will have a selective advantage over smaller ones (K-selection). Resource use has been shown (Schoener, 1983) to be a major factor in influencing body size. A classic example is the ability of large ungulates to digest fibrous foods (e.g., grasses) because of the longer digestive tract that arises from allometric scaling (Demment and van Soest, 1985). This is also true in many other animals, e.g., primates, where the gorilla is more folivorous than the chimpanzee, which is more reliant on fruit and other high-quality foods. Other monkeys are less folivorous still, eating insects and even higher-density foods (Shea, 1988). On islands where resources are depleted, larger animals become "dwarfed" (Lomolino, 1985), confirming the importance of resource use. However, smaller animals actually increased in size on islands, because they were freed from competition of the formerly larger competitors. This example reiterates that there is no "intrinsic" advantage to one size or another. Rather we must consider the animal's ancestral size (previous state) and the new configuration of forces in its new environment in each case of size change (see McKinney, 1990, for further discussion).

The second size-increasing situation encompasses the direct influence of extrinsic, often abiotic, factors, particularly temperature. Cold temperature, as we have discussed above, is positively correlated (Bergmann's rule) with increases in body size (Davis, 1981; Koch, 1986).

6.4. Adaptive Significance of Small Body Size

We shall argue later that many cases of progenesis can be interpreted as adaptations to particular life history strategies. This is **indirect size selection** (small size as an incidental product of selection for new reproductive strategies). However, in some cases the evolution of small size itself has a functional advantage. This we call **direct size selection**. Among marine invertebrates, so-called "dwarf" faunas have been documented from many parts of the fossil record. These faunas may comprise whole suites of brachiopod or molluskan taxa (Snyder and Bretsky, 1971; Mancini, 1978) and are frequently found in sedimentary environments characterized by soft, unstable substrates. Mancini (1978) considered that the small body size, arising from progenesis, in oysters and ammonites in the Late Cretaceous Grayson micromorph fauna allowed occupation of the unstable environment. Surlyk (1972) has shown that many very small, progenetic Late Cretaceous brachiopods occur in chalks that were similarly very soft and unable to sustain the weight of large shells. These brachiopods, by virtue of their progenetic state, lost any ancestral attachment mechanism, such as a pedicle. However, it has been suggested (Snyder and Bretsky, 1971) that the small body size indicates that such forms were also r-strategists because the environments which they inhabited were immature and unstable, fluctuating in time and space. (These suggested causes are not exclusive, and all may have been at work.)

One of the more important adaptive features of small body size is the reduced risk of predation (see Chapter 5) because in terms of the predator's expenditure of energy it is more profitable to attack larger prey. Selection pressure will therefore favor the survival of forms that reproduce earlier at a small size (progenesis). Selection for small body size may also occur in predators in response to changing availability of prey species, resulting in the attainment of different adult feeding strategies. This has been well documented in a range of Australian elapid snakes. For instance, the venomous dugite, *Pseudonaja affinis affinis*, that lives on the mainland in Western Australia is relatively large (up to 1.8 m in length). By comparison, a form that lives on Rottnest Island, *P. affinis exilis*, attains a maximum length of less than 1.2 m. The larger mainland form undergoes a dietary change during ontogeny, from small lizards when young, to small mammals and frogs when a large adult (Storr, 1989). The island form feeds on lizards throughout its life. Whether the smaller size of the island form arises from progenesis or from reduced growth rate is not known. However, it is likely that the absence of small mammals from the island has resulted in selection for adults that by virtue of their smaller size are able to subsist on the ancestral juvenile diet of lizards throughout their life.

A similar pattern has been observed in the black tiger snake *Notechis ater niger* on Roxby Island, South Australia. On this island the snake attains a maximum snout/vent length of about 0.85 m, which is little more than half of that found on other offshore islands (Schwaner, 1985). On these nearby islands the larger snakes attain a maximum snout/vent length (on Franklin Island) of almost 1.4 m (Fig. 6-30). Indeed, the range in maximum sizes of tiger snakes on these islands exceeds the range of maximum sizes found in mainland populations. Prey species, such as muttonbirds, sticknest rats, and bandicoots, are also larger on these islands than on Roxby Island. On Roxby Island, lizards are the primary source of prey for even the largest of the adults. As with the situation with the dugites in Western Australia, even the largest of the tiger snakes that feed on large mammals and birds as adults feed on lizards as juveniles (see Fig. 6-30). Experimental studies indicate that these differences are adaptive, in that snakes of the same size from Roxby Island and Franklin Island, when fed the same weight of food, grow at 10 and 30 mm/month, respectively. Schwaner considers that these different growth rates arise from the action of regulatory genes controlling maturation at different body sizes. Shine (1978) has shown that in elapid snakes growth rate relative to size at birth is higher in large species than in small ones. Furthermore, Shine (1978) has shown that there is a positive correlation between snout–vent length and onset of maturity (Table 6-1). In other words, the small Roxby Island forms are progenetic, compared with the ancestral (mainland) forms, while those from Franklin Island are hypermorphic. The selection factor has been size of prey: the target, body size. Peters (1983) shows a strong, general correlation between predator and prey body sizes in all animals. The governing cause is cost/benefit energetics of "capture/nutrition."

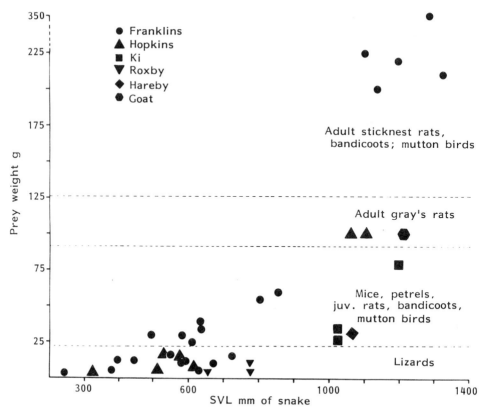

Figure 6-30. Relationship of snout-to-vent length (SVL) to weight of ingested prey in field-caught tiger snakes from offshore islands of South Australia. Reproduced from Schwaner (1985), with permission of the author.

6.5. Size Change in Local Growth Fields

So far we have been concerned with the adaptive significance of body size as a whole. However, local growth fields may undergo size change that is dissociated from body size. This can occur by any of the heterochronic processes, although local hypermorphosis and progenesis may be less common processes than the others, because organ growth offset is often under "global" (hormonal) control (tied to other organs). Thus, local size changes can occur where there is no change in body size between ancestor and descendant, or alternatively where body size is also changing.

While global hypermorphosis can play an important role in extending preexisting growth allometries, dissociation of local growth fields is sometimes necessary to produce a local trait that is larger than otherwise attained by global hypermorphosis alone, or, to reduce in size a particular structure, which if it continued on the ancestral allometry would produce a maladaptive structure. Into this latter category can be placed the evolutionary pattern of the long neck of many Triassic prolacertiform reptiles. Tschanz (1988) considers that this group was aquatic, on account of reduced cervical musculature and stiffened neck. The phylogenetic development of the family shows a trend toward

Table 6-1. Snout–Vent Length, and Probable Age, at the Attainment of Sexual Maturity (Time of First Mating) of Six Species of Elapid Snakes[a]

	Males		Females	
	Snout–vent length (cm)	Age (months)	Snout–vent length (cm)	Age (months)
Unechis gouldii	26	12/24[b]	23	12
Hemiaspis signata	29	12/24	35	12/24
Hemiaspis daemelii	30	<12	46	12
Austrelaps superbus	45	12/24	49	24
Notechis scutatus	65	24	65	24
Pseudechis porphyriacus	78	19/31	88	31

[a] From Shine (1978).
[b] Two ages, separated by a slash, indicates that a proportion of the population fails to mature at the earlier age; e.g., 12/24 means some mature at 12 months of age, but others not until 24 months of age.

increasing the neck length. This occurred by a combination of increased length of each cervical vertebra and also by phylogenetic increase in vertebral number (Tschanz, 1988). Early forms such as *Prolacerta* possess 8 cervical vertebrae that are only moderately elongated compared with later taxa. Extreme neck length was attained by the geologically youngest prolacertiform, *Tanystropheus longobardicus*, where the neck is more than half the total body length and comprises 12 very elongated vertebrae (Fig. 6-31). Polarity of evolutionary characteristics which demonstrates this trend is also shown by the oldest species of *Tanystropheus*, *T. antiquus*, which has 9 very elongated cervical vertebrae. Wild (1973) considers that the neck elongation in this genus occurred not only by meristic increase in number of cervical vertebrae, but also by pronounced allometric growth relative to absolute body size of individual vertebrae.

However, not only do prolacertiforms such as *T. longobardicus* show very pronounced positive allometries but even some earlier forms with relatively shorter necks, such as *Macrocnemus bassanii*, do as well. Rather surprisingly, comparison of allometries of individual vertebrae within the two species shows both a slightly higher averaged allometric coefficient for neck length and a greater variability in *M. bassanii* than in the geologically younger *T. longobardicus*. Tschanz (1988) has postulated that the relative decrease in variability of allometric coefficients in *T. longobardicus* may be a function of the larger body size attained by this species (which reaches 6 m, compared with *M. bassanii* which does not exceed 1 m). The relatively longer neck of the younger *T. longobardicus* was due to allometric hypermorphosis, allowing a longer positive allometric growth period. Tschanz argues that the relative reduction in allometric variability

Figure 6-31. Reconstruction of the Triassic prolacertiform reptile *Tanystropheus longobardicus,* showing the effect of extreme dissociation of local growth fields to produce a greatly elongated neck. Reproduced from Tschanz (1988), with permission of the Palaeontological Association.

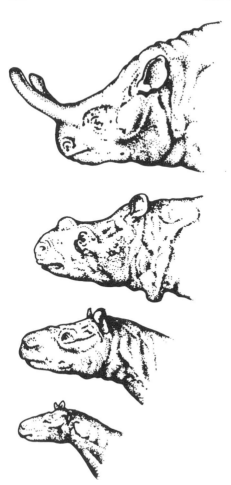

Figure 6-32. Early (lower) and later (upper) reconstructions of the head of titanotheres, illustrating increase in body size and peramorphic development of horns. Hypermorphosis was insufficient on its own to produce the large horns. There was also acceleration in this local growth field. Redrawn from Hersh (1934).

in the later taxon was a functional necessity. Had neck vertebrae in *T. longobardicus* attained the allometric coefficients found in *M. bassanii,* and the animal had still grown to 6 m in length, the resultant neck length would have been *too* long to be functionally efficient and would have become an adaptive liability. The allometric coefficients that it attained, which resulted in the neck taking up about half the animal's entire length, were at its adaptive limit. Above that a longer neck is likely to have been maladaptive, largely on account of the poorly developed cervical musculature. Thus, even though the evolution of traits, such as in this case a long neck, can occur as the result of scaling up of body size, dissociation of the trait can also occur to optimize its functional potential.

The same process appears to have operated in the opposite way in titanothere evolution, for selection on horn size was only made on individuals that dissociated horn allometry and onset of development from phylogenetic increase in body size. Originally interpreted as a classic example of hypermorphosis (Hersh, 1934), McKinney and Schoch (1985) reanalyzed Hersh's data and concluded that extension of the ancestral positive allometric relationship between horn size and body shape by increase in body size was insufficient to produce the large horns seen in *Brontotherium* and *Allops* (Fig. 6-32). The data indicated that there was also allometric acceleration (Chapter 2) in this, and some other traits. Moreover, Oligocene forms show allometric predisplacement of the horns. In other words, compared with their Eocene progenitors, horn development was initiated at a smaller body size in ontogeny. Had this not occurred, horn length would only have been about one-third of that actually attained. Thus, allometric predisplacement, acceleration, and hypermorphosis combined to produce large descendants with proportionately much larger horns in the Oligocene. The functional importance of larger horns can be assessed by considering the degree of stress that they could withstand. While the stress that the horns could withstand is equal to force/area, increase in body size would result in increased force of impact. An efficient way of compensating for this may have been to increase horn size in order to spread the load (McKinney and Schoch, 1985).

The evolution of smaller- or larger-sized traits may vary dimorphically and result in a reduction in competition for resources between the sexes. Dissociation of head growth from body growth has been recorded in species of North American garter snakes, such as *Thamnophis sirtalis parietalis* (Shine and Crews, 1988). Because of the action of testicular androgens early in development, inhibition of growth of the head occurs in males, resulting in smaller heads (shorter jaws) than in females. Body sizes are similar in males and females. Androgen administration occurs early in ontogeny and initiates different growth rates between males and females. Interestingly, castrated males develop heads proportionately *larger* than females. Whether there has been a phylogenetic decrease in male head size, or an increase in female head size is not known. Shine and Crews (1988) suggest that dimorphism in head size was selected in these snakes because it allowed each sex to exploit different foraging niches [a similar interpretation has been proposed for the same phenomenon in birds (Darwin, 1871; Selander, 1966), lizards (Schoener, 1967), and other snakes (Shine, 1986)]. Female garter snakes are known to take larger prey than males, by virtue of their larger jaws. Alternatively, the smaller head size might allow males to forage in different areas from females. The hormonal environment clearly plays a crucial role in early development. And it is likely that perturbations in this area at a critical time early in ontogeny play a highly significant role in affecting not only the whole organism, but also specific traits. The phenotypic and thus ecological, behavioral, and genetic consequences can have a profound influence on phylogenetic development.

Size changes in local growth fields are likely to have played at least as important a role in plant evolution. The modular construction of plants means that local growth fields, such as leaves, shoots, fruits, and roots, can undergo heterochronic changes quite independent of one another. However, there is some evidence that developmental changes in one module can directly affect other modules by a "knock-on" effect (Guerrant, 1988). One example of this involves what Silvertown (1989) has termed "the paradox of seed size and adaptation." Much work over the last few decades has established a firm positive correlation between seed size (usually determined by weight) and habitat. For instance, woodland herbs have larger seeds than comparable species inhabiting open habitats. This has been interpreted as an adaptation to different regimes of aridity or shade: large seeds are adapted to shaded or dry habitats, as their larger reserves allow establishment in these less-than-optimum conditions (Foster, 1986). As Silvertown has pointed out, research under laboratory conditions has long shown that seed size is apparently constant within species. The explanation was that stabilizing selection for a particular seed size limits the number of seeds that a plant can produce. Such studies have tended to view the evolution of seeds in isolation from the rest of the plant.

However, as Guerrant (1988) has stressed in his study of the role of heterochrony in plant evolution, developmental pathways followed by one module, or one set of modules, may directly affect others. For instance, Sinnott (1921) showed that the size of different plant organs (seeds included) was related to the size of the meristems from which they develop. Similarly, a close allometric relationship can exist between different sets of plant "modules." Primack (1987) has found that leaf length (which is positively correlated with plant height) was correlated with seed length. Likewise, seed size is also directly correlated with height on the stem (Waller, 1982).

From an evolutionary perspective it might be as plausible to argue for a nonadaptive approach to seed size. However, Primack (1987) has noted that a larger seed size is associated with lower rates of germination and, more significantly from the perspective of heterochrony, with lower relative growth rates of seedlings. Perhaps these are the targets of selection. Recent studies have shown that seed size in wild plants, rather than being conservative, is phenotypically plastic. Rather than being of direct adaptive significance, seed size may not be the actual target of selection. Close correlation between seed size and flower size or leaf length or plant height allows for a range of possible adaptive scenarios. Some of these features are most likely being carried along piggyback fashion. But unraveling the true targets of selection will likely be a long and tortuous path.

6.6. Life History Strategies

Perhaps one of the most important developments to come out of studies of heterochrony over the last decade is the linking of heterochrony with life history strategies. It has been argued that many traits that have evolved by heterochrony did so not because they were targets of selection, but because they were incidental *morphological by-products* of heterochronic changes induced by selection for particular ecological strategies.

Various aspects of life history strategies have been combined in the concept of r and K selection, originally proposed by MacArthur and Wilson (1967). An **r-selected** population is one that inhabits an environment that is either unpredictable or ephemeral. While the population may experience benign resource-rich periods of rapid population growth, free from competition, these are interspersed with very harsh periods during which unavoidable mortality occurs in the population. Thus, mortality rates of both adults and juveniles are highly variable and unpredictable, and often independent of population density and the size and condition of the individual concerned. The predicted characteristics of r-selected individuals are: small size, early maturity, semelparity (single reproduction), large reproductive allocation, and high numbers of small offspring.

A **K-selected** population occurs in a habitat that is constant or predictably seasonal in time. It suffers little random environmental fluctuation, so a crowded population of fairly constant size is present. Competition among adults is high, this largely determining the adults' rates of survivorship and fecundity. In this situation the young must also compete, resulting in few opportunities for them to become established as breeding adults. The predicted characteristics of K-selected individuals are large size, delayed reproduction, iteroparity (repeated reproduction), low reproductive allocation, and few, large offspring that often require extensive care.

The r/K concept has often been criticized, particularly with regard to its oversimplistic (dichotomous) nature. However, no other generally accepted alternative theories have been proposed. There is little doubt that many biologists find it a useful scheme. It is perhaps the very simplicity of the theory that makes it acceptable, though perhaps not valid in all situations. For instance, Stearns (1976) has shown that 18 out of 35 studies conformed to the model. Reasons why some life histories do not fit the model include the fact that reproductive cost may be far greater than its corresponding allocation and also that demographic forces beyond the r/K scheme may be important. For instance, the relationship between residual reproductive value and size can be an important factor in life history strategies. Thus, among small juveniles, there is a premium on large size because of intense intraspecific competition. The habitat will also be size beneficial, with a premium on small size. At larger sizes the habitat may become size-neutral, then eventually size-detrimental. Another problem with the r/K scheme is that it focuses on traits produced in response to competitive and physical factors, whereas there is growing evidence (e.g., Sih, 1987) to suggest that different predation regimes also influence life history strategies (a more complete discussion of the utility of the r–K continuum is found in Chapters 7 and 8).

In proposing a direct relationship between heterochronic processes and the r/K scheme. Gould (1977) suggested that many of the characteristics associated with particular processes were the same as those associated with either one end or the other of the r/K continuum. He proposed that *progenesis might be linked with r-selected strategies* as progenetic forms tend to have short life spans and thus rapid cycling of generations; small body size; and large litters. *Neoteny could be linked with the opposite, stable K-selected environments,* characterized by strong selection pressure for organisms with long life cycles; that attain large sizes; and have small litters (Fig. 6-33).

Not only progenesis and neoteny can be tied to r–K strategies. By delaying onset of maturity, and leading to large body sizes, *hypermorphosis is a major mechanism for the attainment of K-strategy characteristics.* Likewise, the counterpart to neoteny, *acceleration will be part of r-selected strategies.* In an analysis of ontogenetic development in a number of lineages of Eocene echinoids from Florida, McKinney (1986) found that 15 of 17 lineages showed trends of increasing size as they adapted into deeper water. This implied that selection was targeted on factors related to body size as the echinoids evolved into a more stable, deep-water, K-selected environment (Fig. 6-34). These factors were either larger size *per se,* or later timing of maturity or faster rate of growth. Minor local shape changes were apparently brought about by a mosaic of heterochronic processes. A similar relationship has been observed in Paleozoic bryozoans. Pachut and Anstey (1979) and Pachut (1989) have shown that colonies from low-diversity zones tend to be paedomorphic, whereas those from high-diversity zones are peramorphic. Global allometric progenesis has been correlated with onshore, more unstable sequences in some Paleozoic bryozoan assemblages (Pachut, 1989). Furthermore, diversity is also low.

It is probable that many of the allometrically progenetic forms that have been described (see McNamara, 1988b, for review) may have been r-selected. The fossil record provides some support for the view that these forms evolved in unstable environments by early maturation that led to small size, frequently associated with high fecundity. As we have noted, many allometrically progenetic bivalve faunas (Snyder and Bretsky, 1971) were probably r-strategists, inhabiting unstable environments. Likewise,

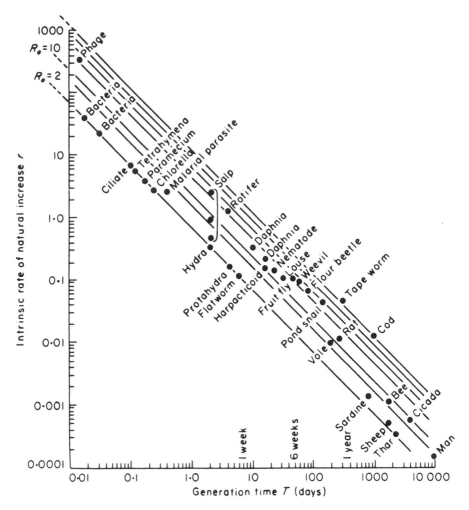

Figure 6-33. The relationship of the intrinsic rate of increase and generation time for a variety of organisms. Diagonal lines represent values of R_0 from 2 to 10^5. Reproduced from Heron (1972).

the small body size in oysters and ammonites in some Late Cretaceous micromorph faunas (Mancini, 1978), that might have been attained by progenesis (McNamara, 1988b), were adapted to soft, unstable substrates. Many small, apparently progenetic brachiopods similarly inhabited unstable environments (Surlyk, 1972). High fecundity among progenetic taxa may be shown by some allometrically progenetic species of edrioasteroids, such as *Timeischytes* (Sprinkle and Bell, 1978), that often occur in large concentrations. Many small trilobites that have been interpreted as being progenetic also tend to occur in relatively high numbers and with low diversity (McNamara, 1983a). An important caveat for this interpretation of size as a reflection of life history strategies, is that any incidental morphological change cannot be grossly maladaptive.

The overwhelming concern with interpreting the adaptive significance of every morphological trait has perhaps tended to obscure the true "trade-off" nature of the targets of selection. A change in body size that is favored by selection may result in a morphological change to a trait that is slightly deleterious. Selection will determine the compromise between these two opposing changes. As Mayr (1983) has observed, more than 100 years ago Wilhelm Roux recognized this as the *struggle of parts*. The phenotype is a compromise between competing developmental interactions [having originated via cell competition *sensu* Buss (1987)]. One way to assess this form of "trade-off" is to examine the ecological strategies of relatively closely related taxa in order to try and separate the relative significance of traits as targets of selection, from life history strategies. Hafner and Hafner (1988) have attempted to do this for the kangaroo mouse (*Microdipodops*) and kangaroo rat (*Dipodomys*). Found in the deserts of North America, these geomyoid rodents both share the same characteristics of an enormous head, huge hind feet, large eyes, and long tail (Fig. 6-35). It has long been argued that these features were specific adaptations to desert living (Mares, 1983): the large head counterbalancing the rodent as it hopped through the cool desert night air on its huge sand-paddle feet, searching for food with its large eyes and steering with a long rudderlike tail. However, Hafner and Hafner (1988) have argued that it was not these structures that were the primary foci of selection, but the ecological strategies of the two rodents. Although the two genera share similar morphological characteristics, their life history strategies are quite different.

The small size and retention of ancestral juvenile traits by the adult kangaroo mouse are likely to have arisen by progenesis, whereas the larger kangaroo rat, the adult of which shares the similar ancestral juvenile characteristics, probably arose by neoteny. The progenetic kangaroo mice have all the classic characteristics of *r*-selected animals: small body size, short life span, large litters, and they live in an unstable ephemeral desert environment. Kangaroo rats, on the other hand, possess features associated with animals situated near the *K* end of the *r–K* spectrum of life history strategies: long life span, slow development, long gestation period, enlarged brains, and small litters. It is these life history strategies which are more likely to have been the principal targets of

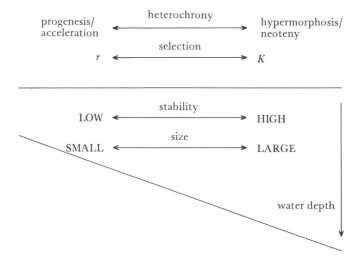

Figure 6-34. Proposed relationship between heterochrony, life history strategies, and water depth.

Figure 6-35. Heterochrony in geomyoid rodents. Upper drawing is of the hypermorphic large pocket gopher, *Orthogeomys* (body length up to 350 mm); middle drawing is the progenetic kangaroo mouse, *Microdipodops* (body size up to 77 mm); lower drawing is the neotenic kangaroo rat, *Dipodomys* (body size up to 200 mm). Redrawn from photographs in Walker (1964).

selection, rather than specific morphological features. This is not to say that the morphological characteristics of the two forms were of no adaptive significance—far from it. But they are likely to have played less significant roles as targets of selection than life history strategies. For instance, the large head possessed by both kangaroo mice and kangaroo rats, relative to body size, is an outcome of the negative allometric trend in head growth that is generally found in rodents. However, there is some degree of dissociation in this and other traits in some genera. The rodent genera *Pedetes* and *Jaculus* have a larger head than would be predicted from simple allometric regression; while in *Scirtopoda* and *Notomys* it is relatively smaller. Thus, localized neoteny and acceleration of traits is a confounding factor in the evolution of morphological traits in rodents, as with so many other groups that we have looked at. Even with the overall neotenic kangaroo rat, features such as hypertrophied auditory bullae and elongated tail are relatively hypermorphic, not having been subjected to the reduced rate of development that occurred in other traits. It is interesting that the long tail of kangaroo mice and kangaroo rats was achieved in different ways. The kangaroo rat has fewer, but longer caudal vertebrae than kangaroo mice, suggesting that acceleration of rates of development of individual vertebrae occurred, whereas in kangaroo rats there was an increase in actual numbers of vertebrae.

The argument that life history strategies have been the main targets of selection in geomyoid rodents is strengthened by examining other genera. Not all of them are paedomorphic. Some the Hafners consider to be peramorphic, having extended their early rapid growth period by delaying the onset of sexual maturity. The most striking of these is the pocket gopher, *Thomomys*. This hypermorphic rodent has ended up with a heavily ossified skeleton, strongly fused skull, and large size (Fig. 6-36). In other words, it has developed "beyond" the normal shape and size attained by a generalized rodent. Like the kangaroo rat, it is adapted to a K-strategic environment, but it was attained by a quite different heterochronic process, hypermorphosis, rather than neoteny.

In all cases discussed, there is *considerable agreement* with the predictions of *r–K* theory on selection of heterochronies. More testing is badly needed, especially in living groups where ontogenetic age and environmental conditions can be determined.

6.7. Fertility Selection

One of the basic tenets of the neo-Darwinian view of natural selection is that it is overwhelmingly dependent upon differential survival (Manly, 1985). In recent years

Figure 6-36. Comparison of skulls of (A) adult pocket mouse, *Chaetodipus,* (B) neonatal pocket gopher, *Thomomys,* (C) adult pocket gopher, *Thomomys,* and (D) adult pocket gopher, *Orthogeomys.* Note the similarity between the juvenile pocket gopher (B) and the adult pocket mouse (A). Reproduced from Hafner and Hafner (1988).

  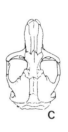 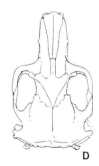

A   B   C   D

there has been an increased interest in other modes of selection, in particular sexual selection (the nonrandom differential ability of phenotypes to acquire and retain mates) and, of particular interest to us here, fertility selection. This is the nonrandom differential rate of offspring production among phenotypes (Travis, 1988). Intensities of fertility selection may be as high as those for differential survival (Endler, 1986). Differential fertility among phenotypes can arise because some female phenotypes are more fecund than others; or because they reproduce more frequently. One of the major factors in more frequent reproduction is earlier maturation. Thus, as we have argued in the previous section, *progenetic forms often show higher levels of fecundity,* and in some cases it may be that this higher resultant level of fertility is the prime target of selection. From a quantitative genetic aspect, small phenotypic differences in the timing of first reproduction can be magnified into large fitness differences that reflect back on the gene effects as a strong epistasis in fitness (Travis, 1988).

Another aspect of fertility selection is the enhanced virility of the male. This can arise from changes in the male's life history strategy that allow a more effective fertilization of the eggs. Possibly one of the most stunning examples of this is the extreme progenetic state of male prosobranch gastropods, such as *Enteroxenos oestergreni* where the gastropod male is so precociously developed that it is little more than an enlarged male gonad which attaches itself to a special male receptacle *within* the female's body (Chapter 4). In such situations it is the extreme small size that is the major target of selection, as it allowed the male to live within the female and fertilize her directly.

Heterochrony that produces such changes in the timing of maturity can play a crucial role in "sympatric" speciation. This can occur and be accompanied by little or no shape or size changes between descendant and ancestor, only changes in the timing of reproduction. Such a situation has occurred in the last 200 years during the evolution of races of the apple maggot fly *Rhagoletis pomonella* (Feder *et al.,* 1988; McPherson *et al.,* 1988; Smith, 1988; Barton *et al.,* 1988). Originally native to hawthorn trees in North America, this fly has spread and infested apples, cherries, roses, and pears. Behaviorally, offspring from apple trees are more likely to lay eggs on apples than are descendants of hawthorn flies. Significantly, there are differences in timing of onset of maturity of the fly on different hosts. Under laboratory conditions the "ancestral" hawthorn fly takes between 68 and 75 days to mature, while the apple fly takes only between 45 and 49 days. In this respect it is progenetic. However, flies that infest fruit of the dogwood terminate diapause and become mature much later, between 85 and 93 days (Smith, 1988). Smith considers that the timing of maturity, which is brought on by the termination of diapause, corresponds to the period of maturation of the fruit of the host species. Because apple fruit mature mid to late summer, hawthorn early autumn, and dogwood mid to late autumn, timing of maturity varies correspondingly in the different flies. The effect of this is not only the establishment of behavioral barriers to gene flow, but also, and perhaps far more significantly, the establishment of developmental barriers, as mating times will vary between the flies on different hosts.

Each race has been found to be genetically distinct (Feder *et al.,* 1988; McPherson *et al.,* 1988). Thus, in classic terms we are dealing with sympatric speciation. There is no geographical barrier to speciation, just behavioral and developmental barriers. Such sympatric speciation, which obviates the need for a prolonged period of geographic isolation, may account for rapid evolutionary radiations. Barton *et al.* (1988) cite the example of 300 species of leaf miners in Britain and 750 species of fig wasps, each species only being able to grow within its own species of fig. Such host shift (which will often necessitate developmental changes in order to coincide with the timing of fruiting of the host species) provides an effective barrier to gene flow.

The timing of metamorphosis in amphibians and insects, i.e., the transition from juvenile to adult state, can vary widely among individuals and between populations (Semlitsch et al., 1988), and can influence adult body size. This may have a direct impact on other factors, such as adult survival, age at first reproduction, and fecundity (Moeur and Istock, 1980; Prout and McChesney, 1985). Adult body size in amphibians is thought to be related to mating success and to the number and size of eggs produced by females (Semlitsch et al., 1988). For these animals, attainment of relatively larger body size will have the effect of improving species fitness (in contrast to the improved fitness attained in some parasitic gastropods and other invertebrates by attainment of much smaller male size).

A parallel situation can occur in insects. Delay in onset of maturity can lead to larger body size, which increases female fecundity levels. Extrinsic factors may play a direct role in this. For instance, the mayfly Baetis alpinus that inhabits high mountain streams in the Pyrenees attains maturity at different ages depending on altitude and temperature (Lavandier, 1988). At an altitude of 1920 m the temperature range is 0–11°C and the life cycle takes 1 year. At an altitude of 2190 m, where the temperature range is 0–7°C, the life cycle takes 2 years to complete. Subimagos (the stage immediately after metamorphosis) are larger in the upstream population. As fecundity is positively correlated with body length, the upstream population has higher fecundity levels.

Chapter 7

Behavioral and Human Heterochrony

You scream, scratch, and throw coconuts, apparently, and then, having another inch or so on top of the skull, you produce the dialogues of Plato, Macbeth, the Ninth Symphony, and the catalogue of the Museum of the Royal College of Surgeons.

In short, when the skull is the right size, it immediately begins to put itself in museums. I am sorry, but there is something here that eludes me.

J. B. Priestley, Among the Glass Jars

J. B. Priestley was a metazoan whose unique genomic combination was created in 1893. These genes programmed a series of cellular mitosis, migration and differentiation which resulted in a cellular assemblage with an exceptional proficiency in the emergent property called "playwriting" by other metazoans. Due to limitations on the number of cell cycles in cellular constructs of his complexity, Priestley died in 1984. His component atoms (mainly C, H, O, N, and P) now participate in the biogeochemical cycle, where most are currently active in the photosynthetic pathways of local metaphytes. Ancestors of these metaphytes, probably green algae, adapted to land over 400 million years ago.

M. L. McKinney, "Now You Know: A Scientist Explains" (unpublished)

1. INTRODUCTION

One of the main themes of this book is that viewing evolutionary change from a heterochronic perspective can often provide much insight, in particular by showing relationships among ontogenies of related groups. We feel that this potentially powerful explanatory view has been largely neglected because of needlessly intimidating terminology and convoluted classifications. Perhaps nowhere else has this neglect been more strongly felt than in the study of evolutionary changes in behavior. Such changes are obviously tightly intertwined with morphological evolution (as both cause and effect) and evolutionary insights about behavioral changes are difficult to come by.

Yet behavior undergoes a predictable sequence of changes during ontogeny in all organisms. As a result, there is great potential for change in timing and rate of that sequence. More importantly, such a behavioral "sequence" is actually composed of many component behaviors which often covary and can therefore undergo differential changes in rate and timing just as morphological traits do. This can produce some fascinating results (behavioral "mosaics" and "dissociations" of behavioral traits) that are only now being documented and explained as the heterochronic changes that they are. This should surprise no one since all behavior must ultimately have genetic roots of some kind and any genetic change is subject to change in timing or rate. Such changes should be most clear-cut in behaviors that are most directly linked to genetic "programming," i.e., reflex and instinctive behaviors. This seems to be the case, as discussed in the first part of this chapter.

In the larger, second part of this chapter, we focus on heterochronic changes in "higher" forms of behaviors, i.e., learning and reasoning. The changes here are less direct but are clearly crucial, as shown in human evolution, the most extreme case. We will show that heterochrony has been the proximate causal process behind all three kinds of changes that led to human adaptation through culture: (1) enlargement of the brain, producing more neurons used to store and process more learned information; (2) the multitude of morphological correlates, such as bipedalism, that coevolved with cultural adaptations; and (3) the life-history changes that affect such key traits as fecundity, the duration of the learning period, and the longevity of older people which allows them to pass on knowledge to younger ones. It speaks highly of the evolutionary potential of heterochrony that all three of these seemingly disparate traits are largely interrelated by one simple heterochronic process: hypermorphosis. We emphasize: the general process is *not neoteny* as has been repeatedly written in many books and articles. This is one of the most widespread bits of misinformation in the history of science. Considering its importance to the evolution of our own species, we hope to correct this by stating the facts as clearly as possible in the latter part of this chapter. But to state it succinctly here: the large human brain, our long learning period, and our large body size (the second largest of 100-plus primates) are the direct result of growing *longer* in all phases of growth (delayed offset or hypermorphosis). It is *not* the result of growing slower (neoteny). Indeed, our rates of growth are (very) generally on par with those of the chimpanzee; we simply experience those rates for longer periods of time and so grow larger, and we live longer than our ancestors.

Aside from reviewing the heterochronic evidence found in the fossil record and living apes for human evolution, we will also discuss some applications of heterochrony for explaining aspects of human races and human sexual dimorphism. We close with a brief look at the role of behavior in evolutionary rates.

2. HETEROCHRONY OF BEHAVIOR

A quick look at most of the literature on animal behavior will show that little space is usually devoted to the ontogeny of behavior, much less a "specialized" topic such as heterochrony. Fortunately, this has begun to change in the last decade [Burghardt and Bekoff (1978) was an important stimulus]. This added dimension will contribute greatly to the utility of behavioral traits in reconstructing phylogenetic events. For instance, the historical pathway that led certain moths to evolve from drinking flower nectar to sucking blood involved a number of intermediate behaviors, such as drinking thick-skinned fruit (Alcock, 1983). Instead of focusing on adults of the intermediate species as is often done in such reconstructions, observation of ontogenetic sequences of the intermediates and end-member taxa would be very useful. Perhaps "terminal addition" of behaviors occurred and intermediates and bloodsucking forms go through the nectar drinking stage in earlier ontogeny. Or perhaps the change was effected earlier in ontogeny and "recapitulation" does not occur. Such observations not only provide more information for phylogenetic reconstruction but also provide crucial insight into the ontogenetic mechanisms which produced the variations and the external conditions which selected them.

As it is most easily studied, we begin with an ontogenetic view of "programmed" behavior. Then we move on to heterochronic aspects of more complex learned behavior. In so doing we will try to clarify what is meant by these two end-member types of behavior.

2.1. Heterochrony of "Programmed" Behavior

"Programmed" behavior (such as reflexive and instinctive actions) may be defined as a relatively simple innate behavioral response (a very relative term given how "complex" such programmed behavior as prey handling can be) to a given environmental stimulus. Usually these are attributed to "lower" organisms. However, since neurological evolution has been basically conservative and accretive, many such behaviors are retained by "higher" creatures. Thus, many "emotional" responses in humans are rooted in the R-complex or "reptilian brain" and the limbic system which has been added to it; "higher" thought results from the most recently added evolutionary product, the neocortex (review in Masterton, 1976).

Given the prominent role of amphibians in classical heterochronic work, they would appear to be best placed to examine heterochronic effects on innate behavior. Yet, as Stehouwer (1988) has recently noted, "there are few systematic laboratory studies of the development of amphibian behavior . . . all of which focused quite narrowly on mechanisms of locomotor development" (p. 384). However, field studies on frogs have been done, and these suggest, not surprisingly, that the behavior of tadpoles is greatly different from that of adult frogs. In particular, the tadpoles are much more active than frogs which mainly sit motionless waiting for prey. More importantly, Stehouwer (1988) himself provides an interesting laboratory study which precisely documents the changes

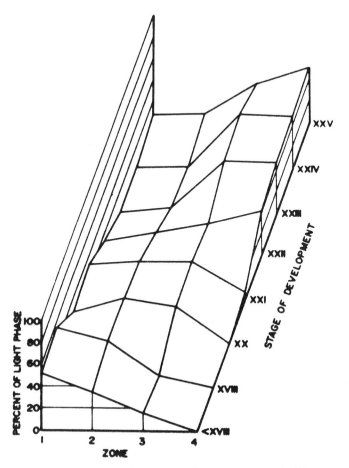

Figure 7-1. Habitat preference of developing frog, measured as time spent in each of the aquarium zones. Zone 1 = deep water, zone 2 = shallow water, zone 3 = transitional, zone 4 = land. Modified from Stehouwer (1988).

in behavior which accompany changes in morphology. Metamorphosis occurs between stages XVIII to XXV and it is during this time that a suite of pronounced behavioral changes occur, in synchrony with pronounced morphological changes. Of special note is the change in habitat preference (Fig. 7-1). This study is only a first step toward detection of behavioral rate and timing changes. The direct heterochronic questions, such as: can behavioral and morphological changes be decoupled or can certain behavioral traits be decoupled from one another?, must await further study of this group.

For a variety of reasons, the literature on the behavior of other animals is much more complete. Beginning with the classic works of Lorenz for example, birds have been under intense scrutiny so it is no surprise that some of the best direct information on behavioral heterochrony (meager as it is) is found here. Among Lorenz's many studies (e.g., Lorenz, 1965, 1970) are those which document the evolution of displays and preening behavior in many species of ducks. There is a clear pattern of different degrees of elaboration among species and at least some of it appears to be traceable ontogenetically. Unfortunately, there have been few systematic studies of the phylogenetic aspects of such ontogenetically changing behaviors. Many tantalizing fragments of observations indicate that such studies would be very productive. For instance, Friedmann (1929) long ago reported that in cowbirds the courtship displays of one species are similar to those of another species at an earlier stage of ontogeny. In many birds, courtship displays and song development have strong learned components. However, cowbirds seem to inherit much or all of this behavior directly. For instance, inexperienced females will automatically assume a precopulatory position, even if they have never heard a male before, while young males can sing the appropriate sounds, having never heard them (King and West, 1977).

Even without systematic studies, it is possible to pick out key heterochronic information from breeding experiments. For example, "Oedipus behavior" in geese occurs when a wild goose mates with a domestic one. Male offspring of this union attempt to mate with the mother because domestic geese carry genes for earlier sexual maturation (progenesis) while wild geese carry genes for later offset (hypermorphosis) of mother-following behavior. When both of these genes are transmitted to the same male offspring, the poor confused fellow is compelled to follow his mother while at the same time having other compulsions to mate with the nearest available female (Harrison *et al.*, 1988). This example makes the crucial heterochronic point that behavioral traits can indeed be decoupled in rate or timing of development, just as local "dissociations" can occur in morphological growth fields.

One of the few systematic studies of behavioral heterochrony has recently been completed by Irwin (1988) whose excellent study analyzed the ontogeny and phylogeny of bird song. She found widespread evidence of behavioral heterochronies which could be used to help reconstruct phylogeny. Bird song development generally follows von Baer's law in that it proceeds from early, generally distributed stages to later, specialized stages. She also found general correlated behaviors that changed as "suites," such as correlations between song repertoire size and mimicry. This is a behavioral analogue to the morphological covariation that is often mentioned as a "constraint" on evolution (Chapters 6 and 8). Here, behavioral change is constrained through linkages to other behaviors so that one cannot change without the other, perhaps slowing behavioral evolution in some cases and/or "dragging" non- or maladaptive traits along.

Also of interest is the phylogenetic information found in Irwin's ontogenetic comparisons. As shown in Fig. 7-2, song development is characterized by a definite sequence of changes. By comparing the ontogenies of certain species to such a sequence, she found, for example, some groups that are clearly paedomorphic, in singing continuous songs and other traits (retain mimicry and have many syllable types). Even the type of heterochrony leading to the paedomorphosis was discernible in some cases. The mimids are apparently neotenic for song development. They never reach the adult stage of crystallized song. Instead the song develops slowly throughout life, ever-changing until they die. They are like *perpetual juveniles* for song development, continually learning new syllables and dropping old ones. In contrast, the reed warblers appear to be progenetic. Their songs crystallize at a relatively earlier stage than most and no new ones are

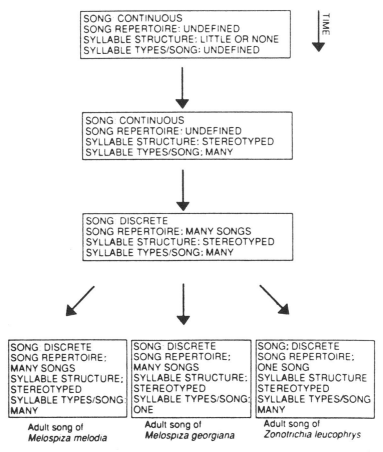

Figure 7-2. Ontogenetic stages of song in three sparrow species. From Irwin (1988).

learned thereafter. (As in many cases the exact ancestor–descendant relationships are unknown. We can only say that such cases are "relatively" paedomorphic, as discussed elsewhere. Perhaps the more specialized groups were derived from the "paedomorphs" by peramorphosis of behaviors. Nevertheless, the important elements of the relative changes are clear. Only the direction of the polarity is unknown—see Chapter 5.) Just how many other behaviors covary with these rate and timing changes in song development is uncertain. However, such considerations provide a fascinating prospect for future research in assessing the role of developmental constraints on behavioral evolution. Obviously some behavioral "decoupling" is possible as shown not only by the Oedipus example but also by Fig. 7-2 (compare adult sparrows). The question of how much is possible awaits more data.

This songbird example includes learning as an important part of song development, unlike the cowbirds. However, even here there is a strong, perhaps overriding innate component. A song "template" is inherited that is only modified through learning. For example, a white-crowned sparrow made deaf at birth can still sing a rudimentary song, with much of the same characteristic ontogenetic sequence of Fig. 7-2 (Konishi, 1965).

Further, even the act of learning is under strong direct genetic control since learning is almost always restricted to often very limited "critical periods" often determined by hormonal "programming," as discussed shortly.

Invertebrates often exhibit "programmed" behavior to an even greater degree than birds and we would expect similar kinds of heterochronic changes. The rampant heterochrony documented in invertebrates so far (Matsuda, 1987) has focused mainly on morphology but it is clear that accompanying changes in behavior also occur. For instance, in the red carpenter ant, behavioral studies have found that nestmate recognition cues (and thus the recognition response) are altered with age (Morel *et al.,* 1988). Hallas (1988) has recently used the ontogeny of web-building behavior to infer phylogeny and adaptive strategies of three species of jumping spiders. Their repertoires were generally species-specific, modified from some common ontogeny. She concluded that the ontogeny of behavior patterns is as potentially valuable as morphology in tracing phylogeny. Of special note are the many heterochronic transformations shown by social insects, wherein larvae can be induced to metamorphose into a number of different adults depending on colony conditions, diet, and so on as motivated by simple hormonal changes (see Chapter 3). Clearly there are also behavioral changes involved when delayed offset of larval metamorphosis causes a larger soldier ant to be produced instead of a small worker ant (Wheeler and Nijhout, 1981). The different groups of termites and their often varied caste systems must involve accompanying behavioral heterochronies as well (Roisin, 1988).

2.2. Heterochrony of Learned Behavior

No behavior is truly fully "programmed" or fully learned. These are end-members of convenience, with genetically determined innate behavior ("nature") at one end and environmentally determined behavior ("nurture") at the other. All behaviors have varying amounts of both in that learned behaviors must ultimately have genetic roots and innate behaviors usually need some kind of environmental stimulus, even if it is only the body's "internal" environment. We repeat these well-known distinctions because they point out that to understand the role of heterochrony in the evolution of learned behavior, we need to focus on the genetic components of it since it must be heritable. Since learned behavior is dependent on a large number of neurons devoted to the storage and processing of information, the most basic role of heterochrony in the evolution of learning has been in the production of ever larger numbers of neurons and the connections among them. This major aspect will be discussed shortly.

However, it is important to first point out that, even given these available neurons, learning is a sequential process that shows both rate and timing elements of its own. For example, the process of learning the grammar and vocabulary of a language is a step-by-step process that therefore has its own intrinsic rate of growth and completion (offset time). Such a rate or time is no doubt very far removed from the action of genes although even here the remoteness is not complete. For instance, one might argue that children with more neurons in certain areas learn language faster. Under even more direct control of genes is the onset and offset timing of such learning. Thus, the learning of language (to continue the example) has definite developmental timing aspects in many ways. Infants up to the age of a few months are able to discriminate among all phonetic sounds. However, they soon begin to lose this ability and their sensitivity narrows until by the age of four they can no longer discriminate sounds outside of their own native language (Werker, 1989).

Many learned behaviors exhibit such **critical periods,** defined as that limited time when an outside stimulus is required for development (see Scott, 1978, for review). Nonhuman examples include socialization in dogs, which must occur from about 3 to 12 weeks (Scott, 1978), and birds. The main points are that the timing of these periods can vary greatly, as in the bird species noted above, with major effects on what is learned and how much, and that this timing is often under direct "internal" control. For example, castration of a juvenile chaffinch delayed song learning until testosterone was artificially administered in the bird's second year (Nottebohm, 1978). Such critical periods are particularly important in our own evolution in that much of our vaunted cultural abilities (e.g., language learning) have resulted from delay and prolongation of the basic primate critical periods. Indeed, critical periods are a prime candidate for heterochronic change in many species.

More recent work has modified the critical period concept to include those many behaviors which have large innate and learned components (Aslin, 1981). In these cases, the outside stimulus is not totally necessary for the development of the behavior, but only accelerates it or allows it to reach full development (Fig. 7-3). Language acquisition seems to follow a maintenance/loss pattern wherein a capacity is innately programmed but is lost unless continued learning occurs. In all cases, the onset of experience is susceptible to timing constraints, many or most of which are most proximally related to hormonal events.

2.3. Heterochrony and Neural Bases of Behavior

To better understand the role of heterochrony in all behaviors, innate or learned, we return to the cellular view which we have promoted throughout this book. The focus will thus be on neurons. In this section we present a brief look at brain development in order to compare heterochronies in innate versus learned behavior. Later, in discussing human evolution, the subject will be raised again.

When one views "programmed" behaviors, the neural basis seems to hinge on the origin and number of nerve cells that are quite specific about the behaviors that they motivate and the stimuli that initiate them. An elegant study by Gurney and Konishi (1980) demonstrates this principle in zebra finches. Males produce a complex song in order to court females. The neural basis of this behavior is found in a chain of distinctive neural elements that constitute the "song system" leading to the vocal organ. In males these elements are much larger and more numerous. The genetic underpinnings of this development can be traced to the production of estrogen in the embryonic gonads of the male (in birds the males are XX and the females have the Y chromosome). Either directly or indirectly, the estrogen thus stimulates cell growth and differentiation in critical parts of the brain. While some learning is involved in the song of this species, the underlying

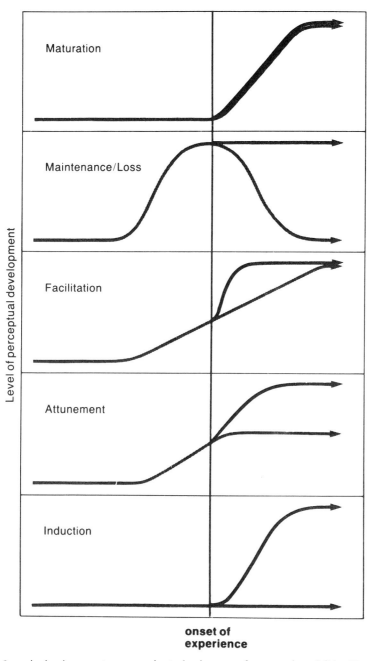

Figure 7-3. Several roles that experience may play in development of perceptual capabilities. Upper arrow in each graph shows development after onset of experience; lower arrow shows development if experience does not occur. At one extreme, in maturation, complete, or nearly complete, development will occur whether or not experience occurs. At the other extreme, induction, development will not occur at all if experience is lacking. In intermediate kinds, experience is necessary for some level of development. From Aslin (1981).

template is inherited and appears to lie in these particular neurons. Ostensibly a molecular "programming" is involved, much as memories are stored in molecular arrangements. In innate behaviors the arrangements apparently are constructed directly by genes.

In terms of heterochronic changes in cellular self-assembly (Chapter 3), we may view changes in innate behavior as changes in nerve cell multiplication or movement. This results in the net addition or subtraction of neurons in certain areas of the brain, neurons that are "programmed" to respond in a certain way to certain stimuli. Thus, local heterochronies such as late offset or acceleration of neuron mitosis could carry out enhancement of a given behavior.

Shifting our focus to the other end of the gradient, learned behavior, neuron numbers are critical here as well. Information from experience is stored in neurons and obviously the more there are, the more can be stored. However, connectivity among the neurons assumes a greater importance in learned behaviors as the greater amount of information is useless unless it is "processed" by association and communication among the cells. In terms of information theory, the value of any data bit is context-determined: only in the context of other neural connections (i.e., other stored data bits) does any new information have utility.

In the last 15 years, neurobiologists have come to realize that these crucial connections among neurons are not "hard-wired" (predetermined) in a precise way during ontogeny. Rather they undergo a twofold process of **neural competition and regression** (see Clarke, 1985, and Purves, 1988, for reviews) that may be described as follows. The adult's complement of neurons is established at a very early age (even in our much-delayed human development, all such brain cells have formed by 5 months after conception). Thereafter, brain (and nervous system) growth is limited mainly to the growth of supportive tissues (glial cells) and growth in size of the neurons through extension of axons and dendrites. Axons carry a regenerative electrical signal (action potential) to sometimes very long distances. The shorter dendrites are profuse branches that passively receive innervations. These extensions grow toward "target neurons" apparently guided by signal molecules, perhaps in the form of a concentration gradient. Much current neurological research is in this area. For example, Gehlsen et al. (1988) have recently identified the basement protein laminin as a guiding signal. As discussed later, timing and rate changes can occur in such axon and dendrite growth.

The main point is that more neurons and extensions originally form than are ultimately needed and those not forming functional circuits die and wither away. In some cases, up to 80% of a cell population may die (Rakic and Riley, 1983). As Bonner (1988) has discussed, this process has evolved because it allows large numbers of neurons and connections to develop without the need for precise gene specification of each event. Such "hard-wiring" would need to encode far too much information and, additionally, even if it were physically possible, the inflexibility to inevitable perturbations would be perhaps insurmountable. Also, in neuron death, errors in connectivity are removed. Purves (1988) argues that the major proximal cause of neuron survival is "trophic" interactions by which neurons are sustained by intercellular molecular messages. Those with insufficient connections therefore die. It is interesting that this neural competition and regression, dovetails elegantly with the ideas of Buss (1987) that cell competition and selection are behind the evolution of complex life. In any case, the process is clearly highly successful, most evident in the human brain which has over a trillion brain cells, many cells having over 10,000 connections each. Heterochronically, this discussion leads to the notion that changes in the rate and timing of neural development may not only affect the number of nerve cells but may alter patterns of connectivity as well.

This obviously oversimplified discussion may give the misimpression that "rate" or "timing" genes for neuron mitosis or dendrite growth may underlie all changes. While this may be true in some cases, the real complexity of brain development indicates that many kinds of gene changes can affect the timing of neuron and dendrite development. A sampler of potentially affected processes from Welker's (1976) overview of brain development would include: maturation of receptors, motor endplates, and synaptic conglomerates, all of which are dependent on localized biochemical contacts; gradients of differentiation in neuroblasts which result from (presumably) biochemical signals; sensitivity to neuroelectric stimulation; and many others.

Synthesis: Heterochrony and Brain Sculpting

In Chapter 3, we emphasized that development may be profitably viewed as a process of cellular self-assembly. In particular, the three activities of cell mitosis, migration, and differentiation cover the major aspects of all that is involved (sensibly enough: changes in number, position, and kind of cells subsume just about everything). The same logic holds true as well for neuron behavior during development with the addition, mentioned above, that cell death (regression) plays a key role in determining cell number, as well as the communication among the cells. Therefore, changes in rate and timing of cell death must be considered in assessing the role of heterochrony in affecting brain development (this is also true of some somatic development). This view is not only our own, having been extensively reviewed in a heterochronic context by Finlay *et al.* (1987) to whom the interested reader is referred. As with development in general, brain growth can be changed in both local and global ways through these changes in cell assembly; the addition of cell death as a major process creates an even greater reservoir of potential change as we shall illustrate.

For neurons directly participating in programmed responses, cell death is a way of carrying out local changes in cell number that can affect response. For example, in the above case of the zebra finches, the increased size of the male song system results from differential cell death that is hormonally regulated. Males and females begin with local neural structures of similar size but there is less cell death in males (Konishi and Akutagawa, 1985). Numerous other examples are known of this kind of local sculpting of innate behavior, such as more motoneurons in finch males in areas involved in copulatory behavior (see Finlay *et al.*, 1987, for examples). The main point is that there are numerous heterochronic changes that can lead to this kind of local change. Not only can local mitotic rates and offset/onset times be involved but because of cell competition for connectivity, rates and times of axon growth are critical in determining how many neurons are trophically sustained and thus survive. For instance, time of generation of cortical cells precisely predicts their eventual position and adult function in rhesus monkeys (Rakic, 1974). Timing differences in axon outgrowth of projecting fibers in retinal ganglion cells in the hamster appear to help control eventual morphology and function (Frost *et al.*, 1979). They are determining factors in which neurons survive and which do not. A key, very recent study (Davies, 1989) has just shown that, in chick cranial neurons, differential growth rate to target cells among axons is intrinsically regulated. Hence, they are highly amenable to genetic control.

Of course, local neuron sculpting is crucial in learned behaviors too. Localization of brain function occurs in many parts of the cortex, such as Broca's area for speech in humans (see below). It seems likely that progressive evolution of language in humans has occurred through such change. However, in the case of learned behavior we must consider two additional kinds of cell assembly changes as being of major importance: (1) global changes in overall cell number, and (2) cell connectivity.

Animals with enhanced learning abilities have more neurons in many parts of the brain, in particular the cortex. Heterochronic changes in rate or duration of mitosis can lead to regional or global increases in cell number in many ways. For example, change in maturation time via manipulation of thyroxine levels in rats results in major changes in rate of neuron origination, duration of axon outgrowth, and even cell death (Lewis *et al.*, 1976; Lauder, 1978). In a similar vein, gerbils have about twice the number of retinal ganglion (and other) cells as hamsters, largely because the gerbils' gestation is nearly twice as long (Wikler *et al.*, 1986). As we shall see, a similar process is operating to create the greater number of cells in our own large brains.

Because of the nature of the competition–regression process, *connectivity alterations accompany changes in neuron number.* In general, systemwide cascades of connectivity changes are promoted by the mitotic process itself. Increases in the number of muscle cells provide more targets for the neurons which innervate them, as experimentally shown in the chick limb (Yip and Klein, 1984). However, more than just correlative connectivity changes can occur; patterns of connectivity may be altered. In the gerbil example, the larger number of neurons means that more cell axons are competing for access to central target cells and they are competing for a longer time due to prolonged growth. This will have a major impact on how the axons are sorted and how the connectivity is segregated. Longer periods of axon competition lead to greater amounts of axon sorting and can have major effects on channels of communication among neurons (Frost, 1984).

In summary, local, regional, or global changes in neuron number and connectivity can result from simple changes in timing or rate of neuron assembly. The addition of cell death as a major process adds another dimension to the flexibility of changes. This is best seen in large, highly connected brains such as our own where rapid growth and extensive, diffuse connectivity occur early, only to be modified by later changes of neuron competition and regression. Perhaps the best summation is provided by Finlay *et al.* (1987, p. 114): "Alterations of timing in developmental events may produce different initial states for the competition to proceed from or different durations for the competition to occur in . . . ." Resulting changes can affect innate behavior in the number of programmed neurons and their connectivity to stimuli. It can affect learned behavior via the number of neurons available for information storage and connectivity for associative activities in information processing.

2.4. Behavior and Morphology

The view of brain sculpting and neuron assembly is a frankly reductionist one that has some heuristic value in relating the mechanics of cell development with emergent qualities like organismal behavior. However, for heterochrony to be more than simply a taxonomy of labels describing rates and times of cellular differentiation, mitosis, and migration, it must also be applied to emergent processes at higher levels of development. In other words, it must help explain why some cell configurations occur and not others. Evolutionarily, this means we are looking at the issue of constraints. Organisms (or brains) are not free-moving colonies of cells which move around randomly.

We discuss this in more detail in Chapters 6 and 8 but our point here is that the study of constraints is an empirical issue which can only be resolved by study of the specific developmental system in question. In the context here, of behavior, we must look at correlations among neuron multiplication and connectivity and, more visibly, behaviors, the last being the phenotype on which selection truly acts. For example, the pygmy chimpanzee is not only paedomorphic in morphological development, but exhibits generally paedomorphic behavioral patterns as well (Shea, 1988). Similarly, dog breeds consist largely of heterochronic variants and behavioral heterochronies may well correlate with the known morphological ones characterizing different breeds (Wayne, 1986). Correlations and dissociations among "local" behaviors may be expected as well. For instance, a descendant adult could show juvenile behaviors of "playfulness" while developing adult patterns of courtship displays (recall the Oedipus behavior in geese). Such mosaics have rarely if ever been couched in heterochronic terms. They would require much genetic programming of the behaviors. Given the strong linkage between behavior and morphology, we also expect that *behavioral suites would also show ontogenetic linkage with local morphological traits* (e.g., bird beak morphology with feeding behavior). De Ghett's (1972) study of ontogenetic changes in behavior and morphology in the Mongolian gerbil seems to bear this out in revealing discrete clusters of development (Fig. 7-4). Thus, the concept of heterochronic changes in dissociated "growth fields" (Chapter 3) can be extended to include behavioral traits as well.

Such morphology–behavior linkage is often a complex causal feedback that often defies any simple explanation. It is often stated in evolutionary texts that change in behavior precedes change in morphology. However, it is also true that a heterochronic change in morphology may precede change in behavior. For example, it was long thought that feeding behavior of various finch species had modified beak sizes for optimum intake. However, it was later shown that the finches in fact learn through experience which seeds are most suited for their particular beak (Kear, 1962). In many cases, both morphology and behavior change each other in a positive feedback loop. But for whatever reason, the manifold developmental time and rate correlations among morphology and behavior are real, with genetically controlled underpinnings, and are subject to selection. A heterochronic view of such changes not only provides a good descriptive framework, but has explanatory value in seeking to understand the selection behind such correlations and their strength in constraining or guiding evolution.

2.5. Behavior, Morphology, and Life History

Heterochronic changes can occur in any of three basic realms of an organism's ontogeny: behavior, morphology, and life history (at the cell level: neuronal, somatic, and gonadal cells, respectively). This is reviewed in Chapter 8 but for now, having just

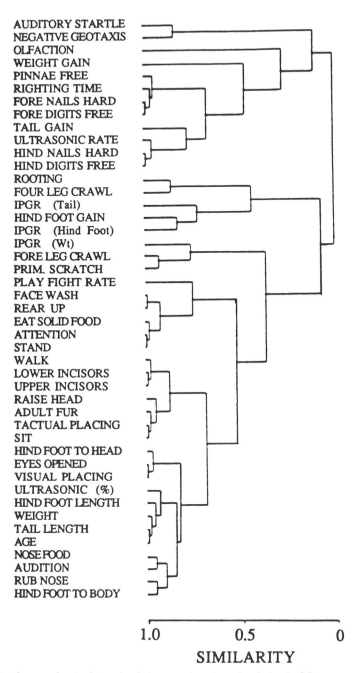

Figure 7-4. Dendrogram showing hierarchical cluster analysis (complete-linkage) of the ontogenetic associations between behavior and morphology of the Mongolian gerbil. Redrawn from De Ghett (1972).

discussed behavior and morphology, we turn to cases that include life history as well. Lawton and Lawton (1986) show that young males of social birds tend to retain paedomorphic plumage and other morphological characteristics for some time after reproductive maturation. Selection has favored this decoupling of morphology and gonadal maturation because paedomorphic males do not elicit attack responses from older males; they are viewed as nonthreatening. On the individual level, this is advantageous to the young male who is able stay in the area to gain experience observing older males and (sometimes) manages to copulate with defended females. Further, as Lawton and Lawton argue, on the social level such nonaggressive behavior is crucial for the integrity (and evolution) of the social structure of many social birds. In contrast, young males of nonsocial birds show this "shape" retardation much less. (In an interesting extension of the r–K hypothesis of heterochronic selection, discussed in Chapters 6 and 8 of this book, they also argue that social birds, with paedomorphic young males, tend to inhabit stabler environments.) Geist (1971) has revealed similar findings in mountain sheep, where young males look like females (which in many species, including our own as shown below, are in many ways simply paedomorphic males). This reduces aggression and enhances social stability. Gould (1979) has discussed Lorenz's idea that juvenile animals have rounded "cuddly" features to help stimulate care behavior from adults. The reduction of aggression toward young paedomorphically "shaped" males would seem to be a subset of this.

Jamieson (1989) has argued that helping at the nest is an "unselected" heterochronic by-product of communal breeding in birds. Where many birds aggregate, some will inevitably be in young, nonbreeding stages and in proximity to nestlings. He argues that these nonbreeders are "prematurely" stimulated (predisplaced = early onset) to feed the nestlings since they evoke the care response. In noncommunal birds, this response is not stimulated until parenthood since they see no nestlings until they have their own.

Our last example is fascinating in being the only known case where nestling birds incubate other eggs in the same nest (Tarburton and Minot, 1987). The Australian white-rumped swiftlet couple will incubate only one egg. Upon hatching, the nestling will incubate his nest-mate until it hatches. The authors interpret this as a heterochronic change in that the first nestling is exhibiting adult behavior (early onset). The proximal mechanism is speculated to be a hormonal change, which is the cause of such behavior in adults (Brown, 1985). Ecological selection for this behavior seems to lie in the swiftlet's exceptionally long nestling period which requires a great deal of care.

3. HUMAN HETEROCHRONY

Heterochrony has particular relevance to human evolution. Not only has the process played a major role in the origin of our species, but the study of heterochrony itself has many roots in assertions and speculations about human evolution. Unfortunately, many of these speculations are wrong, emphasizing neoteny as the "general" heterochronic process behind human evolution. Worse, this view has become widely disseminated, not only in the popular press but in the scientific literature as well. We will show that it is wrong on two counts: (1) there is no single heterochronic process that accounts for all of human evolutionary change, although there is one process that accounts for much of it. However, (2) this process is not neoteny (retardation of growth rate), it is hypermorphosis (prolongation of growth). [See also Shea (1988) who has documented and argued similarly on this and other aspects of primate heterochrony.]

The problem began in 1885 when the Swiss zoologist Julius Kollman proposed the term "neoteny" (literally: "holding on to youth") to describe the process, observed in amphibians, whereby larvae attained sexual maturity. Observing his own species, Kollman (1905) apparently found enough similarity between human pygmies and ape juveniles to propose that early humans arose from apes through neoteny. Human pygmies represented an extant intermediate stage. In the 1920s the Dutch anatomist Louis Bolk (1929) furthered this general view with his "fetalization" theory, suggesting that most major human features arose from retention of ancestral primate fetal features into adulthood. Racial overtones arose in his works, as we shall see. Yet another major exponent, Gavin de Beer (1958), continued the emphasis on neoteny although he rejected some of Bolk's more speculative interpretations. Most recently, Gould (1977) and especially Montagu (1981) have written major books on heterochrony which have included the notion that humans are at least largely "neotenic" apes. Citations in basic textbooks and articles are, unfortunately, far too numerous to mention.

How could such a major bit of misinformation become so entrenched? The diagnosis of neoteny in humans has been mainly justified by two major comparative features (see Gould, 1977, or Montagu, 1981): (1) a superficial resemblance of humans to juvenile chimpanzees in morphology (especially the "bulbous" skull shape) and behavior (curiosity, learning ability, and other qualities), and (2) humans' "slow" development as seen in their "retarded" birth, maturation, and longevity. We will show that both of these are incorrect for the following reasons. (1) The superficial skull similarity is based upon a greatly enlarged brain in humans resulting from prolonged brain growth (hypermorphosis), combined with other, dissociated growth field changes, especially reduced (neotenous) jaws and dentition (prognathism). "Shape" is an unreliable trait of comparison for just this reason; it can gloss over disparate changes in different parts, subsuming them under a subjective "whole" (discussion in Chapters 2, 6, and 8). Our curiosity and enhanced learning abilities occur because our brains are larger (from hypermorphic growth) and because we are in the juvenile "learning" (primary absorptive) phase for a longer time (hypermorphic delay of maturation), not because we grow slowly (see next point). (2) The constant references to "slow" or "retarded" development are imprecisions aggravated by lack of a clear classification scheme: "slow" is a rate term, but humans do not grow more slowly than chimpanzees (or our ancestors as noted in point 3); we grow for a longer time in each growth phase. This is a delay in life history events, not a "slowing" of them; maturation, like *any single event, is not "slowed," it is delayed* (late offset, i.e., hypermorphosis). "Rates" of processes, until that event, are a separate issue. Indeed, late offset of our fetal stage leads to *faster* rates. It is true that development can be both slowed and delayed, and this is the case in many salamanders, the mainstay of heterochronic studies. This has added to the conflation because slow development (and this is the main point) is not constrained to be late in offset: the rate of a process is not governed by the time of its offset. Finally, (3) the fossil record, in light of our recent

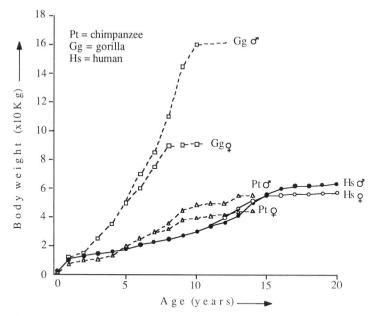

Figure 7-5. Change in average body weight during ontogeny of African apes and modern humans. Ape data from Shea (1983), human data from various sources, especially Harrison *et al.* (1988). Gorillas are accelerated relative to chimpanzees; humans are sequentially hypermorphic relative to the chimpanzee.

ability to assess ontogenetic age of early hominids, indicates that they did indeed mature and die earlier than ourselves, confirming our overall hypermorphic trajectory. All this is discussed in more detail below.

3.1. Hypermorphosis as the Basic Human Characteristic

Much insight into the ontogenetic changes leading to humans is derived from comparisons of human ontogenies with those of our closest living relatives, the African apes, especially the chimpanzee. We share 99% of their genes (King and Wilson, 1975), having shared a common ancestor as recently as 4–6 million years ago, according to generally accepted immunological and DNA hybridization estimates (Harrison *et al.*, 1988). As shown in Fig. 7-5 the average human body size growth trajectory is characterized by a delay in virtually all phases relative to the apes. Most evident is the later maturation (growth offset). Recall the discussion from Chapter 2 concerning the complex nature of growth curves and, especially, the classification of heterochrony using them. Gorillas generally show simple acceleration of each growth phase relative to chimpanzees (Fig. 7-5; see Chapter 2), as first documented by Shea's (1983) elegant study. The human relationship is more complex because, even though it seems mainly caused by simple, roughly "proportional" delays of each phase (relative to the chimp), the fact that each phase has different growth rates changes the overall curve shape in "stretching" it out. This is a major point since it means that delay (offset) can affect

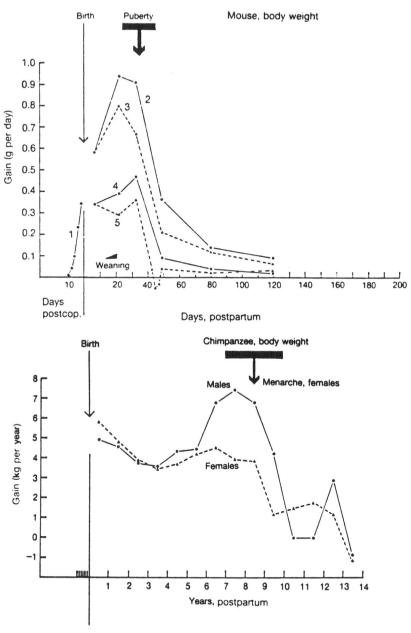

Figure 7-6. Top: weight–velocity curves for males and females of various mouse strains. Bottom: weight–velocity curves for chimpanzees. Facing page: weight–velocity curve for humans. Note progressive delay of adolescent growth spurt. From Harrison *et al.* (1988).

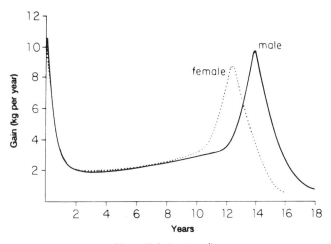

Figure 7-6. (*continued*)

overall (net) "rate" (e.g., adult size divided by years to reach it) by altering the time spent in rapid and slow growth phases (see Chapter 2). The result of "across-the-board" offset delays of each phase is that the hypermorphic species becomes larger since it is spending more time in each growth phase. We use McNamara's (1983a) term, **sequential heterochrony**, for this, where a number of phases (or stages) in a multiphasic ontogeny are affected in a similar way (see also Glossary). In this case, it is *sequential hypermorphosis,* whereby offset of all phases is delayed.

Contrast this with the gorilla's mode of size increase: instead of staying in each phase longer, it accelerates growth during each phase (*sequential acceleration*), staying in it for only about the same time as the chimp (see Fig. 2-9). This says much about the different selective pressures between the gorilla and the human modes of growing larger, discussed shortly. (Recall the larger musk shrew of Chapter 3; its growth curve also reflected hypermorphosis similar to humans, only their curve was simpler; compare Fig. 7-5 to Fig. 3-4.)

This general pattern is seen more clearly in the velocity curves of body growth (Fig. 7-6). Comparing the chimpanzee and human curves reveals that they share the same overall pattern of growth, the roughly "triphasic" pattern discussed in Chapter 2, but with humans delaying the adolescent growth spurt (and sexual maturation) 4 or 5 years later than the chimpanzee. Note also the mouse curve, showing very little time passing between birth and puberty. Primates in general have progressively altered the basic mammalian growth pattern by prolonging the time between birth and puberty. This hypermorphic progression has extended into the primates as a group, being a major factor in the progressive differentiation of prosimians, monkeys, apes, and finally humans. This is shown in Fig. 7-7: note the "proportional" aspects of the hypermorphosis, whereby the offset of each stage (e.g., gestation, infancy) is delayed (sequential hypermorphosis as a general primate trend). An obvious deviation is the disproportionately short human gestation, which occurs because our rapidly growing brain would make a later birth even more trouble than it already is, due to limitations on birth canal size (this

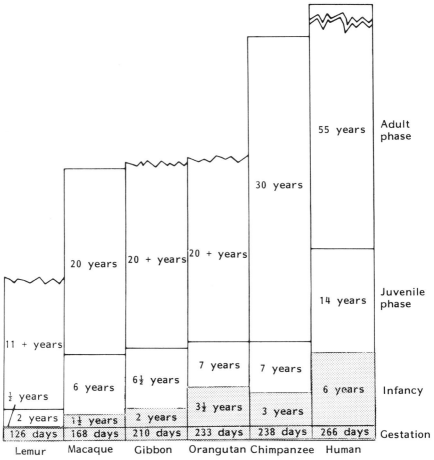

Figure 7-7. Life stages of various primates. Note progressive increase in durations of stages from prosimian to human. From Weitz (1979).

is discussed later). Another point is that this "proportional" offset change of all stages is yet another "structural" relationship: there are often numerous "adjustments" that have been made by intense selection to increase or decrease certain stages disproportionately, e.g., the human gestation.

The proximate cause of this general hypermorphic progression seems traceable to simple change in timing of events in the hypothalamus, which initiates the adolescent growth spurt (Harrison *et al.,* 1988). In any case, for the many workers who, over the years, have sought to "explain" human evolution in terms of a general ontogenetic mechanism, the basal one would have to be the delay of offset of events: hypermorphosis. Human rates are not slower (neotenic) relative to the chimpanzee (compare Fig. 7-6, velocity curve = rate), they are simply extended. Most crucial for the argument of generality of hypermorphosis is that not just body size, but brain size and our extended learning period and life span, are direct products of delay in growth offset as we shall see. These are the things that make us the premier cultural animal.

Evidence for Hypermorphosis from the Fossil Record

The critical reader may justifiably note that the heterochronic observations so far have relied on comparison of human and living ape growth curves and may not be directly applicable to the key issue, that of human growth relative to ancestral human ontogenies. This has been necessary because direct information on fossil hominid ontogeny has been very sparse. (In addition, the many references to humans as "neotenic" chimpanzees, replete with photos of round-headed chimp youngsters, have made it important to compare our growth to apes. We show shortly that this resemblance is superficial.) Fortunately, important new information on early hominid ontogeny is coming to light. This new information based on growth line periodicity in teeth, is a classic example of the overwhelming importance of knowing the absolute ontogenetic age of individuals in making developmental inferences (Chapter 2). The information substantiates the inference of progressive hypermorphosis in human evolution.

The first generally accepted direct human ancestor, *Australopithecus,* dates to at least 5 million years ago. The first known species, *A. afarensis,* was roughly chimpanzee-sized in body (25–50 kg) and brain (400 cm³) dimensions. When this genus was first described by Raymond Dart in 1925, his claim that they were distinct from apes and ancestral to man was greeted with great skepticism. However, evidence for his ideas accumulated as more fossils were unearthed and, in the process, the humanlike features of the group became the focus and therefore emphasized. This culminated with a major work by Mann (1975) who proposed that *Australopithecus* not only had many humanlike anatomical features but shared human life history traits, with extended juvenile development, long parental dependency, and generally delayed ontogenetic events.

Recently, this view has been challenged by a number of workers using more refined techniques (definitive review in Beynon and Dean, 1988). Mann (1975) had used tooth wear and eruption patterns, among other methods, which yield data on relative but not absolute ontogenetic age of the individuals. These newer methods, in contrast, provide absolute age and are thus much more informative. Especially important is the use of incremental growth markings on teeth. Using high-resolution instruments it is possible to see both daily markings in the enamel and circaseptan (7–9 day) cycles in the striae which represent successive enamel-forming surfaces (see Bromage, 1987, for description of methods).

The important conclusion of this newer work is that early hominids had shorter periods of dental development than modern hominids and a less prolonged infancy. They were apelike and not humanlike in their development (Bromage, 1987). For example, the lower M1 molar erupts into occlusion at about 3.5 years of age in *Australopithecus.* This is well within the age range of African apes for that event [chimpanzee mean is 3.26; gorilla 3.5 (Dean and Wood, 1981)] but more than five standard deviations below the human mean of 6 years old (Beynon and Dean, 1988). Even more, the M1 event coincides with the attainment of 95% adult cranial capacity (Beynon and Dean, 1988) indicating that many events, from brain to dental development, were earlier. Independent evidence for this conclusion is provided by recent computerized tomography of the Taung specimen (australopithecine child) which indicates developmental affinities with modern apes (Conroy and Vannier, 1987).

In summary, emerging evidence shows that humans have evolved through a delay in developmental events (hypermorphosis). *Australopithecus* apparently had significantly earlier offset times for many events. In light of past confusion, we reiterate that these data imply that hominid ancestors were maturing and undergoing events sooner (earlier offset) than us, not that they were growing faster. Rates are not the major issue, as discussed above. In fairness, not all workers accept this view of apelike ontogeny in early hominids (Beynon and Wood, 1987), but it is clearly becoming widely accepted: ". . . overall, *afarensis* was ape-like rather than human-like in features such as maturation rates, and newborn body and brain size" (p. 115, anthropology text by Harrison *et al.*, 1988). Note the common, unfortunate mention of maturational "rates" instead of maturational timing. In any case, this could completely revise our view that early hominids needed an extended learning phase in which to pass on their primitive cultural adaptations.

Whether our hypermorphosis has been a steady progressive change, occurred in punctuated, irregular fashion, or at some rate in between is unknown at this time. Given the ready potential for rapid change in heterochronic events (Chapter 3), no possibility can be ruled out *a priori*. However, it is notable that Beynon and Dean (1988) have found indications that early *Homo* was also apelike in development. A recent find of *H. habilis* revealed limb bone proportions much more similar to *Australopithecus* and chimpanzees than modern humans (Lewin, 1987, and references therein). It is also relevant that, unlike pygmy chimpanzees which are almost "purely" neotenic to the common chimpanzee (Shea, 1988), our hypermorphosis has been far from global. Many local deviations from this occur as we discuss shortly. Nevertheless, delay in developmental events is the major underlying process, setting the general course of development, against which all deviations are measured.

3.2. Human Life History, Brain/Behavior, and Morphology

A more detailed view of the generally hypermorphic human trajectory is perhaps best analyzed by breaking it down into the three major components mentioned above: life history, brain/behavior, and morphology. All of these undergo important timing or rate alterations in their ontogeny in the course of human evolution. Also, their ontogenies can be related to the view (admittedly oversimplified) of ontogeny as cellular assembly with gonadal (life history), neural (brain), and somatic (morphology) cells serving as the foci. These three are tightly interlinked into an integrated adaptive suite. Much confusion in the literature has resulted from trying to analyze one aspect independently of the others; they are best seen as an adaptive "package." In the case of humans, this "package" is probably less "tidy" in origin than in most evolutionary lineages (e.g., pygmy chimps) because the special morphological demands of cultural adaptations (e.g., bipedalism, speech) have required numerous "adjustments" away from our ancestral adaptations (arboreality). Nevertheless, this makes the adaptive suite no less integrated. This is nicely illustrated by medical dogma on child development which states that individuals who vary in maturity will do so "as a whole." That is, if maturation is delayed, then skeletal ossification, body growth, dentition, physiological reactions, and even psychological development will lag behind also (Harrison *et al.*, 1988).

Table 7-1. A Partial List of Human Life History Traits Attributed
to Neoteny by Bolk, de Beer, Montagu, and Gould[a]

Late eruption of teeth
Prolonged period of dependency
Prolonged period of growth
Long life span
External gestation
Low birth weight
Fetal rate of body growth during first year
Rapid growth of brain into third year
Late descent of testes
Late development of secondary sexual traits
Late development of reproductive maturity
Lack of estrus period (sexual periodicity)

[a] Compiled by Shea (1990).

3.2.1. Human Life History: Hypermorphosis Everywhere

"Life history" is a commonly employed but vague term. Usually it refers to the timing of traits most intimately involved with demographic features: especially life span, age of maturation, and parental care. Gould's (1977) category of "gonadal" maturation roughly falls in this category since age (timing) of reproductive maturation is one of the focal traits. Often, for a given species, delays in one such life history event are seen in others (sequential hypermorphosis of life history events). In part, this is because each species has an internal "clock" that provides a relative scale during which all events are timed (Chapter 8). However, as discussed above relative to Fig. 7-7 (and in Chapter 8) this "proportional" scaling is very coarse ("structural") and there are many deviations (within the limits of the "structural" relationship) caused by local selection.

For example in humans, a number of life history (and indirectly related) events are delayed relative to ancestral ontogenetic patterns (Table 7-1). We are substantially later in offset of the fetal growth phase, the juvenile period, and all other phases; we are later in the development of many resulting traits: sexual traits, teeth eruption, and so on. We live longer. As noted, we have simply extended the primate trend. Clearly a major consequence of this extension is a larger brain and longer learning period, with older people around longer to learn from (next section).

However, we should also consider the larger context of ecological theory: humans are extreme "K-strategists" in terms of the well-known $r–K$ selection theory. That is, humans occupy a very stable niche (not just because we arose in the tropics, as is often said, but because we control our environment, keeping it stable ourselves). We therefore divert more energy into competitive traits rather than reproduction, with highly developed child-rearing and low fecundity. Most readers will know that $r–K$ theory has been a lightning rod for debate since its proposal in the 1960s by MacArthur and Wilson. Yet, as discussed elsewhere (Chapters 6 and 8), even most of its critics admit the concept is useful; the problem has been its abuse. More importantly here, the concept has recently been rigorously tested in primates and found to be very useful in explaining adaptive patterns among groups (Ross, 1988; Lewin, 1988a, and references therein). Primate species that inhabit unstable environments have a higher potential rate of population increase when body size is adjusted for (further discussion in Section 3.4).

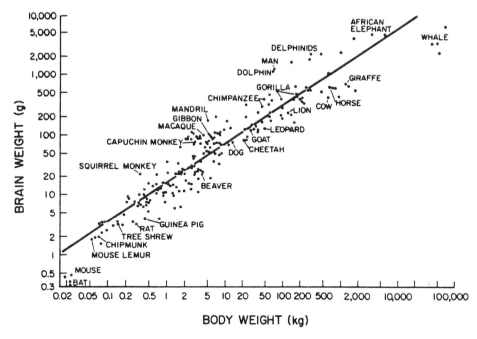

Figure 7-8. Relationship between brain and body weight in a large number of mammalian species. Modified from Purves (1988).

3.2.2. Human Brain and Behavior: More Hypermorphosis

3.2.2.1. Brain Size. At about 1400 cm³ for the average adult, the human brain is smaller than that of some larger-bodied animals such as the whale. However, much of their extra brain tissue is given over to the sheer mechanics of sensation and motor control of the larger body. It is therefore more telling to look at relative brain size, corrected for body mass, if we wish to infer something about the number of neurons devoted to "higher" functions such as learning. The well-known log–log plot of body versus brain weight (Fig. 7-8) does this by allowing comparison of brain size among animals of the same body weight. Thus, primates in general, and especially humans are well above the regression line.

The slope of the regression line has attracted much attention since the high correlation of the two variables indicates a clear pattern figuratively begging for explanation. One school (e.g., Gould, 1975) argues that the slope is about 0.67 as would be expected if body surface area were a determinant of brain size (i.e., amount of area innervated). Another school (e.g., Armstrong, 1983) argues that the true slope is closer to 0.75, as would be the case if body metabolism were the main determinant of brain size. The question remains unresolved (review in Purves, 1988) but the sheer strength of the correlation clearly indicates some strong underlying "structural" relationship. Also of importance is that when only closely related taxa (species, genera instead of the "mouse to elephant" approach) are regressed in the same way, a slope of only 0.2–0.5 results (Lande, 1979; Purves, 1988).

These contrasting scaling patterns apparently reflect the results of different selection regimes. As noted by Gould (1975), and Lande (1979) who used quantitative genetics, selection acting only or mainly to increase body size alone will cause the shallower slope (0.2–0.5) because little correlated brain increase is needed to "keep pace." Where selection is operating to increase brain size as well, the higher slope results. From this we can infer what parts of ontogeny were under selection to create the pattern. Riska and Atchley (1985) have shown that selection acting to increase size in early ontogeny will affect both body and brain while that acting on late ontogeny will affect mainly body size. As discussed in Chapter 3, there is a hormonal basis for this in that IGF (insulinlike growth factor)-II is active during fetal growth and is an effective mitogen on both brain and body tissue (Hintz, 1985). IGF-I is a major mitogen in later ontogeny and is not effective on brain tissue. While much of this is genetically determined, Leamy (1988) shows that the prenatal maternal environment is also a key determinant of brain size.

However, Pagel and Harvey (1989) have recently argued that these genetic and developmental constraints are not the cause of the slope differences. Rather, they say, closely related groups live in more similar environments than more distantly related groups so that the different slopes reflect differences in selection. This is an excellent example of the "extrinsic–intrinsic" debate discussed in Chapter 8: which actually controls evolution more, external selection or internal (genetic, developmental) constraints? In this case, Pagel and Harvey argue that as soon as ecological pressures change, brain size can change rapidly. "Internalists" such as Atchley are not so sure. Atchley is currently doing quantitative genetical experiments that should help resolve the issue (Atchley, 1989, and references therein).

In the case of human evolution, our enlarged brain results from changes in early ontogeny. There are three ways that enlargement ("overdevelopment" or peramorphosis) can occur (Chapter 2): acceleration, predisplacement, or hypermorphosis. In our case it is easy to see why the last occurred: extending the early period of brain mitosis not only resulted in larger brains; the same process extended the learning period in which to use the brain and the longevity of experienced people to pass on knowledge. Acceleration of brain mitosis may have attained the same brain size but the life history suite would have been much different.

The effect of extension of the fetal brain–body growth trajectory is illustrated in Fig. 7-9. It shows that during fetal growth, brain and body weight increase nearly isometrically (slope = 1; Martin, 1983, Passingham, 1985) as IGF-II is active. Around the time of birth, brain growth slows in other primates. In humans, however, the isometric pattern continues for about a year beyond birth as fetal growth undergoes late offset and brain size continues to increase. This is why Portmann (1941) made his now-famous observations that humans undergo a 21-month gestation with the last 12 months occurring extrauterine. It explains why only our babies are helpless ("secondarily altricial") during that time compared to the precocial neonates of other primates. The reasons for this "premature" birth stem from the obstetrical mechanics of female pelvic design with locomotory constraints (Frost, 1987). The head becomes too large to pass through the birth canal although (much less often noted) our body is also exceptionally large at birth. At 9 months after conception the human baby is about 3375 g while even the gorilla only averages about 2110 g at 9 months, when it is also born (Martin, 1983).

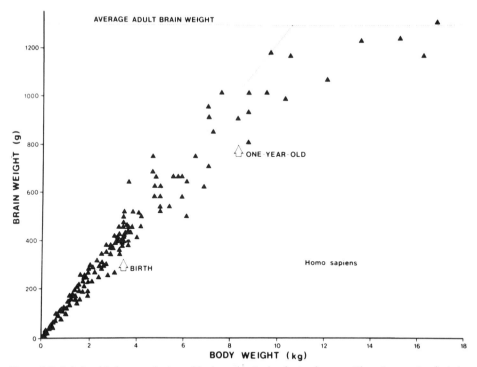

Figure 7-9. Relationship between brain and body weight in developing humans. There is no major deviation from the fetal trajectory until about 1 year after birth. From Martin (1983).

Our large birth size shows that the process of our "extension of fetal growth" is more complex than usually indicated. If the extension of "fetal growth rates" described a single process, our brain and body would weigh about the same as gorillas and chimpanzees at 9 months after conception. Our enlarged brain would result from later growth. Instead, our brains and bodies are both twice as large or larger (relative to the apes) at that time (Martin, 1983). This indicates that both are accelerated (they are still isometric). However, this does not mean that offset delay is not the primary cause. What seems to be happening is that "fetal growth" involves more than one simple rate process: there are a number of multiplicative phases within it, each with its own pace. If we prolong each of the rapid phases ("sequential hypermorphosis" discussed above), we have a net increase when size is measured at a later time (9 months) since more total cells will have been generated. The cumulative nature of such mitotic offset delays is what makes them difficult to distinguish from simple acceleration wherein only one process is speeded up (see Chapter 2 and other discussion in this chapter of growth curve complexity and interpretation).

The approximate shape of this early brain curve has been described in a study of human fetuses by Burns *et al.* (1975). They showed that early human brain growth can be described by a simple quadratic equation, brain mass = $29.81 - 0.988$ (days) $+ 0.008$ (days)^2, where days is postmenstrual days. When plotted, this curve has a roughly exponential shape, starting slowly and grading to rapid monotonic increase (Fig. 7-10).

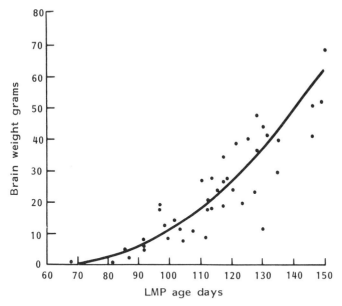

Figure 7-10. Brain weight versus postmenstrual age (LMP), i.e., approximate time (days) since conception. From Burns *et al.* (1975).

(Body growth would have a similarly shaped curve since it is isometric with brain growth.) It predicts brain size very well from conception to birth. Even though "phase delays" seem to be at the heart of causation, a smooth curve results, as transitions among growth phases are delayed in an integrated fashion.

3.2.2.2. Brain Structure. While brain size alone is informative, a finer look at brain structure is necessary to appreciate human heterochrony. Our brains are more than simply larger versions of our ancestors' brains. One reason is that when a nonlinear process, such as brain ontogeny, is extrapolated, the results are more than "scaled-up" versions of the ancestor. Emergent properties are taken on. Another reason is that, like any growth field, the brain is susceptible to "local" alterations (Chapter 3). This would include the brain or neural "sculpting" discussed earlier in this chapter.

Perhaps the most basic feature of our brain structure is that while the human brain is about three times larger than that of other primates of comparable body size, most of our extra tissue is "filler." We have only about 25% more neurons in the cortex (Harrison *et al.*, 1988), that outer half-inch layer where learning and other higher functions occur. The rest is mostly glial cells which serve as connective tissues. Ontogenetically, the cortical neurons all form in the first 15–18 weeks after conception in humans so that our "extra" 25% probably comes from prolonged mitosis at this very early stage. Presumably, ape neuroblast formation ceases earlier. Virtually all of our brain mass, derived from later growth offset delays, results from prolongation of axon and dendrite growth and, especially, multiplication of glial cells that will eventually make up about 90% of the brain's mass. Up to about 4 years of age, there is an increase in the number and size of dendrites in all cortical layers. Even after that, myelination and other processes continue.

Having pointed out gross changes during prolonged cortical growth, we turn to changes at a finer level: localization of function within the cortex. The mammalian brain has been very conservative in its evolution and the prolonged growth of our own brains has generally modified existing patterns. Chemical tracers show that similar circuits link homologous nuclei and cortical areas (Harrison *et al.,* 1988). Each new circuit connection derives from axons and dendrites of neurons within already established circuits (Welker, 1976). Therefore, growth changes involve the proportionate sizes of various areas and the amount of association among them. In heterochronic and cellular terms, this refers to variation in local neuron numbers (multiplication) and axonal–dendritic connectivity growth, respectively. A good example of both local enlargement and changed connectivity is Broca's area in the human brain, which is intimately involved in speech and language. Located in the prefrontal cortex, it is greatly enlarged compared to its homologues in other primates. Functionally, it is involved with face and mouth movements during feeding in apes. Vocalizations are under limbic control. In contrast, our own vocalizations are integrated into neocortical circuitry including of course the enlarged Broca's area (see Lieberman, 1984, for review).

In sum, it seems likely that there are two basic heterochronic scales in changes of structural organization. In the first, "global" alteration of the entire brain can result in local changes from extrapolation of local allometries. Importantly, these extrapolations may be "nonlinear" as noted. For instance, Finlay *et al.* (1987, p. 113) note that "prolongation of development could exaggerate distinctions in connectivity patterns of different cell types" in the brain. As discussed above, this is because changes in timing can alter axonal and dendritic growth through either addition (e.g., more segregation of bundles) or subtraction (e.g., neuron death). Clearly, this is a very crucial point because there is more to being "intelligent" than having more neurons: connections among them must increase as well since information is "context-driven."

The second heterochronic process involves purely local "neuron sculpting" through selection for specific mental properties. To return to Broca's area, selection for improved language capability has acted to enhance that behavior in human evolution. At least some of the enlargement and cortical integration of Broca's area is not likely to be due to simple allometric scaling of overall brain size. Indeed the motor area devoted to the tongue is much larger than that of the entire leg. An example of local loss would be our reduced olfactory area. There must have been some alteration of genetically controlled timing and rate aspects of neurogenesis and connectivity on a local scale. As discussed elsewhere, the application of heterochronic terms to local changes may seem unnecessary or unhelpful. Yet it is well known that timing and rate changes are critical in brain development. Indeed the cortex shows two clear temporal gradients of growth, with one sequence relating to general brain development and the second concerning the order of functional localizations (Tanner, 1978). Interestingly, these maturational sequences occur in functional units rather than geographical areas of the brain (Yakovlev, 1967). Welker (1976) has described how large neuroblasts differentiate first, followed by gradients of differentiation among neuroblast populations. Within any region, different neuroblasts may selectively undergo differentiation, migration, multiplication, and death according to a strict sequential timetable. Holloway (1979) has discussed the "unique timing of embryological and all further ontogenetic development of brain processes: that is, myelinization, neural nuclei, and fiber tract maturational interactions . . . that results in species-specific patterns of maturation of different parts of the brain" (p. 62). Clearly there is much potential for local change in timing via quantitative shifts among components. Such local change is homologous to the "growth field" concept discussed in Chapters 3 and 8, except that neural tissue instead of somatic tissue is involved.

3.2.2.3. Human Behavior and Organizational Hypermorphosis.

The application of heterochronic concepts such as "terminal addition" or "overdevelopment" (peramorphosis) to physical traits is fairly straightforward. Limbs get longer from growing faster or for longer times; at the cellular level, cells multiply faster, and so on. However, when applied to behavior, things are not so straightforward. Can we say that humans are "under-" or "overdeveloped" for certain behaviors? In the past, some have tried to account for our "playfulness" or inordinate curiosity as part of our generally neotenic origin as discussed below. However, as we have reiterated (perhaps ad nauseum), our evolution has occurred mainly through delayed offset (hypermorphosis) of the juvenile stage. Because hypermorphosis is a process of "overdevelopment" (peramorphosis), whereas neoteny is a process of "underdevelopment" (paedomorphosis), as in the perpetually juvenile songbirds discussed early in this chapter, the distinction is critically important for understanding human behavior. Are humans "perpetual juveniles" [as Montagu (1981) argues], or do humans (as we believe) simply delay the offset of juvenile behaviors, but eventually behaviorally mature through the same stages as our ancestors? The issue is obviously complex and needs rigorous study. We strongly encourage ethologists and social scientists to give heterochronic processes a close look as an organizing principle and source of generating testable hypotheses about behavior. In the meantime, we would be remiss not to give some brief consideration to human behavior in a heterochronic context.

A basic overview of how neuronal changes may translate into behavioral changes was discussed in Section 2. For innate behaviors, the process is somewhat easier to visualize. Selection acts on the behavioral phenotype and alters the underlying neuronal pattern accordingly. Obviously, relating more complex behaviors to neuronal patterns is more difficult. The basic pattern is that from early fetal life onward past puberty, the appearance of brain function is closely related to maturation in structure. As expected, the bases of simpler behaviors tend to form earlier than more complex ones. Thus, fibers of the auditory system essentially complete myelination by 4 years of age while the reticular formation, concerned with attention and consciousness, continues to myelinate at least until puberty and possibly beyond. Regarding heterochronic bases of complex behaviors in general, we agree with Harrison *et al.* (1988, p. 401): "there is clearly no reason to suppose that the link between maturation of structure and appearance of function suddenly ceases at 6 or 10 or 13 years. On the contrary, there is every reason to believe that higher intellectual abilities also appear only when certain structures or cell assemblies, widespread in location throughout the cortex, are complete." The underlying neural basis for maturation of such complex behaviors as personality type, advanced reasoning, and so on are clear: since all the neurons form long before birth, the ontogeny of complex behavior must involve changes (increases mostly) in connectivity. A neuron in an adult can have thousands of connections, but even millions of dendrites occupy very little space. Thus, even though 95% of adult brain mass is attained by 10 years, a great deal of change in the "highest" brain functions can occur in the final years before maturation (perhaps even beyond). Such an extremely multiplicative process ("connection cascades") will almost surely lead to "threshold" effects and emergent properties in the developing behavioral phenotype [as modeled by, e.g., Kauffman's cellular automata (1990) or connectivity patterns from percolation theory, see Stauffer (1985)].

Turning from structure to behavioral phenotype *per se,* Piaget's numerous works on the ontogeny of mental functioning are the best known. They are very popular, especially among educators because they provide a taxonomy of cognitive achievements occurring in a sequence of development. Important here is that his concepts have been applied to the ontogeny of behavior in primates with very interesting heterochronic implications.

Briefly, Piaget (1952 and later works) observes three main areas of intellectual development in human children. The first is the sensorimotor period when the child learns about the physical world. The second period involves social learning and the third is when development turns from outside relations to internal abilities such as imagery and memory. While there are criticisms of Piaget's scheme, they mainly focus on periods two and three. Even the critics generally accept his sensorimotor sequences (Bower, 1974; Box, 1984). This is fortunate for our purposes because the most revealing and best studied developmental comparisons between humans and other primates involve the first period.

In humans the first (sensorimotor) period occurs during the first 2 years or so of life. The mental learning achievements at this time can be subdivided into six series (see Parker and Gibson, 1979, for fuller discussion): sensorimotor intelligence, space, time, causality, imitation, and object concept. Each series in turn undergoes a six-stage sequence of development. Of extreme interest to the heterochronic perspective is that when we observe other primates, we find that they go through each of the same six behavioral series, but generally stop short in the development of that series, at an earlier stage than us. Because the truncation is associated with earlier cessation of mental development, they are "progenetic" to us, or more correctly in phylogenetic terms, we are hypermorphic to them. Moreover, the more "advanced" the primate, the more hypermorphic it generally is in serial development (see Fig. 7-7). Thus, studies on prosimians (Jolly, 1964; Parker and Gibson, 1979) showed that lorises and lemurs have reflex grasping and simple manipulations typical of the first two stages of sensorimotor intelligence but show no evidence for the fourth and fifth stages of object concept. Nor in fact did they show any sign of object manipulation abilities appropriate for the last three stages of any of the other sensorimotor series (Box, 1984). Monkeys are more developed in these behaviors but not as much as the apes. The stump-tailed macaque, unlike the prosimians, completed the last stage of the object concept series (an object is shown and then hidden) and got up to the fourth stage of the sensorimotor series (e.g., pulling things apart). However, they did not show tool-use or imitative behavior characteristic of the most advanced stages. For example, they never reached the last two stages of the spatial and causality series such as placing objects in other objects. Nor did they "experiment" with new objects characteristic of the fifth stage of sensorimotor intelligence. Chimpanzees and gorillas also complete the object concept series but go further in the space,

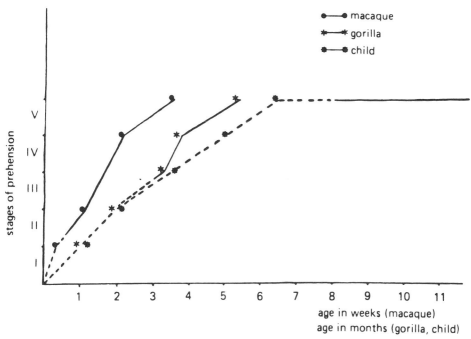

Figure 7-11. Time of attainment of the five stages of prehension for monkey, ape, and human. From Spinozzi and Natale (1986).

causality, and imitation series (Parker, 1977a,b). As is well known, chimpanzees use tools.

The general relationships are shown more concisely in Fig. 7-11 which graphically compares the sensorimotor intelligence development in human, ape, and monkey in terms of the first two stages of Piaget's scheme (the two stages are broken down into five substages). While all three primates ultimately attain the same level, apes and humans are progressively slower about it. Not shown is the key fact, already discussed, that apes and humans ultimately develop beyond the macaque in this series, with the macaque stopping at about the fourth stage. Nevertheless, an important point is illustrated by this elegant study by Spinozzi and Natale (1986): more "advanced" behaviors not only go through the same stages of development but they take longer to do it. Why is this? We have noted that our brains grow faster. Why do our behaviors lag behind? We address this shortly.

First, we turn away from Piaget's scheme to a compilation by Dienske (1986) of developmental observations from a number of sources. The results, shown in Fig. 7-12, indicate a similar pattern to that just discussed but for a much broader range of behaviors. Apes and especially humans undergo delays in the attainment of the monkey's locomotor development. Of special interest is that while most of the changes are proportionately delayed, there is a glaring exception in the play-face or smile which appears much earlier in the ape and especially human baby than it "should." This is another example of *dissociated behavioral heterochrony*. Dienske (1986) suggests as the cause that

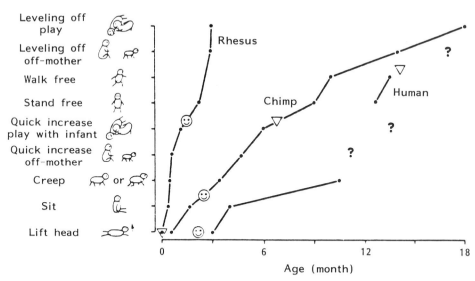

Figure 7-12. Ages at which certain behaviors first appear in monkey, ape, and human. From Dienske (1986).

selection would act to favor smiling in helpless infants because it would encourage maternal care. Thus, the slower the development of motor skills, the earlier such positive feedback signals for maternal care would need to be shifted.

Dissociations aside, the general issue of human hypermorphosis (delay) versus neoteny (rate) is well illustrated in Fig. 7-12. A superficial look at Fig. 7-12 (and Fig. 7-11) would imply that humans are neotenic, with "slower" development. However, changes in offset timing of brain growth phases will also affect brain maturation. Delays in growth offset not only result in larger brains, but "slower" (neotenic) development of behaviors relative to our relatives (and ancestors). The prolongation in the early, rapid brain growth phase results in a larger brain but it also means that the behaviors associated with that brain will be delayed because it is in the earlier phase(s) of organization for a longer time. Since no neurons form after about 18 weeks, our "retarded" behaviors must stem from relatively less axon–dendrite and glial cell growth, i.e., our relative connectivity among the neurons is less. In short, while our brains are bigger at any given time (relative to other primates), the connections among the neurons are proportionately less complete, not only because our brains are in an earlier phase of development, but because there are many more connections to be made now (number of connections probably increases exponentially with increase in number of brain cells, since each cell can have up to 10,000 connections). Put another way, we must distinguish between growth and organization: early human brain development shows more rapid growth but slower organization. The "parts" are growing faster but the holistic integration of them into an "organ" or working system is slowed as a result, because the system is bigger and in need of (perhaps disproportionately) more connections.

This is shown most clearly in that the appearance of most traits (barring smiling) is roughly proportional in its delay, which in turn is proportional to relative brain development (Fig. 7-12). For instance, off-mother and play behavior is of roughly comparable development at 1, 6, and 12 months in monkey, ape, and human, respectively. By Dienske's estimates, the monkey has about 80% of adult brain mass reached at 1 month, the chimpanzee about 65% at 6 months, and the human about 60% at 12 months. Thus, if we look at the onset of these behaviors in terms of *percentage of adult brain size,* there is much more similarity in the time of appearance. This illustrates that the late onset ("slowed" rate) of these behaviors is the direct result of the relative "immaturity" (probably from relatively reduced neural connectivity) of larger brains when compared to the smaller ones at the same absolute ontogenetic time.

As a useful phrase for this, we suggest **organizational heterochrony**. This occurs whenever some composite trait (either behavioral or morphological) shows change in time or rate in "forming," even though its components have appeared earlier. In the human case, we show organizational hypermorphosis: delays in offset of one phase (with us, beginning with the fetal) will result in late onset of any successive phases that are contingent upon the previous one(s). (Further discussion below.)

3.2.2.4. "Advanced" Human Behavior. The late onset of behaviors just discussed has contributed greatly to the myth of human neoteny. Montagu (1981) in particular wrote a book, *Growing Young,* which had the basic premise that we owe virtually all of our cultural capabilities to neotenic retention of juvenile ape characteristics: curiosity, educability, and many others, such as optimism and honesty (Table 7-2). Much of the book is a lament that as we grow older, we lose these "youthful" qualities. More importantly, he argues that we are not "supposed" to lose these qualities: "we are designed to grow and develop in ways that emphasize rather than minimize childlike traits; we were never intended to grow up into the kind of adults that most of us have become" (p. 2). Indeed, he goes on to argue that many of our social problems today result from a loss of such youthful flexibility.

While it is difficult to pin down such behaviors as "resiliency" in Table 7-2, it is clear that the general idea is that of a life-long retention of "youthful" qualities due to a slowed "rate" or even "frozen state" of development. Here we encounter the truly crucial distinction between neoteny and hypermorphosis: the former is a process of paedomorphosis such that the descendant adult (us) *never attains behaviors* possessed by the ancestor. In stark contrast, hypermorphosis is a process of peramorphosis such that the descendant adult *goes beyond* behaviors of the ancestor. As a basic premise, we have noted that prolongation of human brain growth leads to greater brain size; this permits more learned behaviors as there is simply more room to store and process information. We have also noted that "local" changes, such as Broca's area, have allowed alterations beyond or below that of a general increase. By these reductionist quantitative criteria (neuron number), the meaning of behavioral "peramorphosis" is straightforward: as adults, our behavior is more complex than that of apes (and our ancestors) because it is based on more stored and processed information. In more rigorous terms, increasing complexity is usually defined as increases in number of parts and kinds of relationships among them (O'Neill *et al.*, 1986). Our increased behavioral repertoire (the

Table 7-2. A Partial List of Human Behavioral
Traits Attributed to Neoteny by Montagu (1981)

The need for love
Friendship
Sensitivity
The need to think soundly
The need to know
The need to learn
The need to work
The need to organize
Curiosity
The sense of wonder
Playfulness
Imagination
Creativity
Openmindedness
Flexibility
Experimental-mindedness
Explorativeness
Resiliency
The sense of humor
Joyfulness
Laughter and tears
Optimism
Honesty and trust
Compassionate intelligence
Dance
Song

more complex phenotype: more kinds of behaviors) therefore reflects the underlying increased neurons (parts) and connectivity (kinds of relationships).

In short, humans show faster brain growth but delayed mental development. The pivotal point is that this does not mean that we do not go through the same general stages of the ancestral or ape pattern. We have only delayed them. It is not true that we are "intended" to remain mentally youthful: we are in the youthful learning stage longer, but mental maturation continues. Just as with birth, maturation, and death, we have not put off events, only delayed them. This is clearly shown in our deteriorating ability to learn language as we age. More generally, a number of studies have shown that timing of neurological maturation (which we know is delayed and can physically observe) is correlated with timing of advanced behavioral phenotype. Thatcher *et al.* (1987) observed development of the cerebral hemispheres in 577 children from the age of 2 months to early adulthood, by monitoring electroencephalograph data. They found that the right and left hemispheres developed at different rates and ages, but more to the point, that discrete growth spurts appeared in specific anatomical loci at specific postnatal periods. More importantly, the timing of growth spurts overlapped the timing of the major developmental stages described by Piaget. They conclude that this (plus other, similar studies that they cite) strongly favors the "ontogenetic hypothesis of human cortical development in which there is a genetically programmed unfolding of specific corticocortical connections at relatively specific postnatal ages," providing evidence for

theories (such as Piaget's) of sequential temporally "programmed" cognitive development. Epstein (1979) found similar relationships with a focus on IQ (with all its implications for "advanced behavior"). For instance, maturation of Piagetian stages occurs
much sooner in children with higher IQs. In turn, mental age shows developmental
stages that agree with those found in the brain itself. Considering the delayed Piagetian
pattern of our young juveniles compared to ape juveniles (shown above) and all the
evidence for hypermorphosis of brain growth in general, it seems reasonable to state that
humans are peramorphic in behavior, not paedomorphic.

To summarize the discussion so far, we have tried to show that two of the lines of
evidence for neotenic human behavior are wrong: (1) the misconception that we are
physically neotenous led to the wrong assumption that we must be mentally neotenous
too, and (2) prolongation of early brain growth led to larger brains but delayed organization and maturation of infant behaviors. These are not neotenic in origin (but result from
prolongation) and more importantly, the behaviors are only delayed, not permanently
"arrested" as would be the case with paedomorphosis, and in time even extend
"beyond" the ancestral behaviors. That is, human behavior shows *sequential organizational hypermorphosis.*

However, there is still a third line of evidence often cited for human behavioral
neoteny: the retention of educability, curiosity, and other "youthful" primate behaviors
(Table 7-2). To refute them we must attempt to directly account for them in peramorphic
terms. Conclusive "proof" that advanced human behaviors are extrapolations of ancestral patterns is impeded by the very subjective and complex nature of them (Table 7-2).
We would begin by noting that body size, all life history events (Table 7-1), and many
morphological traits are clearly the result of delayed offset so it is difficult to see why
behavior would not be, especially when brain growth itself is clearly peramorphic.
Perhaps the best general explanation is that *most of the traits in Table 7-2 are simply
emergent qualities of a large (peramorphic) brain, capable of retaining and processing more
information,* rather than some vague "youthful" quality. Curiosity, the sense of wonder,
imagination, creativity, openmindedness, and many others simply result from our
greater brain size and associative powers ("intellect"). We retain exaggerated forms of
these qualities into adulthood because we continue to learn through adulthood, just as an
adult ape can continue to learn, but our brains are larger so we learn relatively more than
the ape (although also like him, we learn somewhat less easily as we age). If, as adults, we
have greater areas to store information, it is inevitable that we are going to have more:
"need" to learn, curiosity, and explorativeness (to gain the data); more: need to organize, imagination, creativity (through associations in the data), and so on. Other qualities
such as honesty and trust are difficult to assess scientifically but they would appear to be
emergent necessities (for social cohesiveness) in a complex social system derived from
our peramorphic brains. A related factor is that with brain enlargement, the greater
learning that we experience during childhood has become more important in determining our adult behavior. Thus, we retain these "juvenile" behaviors precisely because the
juvenile period is when we learn so many of them.

There is more than a technical point involved here because Montagu basically sees
human aging today as an "unnatural" deviation caused by social problems. "Psycho-
sclerosis" is his term for the loss of these "natural" youthful (neotenic) qualities. However, the hypermorphic view of human evolution sees aging as a process of delaying but

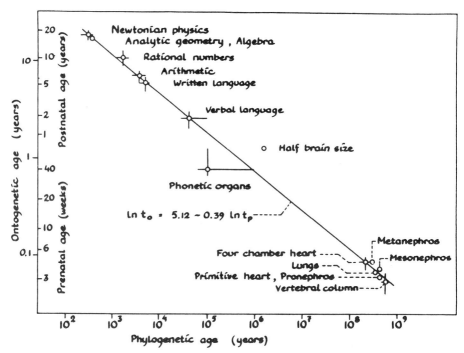

Figure 7-13. Ontogeny versus phylogeny in the human lineage. That is, appearance of traits (biological and cultural) ontogenetically and phylogenetically. Ontogenetic age is measured from moment of fertilization. Both scales are logarithmic. Validity of this graph does *not* mean that ontogeny exactly recapitulates phylogeny; only that the (ontogenetic and phylogenetic) growth of complexity must, be necessity, build upon past contingencies which are therefore often conserved, if sometimes in modified form. From Ekstig (1985).

eventually extending our "natural," ancestral past. The decrease of learning is an inevitable consequence of a finite brain which becomes fuller as we experience life and absorb information. Many learning theorists believe that information in our memories has a "half-life" and there must be turnover for new information. In any case, associative contingencies will build up, making organizational restructuring more difficult (analogous to developmental "hardening" through contingencies of complex growth, as discussed in Chapter 8). We become "set in our ways." This is not as bad as Montagu makes it sound. Strongly held ideas and concepts are very useful, indeed essential, necessities for coping with the world. Being based on past experience, they may become outdated but are important starting points for future change. In other words, they serve as the foundation for the inevitable dialectic between ideas. It is true that we must remain as flexible as possible but there is a price for such flexibility: a readiness to jettison past experience (i.e., in the extreme, a "memoryless" system).

Ekstig (1985) has attempted to extend the concept of peramorphosis from behavior to the ontogeny of knowledge itself. His ideas are based on Piaget's (1970) concept of "genetic epistemology" which theorizes a parallelism between the growth and progressive organization of knowledge in a culture through time and the formative learning of that knowledge by growing individuals (see Fig. 7-13). There are a number of interesting possibilities in this analogy, although like many analogies this one must be viewed with the suspicion of being superficial, false, overdrawn, or even downright fanciful. Perhaps the most insightful aspect of the analogy is that conceptual development does indeed have an ontogeny in which there are rules of assembly. Arithmetic must be learned in order to understand algebra which, in turn, is needed to learn calculus. Further, such an "ontogeny" shows direction in increasing complexity and rate as well. There is a limit to how fast the human mind can assimilate and organize such information. Such growth of organization (complexity) is analogous to the growth of physical contingencies in biological ontogeny (or possibly homologous given that even association of ideas must have a physicochemical basis). In the individual such information growth represents learning of extrasomatically stored and transmitted information, but this information itself originated as cultural innovation and is thus analogous to genetic mutations that act to increase complexity. Note the distinction between this ontogeny of information complexity organized by the mind and the ontogenetic increase of the "mind" itself. It is the latter ("neuron peramorphosis") which precedes and permits the former. What is interesting is that the information assembly itself is "independent" of the specific mind involved. Whether it be a martian or a computer, there are rules of sequential assembly before the next step can be understood (as in the calculus example). More basic information bits must be stored to provide context for the more "advanced" bits. Another example of this is seen in language acquisition where noises (phonemes), syllables, words, and then sentences are sequentially learned, and integrated into the rules of grammar. Thus, Lamendella (1976) concludes on this basis that the development of a child's language recapitulates the ancestral development of language in many ways. However, it is unclear how precise this analogy is. Virtually any process has an "ontogeny" in that it has a beginning and end, and of course contingent events are a major part of, indeed define, all "processes." But perhaps in a *very general way,* "ontogeny recapitulating phylogeny" holds here, as it may with morphology (Chapter 8). While the simplistic application of this phrase has long been discarded, as Gould (1977) has said, evolutionists may have thrown the baby (literally, pun likely intended) out with the bath water.

Perhaps most interesting is to consider the peramorphosis of the brain and information together. Evolutionarily, mental development has dramatically overtaken physical development in an informational context. Biological ontogeny is produced by about one billion bits of information [in a human genome (Calow, 1976)]. In contrast, the human brain can assimilate some 100 trillion bits (Sagan, 1980), about 100,000 times greater. Any insights that can help better relate these two major modes of change should be tried. For instance, we may ask if further peramorphosis can occur in the human brain (or in computers?) and whether this will lead to consequent peramorphosis of information assembly. It is possible that larger brains, with more storage and more connections, could be constructed (either naturally or artificially). The resultant increase in information storage and, especially, association would seem likely to produce a consequent extension in the assembly of unifying concepts. This brings to mind the famous experiments of the cognitive psychologist George Miller who estimated that the (current) human brain can hold at most $7 + 2$ distinct items before its attention. Thus, part of the "intellectual phenotype," resulting from a still further extension of human brain growth, might be for the mind to hold more distinct items before its attention. We find this interesting as an example of how peramorphosis at one level (cellular) translates into peramorphosis at a much higher level (intellect, and even "knowledge"). Further, there is the resulting "peramorphosis" of behavior from such intellectual "overdevelopment," and, even more tangible, the increasingly sophisticated ("peramorphic") technological products.

In sum, we agree with Haldane (1932) who wrote that future human evolution will "probably involve a still greater prolongation of childhood and retardation of maturity." However, we disagree with his (and numerous others') conclusions that this allows the preservation of infantile traits. The *mental prolongation of human youth is only relative,* for it is followed by prolongation of later stages as well, where we not only grow out of the ancestral infantile and juvenile traits but go beyond the brain complexity and conceptual abilities of our ancestors. We also want to point out that while neoteny has been the cause célèbre for not only Haldane but so many others, there has been one important dissenter, at least when it comes to brain ontogeny. Ernst Mayr (1963) wrote that "the brain is ahead in ontogeny in all mammals. Indeed, since human babies have extremely large brains, one might say exactly the opposite of what Bolk has said and state that they have become 'adultified.' "

3.2.2.5. Summary: Heterochrony of Human Body and Brain Size.

Because of the complexity of the issue and past confusion over human heterochrony, we offer a summary of the general body and brain effects of hypermorphosis in human evolution, illustrated in Fig. 7-14. The longer duration (from late offset) of each somatic growth phase (sequential hypermorphosis) has made us generally larger than the chimpanzee, which matures earlier. Note that during intermediate ontogeny, we are actually smaller than the chimpanzee because we delay the adolescent growth spurt (Fig. 7-14A). Superficially, this may appear to make us neotenic but, again, this is not the true mechanism. The gorilla has become larger through acceleration mainly of the exponential juvenile phase, showing little change in time of maturation from the chimpanzee. We suggest that chimpanzee ontogeny roughly approximates that of the early ancestral hominids which were smaller and, as noted, apparently matured earlier than ourselves. This is in keeping with other investigators who have suggested that chimpanzees serve as a reasonable "prototype" for early *Australopithecus.* The pygmy chimp in particular has been singled out for a variety of behavioral and morphological reasons (Zihlman *et al.,* 1978; Susman, 1984). Importantly, this suggestion is based on independent (nonontogenetic) evidence from our own, e.g., similarity in adult body size and lengths of lower limbs between these chimps and early australopithecines (nominally *afarensis*) and behavioral similarity of the chimps to modern humans. This would also agree with the 99% DNA similarity between ourselves and chimpanzees and their apparent status as our closest living relative. Even further, the timing is about right according to the "molecular clocks" which estimate a separation from the chimpanzee line about 5–7 mya (Harrison *et al.,* 1988).

Regarding brain size, the progressive hypermorphosis has caused an increase in adults from about 400–500 cm^3 to 1400 cm^3 in modern humans. The proximate cause is the extension of fetal growth, during which brain growth is isometric with body growth. The hypermorphosis is easier to see with this trait because a simpler growth curve is extended (Fig. 7-14B) than that of body size. In contrast, the effects of delay on behavior are, perhaps counterintuitively, the reverse of brain delay. At the same time our brains

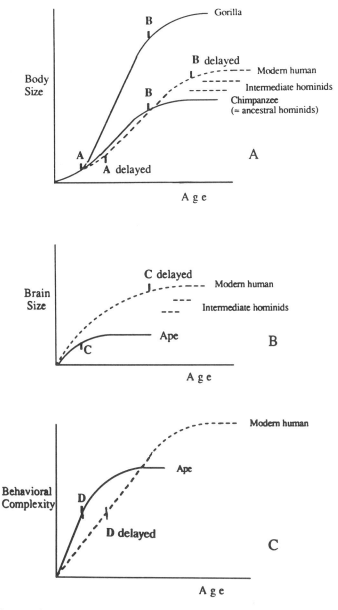

Figure 7-14. Schematic diagrams summarizing heterochrony in human evolution. Timing changes shown are relative to modern apes and ancestral hominids. (A) Body size change. Events A (offset of slow growth phase = onset of rapid growth phase) and B (offset of rapid growth phase = onset of maturation) are delayed in humans relative to apes. Intermediate delays characterize ancestral hominids. (B) Brain growth. Late offset (event C) of brain growth [occurring mainly in early (especially "fetal") phases] leads to progressively larger human brain. (C) Behavioral complexity. Late offset of brain growth leads to late differentiation of neural structures and pathways so that phenotypic behaviors are also delayed. Superficially, this looks like "slower rates" relative to apes (and ancestors) because humans perform homologous simple ("juvenile") behaviors (e.g., event D) later; however, temporary delay of development is the cause and humans eventually exceed them (sequential organizational hypermorphosis).

are growing larger than the ape's, our behavior is temporarily "retarded," although "delayed" is a more accurate term (Fig. 7-14C). This is caused by the delay in brain maturation (organizational hypermorphosis, if delay in offset of the prerequisite events is the frame of reference). Thus, our behavioral ontogeny is similar to that of body size in lagging behind at some stages but becoming eventually "more developed" at each stage (thus, more properly: sequential organizational hypermorphosis), extending beyond our ancestors (and the apes) in behavioral complexity. Superficially, this temporary lag in body and behavior (temporary "paedomorphosis" since we resemble other primate juveniles when we are at older ages than them) may be (and has been) misconstrued as neoteny, but that is not the basal cause. Rather, it is sequential hypermorphosis, delaying but ultimately extending beyond the ancestral pattern. Also, one might be tempted to focus on the late onset and call this sequential postdisplacement. Yet again we believe this is not truly correct: it is the delay in the offset of the preceding stages that is the ultimate cause of the late onset. [See McNamara (1983a) for similar hypermorphic "delays" in sequential formation of trilobite molting stages.]

A more graphical, phylogenetic summary is shown in Fig. 7-15. It is somewhat obsolete in specifics (*Pithecanthropus* is *Homo erectus* and the australoid race does not deserve separate status from "European") but the general outline of phylogeny, brain size, and offset timing is valid. We show it not only as a convenient outline but because it illustrates the wrong reasoning on human "neoteny" ("fetalization"). The delayed age of differentiation (from late offset of growth) is wrongly attributed to "slow rate of growth." The larger brain is also due to late offset of growth but is attributed to neoteny because of the superficial shape of the modern skull to the anthropoid juvenile braincase.

3.2.3. Morphology: Not Only Hypermorphosis

While human life history and behavioral (brain) development are dominated by the single process of delay, human morphology shows a number of local deviations from this underlying process. Perhaps the most striking is our relative hairlessness so that, far from being "overdeveloped" for this trait, we are extreme among all the primates in our neotenic expression of it. This appears to result from the negative allometry of decreasing hair density with increased body size in apes; this led to the consequent development of eccrine sweating to compensate for the loss of the reflective hair coat (Schwartz and Rosenblum, 1981). Many other such dissociations are also discussed by Shea (1988) who provided much of the information for this section (3.2.3). Also see Schultz (1969), an anatomist who has documented many traits where humans show local changes relative to other primates. Such "adjustments" are perhaps no surprise given the demands of cultural adaptations, which are so different from our original arboreal adaptations. We may note two "suites" of deviations. The larger suite has those traits associated with full-time bipedal locomotion; the smaller consists of cranial adjustments related to dietary changes and speech. There may well be other local growth field deviations from the general theme of hypermorphosis; this discussion is not meant to be exhaustive.

Among numerous adaptations to bipedalism perhaps the most central is pelvic restructuring which seems to result primarily from local accelerative processes (Chaline *et al.*, 1986; Berge *et al.*, 1989). Another important change is decelerative (neotenic),

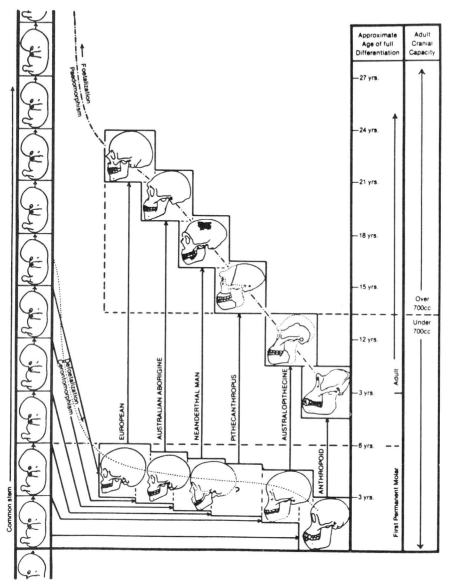

Figure 7-15. "Heterochronic phylogeny" of human evolution as interpreted by Montagu (1981). "Fetalization" should be peramorphosis (hypermorphosis), but age and cranial data are approximately correct. However, aborigines should not be distinct from "European" and *Pithecanthropus* is now *H. erectus*. From Montagu (1981).

causing our skull to sit atop our spine when erect. At birth, head position on the spine in apes and humans is about the same. In apes, this position changes greatly with growth because the growth field in front of the occipital condyles grows faster than that behind them, causing the joint to be shifted backwards. In humans, this forward growth field has been retarded to the same rate as that posterior to the condyles so there is little backward

Table 7-3. Comparison of Human versus Chimpanzee Adult
Male Limb Lengths[a]

Upper arm length	103.6%	Thigh length	144.0%
Forearm length	93.4%	Leg length	161.0%
Hand length	79.5%	Foot length	109.9%

[a] Human values expressed as percent of chimp. From Gavan (1953).

movement of the skull–spine connection (Harrison *et al.*, 1988). However, Moss *et al.* (1982) show that other morphological features of the skullbase cannot be derived from the fetal or subadult morphologies of nonhuman primates via simple neoteny. So, still finer local (growth field) changes must have been operating.

Limb, hand, and foot adaptations to bipedalism show a number of changes which, at least when compared to the chimpanzee, attest to the ready plasticity of these growth fields for simple changes. As shown in Table 7-3, for some linear dimensions, humans are considerably larger (total leg length) and considerably smaller for others (hand length), with a wide variety in between. Heterochronically, these differences arise from a number of processes. The leg dimensions seem to result from simple continued growth (hypermorphosis) of the chimpanzee trajectory (Fig. 7-16) and so do not constitute a growth field deviating from the underlying hypermorphosis discussed above. Manley-Buser (1986) has suggested that our foot development is neotenic (to apes) but these data do not seem to bear this out. It is true that our feet grow at a slower rate, but they grow for a longer time and therefore become larger (Table 7-3). More likely, we suggest, is that foot growth resembles that of upper limb growth which seems to be a combination of prolonged development coupled with the reduced rates of growth (Fig. 7-16). The net result is an organ that is about the same size or smaller than that of the ape's. Perhaps most notable is our relatively smaller, more gracile hands specialized for fine manipulation.

Cranial adjustments to dietary habits have been well studied and, at least superficially, are dominated by neotenic reduction of teeth, jaw, and associated growth fields in humans. de Beer (1930) used it as an indicator of human "neoteny" in general. Brace *et al.* (1987) have documented this reduction particularly well; it is selectively caused by increasing reliance on softer, processed foods. However, this is by no means a matter of "nothing but" neoteny of these cranial features as demonstrated by Bromage (1985). There are a number of even more local deviations within the overall growth field involving changes of deposition and resorption. These are apparently not the simple result of proportional allometric down-scaling. This also seems to be the case in differences in the facial morphology of *Australopithecus africanus* and the more robust species (*A. robustus* and *boisei*). Pilbeam and Gould (1974) argued that the differences were simple size scaling, but Shea (1985b) shows that allometric dissociations are involved. This important finding indicates that dietary differences between the gracile and robust forms led to the different, morphologically localized changes. Finally, we note the specializations of the modern human skullbase and upper respiratory tract: perhaps surprisingly, given the complexity of the apparatus, much of the change may be attributable to a single, general kind of peramorphosis, possibly acceleration or more likely hypermorphosis. Thus, human infants resemble adult apes in these characters (see Laitman and Heimbuch, 1982, and Lieberman, 1984, for discussion).

3.3. Intraspecies Heterochrony: Human Races and Sexual Dimorphism

The study of human races has historically been fraught with biases so unsubstantiated and (now) obvious that it seems almost comical in retrospect. Much of this "science" was intended (consciously and unconsciously) to objectively "prove" the more advanced evolutionary status of Caucasoids (Gould, 1981). Thus, Negroids were said to be more apelike in features such as a greater prognathism. Given the major role attributed to neoteny in human evolution, this rationalization soon became an effort to see which races were more neotenic (i.e., "evolved"). Thus, Kollman (1905) saw African pygmies as intermediate to apes and humans. Bolk (1929) wrote a major article which specifically considered the role of neoteny in racial development of hairlessness, brain size, prognathism, pigmentation, and other "fetal primate" traits supposedly derived from "retardation" of ape ontogeny. The basic idea was that modern human groups could be ranked along this trajectory of fetalization (a heterochronic "Great Chain of Being"). Unfortunately for the devoted Caucasoid racist, such a line of reasoning proved tortuous since, for some traits, Caucasoids are more "apelike," that is, less neotenic (e.g., more abundant body hair). Montagu (1981) has compiled and discussed a list of "neotenic" racial traits, listed in Table 7-4. As already discussed, many of these are not neotenic but hypermorphic (e.g., larger brain of the Mongoloids). However, some are neotenic (less hairy, small ears). Once we realize the complexity of the heterochronies involved, i.e., numerous local dissociations, the whole justification for such a compilation becomes very questionable. For the most part, this kind of bookkeeping provides little insight, yielding mainly a catalogue of morphological adjustments to the race's original adaptations, or perhaps genetic drift. As an exercise in adaptationism, such accounting may be of interest. For example the light skin of Caucasoids appears to be an adaptation to increase vitamin D production in the decreased sunlight of northern Europe (Brues, 1977). And it is true that one race, the Australoids, is often labeled as generally "least derived" morphologically, having prominent brow ridges and prognathism (among other things).

There is one case where a single underlying change does serve to distinguish a human group. The African pygmy is characterized by slower growth, or neoteny (certainly this group is derived). As discussed in Chapter 3, this seems to be caused by decreased levels of production of IGF-I (Merimee *et al.,* 1982) which leads to a lack of the normal human pubertal growth spurt (van de Koppel and Hewlett, 1986). [However, very recent work by Gerhard Baumann, M. Shaw, and T. Merimee (in press) indicates that defective growth-hormone receptors are also involved.] While their small size may be adaptive in some respects (Hiernaux, 1977), it is clear that many of their allometric distinctions are simply by-products of the body size change. Hiernaux (1977), among others, has suggested for example that their relatively long distal extremities and short hindlimbs reflect specific selective forces. However, anthropometric data collected by Shea and Pagezy (1988) indicate that the large majority of such differences reflect simple body size scaling correlates.

However, in all cases, there is no evidence for any significant racial variation regarding the heterochronic traits that make us truly human: hypermorphosis of brain growth and learning ability. All races are separated by a mixture of various minor heterochronic differences; but the basic trajectory of hypermorphosis of neural development

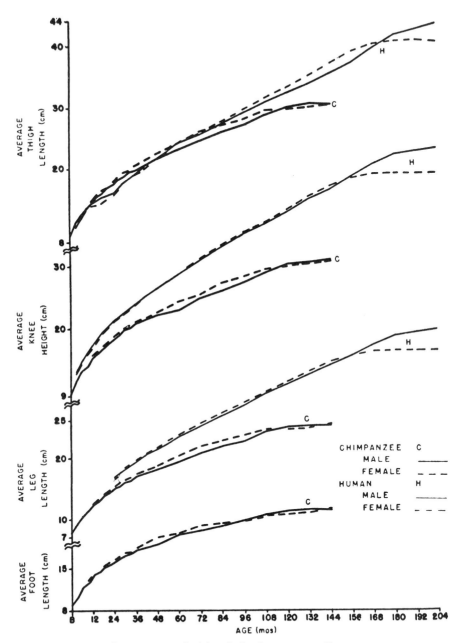

Figure 7-16. Ontogenetic change in average limb lengths for chimpanzees and humans. Discussion in text; see also Table 7-3. From Gaven (1953).

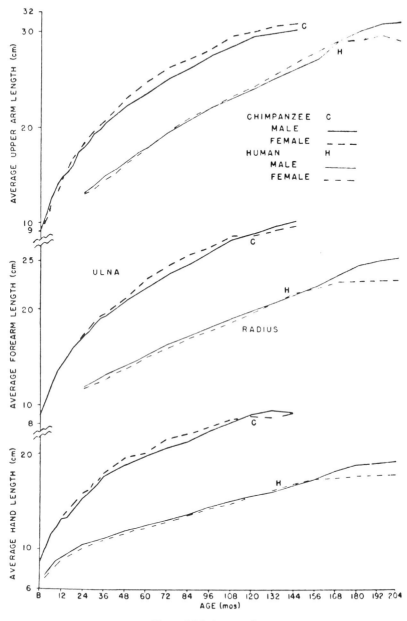

Figure 7-16. *(continued)*

and learning stages (or any life history events) is not known to be significantly different in any racial ontogeny.

Another area where heterochrony is important is human sexual dimorphism (some of this discussion is from Shea, 1988). In heterochronic terms, human (and ape) females have a similar growth pattern but show an earlier offset time (Figs. 7-5 and 7-6). They are

Table 7-4. Supposedly "Neotenous" Traits Which Distinguish Races Listed from Caucasoids[a]

Mongoloids	Negroids	Bushman–Hottentot
Larger brain	Flattish nose	Light skin pigment
Larger braincase	Flat root of the nose	Less hairy
Broader skull	Small ears	Large brain
Broader face	Narrower joints	Round-headed
Flat root of the nose	Frontal skull eminences	Bulging forehead
Inner eye fold	Later closure of premaxillary	Small cranial sinuses
More protuberant eyes	sutures	Flat root of the nose
Lack of brow ridges	Less hairy	Small face
Greater delicacy of bones	Longer eyelashes	Small mastoid processes
Shallow mandibular fossa	Cruciform pattern of lower	Wide eye separation
Small mastoid processes	second and third molars	Median eye fold
Stocky build		Short stature
Persistence of thymus gland into		Horizontal penis
adult life		
Persistence of juvenile form of		
zygomatic muscle		
Persistence of juvenile form of		
superior lip muscle		
Later eruption of full dentition		
(except second and third		
molars)		
Less hairy		
Fewer sweat glands		
Few hairs per square centimeter		
Long torso		

[a] From Montagu (1981).

thus "progenetic" to males in a nonphylogenetic sense (called "bimaturism," see Glossary). In humans and other primates, many intersexual limb proportions are simple allometric correlates of smaller female body size (Wood, 1986). For example, males have longer legs relative to trunk length simply because males spend a longer time in the prepubertal growth phase, when legs grow relatively faster than the trunk (Harrison *et al.*, 1988). Many other female traits, the more gracile skeleton, narrower joints, and so on, are also simple correlates. However, dissociations from strong local growth field selection do occur, not surprisingly in the pelvis (Leutenegger and Larson, 1985). The proximal cause of wider female hips and other traits is due to specific stimulation of target cartilage cells by estrogens. Montagu (1981) discusses a number of other female paedomorphic traits, including skull features, but attributes the paedomorphosis to neoteny instead of progenesis. Given the frequent covariation of behavioral with morphological heterochronies, it would be of interest to investigate whether male–female differences in human behavior may be (at least partly) related to the female's "progenetic" status.

Similarly, the ecological and sexual selective pressures affecting the different maturation timing mechanisms between sexes are of great interest and deserve much more study than they have received. For instance, the extreme sexual dimorphism of the gorilla versus the lesser degree seen in the chimpanzee is almost certainly related to the very different social groupings and sexual behavior of the two species. (Gorillas basically have a harem system while chimpanzees are much more promiscuous.) Our own sexual dimorphism has been affected by our long life span, long parental care, and possibly a polygynous mating system where adolescent sterility was an advantage for females but not males (Ford and Beach, 1952).

3.4. Selective Forces and Human Heterochrony

The basic pattern that emerges from the above is that the primary developmental mechanism underlying much of human evolution has been hypermorphosis. In the adaptive triad of life history, behavior, and morphology, delay in developmental timing has played the central role in the first two and part of the third. Of course, diagnosing the developmental mechanism is only part of the task of reconstructing evolutionary paths. The second half is to isolate the selective (external) forces that have acted upon these genetically (internally) produced phenotypes. Specifically, why have hominids with those phenotypes been more successful?

A vast literature has grown up around the subject of human evolution and much of it is more confusing than enlightening. One main reason is that workers often insist on focusing on only one aspect of the total human adaptive suite. Thus, there is a great deal of argument over the function of this or that tooth or clavicle with little reference to the whole. The biggest drawback of this piecemeal approach is that cause and effect become major bones of contention when in fact covarying suites of traits change in unison under selective pressures that affect them all. This has been stated by others and is probably obvious to many readers but the holistic approach is explicit in the heterochronic view. We are forced to look at timing and rate changes of ontogeny as a whole, not only of morphology but behavior and life history as well. In short, we must look at ontogenetic covariation in the context of ecological processes.

We suggest that the basically hypermorphic human adaptive suite represents an almost end-member example of K-selection. Large body size, low fecundity, complex behavior, long life span, extended parental care, and other human features are classic K-selected traits, as the r–K continuum is usually defined. It has become fashionable to focus on the many shortcomings of the r–K framework, but its widespread use indicates that there must be something to it as we have already noted (see Chapters 6 and 8). Most important, an elegant study by Ross (1988) has recently provided sound empirical evidence that r–K not only "works" but is a good predictor of primate life history patterns in particular (see also Lewin, 1988a, and references therein). While primates are relatively K-selected compared to many mammals, Ross showed that within primates themselves, some are more K-selected than others. As expected by the theory, these latter inhabit more stable environments. Clearly, hominids fall in this category as well so that, in seeking selective forces in our evolution, we must consider this strong underlying background of environmental stability. Indeed, this logic has recently been given even more support by Pagel and Harvey (1988) who showed that K-selected mammals clearly have larger relative brain sizes. In a life history sense, larger brains correlated strongly with longer gestation lengths and was only weakly or not correlated with maternal body size and metabolic rate. Thus, brain size is directly linked to life history strategies, largely as a product of stable environments and low mortality (see Lewin, 1988b, and references therein for discussion). One key insight provided by this approach is that culture, the truly central theme of our adaptive suite, not only arose in this stable environment but has acted, in a positive feedback manner, to increase that stability. As our species has exerted ever-greater control over the environment, it has become progressively even more predictable. It is no wonder that developmental mechanisms that enhance these traits have been progressively selected so that we have become extreme K-selectionists.

This scenario seems straightforward enough but a crucial missing part is the attainment of a threshold high enough to initiate this positive feedback process. In other words, what selective forces led to an organism with a brain large enough and other associated traits to begin to rely so heavily on learning? Evidence on this is found in patterns of brain size and life-style in living primates and other mammals. Clutton-Brock and Harvey (1980) showed that primate relative brain size decreases as the proportion of leaves in the diet increases. They suggested that a larger memory is needed to exploit fruits, which are relatively more scattered and patchy in occurrence than leaves. A similar pattern has recently been found in myomorph rodents (Mann *et al.*, 1988). Folivorous groups average only about two-thirds the brain size of granivorous, insectivorous, or generalist groups of the same body weight. Hence, we may infer that nonfolivory played a role in our ancestors reaching the threshold. An important additional factor has been extensively discussed by Gibson (1986 and references therein) who has shown that primates regularly eating foods that need to be extracted from the environment have relatively larger brains. Such "extractive foods" include such diverse diets as nut-meat, pod seeds, termites, snails, and many others. While they are more trouble to extract, they are generally higher in nutrition and more available throughout the year compared to more readily eaten foods such as berries or leaves. Omnivorous diets of extractive foods lead to the largest relative brain sizes of all, as opposed to those primates which specialize on just one or a few types. Hence, it may be that dietary habits of extractive omnivory and nonfolivory played a major role in the development of brains large enough to reach the threshold.

However, in addition to subsistence behavior, another school of workers stresses the demands of social life as a main factor in the evolution of learned behavior (see Box, 1984, for review). Unlike objects, the social environment is reactive so that there is a strong likelihood of powerful positive feedbacks ("arms race"). Individuals able to communicate more effectively, make alliances, and even "outthink" other individuals in the group (e.g., in bargaining) may be at a considerable advantage (Byrne and Whiten, 1988). This is sometimes called "Machiavellian intelligence." This may also include effects of group selection (Grant, 1985). Social groups whose members are more able to cooperate and innovate may be at a selective advantage in the long run. While this form of selection is disfavored for biological evolution, it may occur where cultural transmission is important because it does not rely on individual reproduction for diffusion. In any case, the point is that other humans became a major selective force in the survival of any one individual; they became the greatest threat to life and the greatest asset. This social factor has increased in importance as other aspects of the environment came under ever-greater control. This could also explain the problem of human "overintelligence" articulated by A. R. Wallace. How could brains evolved in a hunting and gathering society be preadapted to our high technology and information age? The intellectual demands of coping with the evolving social environment may have been great enough to require a very high aptitude for learning in general.

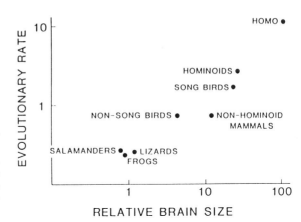

Figure 7-17. Rate of evolution versus relative brain size (adjusted for body size), showing increasing rate with behavioral complexity. Evolutionary rate was calculated from lengths of eight morphologic features, method and date described by Bateson (1988). From Bateson (1988).

While there is much controversy over the relative selective strengths of the two forces just discussed (social versus subsistence forces) in promoting evolution of learned behavior (Box, 1984), these forces are not exclusive and there is little doubt that both played a role. Indeed, the traditional scenario of human evolution invokes both. Our hominid ancestors, having taken up life in the African savanna, took up new (and more challenging) modes of subsistence, going from dominantly vegetarian to omnivory (including food extraction and hunting). Hunting, and new challenges of predator avoidance, would have simultaneously increased selection on cooperation and other forms of social behavior.

The role of hypermorphosis in life history, brain size, and learning aside, selection on morphologically local developmental changes emphasized "adjustments" that, while often nonhypermorphic, were largely determined by that more central suite of adaptations. The above-mentioned dissociations in speech, bipedalism, finer manual powers, and others all reflect pressures associated with culture and learning. The fact that we have so many should not be taken to mean that such heterochronic mosaics are common in evolution. The different demands (and consequent morphological refinements) of our adaptive cultural trajectory, compared to those of our arboreal ancestors, are likely to have been much more extreme than that of most groups, as illustrated by the pygmy chimpanzee. This smaller version of *Pan* is almost globally neotenic relative to the larger form.

4. CODA: HETEROCHRONY, BEHAVIOR, AND EVOLUTIONARY RATES

In Chapter 5 we looked at the effect of heterochrony on evolutionary rates. In that context, heterochrony was a direct mechanism for carrying out change, either fast or slow. The focus of such change was morphology. It is also important (and often neglected) that behavior can also affect evolutionary rates. As a major developmental vehicle for altering behavior (either innate or by increasing learning capacity), heterochrony also has an indirect effect on evolutionary rates.

There is evidence that, as a very general rule, increasing behavioral complexity leads to an increase in evolutionary rate (Bateson, 1988). This is shown in Fig. 7-17.

Also, Stanley (1979, in press) has shown that behaviorally advanced taxa such as mammals, birds, trilobites, and ammonoids have higher speciation (and extinction) rates compared to foraminifera, corals, and bivalves. The reason for this relationship is probably found in Mayr's (1963) conclusion that "ethological barriers to random mating constitute the largest and most important class of isolating mechanisms in animals." Thus, the more complex the behavior, the more likely that barriers to gene flow will be formed. In addition, we might consider the greater adaptive flexibility of advanced behavior in finding new environments or niches to inhabit, requiring subsequent morphological "adjustments." This would be especially important in advanced learned behavior. Thus, by its role in changing behaviors, especially the increase of learned behavior through delayed brain growth and learning stages, heterochrony plays an indirect role in determining evolutionary rates. As with human intelligence, this is probably a good example of positive feedback in evolution because as other organisms evolve more advanced behaviors, there is ever-greater pressure to increase behavioral complexity. This is seen in the increasing mammalian brain size during the Cenozoic Era (Gould, 1988b). Such positive feedbacks would create ever-increasing evolutionary rates in the participating taxa (Chapter 8). In addition, the more behaviors an organism has, the more there are to phylogenetically change. Complexity, also in positive feedback fashion, inherently promotes increased rates of change and more complexity (Chapter 8).

Chapter *8*

Epilogue and Synthesis
Ontogeny in Evolution

Evolution is the control of development by ecology.
L. van Valen, prominent evolutionary biologist

That's all folks.
P. Pig, prominent peramorphic porcine pundit

1. DEVELOPMENT AND EVOLUTIONARY THEORY

This chapter is an attempt to synthesize the ideas presented earlier in this book. It is to some extent reductionistic because we want to unite a number of disparate ideas and so must reduce things to the basic form to make them manageable. Fuller discussion of most of the concepts is found in earlier chapters. A book of this sort should not only inform but also stimulate further thought and research. To that end, some of what follows is quite speculative.

1.1. A Rebirth of Interest

The role of ontogeny in evolution is becoming a major focus of interest in evolutionary theory (Futuyama, 1988). This renaissance of interest essentially began with the publication of Gould's (1977) major work on ontogeny and phylogeny. This work, of partly "antiquarian indulgence" (Gould, 1988a), seems to have acted as a catalyst on largely dormant ideas. Many would agree that, given the growing data base in developmental biology, its role in evolution was certainly overdue for scrutiny, as noted by many recent authors (e.g., Hamburger, 1980; Raff and Kaufman, 1983; Maynard Smith *et al.,* 1985; Levinton, 1988; Thomson, 1988). At some point, examination of this role was inevitable.

But just why is it that modern evolutionary theory, beginning with the modern synthesis of the 1940s, has neglected development? Why has it emphasized genes and phenotypes of (mostly adult) individuals, with relative neglect of the "black box" of developmental events occurring in between? At least in part the answer must be, quite simply, that it worked. The elegant tractability of population genetical theory, combined with natural and experimental observations, has made great explanatory strides. To this we may add a historical reason: the conceptual abuse of available ontogenetic information by early prominent figures such as Haeckel (see Chapter 1). This encouraged developmental biologists to take up a largely mechanistic, nonevolutionary approach. Simultaneously, evolutionists became very suspicious of attempts to incorporate developmental regularities into evolutionary patterns.

While some authors seem to treat the relative neglect of development in evolution as if it were a lamentable historical accident, the above scenario was probably unavoidable. The complexity of development, combined with its central yet complex connections to evolution, made it almost inevitable that early attempts to relate the two would be irresistable and inaccurate. Further, the ability of genetics and individual selection to explain many salient patterns of evolution made it natural to focus on those aspects, while waiting for the ahistorical methods of embryology to unravel the mechanics of development. Even now, as is evident in this book, science is ignorant about many of the mechanical details, though workers know enough to make more educated guesses.

In addition, the "neglect" of development by evolutionists has probably been over-emphasized. Throughout the postsynthesis period, there have actually been some prominent evolutionary theoreticians who created a high profile for developmental issues. de Beer (1958) was perhaps the most directly concerned with the general issue, but many others brought development to bear. Waddington contributed substantially to developmental quantitative genetics while Mayr, Dobzhansky, and others wrote widely about developmental homeostasis and its evolutionary implications. It is true that much of this was rather vague and speculative, but with good cause, because it largely reflected the relatively depauperate state of knowledge of developmental biology at that time. Thus, Mayr (1963), in his major treatise on evolution, devoted only nine paragraphs out of some 673 pages to developmental questions (Thomson, 1988).

1.2. The Possible Importance of Development in Evolution

Organic evolution, like any change, has two properties of particular interest: the *rate* and *direction* of change, representing the two basic parameters of existence, time and space, respectively. In both cases, a major question has been not only the rate and direction of evolution, but whether the motive force behind each has been largely, or even significantly, "internally" generated. Both properties have been of keen interest to evolutionists for many years. Darwin (and Huxley among others) of course wrestled with the question of evolution's gradualness as do evolutionists today. Long before "punctuated equilibria," various schools of "macromutationists" and "saltationists" were challenging the notion of slow change. Geologically "rapid" evolution may have other causes than rapid internal "jumps" (e.g., intense selection on small populations) but such jumps have been a source of fascination for many. A key point is that as our knowledge of development increases, the feasibility of such "hopeful monsters" (a graphic phrase construed by Goldschmidt himself) seems to be increasing among evolutionists. Numerous developmentally knowledgeable evolutionists (Raff and Kaufman, 1983; Arthur, 1984; Thomson, 1988) have argued for the likelihood of such events (a notable exception is Levinton, 1988) as discussed further below. For instance, major differences in the early development of closely related frogs and closely related echinoids indicate that not all evolution has been via minor, late-acting changes (Thomson, 1988; Raff and Kaufman, 1983; Raff, 1989; see Chapter 3, this volume). As Thomson (1988) noted, "monster" is a pejorative, subjective term. Few seriously argue that the first bird hatched from a reptile's egg.

However, while rates have dominated much evolutionary debate in recent years, interest in directionality is fortunately growing. We say "fortunately" because it seems (to us) to be a more central issue: it is usually of more interest to know where something is going than how fast it is getting there (though this is not to deny that the major interest in rates has really been to explore implications for underlying evolutionary mechanisms, e.g., stasis). Thus, there is a growing literature on trends in general (McNamara, 1990), abiotic versus biotic forces in trends (Ross and Allmon, 1990), coevolutionary trends [Vermeij's (1987) escalation or "arms race"], clade-level trends (Gould *et al.,* 1987), biosphere-level trends (Bonner, 1988), and so on. True to the neo-Darwinian view that most of us at least essentially accept, these discussions emphasize external selection in directing evolution. Yet it may be that internal forces play a major role in that directionality. Obviously it will be nothing as simpleminded as the orthogenetic internal drives suggested in the past, but it may be that emergent "rules" of developmental processes constrain or bias change in certain ways. For example, the very existence and, especially, abundance of heterochronoclines (Chapter 5) argue for the importance of ontogenetic patterns in influencing evolutionary direction.

Recalling the watchmaker analogy of Chapter 5, where development (the "internal" process) does not produce each piecemeal "part" (trait) that is needed, when it is needed, it has a limiting effect on selection (the external process). At the greatest extreme, development would produce no piecemeal units at all, but intrinsically generate "new" forms *in toto,* omitting a role for external selection altogether. The point of this perhaps obvious thought experiment is to outline the borders of reality: the truth is somewhere in between. Between the end-member roles of inconsequential "on-demand supplier" and sole "creator" lies the vast domain of "constrainer." (As in the case of heterochronoclines: ontogeny is often producing highly covariant trait suites; yet even here dissociated, individual traits, or mosaics of them, can be clinically patterned—Chapter 5). The question is where in the middle ground, of weak to strong constraint, development lies. It will vary of course (especially among major taxa, i.e., among developmental programs), but how much, and where is the center of the distribution? An excellent example of this "intrinsic versus extrinsic" debate is the brain–body allometry discussed in Section 3.2.2.1 of Chapter 7. Thus, Pagel and Harvey (1989) argue that brain size changes "freely," whenever (extrinsic) selection changes. Others (e.g., Riska and Atchley, 1985) argue that brain size is linked to body change. This also illustrates that, while it is convenient to discuss rate and direction of change separately, they are linked (though not inextricably). Large-scale developmental change is not only usually "rapid" but also is associated with a large developmental contribution to evolutionary directions as the role of external selection in creativity is diminished.

1.3. Heterochrony and Those "Emergent Rules"

While evolutionists are often embroiled in theoretical debate over the relative role of development in evolution, this is really an empirical issue. Until more is known about what transpires between the zygote and the adult, it is impossible to say how constraining or creative that process can be across evolutionary time scales and in different environments. As discussed below, "internal" development and the "external" environment both exhibit varying rates and directionality of change which can, theoretically, serve as controlling forces of rates and directionality of evolution. Nevertheless, as data accumulate we are at least setting narrower boundaries on our speculations.

Perhaps the most striking impression on anyone familiar with developmental processes is the interactive complexity [to use Thomson's (1988) apt phrase] of them. Further, this complexity occurs on a number of levels, from molecular to tissue scales. The crucial result is that development is an extremely "nonlinear" process with an enormous potential for intricate cascades of change, composed of causal chains traceable through a number of scales. In the words of Kauffman (1983), such a system is highly "nonisotropic"; it is most mathematically tractable to such approaches as bifurcation theory (Chapter 3). This of course is why the study of development (and hence its role in evolution) has lagged behind. It will strongly resist naive reductionism since emergent properties will occur at each level. If we are to ever understand the "rules" governing rate and directional change in ontogeny, a hierarchical approach is essential. The nested branching tree model used in Chapter 3 and below is but a simple start.

1.3.1. Pan-Heterochrony?

Our point is that heterochrony may be an organizing principle for not only describing but also characterizing and explaining developmental "rules." We have shown how it can be applied to a number of levels, and related to the hierarchical levels of a branching ontogenetic tree (Chapter 3). We have done this by extending the usual application (but not definition) of the term "heterochrony." In the past, heterochrony has been limited to classifying large-scale, late-ontogenetic patterns. As a result, it has been used as a largely static descriptive scheme in spite of its definition as a process of change without reference to the spatial or temporal scale involved. Thus, we have sought to redefine heterochrony as any "change in the rate or timing of cell dialogue" (Chapter 3). This means that very early developmental changes (e.g., tissue induction) as well as very late ones (e.g., local growth field change) would fall under its rubric.

This approach arises from: (1) the need to be consistent and nonarbitrary in the application of heterochronic terms (and the confusion that has occurred because people have not), and (2) the general utility of an agreed-upon vocabulary and frame of reference across many scales of ontogeny. Notice that we do not claim that this extended view yields any more explanatory power *per se*. Developmental events can be explained only through direct study. What we are suggesting is that heterochronic ideas, changes in rate and timing, pervade ontogenetic change at all levels, and a common organizing principle and vocabulary are essential, even if simple large-scale ("allometric") patterns are not the result.

Let us review some concrete examples to justify both points. The "classic" cases of heterochrony have involved the recognition of similarities between the adult of one species and the juvenile of another. Of necessity, this means that comparable, late-acting changes must have occurred to separate the two. Thus, in the Irish Elk (Gould, 1974), the well-known axolotl (Raff and Kaufman, 1983), and the titanotheres (McKinney and Schoch, 1985), the evolutionist is drawn to the pattern, resulting from some simple rate or timing change: a number of traits (e.g., horns, body size, or some other traits) become "over-" or "underdeveloped" in one species relative to the other. The simplicity is both obvious and attractive to the evolutionist seeking explanation of process through the study of pattern. But what happens when we shift away from such obvious covariant suites and focus on single traits? What if only horn length increases through faster growth? Can we say it is accelerated? Or is this just more unneeded jargon? If so, then where do we draw the line, since covariance of change can extend from globality to very local suites of traits (as in Chapters 2 and 3)?

At the other end of the ontogenetic scale, consider the apparent arbitrariness of "heterochrony" in early ontogenetic events. Here, instead of "trivializing" heterochrony, applying it to minor events (e.g., local trait acceleration) as above, we may be accused of "aggrandizing" heterochrony by applying it to massive tissue alterations. Yet, Hall (1984a) has shown how changes in rate or timing of cell interactions can have two outcomes. It may affect only the size of the tissue-grade traits involved (what we have called size differentiative heterochronies), or it may result in novel interactions and tissue-grade changes (novel differentiative heterochronies). In "classic" heterochronic jargon, the former trait would be routinely classified as a typical heterochronic "event" but the latter would be "something else" (the noncomparable ontogenetic event of Alberch, 1985a). Yet in both cases, the same cellular events may be the cause: delay in cell migration for example. (Novelty only seems spontaneous; everything has precursors that give continuity and comparability to the previous state.) Thus, current usage is not only arbitrary, it essentially emasculates heterochrony by circular logic: "heterochrony" is limited to simple allometric changes (and thus has little explanatory power or importance as a creative process according to most authors; see Chapter 3) but is defined as only those heterochronic events that involve simple allometric changes! Yet we see here that changes in rate and timing at the cell level can improve our interpretive view greatly by allowing us to trace causal events up through the tissue grade and that the creative power of such events is great where such heterochronic changes occur in early ontogeny. If we examine the example of differentiative changes, is there not an explanatory gain from knowing that both a novel tissue and a simple size change arise from similar changes in rate or timing at the cell level? And formalizing that similarity with a common, shared set of terms?

Certainly the most tentative proposal we have made is the application of heterochrony to biochemical levels. Yet, many workers discuss rate and timing aspects of diffusion across growth fields (e.g., Raff and Kaufman, 1983; Bonner, 1988). Again we confront the arbitrariness problem because in one case, traditional allometric changes may result while in others cellular reorganization may occur, affecting, say, the number of digits (Chapter 3). Both processes involve cellular activities, only in the latter, nonlinear bifurcations result. The problem is aggravated by such terms as "pattern formation," which is heavily used but poorly defined (Arthur, 1984). Nor is recourse to the underlying genetics any help. The simple "rate" or "timing" genes of the past may sometimes cause changes in rate or timing of morphogen activity but often such diffusive changes may come from "structural" gene changes such as affect the extracellular matrix (Chapter 3).

1.3.2. Summary: Justifying the Extended View of Heterochrony

Heterochrony is no panacea and not all developmental events can be fruitfully viewed in rate and timing terms. All change involves time but, for instance, it is not necessarily helpful to view a car crash as having been caused by change in timing (one car had "delayed onset" of the brake, and hit the other). Rather, a more ultimate cause is sought: a drunk driver, with reaction impaired, had delayed reflexes. The alcohol is the condition most usefully identified in explaining the cause. Timing change was more an effect than a cause, though it is useful as a description in the causal chain. Similarly, simple "rate genes" may cause some heterochronies but many result from changes in cell properties and other causes not easily reducible to single causes (as is also the case in some car collisions, e.g., perhaps a wet road is involved above). Thus, in some cases, heterochronic terms may be simply a matter of terminological convenience. Various workers have objected to the use of phrases such as "accelerated digit growth" or "hypermorphic limb." We all tire of jargon but it exists because it is useful to workers who need a precise meaning for commonly observed phenomena. (This is the same logic for the use of "allometric heterochrony"—Chapter 2.) Further, in this case the jargon is neither esoteric nor unmanageable: anyone who studies organisms at any level will observe rate and timing change, and there are only six major types of heterochrony (Chapter 2). Beyond convenience, interpretability of events is enhanced by focusing attention on the cellular level of development and tracing causation across all scales. In this we adhere to Oster *et al.* (1988) who state that "developmental constraints must be understood at the level of the cellular and biochemical interactions that underlie morphogenesis." A specific example of this is the frequent "definition" of heterochrony as a simple extension or truncation of existing developmental "pathways." What does this mean in real terms? In some cases, extension of the ancestral "pathway" of cell migration may lead to simple size change. In other cases, it can lead to tissue-level novelties. If we seek to understand fine-scale phenomena (and their resultant cascading effects through higher levels), we must first have precision of meaning. Phrases such as "pattern formation" or "pathways" are often nothing more than labels of ignorance which is why they are so often the focus of debate.

In the past, heterochrony's attraction has been its comparative view of large-scale covarying trait suites between ancestor and descendant. Specifically, it has been useful in reconstructing three key evolutionary phenomena: phylogeny, adaptation, and mechanism of change. The similarity between ontogenetic stages of different species is a major way to recognize and reconstruct phylogenetic derivations. In the case of adaptation, heterochronically covarying traits either covary because they are jointly adaptive or (of perhaps more interest) some have been "dragged along" with others as parts of a developmental package (Chapter 6). Finally, the ability to identify differences between species as originating from rate or timing of growth provided major insight into the mechanisms of speciation. Our point here is that in extending the scope of heterochrony, none of these traditional aims are compromised. Indeed, they succeed better. Phylogenetic relationships still occur where the ontogenetic stages are not directly comparable. In fact, the complex interactive processes of development probably make such disjunctions inevitable (Chapter 3; Alberch, 1985a). However, rather than despair over such non-comparability, it is better to realize that comparability *does* exist at finer scales: the descendant may undergo delayed biochemical and/or cellular change, relative to the ancestor, at some critical point. As noted above, all novelties have precursors; in this case, the cell-level behaviors of the ancestral ontogeny are comparable up to the point where they deviate. The only difference with this and traditional heterochrony is that it does not cascade upward in any simple extrapolative way. The resulting outcome may also still be of adaptive interest since many such changes may still be functionally important. Finally, the goal of reconstruction of mechanism is obviously greatly aided by our expanded scope. As already noted, we can identify changes at more ultimate causal levels (cellular, biochemical) rather than limiting our view to the gross tissue-grade level.

Cell-Level Selection. A final justification for the expanded (especially cellular) view of heterochrony may be its use in analyzing selective processes. Until now we have focused mainly on developmental production of variation. Selection of that variation is obviously crucial and while the view of selection as a hierarchical process is growing [selection on genes, individuals, species (e.g., Vrba and Eldredge, 1984)], the levels seen in developmental processes have often been conspicuously absent. This has been discussed by Thomson (1988) who points out that any level of life where variation is introduced, sorted, and replicated will undergo selection (and thus evolution). Thus, developmental "pathways," be they biochemical, cellular, or tissue grade, will evolve in the Darwinian, phylogenetic sense. The sorting (selection) is often via "internal" environments, being internal to the individual (as opposed to selection by the extraindividual environment, which acts on the individual as the selective unit).

An evolutionary perspective on this cell-developmental level selection is found in Buss's (1987) treatment of the evolution of metazoan individuals from cell colonies. By focusing on cell-level selection, he sheds much light on how evolution has occurred. In these terms, heterochrony at the cell level arose as the natural by-product of cells vying to survive. Changes in rate and timing of cell migration, mitosis, or differentiation allowed some cells (and cell aggregates) to survive where they otherwise would not. In particular, this led to the formation of different kinds of cell aggregates. At the organismic level this translates into new kinds of variation as the novel cell aggregate is able to exploit new ways of life. Buss (1988) has added further insight into heterochrony by documenting the critical effects of timing of germ-line determination in passing on these cell traits to conserve the diversity produced.

Michaelson (1987) has reviewed the evidence for cell selection: overabundance of cell death and growth, explicitly in terms of the "struggle for existence" in cell societies. Even morphogens (Chapter 3) are, in his view, agents (and targets) of selection. The living "prototype" for this single- to multi-cell transition is the slime mold. Key insights into the origin of cell–cell interactions, gene expression, and morphogenesis are found in an excellent review of slime mold research by Devreotes (1989).

2. THE EVOLUTION OF CONSTRAINT

Discussion of the role of development in evolutionary rates and direction must focus on how greatly variation is constrained. If large changes are feasible, then a mechanism for rapid evolution and internal directional control is available. There has been much debate on the relative importance of internal forces because: (1) terms such as "constraint" are relative, subjective, and hard to define (Bonner, 1982), indeed everything is "constrained" in many ways since all things are limited in change by their state at preceding times; and (2) a point often overlooked, the importance of internal constraint has not been constant; it has increased as complex ontogenies have evolved. Neo-Darwinian proponents tend to emphasize constraint as a way of minimizing the input of development to evolution, whereas challengers to that view point out that "constraint" is a relative term that may leave much room for internal change.

Both points, one and two above, are clarified considerably by Levinton's (1988) discussion of the **evolutionary ratchet** model which states that, since organisms have a long evolutionary history, "the evolution of timing, rates, and localization leads to a complex developmental process what can be disrupted less and less easily as time progresses" (Levinton, 1988, p. 217). In short, there is an evolutionary "hardening" of integration and contingencies within the organism which makes it more difficult to modify development and functional relationships as organisms evolve. From the cellular view of evolution of Buss (1987) and our discussion earlier (Chapter 3), we would say that cellular aggregation and assembly have resulted in increasingly greater amounts of integration and specialization, leading to ever-greater amounts of interdependency in both assembly (ontogeny) and function (physiology).

Levinton (1988) specifies three kinds of "ratchets" that contribute to the overall evolutionary ratchet. These are not entirely mutually exclusive. The *genetic ratchet* refers to new genes which, once incorporated, cannot be easily lost. They can become too integral to the action of other genes (see especially Wright, 1982). The *epigenetic ratchet* refers to the ontogenetic interdependencies that evolve, such as those modeled by the branching tree. Early, major ontogenetic changes affect so many later contingent events that they rarely successfully occur. Finally, the *selection ratchet* refers to features that, once acquired, are not easily lost because of their functional integration in the developing phenotype. For instance, one limb bone, the humerus, is essential to the proper functioning of others, the ulna and radius. (The term "selection" ratchet is questionable, since the other kinds of ratchets also result from selection. It seems to be another case of wrongly emphasizing selection as a process acting only on adults. Perhaps "functional" ratchet is more appropriate.)

2.1. Developmental Inertia and Directionality of Biosphere Evolution

From the ratchet concept we get a better idea not only of what "constraint" means (in biological terms) but how (and why) it has changed. The major implication for rates and direction of evolution is that in the early phases of life's evolution, when "hardening" was not advanced, rapid "jumps" and internal directionality were more prominent than today. However, there is an external component to these "jumps" as well because such internally created novelties are likely to be poorly adapted in many ways (e.g., function, efficiency). Therefore, a considerable amount of open adaptive space was required to permit their survival until external and internal selection could act to "shore up," integrate, and generally better adapt the organism. "Internal selection" refers to the selection of developmental and other contingencies that improve the mechanical functioning of the organism (Whyte, 1965). These of course are the same contingencies that form the "ratchet" since, once established, they are difficult to lose. Katz (1987), in his discussion of the "phylogenetic ratchet," points out that it is almost always much less detrimental to add (a gene, developmental pathway, organ) than subtract since loss will probably interfere with adult function (or later ontogenetic development if it is an ontogenetic "trait" that is lost). Or as Arthur (1984) has put it, since organisms are well adapted to their environment via their components, any mutation which deletes components will have a near zero chance of being advantageous. In addition, any "step backward" which leaves the adult with "juvenile" traits may throw it back into competition with related species which share the trait. In contrast, added components will retain the old, useful components (or ontogenetic contingencies) so that even if the change is only incrementally useful, it will not be opposed by selection. To combine Jacob's (1977) and Frazzetta's (1975) metaphors, evolution is not only restricted to "tinkering" with preexisting ontogenies, but must do so "while the engine is running" (keeping the ontogeny viable at all times). Most additions will occur at the end of ontogeny (minimizing interference with later development) so "terminal additions" (heterochronically = peramorphosis) accumulate through time. Some of these terminal additions create forms that are able to exploit new niches. Since these are additions on a more basic ontogeny (**bauplan** = body plan, or roughly speaking, a phylum architecture), they represent permutations of a more basic morphological theme. This is diagrammatically illustrated in Fig. 8-1.

We are certainly not the first to argue that early metazoans had both internal conditions (less "hardening") and external conditions (less competition, predation) favoring rapid evolution (see Valentine 1980, 1986; Campbell and Marshall, 1987). More controversial is our scenario of the evolution of contingencies. Some critics may accuse us of simply reviving the long-discredited notion of recapitulation, whereby all evolution has occurred by terminal addition (Chapter 1). However, we are not saying that. We have already discussed numerous cases in earlier chapters showing that heterochrony can work to alter ontogenetic events earlier than the terminal ones. Change in rate or timing of cell migration, or cell differentiation early in ontogeny can significantly alter later events, and still be viable. Our goal is to point out that the general trend toward complexification of ontogeny has occurred mainly by addition. To become more complex is to have more parts and more kinds of parts (see below). Addition is the only way to do that. Addition is easiest at the end of a sequential process, but it is not the only way as shown by the success of those earlier alterations just noted. In this, we agree with Gould (1977) who has noted that the past wholesale rejection of recapitulation is a good case of "throwing the baby out with the bathwater," an overcompensation.

Valentine (1986) lists two lines of fossil evidence for this scenario. One, it is when adaptive space is the most open (Cambrian "explosion," invasion of land) that distinctive bauplane evolve. This observation has recently been further enhanced (Erwin *et al.,* 1987) to show that *progressively milder refinement* of each distinctive bauplan occurs as

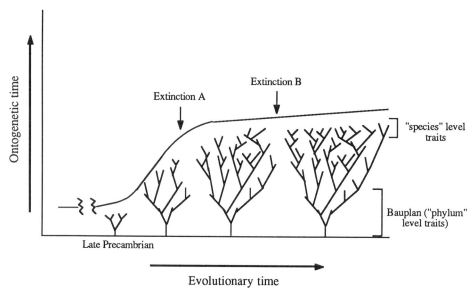

Figure 8-1. Schematic diagram of the accumulation of "contingencies" as the ontogenetic program of one particular bauplan evolves. This "hardening" process (i.e., evolutionary ratchet) is sometimes (but not necessarily) part of complexification [e.g., accretion of new pathways (new contingencies), or "terminal additions"]. Such complexification is exponential at first (multiplicative, positive feedback law of complexity: more parts mean more parts to modify), but soon slowing down, due to the (also) multiplicative nature of both internal ("hardening") and external ("niche packing") contingencies (negative feedback mechanisms). The net result is that a basic body plan is progressively modified into permutations that are progressively more minor: as ontogenetic complexity increases, the number of permutations increases but the ability to change them, and their relative effect, both decrease. Early extinctions (e.g., mass extinction A) will have more overall effect on the evolution of the biosphere than later ones (mass extinction B). See text for fuller discussion.

class- and then ordinal-level differences (i.e., ontogenetic alterations to the taxonomist's eye) originate (Fig. 8-2). Two, even when diversity is low (e.g., Early Triassic), no new bauplane will evolve if bauplane have not been eliminated. It is easier and faster to fill adaptive space by simply modifying the existing basic body plans (see Erwin *et al.*, 1987). Importantly, there is also some theoretical evidence from population genetics.

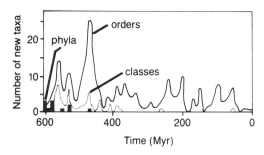

Figure 8-2. Pattern of diversification of families of skeletonized marine animals. Graph shows number of new phyla, classes, and orders that appeared in each time interval. Late Precambrian–early Cambrian is circa 600 Myr. From Benton (1988b).

Arthur (1984) shows that a kind of selection, called **n-selection**, may explain the processes acting on these early forms. In this case, the n stands for net reproductive rate which is the basis of selection. Instead of competitive ability in predation, feeding, and other aspects of energy gathering and use (the basis of most current Darwinian selection), selection favors those with higher n. Not surprisingly, the ecological conditions under which n-selection is most feasible are in small, isolated newly founded populations. However, since the genes and other internal systems are not yet well coadapted, they must remain isolated for some time. This environment and small population also help resolve the problem of "who does the monster mate with." For instance, if a mutation occurs early in the germ-line cell, then a large fraction of gametes will have it and will appear in many offspring. Also it may be recessive, in which case it will be delayed a generation and then may appear in a number of individuals. After a number of such n-selected bauplane become adapted, a new kind of selection, **w-selection**, takes over. This is the traditional Darwinian selection, based on competition between individuals, where other factors besides net reproductive rate are important in determining fitness, especially the general ability to gather and use energy, as noted above.

Finally, there is evidence for the evolutionary accumulation of contingencies (= the bifurcations of Fig. 8-1) in the well-known "recapitulation" of embryology: "higher" organisms (e.g., humans) go through stages bearing some similarity to the ontogeny of "lower" organisms (e.g., fish). Contrary to Haeckel (Chapter 1) these early stages are not simply adults of ancestral ontogenies (as would occur in "pure" terminal addition). However, their very similarity reflects the ontogenetic conservatism dictated by the evolutionary accumulation of the contingencies we have just discussed. The shared similarity at early stages, and subsequent complexification (in "higher" organisms) at later stages argues for a *general* (coarse) vector of complexification by mostly late-stage, ontogenetic accretion. Thus, two key ontogenetic patterns have evolved: (1) the more closely related two groups are, the later their ontogenies (contingencies) tend to diverge, and (2) the more "complex" a group is, the more "pathways" (contingencies) it will have added on, usually toward the later stages of development.

The net result of this process of accumulating developmental contingencies has been the accumulation of more complex ontogenies through evolutionary time. The total number of phyla has decreased from about 100 to 30 but the number of permutations based upon those 30 body plans has been enormous (Fig. 8-2). Of course, this has been an additive (as opposed to a zero sum) process, with simpler permutations being conserved. Therefore, the net diversity of the biosphere has increased considerably since the Cambrian. Whether it continues to increase, or has reached a diversity equilibrium (or asymptote) is a matter of debate (Valentine, 1985). Gould (1988b) has pointed out that such "trends" of increasing variance are common in evolution. In this case, the variance increase is asymmetrical: due to the laws of physics, open systems far from equilibrium (e.g., life) must start off small and simple and build "upward." However, we should recall Maynard Smith's (1972) point that such "nowhere to go but up" theories are incomplete. They posit that (and, more importantly, may explain why) things start off simpler, and that they therefore can *potentially* become more complex. But such theories do not specify what *actually causes* them to change toward more complexity (or greater size, or whatever state variable is changing). In this case, we have already noted the explanation of Katz and Arthur (see also Bonner, 1988) that any genetic/developmental additions that are randomly added will sometimes (albeit rarely) be adaptive (and importantly, the counteractive tendency to lose them will be, on balance, less successful). From the biosphere point of view, this adds up to life which is constantly finding new ways of doing things and "creating its own niches," sometimes by becoming more complex. Thus, there is evidence that the adaptive landscape has added many more peaks, caused by progressively greater amounts of resource and niche partitioning ("niche packing") through time (Bambach, 1985, 1986; Sepkoski, 1988).

The evolution of the ontogeny of only one bauplan is depicted in Fig. 8-1 but we may consider it representative of many. Couched in terms of the "tree model" of Chapter 3 (Section 6), terminal addition occurs via peramorphosis involving one of three possible heterochronic processes—acceleration, hypermorphosis, or predisplacement (Chapter 2). The human brain, the most complex organ, with perhaps the most complex ontogeny (of an organ), is a good example, having arisen from hypermorphosis of neural growth (Chapter 7). It is also an excellent example of accretive growth, retaining more primitive functions while adding others. Finally, it is also an excellent example of the nonlinearity of terminal addition. The emergent properties of the human brain reflect a highly nonlinear output derived from a linear input. The net sum of such additions to many organs would represent the growth of peramorphic ontogenies shown in Fig. 8-1. (Again, this is only a general, reversible vector of increasing maximal complexity; we are not reviving recapitulation.)

Regarding the biosphere itself, the general pattern of Fig. 8-1 is roughly logistic, with a very long lag phase as cells evolve into units which will eventually have the ability to: (1) specialize and (2) become integrated. Once this occurs, in the late Precambrian, a critical threshold is reached and cellular assembly begins with all its multiple options, which increase in positive feedback fashion (more complex = more permutations possible). In terms of hierarchical sorting (Vrba and Gould, 1986), this threshold ("Cambrian explosion") occurs when the individual metazoan, in addition to its component cells, becomes an emergent target of selection (Section 1.3.2). Increased "hardening" (internal negative feedback) and filling ecospace (external negative feedback) soon causes a decrease in the rapid growth of diversity. The greater permutations of multiple options become more and more counterbalanced by the increased contingencies inherent in development of them (see below).

Mass extinctions obviously play an important role in this evolution of more complex ontogenies. The "randomness" of catastrophic selection has been much discussed as a non-Darwinian process (Levinton, 1988). As illustrated in Fig. 8-1, early extinction events could be expected to eliminate more basic body plans than later ones even if both killed off the same percentage of species. Later events would affect large numbers of ontogenetic permutations on each of the 30 or so basic bauplane but, unless there was strong bauplan-related selectivity, some of each bauplan would survive. There is no evidence that such selectivity occurred, including at the massive end-Permian event (Erwin, 1989).

In sum, we have argued that two non-Darwinian processes set the trajectory and have "aimed" the basic directions of the evolution of life (or, in still other terms, they have set major initial boundary conditions determining later evolution). The first process is the "random" *internal* (*developmental*) *creativity* of early metazoan body plans. The second, just discussed, is the "random" *external elimination of body plans by catastrophic selection,* especially in the early stages of metazoan evolution. In both cases, the adjectives "random" and non-Darwinian are used to indicate change (directionality) outside that caused by natural selection of incremental variation by the environment acting to optimize fitness in that environment. As we shall argue next, this latter, deterministic, external directionality of Darwin became important only as complex ontogenies evolved.

Rise of Minor Directional Control by External, Deterministic Selection

As complex ontogenies accumulated, they became more and more constrained by their past. There was less plasticity for major modification. There were more traits to modify, but these could be modified in progressively less profound ways. If alteration (e.g., heterochrony) occurred too early, then later developmental contingencies would likely cascade into a negative impact. It is true that "novel" differentiative heterochronies can occur (Chapter 3 and elsewhere), even today, but "novel" is a relative term and it seems very likely that the novelty producible from a chicken is considerably less "novel" than that producible from a poorly integrated, simple metazoan ontogeny of the late Precambrian (see discussion below on rates). The key point is that the Darwinian paradigm, generated and studied largely through observation of living organisms, has naturally been inclined toward environmental selection of "incremental" variation. At this point in time, the "epigenetic ratchet" has become so "hardened" that few, if any, viable truly novel forms are produced. Even if they were (e.g., by the simpler ontogenies that persist today), the adaptive landscape is so full of highly adapted, fully integrated (internally coadapted) forms that they have little competitive chance (i.e., conditions for n-selection are not present).

Neo-Darwinists rightly use this current state of biological affairs to argue for the predominance of environmental selection in determining the directionality of evolution. Yet the irony is that this argument is inverted. The highly evolved developmental "constraints" which limit internal creativity also guarantee the predominant determination of evolutionary direction by development. Environmental selection can only act upon the relatively minor alterations (e.g., rate and timing) that these "constrained" developmental systems permit. As ontogenetic ratcheting proceeds, the "inertia" of past changes becomes greater and greater so that while environmental selection is more and more of a determinant of direction, it is operating on ontogenies that are less and less open to significant change. It is as though the former pilot of a plane has handed over the (increasingly less effective) controls to another pilot, after building up a great deal of velocity (which includes both speed and direction) so that no drastic course changes are possible. The new pilot is "in control" but he is greatly limited by the course choices of the previous one. The primary controller of the outcome is the one who determines the initial course.

This can be summed up by pointing out the subtlety of the term "constraint." In the usual sense of the "ratchet" we see it as a limitation on developmental change (and hence evolution). Yet to the physicist, life *is* constraint (Fox, 1988). Maximum entropy, the antithesis of life, represents maximum unconstraint. The more complex an ontogeny is, the greater its negentropy (information content) and the greater its "contraints" (i.e.,

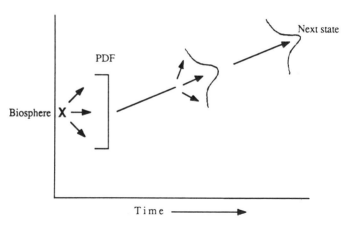

Figure 8-3. As developmental (e.g., "hardening") and ecological contingencies (e.g., "niche packing") accumulate among its components, the trajectory of the biosphere as a whole likewise becomes more constrained. This figure is highly schematic but tries to illustrate this process by plotting a biosphere "trajectory" through time, where the y axis is some metric (e.g., multivariate vector) of any number of "state" variables that describe the biosphere. PDF is the probability density function, i.e., the likely value of the next state. The main point illustrated is that through time, the PDF becomes narrower (more peaked) as the next state becomes progressively more constrained by the last from the buildup of the "internal" (to the biosphere) contingencies.

a contingency = a bifurcation of Fig. 8-1, or in binary language, the growth of information since 1 bifurcation = 1 binary data bit: the more contingencies, the more information). The growth of maximal complexity (increasing negentropy) in the biosphere must, of necessity, proceed by growth of constraint.

A critical question is, if life becomes progressively more constrained in its ability to generate new forms, can it become so constrained as to become unable to cope with major environmental change? This may be somewhat true of the most complex ontogenies, although they would presumably be replaced by complexification of surviving simpler ontogenies if massive extinction of complex ontogenies occurred. [Indeed, it is a basic rule of conservation biology that large, complex organisms are usually the most likely to become extinct in an environmental perturbation (Pimm *et al.*, 1988).] However, a major ameliorating process here is that the biotic environment with which complex organisms most directly interact is also undergoing evolution of constraint. For instance, mammalian predator–prey interactions consist of groups that have long histories of ontogenetic hardening (and shared the same "hardening" history before they diverged). That such interactions form a major force in the groups' evolution is shown by the ungulate–carnivore "arms race" of progressively increasing brain size in both groups (Jerison, 1973).

The general result of the growth of developmental inertia has been to make the biosphere, as a whole, less amenable to change (Fig. 8-3). Most evolutionists tend to think in terms of individual species but change of the biosphere is obviously contingent upon the plasticity of its components. As the developmental constraints of each species accumulate in the pool of the biosphere, the trajectory of the biosphere itself becomes more and more determined. Any number of the "state" variables that compose the biosphere, from the DNA code to embryological cleavage patterns to community organizations, become progressively more constrained, thus constraining the collective global unit itself, the biosphere. Obviously, even within these constraints, many events are still permitted. Indeed, a catastrophic external perturbation could, theoretically, decimate the biosphere so severely that even now some bauplane could be lost. However, such an event has to be truly massive (e.g., nuclear holocaust scale) and even then many previous contingencies would be retained (e.g., microbes with the DNA code) to impart direction on the recovering biosphere.

2.2. The Evolution of Evolutionary Rates

The evolution of ontogenetic constraint not only "hardens" the directional trajectory of a group's evolution but also, by definition, affects (decreases) the rate at which it evolves. In discussing rates of evolution, it is useful to distinguish between **breadth** and **depth** of change. These concepts were introduced in Chapter 3. **Breadth** of change refers to the number of traits (or cells, or growth fields, or some other metric of relative morphological amount) that are altered. **Depth** of change refers to the degree to which they are altered. For example, a descendant species may have many traits that are only slightly reduced (e.g., near global progenesis), or only one or two traits (e.g., limbs) that are greatly reduced. The former exhibits great breadth but little depth of change while the latter shows the converse. This is illustrated in Fig. 8-4. At the cellular level, we would be referring to the change in number of cells within a growth field (depth) and the number of fields affected (breadth). The traditional focus of morphological evolutionary rates has mostly been on depth. For instance, Haldane's (1949) unit, the "darwin," is based on proportional changes in a trait or "structure."

The evolution of more complex, constrained ontogenies has had two major effects on rates of evolution: (1) *decrease in number of "saltations,"* i.e., very rapid changes, of very high depth and breadth (many traits change radically), and (2) *increase in overall rates of change.* At first glance these two results may seem at odds. The first result has already been discussed above, in some detail. Accumulation of ever more contingencies means that earlier changes become progressively more remote because they propagate into later contingencies (Fig. 8-4). The second result is perhaps more subtle but simply arises from the fact that more complex phenomena (in this case ontogenies and the resultant adults) have more "parts" and more kinds of parts (O'Neill *et al.*, 1986). Thus, vertebrates have over 100 cell types with many trillions of cells while invertebrates, protozoans, and "simpler" forms have progressively fewer cell types (Bonner, 1988). Further, in a very general way, this increasing cell pattern is reflected in other levels of the morphological hierarchy, in a larger genome ("functional" genome, not "junk" DNA) and more anatomical complexity at the tissue level. At whatever level is focused on (cell, tissue, organ), the more traits ("parts in the system"), the more possibility there is for alteration. This is a basic property of evolving systems, noted by Halle (1977) and many others. *As a system becomes more complex, the number of possible permutations on it increases, "in proportion" to the complexity.* DeAngelis *et al.* (1986) have noted this as a classic example of a positive feedback process wherein cumulative change begets ever faster change. This principle is not limited to morphology; recall in Chapter 7 that more

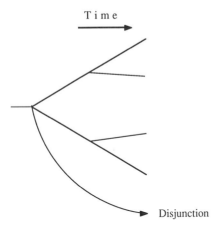

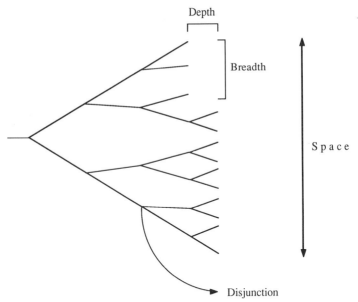

Figure 8-4. Branching tree diagram illustrating the notions of breadth and depth of ontogenetic evolutionary change. Breadth = number of "parts" (cells, traits) affected, depth = degree of effect (amount of change in onset time, rate, and so on). Comparison of top ("simple" ontogeny) and bottom ("complex" ontogeny) tree shows how the same change (same breadth and depth)—e.g., disjunction—has much more relative effect on simpler than on more complex ontogeny. Thus, as evolutionary rates increase in more complex ontogenies (more parts to change), the relative overall change decreases. Also illustrated is the way that "hardening" occurs, in an exponential fashion: the number of "terminal" traits (end branches on right) increases much more slowly than the number of intermediate branches from zygote (first branch on left) to the terminal traits. Fuller discussion in text.

complex behavior has resulted in ever-greater rates of evolution (more behavioral traits = more traits to change, with that much more pressure on morphology).

Evidence that complex ontogenies actually do show more rapid change has been found by Stanley (1979, 1990): more "advanced" taxa (e.g., mammals, birds, trilobites, ammonoids) have higher rates of evolution than "simpler" forms (foraminifera, corals, bivalve mollusks). He also found that animals with narrower niche breadths have higher rates. This may be partly because more morphologically complex organisms are more "specialized," as is sometimes argued. Stanley focuses on behavioral complexity as a main cause but: (1) on a gross scale, behavioral complexity correlates, roughly, with morphological complexity, and (2) the same "rule" holds for behavioral traits anyway: the more behavioral traits, the greater the chance (number of permutations) for modification (and therefore consequent morphological change). It is interesting that when Schopf et al. (1975) found that rates of evolution were correlated with morphological complexity, they interpreted this as possibly reflecting a procedural artifact. One's perception of evolution, they said, is affected by the relative number of traits a group has. We would argue on the side of Halle (1977) that greater complexity gives rise to more possibility for change and that there is thus a real basis for greater rates in more complex organisms. Of course, the big question is how much the human mind subjectively categorizes ("averages out") these changes in making taxonomic decisions. Clearly human artifacts are involved but there is a real basis for the pattern in the positive feedback of increasing complexity.

The key to this last point is that, while more parts make more change likely, they also diminish the relative impact of each change. For instance, in Fig. 8-4 let us say that successful alterations that occur earlier than two contingent events (branches) from the adult phenotype are very unlikely in any ontogenetic sequence. Too many later events are altered for a viable result. In a complex ontogeny, the affected growth field will be much smaller relative to the whole, compared to that of a simple ontogeny. In short, the *relative* breadth of change will be much less. Therefore, even though we may expect more permutation of parts in more complex ontogenies (because there are more parts to change), we may also expect these permutations to be of smaller relative effect. The logical extrapolation of this explains why simpler ontogenies are more likely to experience greater "saltations." Any changes have greater relative impact (in addition to fewer contingencies, i.e., less "hardening," to reduce chances of success). To repeat ourselves, one is therefore more likely to get true novelty from *Tribrachidium* (a late Precambrian fossil from the Ediacara Formation) than a chicken.

2.3. Summing Up: Evolution of Constraint

Let us sum up with a branching tree, to illustrate the all-important "hardening" or "ratcheting" process itself. If we consider each terminal branch to represent some adult "trait," then an ontogeny (i.e., "tree" sensu Fig. 8-4) with 32 terminal branches will have 8 times more traits than a simple ontogeny with only 4 terminal branches. However, the number of "pathways," i.e., the number of total branches in the cascade, will be 62 in the complex ontogeny and only 6 in the simpler one. The complex organism therefore has over 10 times more pathways producing only 8 times the "traits." Similarly, the number of branch points, or contingent events, will be more than 10 times (3 versus 31). This mathematical pattern points out how any complexifying system must accumulate a progressively greater burden (of contingencies) relative to the end result. It also shows that "hardening" is not an arithmetic, or linear, process. It increases even faster than the growth of complexity in the adult phenotype.

We may use this (obviously oversimplified) discussion to exemplify the main points, making the following predictions: (1) simple ontogenies would at first find it relatively "easy" to make "rapid" changes of high depth and breadth because of their low burden of contingencies; also successful "jumps" would be aided by the "open" adaptive landscape as discussed earlier; (2) any random terminal additions would rapidly accumulate on these "novel" disjunctions because contingencies "branch" multiplicatively, accelerating in a positive feedback process; (3) after this "exponential" growth phase of diversification, an asymptote would emerge as the rapidly increasing effect of external and internal contingencies (external = "filled space"; internal = "hardening") was felt. "Saltations" would "quickly" become reduced in "size" (breadth) and, although change would be more abundant, it would involve more terminal events of progressively more minor relative effect. This scenario, diagrammatically shown in Fig. 8-1, is what the fossil record generally seems to indicate (Valentine, 1986; Erwin *et al.*, 1987; Campbell and Marshall, 1987).

3. HETEROCHRONY AND CLADE DIVERSIFICATION

We have said that as developmental "hardening" increased, the role of environmental selection on incremental variation became progressively more important. However, this selection was acting on developmental programs (bauplane) which were already well established (i.e., whose trajectory was set), in large part by intrinsic processes. This ought to be kept in mind in discussions of the relative roles of intrinsic versus extrinsic processes. In any case, it is clear that environmental determinism played an increasingly larger and more direct role in the origination of taxa. Innumerable clades, especially from ordinal to genus level, and certainly many species, must have originated from orthodox Darwinian processes. But has there been any substantial role for heterochrony since the initial creation of body plans? (This question does not of course necessarily address the "punctuated-gradualism" question since environmental sorting of genes can be relatively rapid or slow, depending on gene pool size and intensity of selection.) If development has become so hardened that only incremental variation was produced, then all species and higher taxa ("clades") would originate from the accumulation of small changes. We have noted elsewhere that very local rate and timing genes do indeed exist and can be treated in quantitative genetics sorting (Slatkin, 1987). But this in itself is not very interesting, giving only names to growth in local fields. However, we have also said that some substantial changes may occur due to hierarchical effects, especially early ones, or late-acting changes in hormonal (integrative) systems. "Global" or at least large-scale heterochronies do continue to occur, as evidenced by many cases throughout this book (e.g., chimpanzees and gorillas, Chapters 2, 3, and 7). Accordingly, in this section we will try to show that the contribution of heterochrony to clade diversification seems to range from that of a passive supplier of incremental variation to that of a more active determinant.

3.1. Paedomorphs: Not Necessarily the Best Ancestors

A traditional, major candidate for strong developmental input into clade diversification has been global or large-scale paedomorphosis (i.e., neoteny, progenesis, or postdisplacement); discussion is in Chapter 5. de Beer (1958), for instance, argued that paedomorphosis has more evolutionary potential than peramorphosis. Perhaps his best-known example is that insects arose from paedomorphic millipedes via loss of segmentation. Such examples (he said) occur because peramorphosis produces more specialized, "overdeveloped" structures while paedomorphosis produces "underdeveloped" ones. In branching tree terms, peramorphosis therefore generally burdens the ontogeny with still more terminal additions (contingencies) while paedomorphs have some of them removed, "making room" for other kinds of additions (i.e., further evolution). Usually, such large-scale paedomorphosis has been associated with smaller size. For example, Gould (1977) used the peramorphic giant "Irish Elk" as an example of a poor candidate for evolutionary potential. Stanley (1973) explicitly discussed the specialized "allometric constraints" of large size which make large organisms less likely to be ancestral to major clades. This association of small size and paedomorphosis is probably often valid but should not be regarded as invariable. "Underdevelopment" via slow growth, early offset, or late onset will often apply to body size, but need not, because of dissociation. Amphibian neotenes, for example, can retain juvenile shape and other traits while growing to larger size than the ancestral adult (Chapter 3).

Levinton (1988) has cogently argued that, while paedomorphs (or at least small organisms) are indeed the most common ancestors to evolutionary, clade expansions (e.g., Stanley, 1973; Gould, 1988b), the traditional explanations just outlined are wrong. He argues that rather than reflecting greater evolutionary potential, paedomorphs preferentially survive environmental perturbations. They are therefore the predominant forms left to give rise to any ancestors at all. In short, it is the preferential *extinction of advanced forms* (his phrase) rather than the preferential *potential of unspecialized forms* that lies at the heart of the observed pattern. To be correct, his arguments must show that: (1) "advanced" (peramorphic) forms must have no less evolutionary potential than unspecialized forms, and (2) peramorphic forms must preferentially become extinct.

Four lines of argument support point one, the idea that peramorphic forms (on average) have no less evolutionary potential than paedomorphs (see also Chapter 5, Section 4.4.1). First, as already noted, not all evolution occurs by terminal addition to ontogeny. While we have emphasized the role of terminal addition (and consequent constraints) as a general vector of evolutionary directionality, many changes have obviously occurred through earlier changes in ontogeny, noted above. Therefore, peramorphic ontogenies, with more terminal "events," may still form an ancestral base where change occurs earlier in ontogeny. Second, some cases of paedomorphosis result in reduction to the point of trait loss so that evolutionary potential actually seems to be decreased rather than increased. For example, neoteny in the cave salamander *Typhlo-molge* leaves them blind and depigmented. This hardly seems to be a promising evolutionary progenitor. (Yet, this is not to say that evolutionary potential is eliminated. The loss of legs in some lizards has led to the highly successful radiation of snakes. The same may be said of many parasites. Thus, while loss may not lead to greater potential than gain of a trait, it clearly does not eliminate potential altogether.)

This second line of reasoning is the converse of the third line: peramorphosis ("gain") can open up new niches, i.e., new areas in the adaptive landscape, which can then be occupied by descendant groups. Levinton (1988) cites the example of pteryoid bivalve mollusks which pass through a shelled scalloplike phase but, as adults, become cemented to the bottom in oysterlike fashion. This latter, cementative phase apparently arose from hypermorphic (or some other kind of peramorphic) events, "beyond" the scalloplike ontogeny. Should modification continue, it is possible that a wide variety of cemented forms adapted to various substrates might evolve. Levinton cites other examples, but a quite compelling example is found within the cranium of the primate reading this book. As detailed in Chapter 7, the bulk of our cultural adaptations (brain size, brain complexity, longevity, and so on) originated via hypermorphosis. The extreme success of our line in colonizing a vast range of habitats has resulted in subspecies-level differentiation, that in time may have led to numerous descendant species (and could yet, if humans expand outward into space). In such cases, by having "more" of something, the ancestor was able to open up major new adaptive space, beginning a clade composed of minor permutations (often of continued peramorphosis) of the ancestral peramorphic theme. In paedomorphosis the morphology produced has already been tried and tested in the ancestor, but in peramorphosis the extension of ontogenetic pathways may produce entirely novel morphologies opening up new adaptive space. In addition to all this, some cases of clade-starting "paedomorphs" are actually peramorphs. For example, the oft-cited "paedomorphic" origin of flightless birds is really peramorphosis: the massive, powerful legs are hardly "underdeveloped" (Chapter 5). They resemble the ancestral juvenile only in "shape," a misleading criterion for diagnosing heterochrony (Chapter 2). The same is true of our supposedly "paedomorphic" human brain, which is really peramorphic (Chapter 7).

The fourth line of argument points out two logical errors that often seem to, superficially, indicate that paedomorphs have more evolutionary potential. The first logical error occurs because it is continued peramorphosis that often fills up the major areas of adaptive space (e.g., radiation of apes and humans as mainly peramorphic primates—Chapter 7). Therefore, the ancestor itself *by definition* has less developed ("paedomorphic") traits relative to its descendants. Thus, it is easy to assume *ex post facto* that paedomorphosis lies at the origin of the radiative trend. However, if we are searching for the heterochronic processes which created the ancestor itself, then this is misleading. We must identify the relationship between the clade ancestor and its own immediate progenitor species (its own ancestor). As in the bivalve example above, the clade ancestor may be peramorphic. The second logical error derives from the assumption that paedomorphosis "makes room" for specialized adaptations by removing or reducing preexisting specializations. Yet, as we discussed above, change in ontogeny is not a "zero sum" process whereby something must be subtracted before something can be added. It is instead basically an accretive process of "tinkering" wherein old ontogenetic pathways are often conserved. There is thus no *a priori* reason why an ontogeny with something reduced or lost has any more potential than one that does not.

This whole "paedomorph as best ancestor" idea is actually part of a more general philosophical and semantic argument with a long pedigree. For instance, the anthropologist Elman Service (1971) proposed the "law of evolutionary potential" which said that specialized societies have less potential. The universal fallacy in such arguments is that "specialization" is almost always "explained" post hoc, usually via poor logic. For example, Service says that the past social responses limit future choices (a truism if there ever was). The same is true of ontogenetic "decisions" (changes). The strong element of tautology in such arguments is no doubt familiar to most readers and anyone who has tried to "explain" historical events. While it is true that past events determine future ones, the problem is that events which "limit" the future in one instance may open new opportunities in others. For instance, the "specialized" trunk of an elephant could just as well be seen as a progenitor for other organs, as well as a "terminal" end product. It all depends on the conditions elephants encounter in their evolutionary future. Conversely, the same is true of leg loss in lizards (to become highly successful snakes). The potential of these intrinsic events (i.e., intrinsic to the system changing, be they cultures or species) is *context-dependent* (context = "environment"). Therefore, predictive generalizations based only on intrinsic properties are incomplete and of dubious value.

It is therefore very difficult to say whether an ontogenetic event (e.g., paedomorphosis) "intrinsically" has more potential than another event. The best approach to resolve this would be to document what occurs ex post facto: do an empirical study of the role of heterochrony in a number of clade originations in the fossil record. McNamara (1988b) has come closest to doing this with his massive compilation of heterochrony in the fossil record. He found that, of 272 documented examples of heterochrony, 179 showed paedomorphosis and 93 showed peramorphosis. A culled best-documented subset showed 90 cases of paedomorphosis to 65 of peramorphosis. However, while this gives an idea of the relative frequency of the two types of heterochronies, it does not directly tackle the issue of where the greatest amounts of novelty and clade origination occur. Nevertheless, tantalizing tidbits are glimpsed: he reports that trilobites show a clear predominance of paedomorphosis early in their history (Cambrian), switching to more peramorphosis later. Other examples are discussed in Chapter 6, but the general pattern seems to be that the relative importance of paedomorphosis and peramorphosis varies not only over time but among groups. Further studies of this and other data sets could help resolve this debate once and for all. As always in dealing with the fossil record, artifacts will have to be carefully dealt with. There is the difficulty of diagnosing heterochrony in fossils (although pera- and paedomorphosis can be diagnosed if "loosely" defined—Chapter 2). Also, many paleontological papers have misreported the heterochrony seen because of terminological confusion, so that compiled data must be verified (McNamara, 1988b).

Finally, we reiterate our main point: we are only saying that paedomorphs may not, on average, have more evolutionary potential than peramorphs, not that they necessarily have less. Both Gould (1977) and Hanken (1985, 1989) have shown that paedomorphic miniaturization can give rise to morphological novelty. Hanken (1989) in particular has demonstrated an important innovative mechanism: spatial "repatterning" of developmental pathways as a scaling effect of miniaturization (Chapter 3).

3.2. Small Organisms: The Most Common Ancestors

Having argued that paedomorphs do not necessarily have more evolutionary potential than peramorphs, it remains to address the fairly well documented notion that small organisms form the ancestral stock of most clades (Stanley, 1973; LaBarbera, 1986; Gould, 1988b; McKinney, 1990). As noted above, Levinton (1988) equates small size with paedomorphosis and argues that small paedomorphs form the basis of most clades, in spite of not having more evolutionary potential, because peramorphs suffer more during extinctions. Therefore, it is mainly the paedomorphs that are left to reexpand. However, there is a major problem in using this "theory of extinction of advanced forms" to explain the pattern. As discussed throughout this book, small size does not by any means necessarily denote a fully global paedomorph. It is quite possible for a small (size paedomorphic) organism to be peramorphic for some or many key (high evolutionary potential) traits. To use Levinton's (1988) own example of the attaching scallops, there is no reason why many cannot remain small and still take up that life-style with peramorphic traits. In addition, small size does not mean that the organism is more unspecialized. First, on the scale of many biological comparisons, terms like "advanced" and "specialized" are highly subjective to begin with. Second, and more important, the widespread idea that small organisms are inherently "less specialized" is very flawed. Semantics of "specialization" aside, many small organisms are obviously highly specialized in food-gathering, diet, and other aspects of life-style (e.g., parasitic gastropods of Chapter 4). Indeed, larger animals generally have a more generalized diet because their digestive systems are longer and can handle poorer-quality foods along with high-quality kinds (Demment and van Soest, 1985). If we focus on anatomical specializations for large size itself (e.g., locomotion), we find that it is based mainly on "mouse to elephant" logic that has no sense of scale. For example, Stanley (1973) and others have argued that an elephant is a poor ancestor because of its "allometric constraints," i.e., specializations for large size. This ignores the heterochronic ease (Chapter 3) of rapid dwarfing, but even ignoring that, this kind of gross comparison does not shed any light on whether a large mouse (for example) is a better ancestor than a small mouse. The scales involved are dramatically different and, like the "bacterium to whale" generalization of Bonner about life history discussed in Chapter 2, ignore the much smaller size ranges most relevant to comparisons of potential among related taxa ("vertical allometries"— Chapter 2, see Fig. 2–14).

We suggest that a better explanation of clade origination from small size is found in two well-established ecological generalizations. One is that *small species are much more common* in most clades, resulting in a right-skewed size frequency distribution (e.g., see Bonner, 1988). This seems to occur because small organisms subdivide the environment more finely (May, 1978). For example, only about 9% of mammals are large [over 30 cm head length (van Valen, 1975)]. As LaBarbera (1986) has rightly suggested, this may well mean that clades originate from small species simply because small species are more common; they would do so even if any given species had an equiprobable chance of being a clade-starter. An important modification to this widely held size–frequency relationship has been recently reported by Dial and Marzluff (1988). Assembling data from 46 clades of various sizes and organisms, they show that the very smallest size is not the category of each clade with the most species. Most fell between the 16th and 40th percentiles (1% being the smallest) indicating that smaller categories are still the most diverse but there is certainly no monotonic relationship.

In any case, the higher diversity of smaller species could conceivably eliminate the need to posit preferential extinctions, except that many high-level clades (e.g., orders) do seem to originate and proliferate most rapidly after major extinctions. These extinctions seem to impact large animals more heavily [e.g., dinosaurs at the end-Mesozoic (Bakker, 1977)], and the remaining small ones diversify (e.g., mammals in the early Cenozoic Era). This is to be expected, being based on a second ecological generalization: *large species will suffer disproportionate losses during major ecological disturbances.* This is well established theoretically since large organisms tend to have lower fecundity, grow more slowly, individually need more food, have lower population densities and other characteristics that make them more sensitive to perturbations (for recent review and rigorous mathematical treatment see Pimm *et al.*, 1988).

In sum, we suggest that both factors are operating to make small organisms prominent as clade ancestors: (1) during "normal" times, clades (mostly lower level, such as family and genus level) originate mainly through small ancestors simply because there are many more of them; (2) following major extinctions, smaller organisms preferentially survive and the emptier ecospace allows even greater amounts of clade radiation so that higher-level (e.g., family ordinal) clade differences evolve. In both cases, as descendant radiation occurs, habitats which promote selection for large size are inevitably encountered (such as habitats with cooler temperatures; see Chapter 6). Since the already small size of the ancestor precludes much more size decrease, a trend of **cladogenetic asymmetry** results, or what Gould (1988b) calls an increase in variance since the overall size range expands (Chapter 6; further discussion in McKinney, 1990). This of course forms the basis of one of the fundamental generalities of paleontology, Cope's rule of size increase.

3.3. Shape, Size, Life History, and Macroevolution

The above discussions, while perhaps gratifying in their apparent unification of fossil evidence with ecological theory, have, at first glance, effectively emasculated the explanatory power and role of heterochrony in the origin of species and higher taxa. de Beer's speculations about paedomorphs being better potential ancestors have been strongly questioned. Levinton's attempt to salvage the notion by synonymizing paedomorphosis ("unspecialized") with small size seems too simplistic. Instead we seem to be left with the evidence that "random" selection for small-sized individuals (during both normal times, by virtue of their abundance, and catastrophic times, by virtue of their greater resilience), and not some heterochronically derived plasticity or morphology, is the source of their evolutionary "potential." This "randomness" is a critical point in its own right, yet, in a book on heterochrony, we should focus on the developmental aspect. When we do so, we find that heterochrony is just as involved as it was before, only, as is often the case, the situation is not as simple as we would prefer it. A main complication is the same one that has distracted many thinkers on heterochrony: they tend to think of heterochrony only in "global" terms of many covarying traits. Thus, de Beer speaks mainly of global paedomorphs (and radical changes). Yet, we have shown in many parts of this book that heterochrony can affect various parts of morphology separately, including the detachment of size from "shape" (and local traits from each other).

With this in mind, we propose the following scenario to explain the role of heterochrony in clade origination: (1) those ancestors that give rise to clades are mostly small because of the numerical abundance and/or extinction selectivity reasons noted; (2) during "normal" (non-mass extinction) times, those small ancestors that give rise to the clades are often innovators through some kind of morphologically *local* peramorphosis, which permits the (still small) ancestor to exploit new modes of life (e.g., cementation in scallops), although innovation through paedomorphosis may be equally important (Hanken, 1985, 1989); during postextinction expansion such innovation would seem to be less important since some competition and predation may be removed; (3) as expansion and diversification occur, some descendants will come to inhabit environments where larger size is selected for; this size increase will occur via peramorphosis of size (e.g., acceleration of body size) and, often, peramorphosis of shape as well (in which case peramorphosis is global). This scenario is essentially the same as that suggested by Dial and Marzluff (1988), who provide further discussion and evidence.

3.3.1. Shape and the Measurement of Constraint

Figure 8-5 attempts to depict the allometric trajectory accompanying the size increases just described. In some cases, peramorphosis will involve a number of covarying traits, which are simply extrapolated along the ancestral trajectory. This particular lineage in the clade would form a peramorphocline (a global peramorphocline if all, or

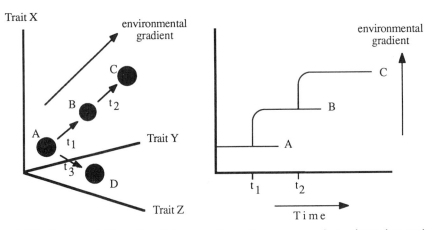

Figure 8-5. Left: schematic illustration of allometric ("shape") changes in a heterochronocline, with the ancestor, species A, giving rise to species B and C. These latter two species migrate along an environmental gradient and the "preaimed" coadapted allometric growth patterns are extrapolated accordingly (e.g., overall size increase in stabler regimes). Traits X, Y, Z can be "standard" organ size measurements but also behavioral or life history traits. Also, only three axes are shown for simplicity but *n* axes are possible. In the case of descendant species D, a decoupling of covariation has occurred, as selection has favored proportions not in keeping with the "preaiming" of the ancestral ontogeny. Right: the cladogenetic depiction of the heterochronocline, denoting times when speciation occurred. The overlap of species ranges is not always found, especially if an "external" forces (e.g., predation) is driving the cline (as discussed in Chapter 5).

most, traits were extended in growth). Such a heterochronocline would usually follow some environmental gradient, and can be "driven" in a number of ways: migration into environmental gradients, "autocatakinesis" (competition from the ancestral group "pushes" the descendants outward), or predation pressures (Chapter 5 for discussion).

Whatever the driving forces for such an expansion, a major point is that such ontogenetic–environmental extensions (or truncations) usually form *suites of coadapted traits*, i.e., they are "preaimed." Some workers (e.g., Jablonski *et al.,* 1986) have suggested that ontogenetic channeling up and down such trajectories may represent a form of internal directional determination of evolution, but a large part of this directionality is already "preaimed" by external selection (see Chapter 6). This is not to say that internal factors have no input in control. As noted earlier in this chapter, the question is how much input they have. In this case, a concrete illustration of this usually abstract concept is shown: how much can the trajectory deviate if external factors "push" on it? It may be that internal forces often "lock it in" (e.g., "hardening of contingencies," as may be genetically exemplified by pleiotropy) to cause trait covariation that results in suboptimal (as gauged by effectiveness in the external environment) traits when extrapolated (see also McKinney, 1988b). In such a case, internal contraints are obviously very important. Alternatively, such trait complexes may be readily broken up when the need arises, reducing or eliminating the constraint on selection. Another key point is that such covariation is not limited to morphology: behavioral traits can covary with one another or morphology (Chapter 7) while life history traits (e.g., timing of maturation, death) can covary as well (with each other, morphology, or behavior). Thus, the axes of Fig. 8-5 are not restricted to morphological parameters, or of course just three axes.

Such phenotypic covariation is best studied by the methods of quantitative genetics which, when applied to living lineages, can sort out the genetic covariations underlying the phenotypic ones. Many of us, paleontologists especially, tend to regard them as the same but as Lande (1979) and others (Cheverud, 1982) have noted, patterns of phenotypic change are not necessarily a valid guide to underlying genetic patterns. However, recent work indicates that such assumptions of equivalency (or similarity) may often be justified (Cheverud, 1988). In any case, the main point is that covariant suites at both the genetic and phenotypic level are made and broken through natural selection and the "flexibility" of such dissociations is at the very heart of the constraint issue. They represent trade-offs between the advantages of "piecemeal" fine-tuning by selection versus those of having preintegrated parts for rapid change when the environment demands it (the watchmaker analogy of Chapter 5). For example, to return to the illustration (Fig. 8-5), in another lineage (D) in the clade, the peramorphocline might not be global. Covariation might be broken up by changes acting on the various rate and timing genes of local growth fields. In this second case, body size or other trait increase may or may not occur when other traits increase so that only some traits are peramorphic.

Thus, the constraint debate will in large part be resolved by the continued work of quantitative geneticists who can experimentally observe changes in genetic and phenotypic covariation under controlled selection [good reviews by Cheverud (1984) and Riska (1989)]. Models show that phenotypic trait covariance can, as one would expect, result from adaptive pressures (external forces) but may also reflect shared development histories (internal forces; Riska, 1986). This is an incipient field with not enough data gathered yet to determine relative influences of external versus internal forces, although some relevant results have been reported. For instance, two excellent studies, on rats

(Zelditch, 1988) and the armyworm (Pashley, 1988), have shown that developmental decoupling of traits can occur rather easily. (See also Chapter 4 for discussion of different kinds of developmental programs, some of which are quite "flexible" in this way). This decoupling is responsive to the age at which selection occurs and can be passed on genetically. This indication that phenotypic and genetic covariation appear to change during postnatal growth, reflecting changes in rate and timing processes during ontogeny (see also Atchley, 1984), and can be readily affected by selection, would provide flexible means of change and would imply a minor role for developmental constraints in evolutionary directionality in these trait complexes. Heterochronically, we may visualize growth fields (traits, organs) which, during development, have considerable flexibility in coupling or decoupling rate or timing changes to those of other fields. As Atchley (1987) has said (and illustrated with jaw ontogeny), each field (e.g., jaw component) has its own ontogenetic trajectory which may be separable into its own rate and timing changes. Such covariation (or lack thereof) need not be limited to morphological shape but may include behavioral trait covariance and timing of reproductive events, as noted earlier.

Because constraint is perhaps the major issue in the role of development in evolution, we should pause here briefly to consider it in the context of the above discussion. Numerous authors (Oster and Alberch, 1982; Maynard Smith *et al.,* 1985; Oster *et al.,* 1988) have defined developmental constraint as a "bias in the expression of phenotypic variation due to properties of the development process." Alberch (1982) notes that such biases can result in discontinuous and clumped "morphospace." Our discussion of covariation serves as a reification for these abstractions in that common allometric extrapolations, illustrated above, will occupy more-or-less continuous morphospace. It is a bias that has not resulted in any great disjunctions. Of course we must distinguish between a bias in what the genome (or developmental program) is capable of producing now and what it is capable of evolving into. This means that *constraints themselves evolve* which is a major reason why the concept is so slippery. We are dealing with potentialities, yet the potentiality of an evolving ontogeny will itself change, depending on events not only in its own contingent chain but in the outside environment: a mass extinction, as we have said, may make an otherwise unviable organism competitive. (Again, this is the issue of context-dependency.)

This "allometric" view of clade expansion is not meant to indicate that it is the only mode of change. It only means that the "preaiming" and environmental-trait suite covariation discussed above, and the smooth changes permissible by decoupling of loosely integrated systems, make disjunctions uncommon. This is all the more likely since sorting of the rate and timing genes can be carried out in gradual Darwinian style (Slatkin, 1987; McKinney, 1988a,b). These would be the polymorphs and other kinds of intraspecific "heterochronic morphotypes" discussed in Chapter 4. However, major developmental disjunctions are not prohibited and may well occur. We have discussed such "novel differentiative" heterochronies elsewhere (Chapter 3). Such heterochronies may in fact lead to the originator of the clade (e.g., the ancestor of Fig. 8-5), whose descendant species may thereafter diversify via less radical (e.g., "allometric") heterochronies.

3.3.2. Size and Life History

So far the focus has been on morphology: shape and body size. This is expedient because both are visible and easily measured (at least approximately). Additionally, the observable evidence emphasizes the extreme importance of body size itself as a selective target (Chapter 6). Clade origination and proliferation, and even lineage anagenesis often involve body size change. This has been documented by Stanley and Yang (1987; see also Gingerich, 1985; LaBarbera, 1986) who argue that most evolution involves change in body size. Much of the rest (shape changes) consists of correlated allometric by-products (which may or may not be optimally adaptive, as just discussed). However, to fully understand the causes of evolution, we must broaden the focus, realizing that size is only one aspect of a suite of life history characters. Other aspects include timing of differentiation, energy intake and storage, and especially reproduction. To most workers, the topic of life history immediately brings r–K selection to mind. We have discussed (Chapter 6) how r–K theory predicts that large body sizes will tend to be associated with stable environments and small sizes with more unpredictable ones. Given the large amount of criticism against r–K, it is worth reiterating here that the r–K "continuum" is obviously an oversimplification of nature. It is rarely noted for instance that "r–K" is only a special case (density-dependent selection) of a much larger body of demographic theory and classification (Boyce, 1984; Begon et al., 1986). However, the extreme resilience of the theory and its continued use by a large proportion of ecologists are indicative of its basic value and validity [and there is empirical support, e.g., Wootten (1987); see also Chapters 6 and 7]. Even most critics, who have focused on its negative points, admit its utility when applied explicitly as a model of density-dependent selection (Boyce, 1984). Along with increasing body size, other life history correlates with increasing environmental stability are delayed maturation and greater longevity, which lead to fewer, but more "competitive," offspring, and increased parental care (review in Begon et al., 1986). We reemphasize that bigger size does not always mean longer growth (delayed maturation) among closely related species; this is only a "bacterium to elephant" relationship (Chapter 2 and elsewhere). Global acceleration may increase body size as well.

The role of heterochrony in life history was most explicitly articulated by Gould (1977) who hypothesized, and presented evidence for, the association of progenesis with unstable (r-selecting) environments. Small size, early reproduction, and reduced longevity are all products of early timing of both maturation and death (as discussed in Chapter 7, life history events tend to correlate: delay or early onset or offset in one event often occurs in others as well). Such early maturation can greatly increase fecundity (Gould, 1977). Conversely, one would expect to see a trend toward hypermorphosis as descendant species migrated into more stable habitats. Larger size from delayed maturation (more growth), and increased longevity would be predicted. This general correlation of size and environmental stability seems to have stood up favorably under further study. Echinoids (McKinney, 1984, 1986), bryozoans (Pachut, 1989), corals (Foster et al., 1988), trilobites, ammonites, and brachiopods (last three in McNamara, 1988b) show environmental stability–morphology relationships predicted by the r–K theory. Hallam has published a number of excellent studies recording similar body size–stability associations in bivalves and ammonites (see Hallam, 1990, for review and references). (Further discussion of this r–K relationship is in Chapter 6, especially Section 6.6.)

This stability–size relationship corresponds well with the origin of clades at small size and their proliferation into size-increasing environments, suggested above. It is therefore tempting to use r–K selection as part of a general explanation for the origin of clades. It is particularly tempting when we consider the pattern of **nearshore innovation** reported for the origin of so many marine clades (Bottjer and Jablonski, 1988 and references therein). The greater unpredictability ("r-selectedness") of shallower waters was used to equate increasing water depth with increasing stability in many of the above paleontological studies. Indeed, McKinney (1986) explicitly suggested that the unpredictability of shallow water promoted paedomorphosis via r-selection and would account for nearshore innovation.

However, while it makes for a good story, we hasten to add that this suggestion is an oversimplification and needs more scrutiny. For instance, we have already noted that paedomorphs are not necessarily the most novel ancestors. More importantly, it needs testing with better data. By better data, we mean that size alone is insufficient to pick out heterochronic events. As often pointed out in this book (especially Chapter 2), a knowledge of ontogenetic age (lacking in the paleontological studies cited) is essential for a good understanding of true heterochronic events. This includes life history events such as age of reproduction and death. Also, we need more information about the environment in such studies. Without these improvements, all we have is a body size–depth relationship which, while very interesting, is not enough to build a true r–K model on. For instance, Gould (1977) also suggested that neoteny should occur in stable, competitive environments. Yet neoteny is indistinguishable from progenesis in the fossil record since both produce paedomorphs, one by growing slower, the other by stopping growth sooner. And, as we have said, larger size in itself does not prove a longer growth period. Such troubles are compounded given the incompleteness of the r–K itself as only a partial demographic model. For example, what about stressed environments that occur nearshore?

Our goal is not to discourage the use of the size–stability model, but to encourage further testing. In fossil studies, we encourage the use of growth lines and other means of ontogenetic age assessment to gather more complete data on life history timing. Also, environmental parameters should be better documented. Actually, both age and environment are more easily studied in living organisms so it is surprising that almost no such studies have been carried out. The one that we know of (Harris, 1987) has indeed supported Gould's prediction of paedomorphosis in a fluctuating environment. However, this used the ever-studied salamander and more types of taxa need to be looked at too. The importance of a much larger, age-based data set is seen in contrasting results of frog studies. In the wood frog (*Rana sylvatica*), developmental timing has a high heritability (Berven, 1987) while in the frog *Hyla crucifer*, developmental time has low heritabilities (Emerson *et al.*, 1988). As a result, environmental selection effects vary considerably.

3.4. Summary: Heterochrony and Clade Origination

The main points in this section can all be combined to yield a tentative, general picture of how clades originate. We have noted that small organisms tend to be the clade originators. An earlier line of argument (e.g., McKinney, 1986) would have linked this fact to the r–K notions above to state that small organisms are size (and often shape) paedomorphs, selected for in unstable environments via progenesis. In addition to a gratifying fit with Gould's (1977) ideas on progenesis and r-selection, this would agree with Bottjer and Jablonski's (1988) evidence for shallow water innovation, followed by offshore migration. This would occur via largely late-acting heterochronies, within the developmental ("allometric") constraints discussed above. These would often be "preaimed" extrapolations based on environment-trait suite coadaptations. Where environmental gradients occur, heterochronoclines may result as expansion continues and coadaptations are extrapolated. Finally, this scenario dovetails with yet another pattern, Cope's rule, in that subsequent offshore migration into stabler environments leads to larger-sized descendant species (via hypermorphosis and other forms of size peramorphosis) through time.

We believe that this scenario is viable as a tentative, testable first approximation of the process of clade origination. It is particularly satisfying in its combination of developmental mechanics with ecological selection. However, in this book we have tried to make it more realistic by bringing in two complications that the scenario must explain. The first complication is that small-sized ancestors would be expected, even without the r–K scenario, because: (1) small species are much more common, and (2) small species tend to preferentially survive major extinctions. Yet, these also do not invalidate the scenario above because they are not exclusive to it. That small species are more common and survive extinctions better does not address where small species originate, live, what habitats they are adapted to, and by what developmental mechanisms they are produced. Clearly, small species are not randomly distributed. Thus, incorporating this first complication into the original scenario, we can say that small species are most common in unstable (r-selected) environments and that this is where they originate (perhaps often through size paedomorphosis of larger ancestors). Further, when major extinction events occur, the small organisms, being preferentially left (in part because they are "preadapted" to unpredictable change), will therefore be preferentially found in unstable environments where they are disproportionately common to begin with. The second complication is that paedomorphs are not necessarily more commonly clade originators: they are not necessarily more novel, nor is paedomorphosis necessary to create "ontogenetic room" to permit later evolutionary modification. Thus, while paedomorphosis of size is necessary to explain size decrease, the role of paedomorphosis as necessary to create novelty in the smaller forms may be questionable [although it could be important as shown by Hanken (1985, 1989)]. Therefore, the scenario should allow for the possibility that the small ancestors may also become novel via some kind of local peramorphosis, as in the bivalve example, or even that no real novelty is always involved, only reexpansion in the case of postextinctions.

In sum, the scenario summarized above still stands as a testable explanation of observed patterns, even when complicating observations are included. It provides an exciting possibility of uniting developmental and ecological processes to account for major evolutionary patterns. We were thus surprised to read that Jablonski and Bottjer (1990) "discount" the role of heterochrony in onshore–offshore trends. In addition to the above, we argue that their conclusion is premature for the following reasons. (1) They cite just two fossil cases (both by the same person and of the same animal group) even though McNamara (1988b) has compiled over 250 cases of heterochrony in the fossil record (many of which agree with our ideas). (2) Both of their cited cases report "neoteny" nearshore, so they say the progenesis-nearshore idea is unsupported. Yet as we have noted repeatedly, heterochrony cannot be deciphered without age data. Further, the two cases of neoteny cited may thus agree with the scenario because both neoteny and progenesis are processes of paedomorphosis, which are indistinguishable here since age data are lacking. (3) Finally, as this book has tried to document, heterochrony is the major mode of evolutionary change, whether it be minor allometric alterations or changes in body size or profound innovation. To "discount" a role for heterochrony, whether or not one agrees with our specific proposed scenario, is to ignore the great explanatory power of combining developmental with ecological processes.

4. DEVELOPMENTAL, ECOLOGICAL, AND EVOLUTIONARY CASCADES

We have argued that evolution has created increasingly complex ontogenies characterized by a "hardening" of contingent interactions. This constraint on development has led to a similar hardening of ecosystems and even the biosphere because as their components become constrained, so do these larger entities. Just as contingencies accumulate in number and kinds of cell interactions (in ontogenies), so do contingencies accumulate in number and kinds of interactions among the ontogenies (not just adult organisms as is often implied) composing an ecosystem. Obviously the latter is a much "looser" arrangement, not being as highly orchestrated (in spite of past analogies to ecosystems as "superorganisms" and succession as an analogue to ontogeny). Nevertheless, a large number of contingent interactions are required for any ecosystem to operate in even a roughly stable homeostatic fashion for any period of time (predator–prey relationships, plant–herbivore interactions affecting carbon turnover, and so on). The point is not only that ecology affects the evolution of ontogeny, but also that the much less-often noted reverse is true: *ontogeny affects the evolution of ecology* (ecosystems). Ontogenetic changes that affect interactions among organisms (other "ontogenies") can have cascading effects on the ecosystem in the same way that cellular contingencies can cascade in developmental systems. Thus, a change in a predator may affect its relationship with its prey. In most cases, these changes (and the resulting "cascades") are gradual, as exemplified in the mammalian coevolutionary (carnivore–herbivore) brain size increase, noted above. In this case, the role of selection is very strong as each of the major interacting components (mammals) is developmentally constrained. Thus, the sudden evolution of a "super-predator" is virtually impossible. Similarly, the "niche packing" within communities that has apparently occurred through evolutionary time (Bambach, 1985, 1986; Sepkoski, 1988) probably arose at least in part via selection on ontogenetic variation for refined interactions (increased "specialization"). In such situations many of the biotic elements of an ecosystem essentially evolve together (albeit "loosely" integrated). Futuyma (1986a) has termed this **diffuse coevolution**. This results in higher taxa "tracking" each other, e.g., increasing carnivore-herbivore brain size, or the "arms race" (predatory versus escape capabilities) among the higher taxa of marine invertebrates (Vermeij, 1973, 1978, and especially 1987). Ecosystems are thus metastable over long time periods. In the process, some ontogenetic changes in the component species are being slowly "amplified" (favored by positive selection) while other ontogenetic changes (most of them) are being dampened out by negative selection. [The process of "niche packing" itself is also a positive feedback process: as species accumulate in an ecosystem, the number of possible interactions (permutations) increases nonlinearly; taxonomically, this would be manifested in the rapid initial rise of the hypothetical curve in Fig. 8-1.]

From this it follows that if abrupt ontogenetic changes could occur, they could potentially cause rapid cascades throughout the system, as opposed to the slow, diffuse ("metastable") coevolution just described. Gray (1988) has discussed a useful phrase applicable to this: **ecological and evolutionary cascades**. Humans produce these artificially when we introduce foreign species into native ecosystems (e.g., placentals into Australia). In such cases, the developmental "abruptness" arises because the novel ontogeny evolved elsewhere over a long period. The interesting question so central to heterochrony is: can development produce such abruptness on its own? If so, we would have the theoretically fascinating possibility of a tiny change at a very low level (a molecule in one gene) leading to a large developmental cascade, which in turn would lead to a large ecological (and hence evolutionary) cascade. Thus, major evolutionary "branch points" could be traced to "quantum"-level atomic changes and, perhaps, true randomness. This is the biological equivalent to the common example in nonlinear dynamics that a butterfly's flapping wings can change the weather through amplification (nonlinear effects). The point here is that in systems with a memory (i.e., heredity), amplification may reach truly extreme levels. We have noted that such "jumps" were more common in early developmental systems; yet at the same time, the ecosystems comprised by these interacting ontogenies were possibly less tightly integrated. To what degree would this "dampen out" a potential ecological cascade if abrupt changes in ontogenies (and thus interactions) could occur?

5. INTERNAL AND EXTERNAL TIME

Few subjects are more intoxicating than time. All things change with time; entropy has seen to that. The arrow of complexity of ontogeny is life's way of rebuilding itself, brick by cellular brick, in the face of certain death. Likewise the directional complexity of the biosphere has been built brick by brick, via accumulation of ontogenies, some of which become progressively more complex as random changes are conserved.

While we should continue our successful scientific efforts to reduce reality to understandable bits, it is also useful to step back and take in the larger picture. Consider the description of ontogeny as cellular assembly. We have discussed how heterochrony is change in rate or timing of the ancestral assembly process. Traditional discussions of ontogeny in evolutionary terms (where it is discussed at all) tend to see ontogeny merely as a vehicle for delivering an individual (cell aggregate) to reproductive fruition (adulthood) at the most advantageous time. Even where life history analysis is used, it usually focuses only on one or a few major punctuations in the individual's life: birth, maturation, death. Yet obviously the whole ontogeny must be considered if one is to fully understand the selective forces that designed it. The rate and timing of all events, both internal (intrasomatic) and external, during each second of life, from conception to death, are equally a part of an individual's struggle for existence. That is, death as an

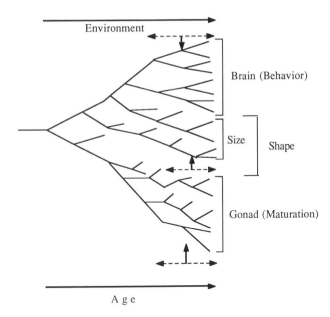

Figure 8-6. Schematic morphogenetic tree diagram illustrating the developmental contingencies in the growth of the three primary aspects of an organism: behavior, morphology (size and shape), and life history. In cellular terms (heterochrony arising as change in rate/timing of cellular assembly, at least in part from cell-level selection), these would generally refer to neural, somatic, and gonadal cells, respectively. A critical feature is the rate/timing "decoupling" of the three aspects, e.g., progenesis of shape while brain growth may be accelerated. (Arrows illustrate possible selection for various offset times.) However, such decoupling is only possible on a relatively fine scale not on a "mouse to elephant" basis (i.e., arrows cannot differ too much among the three). That is, there is a general positive correlation among the three for reasons of "structural" constraints, as discussed in text.

embryo or juvenile, from internal or external events, will reduce fitness just as effectively, indeed probably more so, than death as an adult from external causes (such causes being the usual focus of discussion). Our attempts here to extend heterochrony to early ontogeny and very late ontogeny, and to fine spatial scales, reflect an attempt to broaden the view in this way.

It may be useful to conceive of selection as acting to entrain the rate and timing of ontogenetic events with the rate and timing of events in the external (an individual's extrasomatic) environment. Heterochrony is not just change to adjust the adult morphology to this environment. It is an adjustment of ontogenetic events affecting: (1) the behavior, reproduction, size, or shape of the individual, (2) at any point in its life (Fig. 8-6). Note an important constraint on this "adjustment" of timing, however: a "structural" covariation among behavior, reproduction, size, and shape. Some of this covariation is on a "mouse to elephant" scale such that (for example) there is a limit on how soon a brain may stop growth if body size is permitted to continue growth. Thus, in mammals, body mass accounts for 56% of the variance in age of first reproduction on a "mouse to elephant" basis (Wootton, 1987). There is still much variation from "local" factors (deviation within this gross constraint = "vertical allometries," e.g., Fig. 2-14). In addition, the ontogenetic–environmental entrainment often has a vector wherein environmental gradients lead to gradients of ontogenetic shifts, as manifested in heter-ochronoclines (Chapter 5).

The extent of the ontogenetic approach is clearer when we consider that many organisms change size by over *four* orders of magnitude during their ontogeny (e.g., many fish). This causes radical changes in predation pressure, resource availability, and many other external "events" in their life, all of which have their own deterministic timing and rate controls. Werner and Gilliam (1984; see also Ebenman and Persson, 1988) have presented an excellent review of this concept, the **ontogenetic niche**, in which the shifting of adaptations during ontogeny is considered (discussed in Chap-ter 5).

Next consider that internal events in living organisms strongly tend to scale propor-tionately (review in Calder, 1984). For instance, among mammals, any given individual has about 800 million heartbeats and in general burns about the same number of calories (per unit mass) in a given life span. The major difference is that smaller mammals expend their allotment progressively faster in absolute time because their metabolisms are faster. Thus, in terms of their own "internal" clocks, individuals live about the same amount of time despite the differences as measured by external time scales, which tell us an elephant lives decades and a rodent for a few years at best. Reiss (1989) has carried this view to the greatest extent, analyzing developmental change in such physiological time units (PTUs) and finding considerable evidence for its utility.

Is it valid to view this as nature's way of scaling internal events to match the tempo of external ones? If this were true, we would expect small organisms to be subjected to a more rapid pace of events (see Fig. 8-7). This seems to be the case, in a general way. External biological events at their spatial scale are more closely spaced: smaller animals are more likely to be preyed upon, for instance. Similarly, external physical events are more common: a mild rainstorm to a large mammal may be a life-threatening flood to a small one. As DeAngelis and Waterhouse (1987) have said, "it is almost tautological to say that at sufficiently small spatial scales the dynamics of ecological systems will be short-lived or transient." Since energy drives all events, the scaling of energy with size may lie at the heart of this relationship. Smaller organisms perceive the universe at smaller spatial scales and the temporal events on that scale are driven by energies that are similarly smaller. Such energetic events include the time between life-threatening events (e.g., number of potential predators; rapid drastic temperature change), food and resource encounters, and so on, to which the internally timed events of birth, death, mating, and so on are adapted. The r–K continuum addresses this scaling in an incom-plete way, attempting to link a few internal events (especially reproduction) with exter-nal ones (e.g., "stability" = longer absolute times between events).

Theoretically, it is tempting to apply the concept of *self-similarity* to this scaling. The recent rise of fractals into the scientific consciousness has shown that many natural phenomena are the same at any scale (e.g., lung branchioles). Mandelbrot (1977) has shown that self-similarity is very common in both physical and biological systems. It may be possible that ontogenies are self-similar in many ways, at least within major groups (e.g., mammals). The rate and timing of internal (ontogenetic) events, when normalized for intrinsic time, may be self-similar (e.g., same number of heartbeats), being scaled to the rate and timing of external biotic and abiotic events.

A practical result of this scaling of internal and external time is that body size is

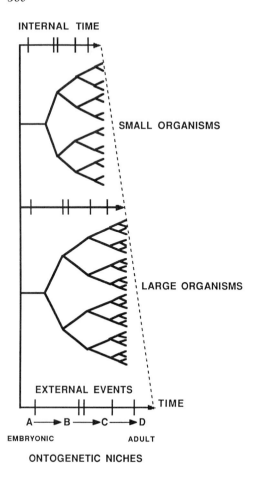

INTERNAL TIME

SMALL ORGANISMS

LARGE ORGANISMS

EXTERNAL EVENTS

TIME

A → B → C → D

EMBRYONIC   ADULT

ONTOGENETIC NICHES

Figure 8-7. A "self-similarity" view of entraining intrinsic time to the timing of external events. Smaller organisms "run faster" (have higher metabolisms) and therefore have faster rates of intrinsic time, and also experience faster rates of extrinsic time because smaller events and smaller objects (smaller units of energy and matter, respectively) occur (and are encountered) more often than large ones. That is, prey, predators, life-threatening natural events, and many other phenomena (illustrated by "random" tick marks) are encountered more frequently (as measured in absolute time) than in larger organisms. If this perhaps fanciful scaling occurs at all, it is only at coarse scales ("mouse to elephant"). As discussed in the text and Fig. 8-6, these are "structural" entrainments, not precise linkages. The ontogenetic niche shifts of a large organism are also shown. These may not scale even in a coarse way because, at least as Werner and Gilliam (1984) argue, major body size change is itself the cause of niche shifts; hence, it will occur less in smaller organisms.

often a better gauge of ontogenetic "age" than absolute age (Hughes, 1984; Sauer and Slade, 1987; Blackstone, 1987a,b; review in Ebenman and Persson, 1988). For instance, if an individual is "retarded" in growth because of a deprived or stressed environment, its consequent smaller size will more accurately reflect its younger "age" than the absolute age which would be the same as a normal individual. The "retarded" form is intrinsically younger because the deprived environment forced a slowing down of internal events (including growth). As a result, ecologists are beginning to analyze population dynamics in terms of size-structured communities (Ebenman and Persson, 1988) instead of the traditional age-structured focus (Charlesworth, 1980). This often yields better results, in part because many events are not age dependent but size dependent, including both external events (e.g., predation pressure) and internal ones (e.g., growth rate). Indeed, body size has been suggested as the major variable structuring food webs (Lawton and Warren, 1988).

Having said all this, there are many caveats. Biological systems are rarely as well behaved as physical ones and the use of the self-similarity idea is clearly speculative. As an extreme extension of it, one might even theorize that large-sized species would have geological durations proportionately longer than small ones. Thus, any given species might then have roughly the same number of generations before extinction (i.e., the same "tenure" on earth in terms of intrinsic time). This would reflect a rough scaling of intrinsic with extrinsic events, including a scaling of external stochasticity. [A quantitative vindication of the recurring temptation to analogize species "ontogeny" with individual ontogeny. Since as least Brocchi (1814), numerous thinkers have viewed each species as having its own intrinsically predetermined life span.] Alas, while this is true for large and small mammals, the opposite is true for foraminifera. Large-bodied mammalian species do seem to persist longer than small-bodied species, but not so with large and small foraminifera (van Valen, 1975). Another problem is that the scaling of intrinsic time to size is strong only on a "mouse to elephant" basis: groups have bent the rules in many ways. Humans are a good example, living over twice as long as we "should," although other examples are not uncommon (Calder, 1984). Statistically, this is seen in that only about 50% of the variance of maximal life span in mammals is explained by body size (Prothero and Jurgens, 1988).

Perhaps the best way to view this is to see "self-similar" scaling of physiological time as an underlying "structural" constraint built into cellular assembly at a very basic level. It takes a certain minimum time to build an organism of a certain size (developmental time). Similarly, metabolism is governed by surface-area considerations and other physical factors (physiological time). Operating to cause changes within these constraints are a large number of local ecological forces ("vertical allometries" discussed above; also Chapter 2). In one case it may be best to become large as rapidly as possible (global acceleration) if predation at smaller size is great. In other cases, large size may be best attained by growing longer at a slower pace (global hypermorphosis), if a long learning period is useful (e.g., humans). Disproportionate changes in some ontogenetic stages could occur within these global changes if very strong selective pressures act at certain stages in the ontogeny (e.g., to prolong the juvenile learning period even longer than proportional time scaling would otherwise allow). But the amount of change would have limits, set by the underlying "structure" of intrinsic time. Such speculations can only be fleshed out by much more work on the life history, ecology, and quantitative genetics of living organisms [e.g., see especially Ebenman and Persson (1988) for models of selection on growth rate].

6. ONTOGENY, HETEROCHRONY, AND EXTINCTION

The focus of this book has been on ontogenetic change as a source of creativity in evolution. Thus, it is mainly a book about how groups originate. However, it is also true that ontogeny plays a major role at the other end of the evolutionary tenure: the extinction of groups. Since evolutionary theory in general has (until recently) focused mainly on origination via study of genes and adults, this intersection of extinction and ontogeny is a ripe area of study, incorporating two relatively neglected aspects of evolution. Even without a great deal of previous work, there are tantalizing indications, both theoretical and empirical, of how crucial a role ontogeny may have in extinctions. Much of this role involves ontogeny as a target of selectivity in extinctions. Ontogenetic traits evolve to meet only local demands. Their conferring of survival in catastrophes is therefore a fortuitous event, pointing out once again the role of "luck" in evolution. For example, theoretical models show that organisms with rapid growth rates and early maturation can rebound more quickly from perturbations (Pimm *et al.*, 1988). This provides a better chance of avoiding extinction because populations can rebound fast enough to reach levels safe from stochastic and demographic causes of further loss. This rate of rebound is partly responsible for the selection against large body size that we have noted above. Large organisms not only have slower rebound rates (longer growth and later matura-tion) but have lower population densities so that their numbers are reduced more pre-cipitously to begin with (Pimm *et al.*, 1988).

Aside from body size selectivity, a second ontogenetic criterion for selectivity that has been studied is planktonism. Kitchell *et al.* (1986) showed that life history tactics can be a key factor in extinction survival of plankton. Chatterton and Speyer (1989) have shown that trilobites with planktonic stages in their ontogeny suffered disproportionate losses during the Ordovician extinction. This agrees with earlier suggestions (McKinney, 1987) that being planktonic as adults or in other parts of the ontogeny seems to increase extinction rates in many kinds of organisms.

There is a strong role possible for heterochrony in both of these selective features. Regarding body size, growth rate and time of maturation are obvious and have been discussed extensively in this book. We have also mentioned the major role of heter-ochrony in producing direct-developing larvae versus other less direct developers in echinoids (Chapter 3). Freeman (1982) also discusses how heterochrony can produce nonplanktotrophy from groups with planktotrophic larvae.

7. SUMMARY OF CHAPTER AND CONCLUDING REMARKS

Darwin's original formulation of evolution is reducible to a kind of algorithm:

$$\text{variation} + \text{environmental selection} = \text{evolution}$$

Variation is often characterized as the "internal" creativity of the system, whereas the environment (divisible into biotic and abiotic elements) is often viewed as the "exter-nal" creativity. Evolution is thus the sum of the dialectical interplay between these two forces. Both forces have stochastic (mutation, catastrophic selection, respectively) and deterministic components.

Traditionally, the "Darwinian" view (first Darwin and now neo-Darwinians) has focused on the external forces of creativity, not for deep philosophical reasons so much as that the production of variation by genes and development was largely a "black box." However, with the growth of knowledge of genetics and, more recently, of the complex process of development, it is time to incorporate them into explanations of evolutionary patterns. It is time to reassess the "internalist" perspective discredited by the old anti-Darwinian schools of orthogenesis and Goldschmidt. In doing so, we must be careful of the dichotomist tendency to overemphasize either external or internal factors. The main question is one of relative input: how much of a role does development play in the direction and rate of evolution? "Constraint" is an agnostic term often employed to describe this role because it implies that development has some control over evolution but does not specify how much. This control arises because development presents only limited and nonrandom raw materials to selection. To some extent such terminology reflects the historical focus on external selection: one could just as well say that *selection acts as a constraint* on developmental (and of course genetic) products.

To answer the question of relative roles, an understanding of the emergent rules of development is necessary. (In contrast, we know much more about rules of selection.) Only then can we know just how much creativity there is when development programs are "tinkered with" and where the internal directional biases are. Obviously, the emergent rules of development must be examined at all levels: molecular, gene, cell, and so on. As heterochrony is the most common means of developmental change, we have tried to explain it in just such multilevel terms. Such examination indicates that heterochronic changes in early ontogeny can lead to highly novel products.

This creation of novelty implies that the internal input to evolution may be quite significant, even today. However, a major point, not often explicitly made in the "internalist–externalist" debate, is that the relative contribution of these two forces of creativity has not been constant through evolutionary time. A major theme of this chapter is that the role of development as a steering force has declined. Ontogenies have become more complex as cells aggregated into metazoans: cells became more diverse, specialized, and integrated in metazoan development, via conserved random mutations. This complexification was sometimes (*not* entirely) a matter of peramorphosis, as new ontogenetic events were added, often late in ontogeny, minimizing the effect of cascades as often found in early ontogenetic changes. This "nowhere but up" complexification occurred because random changes sometimes permitted the metazoan to find "new ways of doing things," ways unavailable to simpler ontogenies. This "niche packing" was at first a rapid process (geologically speaking) and many basic body plans (bauplane) were established in late Precambrian and early Paleozoic times. "Random" ontogenetic creativity was profoundly important at this time as there were many fewer internal and external limitations (i.e., relatively few ontogenetic and ecological contingencies, respectively).

After the initial burst of body plan formation, successful profound ontogenetic changes soon became much less common as internal and external contingencies accumulated (i.e., as ontogenies "hardened" and ecosystems became more "niche-packed"). This led to a major shift in evolutionary creativity, from mainly internal forces (genes, development) to external ones (environment): less profound ontogenetic creativity was possible and, even where it was, truly novel forms were usually eliminated by the negative forces of a highly structured biotic environment. Nevertheless, the basic control of directionality by the internal "inertia" of past changes remained powerful: external directional control is strongly limited to products in which the most basic traits (bauplane) originated from early, largely intrinsic creativity. Diversification became largely (and progressively more) a process of permutations on established body plans. This, along with the increased packing of niches, has led to a "hardening" of ecosystems and even the biosphere as the interacting component organisms have become less subject to change. The evolution of complex ontogenies has also led to more rapid evolutionary rates as there are more parts and pathways to change. Simultaneously, these changes are less profound, as they affect a progressively smaller relative amount (breadth) of the total organism.

Even though hardening has progressed, there is still much internal creative potential in modern ontogenies. Numerically, most of this lies in late-ontogenetic "allometric" rate and timing alterations, but early ontogenetic changes, including novelties, may yet be possible (though not to the level of creativity of early metazoan ontogenies). This latter may be a major source of much "high-level" (e.g., orders, families) clade creation whereafter late-ontogenetic "allometric" changes create the variations on those themes, the "low-level" clades (family, genus levels).

In particular, we have tried to relate clade creation to observed fossil patterns, such as the dominant origin of clades by small ancestors and in nearshore environments. The numerical abundance of small organisms and their preferential survival during extinctions may account for much of this pattern. The developmental mechanism creating such small, sometimes novel ancestors may often be global paedomorphosis. Once produced, such forms would most often be selected for in nearshore (r-selecting, unstable) environments. Alternatively, we have also suggested that local peramorphosis may create novelty. This process is not necessarily associated with small body size (although it does not exclude it). Yet it could still explain the small size of most clade originators because, even if this kind of novelty were random with respect to body size, the numerical dominance of small body sizes in most groups would guarantee that most novelties would appear in small organisms.

Once clades originate, much of the late-ontogenetic (allometric) changes which produce their ramification into lower taxa often show a gradation of ontogenetic traits, called a heterochronocline (there are numerous kinds: see Glossary). Such clines also have a strong spatial, environmental vector. Individual selection producing these clines often operates on small-scale heterochronic variants called heterochronic morphotypes. Selection is not only on morphology (size and shape of both body and local traits) but heterochronic variation of life history (timing) and behavior.

We have also tried to show not only that ecology affects development but that development affects ecology. The "hardening" of ontogenies through time means that after the early Paleozoic, ecosystems probably evolved gradually, with component groups "diffusely" coevolving. However, it is possible that the origination of an "abrupt" ontogeny could, rarely, create ecological cascades. In the early Paleozoic it may be that the greater frequency of "abruptness" initiated such cascades more often, although here the ecosystem elements were more loosely interacting which would tend to dampen such contingent events.

We also discussed how it may be useful to look at heterochrony as providing raw material for selection to entrain the timing of intrinsic, developmental events with that extrinsic event. This arises from the view that each second of an individual's existence is equally important. Thus, behavioral, reproductive, as well as morphological events are entrained (probably not tightly) to environmental events. The "ontogenetic niche" describes the niche occupied at any point during such growth and entrainment. Scaling of internal and external events often centers around body size, which can change over four orders of magnitude during growth in some groups. Thus, smaller organisms are more affected by smaller (and therefore more frequent) biotic and abiotic events. This size scaling means that size may often be a better metric of "internal" age than external, chronological time: the rate of internal events is often reflected in the organism's size. This will lead to size-structured, as opposed to age-structured, populations.

Finally, the role of ontogeny in extinction is even less well examined than its role in origination. There is some evidence that selectivity of various ontogenetic features (e.g., growth rate and maturation time as reflected in body size, duration of planktotrophic stage) has occurred in past extinctions. These ontogenetic features are almost certainly fortuitous in conferring extinction resistance, having been evolved to meet local, non-catastrophic environmental demands.

8. SOME MAJOR POINTS OF THE BOOK

1. An understanding of the rules of development is a major deficiency in current evolutionary theory. Beginning with Darwin, and up through the modern synthesis, the focus has been on heredity (genes) and adult phenotypes, with much less emphasis on what happens in between. Yet these rules of development can strongly bias evolutionary directions and control evolutionary rates. The very existence of heterochronoclines directly illustrates this effect of ontogeny on evolutionary directions, and global heterochronic changes illustrate the effect on rates.

2. Despite much abuse by early workers, the concept of heterochrony is very useful in deciphering these rules of development. This is because change in rate and timing of development is the "easiest" mode of generating change, with minimal disruption of development. Further, once the unnecessary jargon and philosophical obscurantism are stripped away, heterochronic concepts are few in number and simple in understanding, providing a ready classification for description, and explanation, of ontogenetic change. Thus, only three basic kinds of changes are involved: time of onset, time of offset (both combining to make up duration of growth), and rate of growth. Positive (increases) or negative (decreases) changes in each category mean that a total of six major categories exist. Acceleration (faster rate), hypermorphosis (late offset), and predisplacement (early onset) cause peramorphosis ("overdevelopment" in the descendant adult) while the opposite processes (respectively) of neoteny, progenesis, and postdisplacement cause paedomorphosis ("underdevelopment"). Such changes may affect the whole organism (global) or "dissociated" to affect only certain events (e.g., local growth fields).

3. Heterochrony is, by definition, a process of change in time or rate, so can only be diagnosed when ontogenetic age is available. Such changes are most easily interpreted on trait versus age plots. In all trait versus age comparisons, we recommend the use of direct size metrics (linear, areal, volumetric) of organs or characteristics as the trait rather than "shape," which is highly subjective and misleading. Many cases of misdiagnosed paedomorphosis (humans, flightless birds) result from peramorphosis of the "shape" components, e.g., humans resemble juvenile chimps only because our brains have grown so large (peramorphosis) relative to the growth of other cranial growth fields. Also, in looking at trait change with age, the multiphasic nature of most growth curves may be deceptive and should be taken into account, e.g., delay of a rapid growth phase (or a number of phases) may "mimic" acceleration (in true acceleration, no change in offset would occur). Often, the same change occurs in many phases (sequential heterochrony). Change as a function of body size (allometric heterochrony) instead of age is also very useful not only for common descriptive purposes but because body size is a

good metric of "intrinsic" age. Such changes are easily viewed as trait versus body size plots, and can be analyzed using bivariate or multivariate methods.

4. The definition of heterochrony may be broadened to include rate and timing changes of traits at finer spatial and temporal scales than the usual tissue or organ grade now used. While this broadened view may be portrayed as "pan-heterochrony," consider this quantitative-genetics view of Atchley (1987): for each complex trait, there are a number of component parts, each of which often has an independent embryological origin, or "ontogenetic trajectory." When considered as functions of time, each independent trajectory may exhibit different "shapes," because of varying times of initiation, termination, and rates. This allows a more complete picture of heterochrony as a process, instead of just a taxonomy of ontogenetic and phylogenetic patterns. The cellular view is especially useful wherein heterochrony is seen as change in rate or timing of cell interactions, often caused by change in rate or timing of cell dialogue. This can occur by changes in biochemical mediators (morphogens) among cells, or sensitivity to them. Two major categories of heterochronies derive from this broader view: (1) differentiative heterochronies that occur before organ or trait differentiation, and (2) growth heterochronies that occur after differentiation and generally affect only the size (allometry) of organs. There are two kinds of differentiative heterochronies: size differentiative heterochronies (like growth heterochronies) affect only the size of organs, while novel differentiative heterochronies result in new cellular and tissue arrangements. In both differentiative types, change in time of induction is often the precipitating event. Such heterochronies can be diagrammed with "tree models" as a heuristic device for visualizing changes in space-time. Growth heterochronies can sometimes be sequential (multiphasic) and/or organizational (see point 12).

5. Simple "rate" or "timing" genes are not always involved in many heterochronies. Many gene changes can act to change rate or timing of ontogenetic processes at a fine scale to produce the changes just described. For example, alteration of rate or timing of morphogen transmission or reception may occur from change in genes regulating morphogen production, or changes in the "structural" genes determining the physicochemical properties of matrix (or cell membrane) across which the signal is transmitted.

6. The large majority of heterochronies are growth heterochronies, in contrast to the historical fixation with saltations and hopeful monsters (novel differentiative heterochronies). Indeed, growth heterochronies account for much intraspecific variation, such as polymorphism, sexual dimorphism (e.g., bimaturism), and other manifestations of morphological plasticity; this variation we have generally termed heterochronic morphotypes. Gradual sorting of such morphotypes is a major mode of evolution, and may be studied by the application of quantitative genetics to heterochronic genes.

7. The relative input of developmental "constraint" in controlling the rate and direction of evolution has not been constant. Simpler and more loosely integrated developmental programs in the late Precambrian and early Paleozoic allowed developmental creativity to cause greater rates of evolution and determine more of the course of evolution than it has since. Since then, there has been a buildup of continencies in developmental programs (e.g., as illustrated in "tree models"), called "hardening." This comes from the fact that successful developmental mutations usually modify only the least disruptive, later parts of the ontogenetic program, to produce "ecologically new ways of doing things." This "hardening" is the reason that "advanced" taxa "go through" stages

of more primitive relatives (although in a much less precise way than the theory of recapitulation once held). It is also why more closely related groups have more similar ontogenies. As a result of the increasing intrinsic "hardening" of developmental programs and the extrinsic packing of niches (by finding "new ways of doing things") since the early Paleozoic, the earth's ecosystems (and biosphere as a whole) have also become more "hardened" as their component life forms became less amenable to change.

8. As this hardening proceeded, permutations on preexisting themes became the dominant form of creativity. This led to the origination of lower level clades, apparently mainly in nearshore environments, for reasons that are not clear, but seem to involve body size and life history strategies, both of which rely heavily on heterochrony as the mechanism for modification. Contrary to traditional arguments, it is not at all evident, either empirically or theoretically, that paedomorphs generally have more evolutionary ("clade-originating") potential. Preliminary studies of the frequency of heterochrony in the fossil record indicate that the frequency of paedomorphosis versus peramorphosis in a group can vary through time and varies among groups.

9. Once clades originated, diversification often occurred via selection on minor heterochronic variants; this often formed gradational ontogenetic–phylogenetic clines called heterochronoclines. These often have a spatial (environmental) vector tracked by regular evolutionary changes in the ancestral ontogeny. This regular trait change in heterochronoclines can involve only some traits (dissociated, mosaic clines) or be global. Heterochronoclines can be either anagenetic or cladogenetic, depending on whether the ancestral lineages persist. The driving force behind such clines can be either internal (e.g., competition from ancestral groups) or external (predation) to the species in the cline. In the case of an internal driving force, the term autocatakinetic may be used, to denote that the pattern is self-generated.

10. Heterochronoclines are good illustrations of the role of ontogeny in directing evolution after the early Paleozoic: evolution is clearly strongly influenced by the ancestral ontogenetic patterns, but that influence is not radical. The clinical differences observed are largely minor rate and timing permutations. That heterochronoclines form is to be expected given that ontogenetic traits are coadapted to one another and that such coadapted trait suites are in turn adapted to environmental parameters. During the course of its ontogeny, an individual must remain well adapted to its environment. In many groups, especially where major body size changes occur, ontogenetic trait coadaptations entail major adaptive shifts, i.e., the individual will have a number of ontogenetic niches. Extrapolation (or truncation) of environmental parameters can naturally lead to extrapolation (or truncation) of ontogenetic coadaptations to those parameters. That is, ontogenetic trajectories are "preaimed" to some extent, although trait dissociation can often occur where it is advantageous to "break up" coadaptations. Graphically, such changes can be visualized as extrapolations or truncations (or dissociations) on an allometric plot of ontogenetic change. Such coadaptations are not limited to morphological traits (size and shape), but include life history and behavioral traits (i.e., a triad of "targets of selection": morphology, behavior, life history, all strongly affected by change in rate or timing of development). It is often very difficult to tease apart which traits (if any) within these coadapted suites are just being "dragged along" (and may thus be maladaptive or nonadaptive). The degree to which dissociation is possible will strongly determine the role of developmental constraint in influencing evolution (e.g., how many

traits result from just being "dragged along" by developmental contingencies), especially after the early Paleozoic. Further work in developmental biology and quantitative genetics should empirically resolve this long-standing question, heretofore often a point of philosophizing and ad hoc reasoning.

11. The role of heterochrony in the evolution of behavior is a potentially major area of study that has been largely ignored. Yet, it is clear that adult and juvenile behaviors can become altered in the timing of appearance, e.g., adult birds can show behaviors expressed only in the juveniles of ancestors. Further, such behaviors can occur and be altered as coadapted "suites" or be dissociated in the same manner as morphological suites. Heterochrony is also strongly evident in the evolution of learned behavior, such as the mitotic increase of brain cells for information storage and connections (axons, dendrites) for information processing.

12. Human evolution is *not* characterized by neoteny (slower growth), but primarily by hypermorphosis (delayed offset of growth): extended learning, larger brains, increased longevity, large body size, delayed puberty onset, and many other human traits result from delay of onset of all ontogenetic phases (sequential hypermorphosis). The general selective regime for this is roughly characterized as "K-selection," and humans are an extreme form of the same generally hypermorphic/K-selective trend in primate evolution as a whole. Complex human behavior may be thought of as the peramorphic phenotype of brain peramorphosis. Behavior shows sequential organizational hypermorphosis. Heterochrony is also a key mechanism determining human racial and sexual differences; human females are generally paedomorphic (progenetic) to males.

13. We have shown how heterochrony can be interrelated to the following phenomena: genes, cells, development, morphology (both size and shape), behavior, and life history, basic ecological principles, the origin of higher taxa (body plans and lower level clades), nearshore–offshore patterns, evolutionary rates, extinctions, and still other phenomena. As a central process which can interrelate such seemingly disparate phenomena (in nontrivial ways), heterochrony may therefore serve as an important concept to incorporate into any general theory of evolution, which would seek to explain evolution (and interrelate all relevant processes operating) across many (temporal and spatial) scales of reference.

Chapter 9

Latest Developments in Heterochrony

1. INTRODUCTION

Since the manuscript for this book was submitted for publication, a number of important articles have appeared that supplement the first eight chapters of the book. Therefore, we add this chapter (during editing) in order to bring this new material to light. However, rather than being simply "tacked on," this new material is discussed in the context of earlier chapters, on a chapter-by-chapter basis.

2. CHAPTER 2: CLASSIFYING AND ANALYZING HETEROCHRONY

A major dichotomy in diagnosing heterochrony is that between age-based and size-based descriptions. Age-based change is to be preferred, representing the ultimate cause from which size-based changes can be derived. Of course, practical difficulties in inferring ontogenetic age often arise.

2.1. Age-Based Heterochrony

Fitting growth curves to age-known growth series has become a sizable industry in the literature and we have not tried to summarize all such techniques in this book. While these techniques serve as improved descriptors of change, they rarely provide insight into underlying developmental processes. However, we should add to our earlier discussion a paper by Jolicoeur and Pirlot (1988) who have developed a four-parameter curve that provides an exceptionally accurate description of age-based asymptotic growth and complex allometry. A major reason for this accuracy is their recognition that the curve must pass through the origin. This is often neglected in the Gompertz and other curves, which are asymptotic to the origin. This is an error because growth begins at such very small sizes (egg fertilization) relative to later sizes that asymptotic curves greatly misrepresent change in early stages. More importantly (to the heterochronic dichotomy), the authors attempt to relate this curve to size-based allometric manifestations of such growth. This attempt was not so successful in their 1988 paper (being highly complex in solution) but Jolicoeur (1989) has recently provided a more simplified model. This three-parameter model can often provide a relatively simple description of complex allometry. After transformation, the allometry is nonlinear only with respect to an "exponent of complex allometry," D.

While Jolicoeur's recent work provides better ways to describe age-based growth and relate it to size-based (and complex) allometry, Brower's (1990) recent paper reemphasizes the need to continue searching for ways to identify age in fossil (and living) organisms. In an allometric study of 17 species of fossil crinoids, Brower inferred that paedomorphosis was a major mode of evolution in this group. While such inferences have been rampant in paleontology (McNamara, 1988b), the large majority of inferences are questionable because ontogenetic age information is lacking: they are allometric heterochronies which may be produced by a variety of underlying age-based heterochronic processes (Chapter 2). Brower (1990) states that it is possible to infer generalized ages in his study, using his solely allometric (size/shape) data, but we remain uneasy about such inferences. Direct interpretation of growth lines or some other marker of *absolute time* seems to be the only conclusive way to diagnose age-based heterochrony. Without good reason to believe otherwise, we must suspect that any *relative morphological* changes can be produced in a variety of ways.

Obviously, age information is a very difficult goal, but continued study of growth-line and related markers (Jones, 1988) should yield rich rewards. Such rewards extend far beyond evolutionary reconstruction (the identification of ontogenetic–phylogenetic linkages). For instance, paleoecology would become much more potent if it could study age-based population dynamics in addition to our current focus on body-size distributions and size cohorts (for a brief review with references see Cadee, 1988). As Threlkeld (1990) has correctly pointed out, even ecologists need reminders that populations are structured by both age *and* size. Currently, the role of body size has become a focus of interest to neontologists (e.g., Ebenman and Persson, 1988) but both perspectives must be integrated for a truly complete understanding of population and community dynamics.

2.2. Size-Based Heterochrony

Obviously, traditional size-based heterochrony ("allometry") will continue to be of great importance as a descriptor of change, even where age data are lacking. However, even in this relatively simple description of size and shape there have been many errors of technique and interpretation. This is pointed out in LaBarbera's (1989) recent review paper on analyzing body size in ecology and evolution. Among the major errors he discusses (and suggests corrections for) are the use of ordinary least square regression (instead of major axis or reduced major axis), bias due to log transformation, and failure to present raw data so that conclusions can be checked or the data subjected to improved

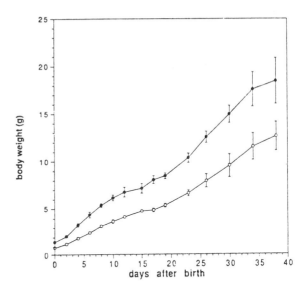

Figure 9-1. Growth curves of wild-type (●) and heterozygous littermates (O). The latter mice show slower growth (neoteny) caused by the inactivated IGF-II gene. Modified from DeChiara *et al.* (1990).

analyses in the future. (He notes that the reduced major axis method is least biased, more efficient, and least sensitive to assumptions about error structure.)

Forbes and Lopez (1989) have recently shown how allometric data can go beyond description and tell us something about underlying processes, even without age data. They present a moving regression analysis to identify bivariate scaling breaks (discontinuities) in ontogenetic trajectories. They illustrate how these can be used to determine critical periods in development. While such periods may not reveal the exact heterochronic processes (being relative ontogenetic shifts), they still provide useful information about ontogeny that is often ignored in allometric studies.

3. CHAPTER 3: PRODUCING HETEROCHRONY

3.1. Biochemical Mediators: Growth Factor Genetics

Among the key biochemical mediators discussed in this book, IGF-II (insulinlike growth factor) is clearly among the most important in controlling growth. This polypeptide is a mitogen for a number of cell lines. DeChiara *et al.* (1990) provide an excellent review as well as the first direct evidence for a physiological role for IGF-II in embryonic growth. Most important, they present considerable evidence on the genetics of IGF-II by introducing mutations at the IGF-II gene locus in a mouse germ line. This gene has at least three promoters and expresses several transcripts during the embryonic and neonatal periods (see numerous references in the paper). Of considerable direct heterochronic interest is that when the IGF-II gene was inactivated by mutation, growth deficiency was produced, producing animals only about 60% the size of normal mice. Yet, in all other respects, including fertility, these small animals were normal. Examination of the growth curves of these animals indicates that the mutants are smaller by virtue of reduced rate of growth (neoteny) rather than late onset or early offset (Fig. 9-1). Not seen in Fig. 9-1 are the prebirth effects of the gene: even before day 16 after fertilization, growth effects are shown. Age of maturation does not seem to be affected, so this is apparently a case of "pure" neoteny. Finally, the neotenic effects shown are for heterozygotes. Here we have excellent documentation of how heterochronic alleles can occur via mutation and be sorted in Mendelian fashion. As we have argued earlier in the book, heterochrony should be "de-coupled" from the uni-generational saltation concept.

In an interesting comparison to the study of DeChiara *et al.* (1990), Xiang *et al.* (1990) used insertional mutations in transgenic mice that also created smaller mice. However, in the latter study, the mutants produced normal amounts of growth hormones but were resistant to its effects. These mutants were allelic to a spontaneous (natural) mutant, the pygmy mouse, so here is an excellent example of how experimental manipulation can provide insight into natural evolution. The authors also point out that similar processes are at work in human heterochronies, with both dwarf and pygmy syndromes, where individuals may be either growth-hormone-resistant or deficient in hormone production.

3.2. Genetics of Heterochrony

In addition to the functioning and genetics of growth hormones per se, there has been some important recent work on the genetics of developmental regulation in general. Paigen (1989) discusses experimental approaches to the study of regulatory evolution. He shows how new methods of molecular genetics (such as the gene insertions just discussed) can provide much information on the evolution of regulation. Far too little is known to generalize at this point, but he gives a glimpse of what can be learned in his study of regulatory evolution in two proteins. For instance, regulatory mutations seem to occur in close proximity to the coding sequence of the regulated ("structural") gene. This forms a "gene complex" with simultaneous and interdependent selection of the regulatory and structural elements in the complex. A main consequence: the entire gene complex segregates and evolves as a unit.

Cavener (1989) discusses the evolution of developmental regulation, focusing on *Drosophila*. He argues that (gene) regulatory elements can arise *de novo* and are not limited to transposing or copying regulatory elements from elsewhere in the genome (as is often suggested). He concludes that until many more regulatory systems are studied, few generalizations can be made. However, he makes the interesting point that the functional properties of a gene evolve somewhat independently of gene regulation. Thus, functional information cannot be used to infer regulatory properties.

There is no doubt that studies such as those just discussed will be a major source of information on the causation and role of heterochrony in evolution. However, aside from this focus on the molecular and genetic level, there is another level that will also be extremely important. These will be detailed studies of developmental variation at the population level. Atchley has been one of the major figures in such population (quantitative) genetic analyses of regulatory phenomena, as we discussed earlier in the book. In a recent symposium (Atchley, 1989 and references within), he discusses this approach, including an excellent paper (with Newman) that uses mammalian development to describe the response to selection in the components of complex morphological structures. "Allometric engineering," discussed below, is another good example of this approach.

Currently, heterochronic research at both the molecular-genetic and population level is very inchoate. However, these are clearly two of the major avenues which must be pursued for a complete understanding of heterochrony in evolution.

3.3. Differentiative Heterochrony: Induction Timing and Cell Fates

In Chapter 3 we discussed how timing change of cell interactions in early ontogeny could lead to dramatic ("nonallometric") changes in cell fates. Ettensohn (1990) has detailed the timing of interactions between primary and secondary mesenchyme cells in the sea-urchin embryo. He found that secondary cells were sensitive to primary-cell signals only for a very brief time, late in gastrulation. Before this time, they were insensitive to the signals leading to a major change in cell fate: some secondary mesenchyme cells formed skeletal tissue instead of the muscle and pigment they would have formed if interaction with primary cells occurred at the normal time.

4. CHAPTER 4: HETEROCHRONIC VARIATION AND SELECTION

Heterochrony from Phenotypic Plasticity

In Chapter 4, we discussed the often-overlooked heterochronies that are produced via ecophenotypic (nongenetic) plasticity. An extremely intriguing example of direct heterochronic relevance was recently reported by Crowl and Covich (1990). They show that the presence of predatory crayfish dramatically alters the life-history traits of the snail *Physella virgata*. In a predator-free environment, snails reproduce at about 4 mm long and live about 3–5 months. However, in the presence of the crayfish, snails grow to twice that size, live 11–14 months, and delay reproduction. Apparently, the larger size, longer life, and delayed maturation thus reflect delay of various stages (*sequential ecophenotypic hypermorphosis*). The proximal mechanism for this is a water-borne signal, a snail protein, that is degraded by a crayfish enzyme. The more ultimate selective reason for the evolution of this response in the snail is the advantage of allocating resources away from reproduction and toward growth. In this case, larger size (and/or delayed reproduction) is apparently advantageous in the presence of predators.

On a broader scale, West-Eberhard (1989) presents an excellent review of phenotypic plasticity, including some references to heterochronic variants. Among the many interesting points she makes is that study of plasticity evokes an appreciation of ontogeny as a *creator* of diversity rather than a "constraint," as it is usually considered. Most evolutionary theorists have focused on canalization and stabilizing selection, leading to questions on how much development restricts evolution rather than how much it provides opportunities. West-Eberhard points out that plasticity not only provides opportunities for local adaptations to temporary (changing) conditions (in which case heterochronic plasticity itself may be selected for), but that plasticity can lead to diversification through the origin of novel traits and altered directions of evolution. This latter occurs because plasticity "creates" new forms which may "take hold" so that specific heterochronic morphs may subsequently become selected for (genetically established).

5. CHAPTER 7: BEHAVIORAL AND HUMAN HETEROCHRONY

Human Heterochrony

In a recent major review of heterochrony in human evolution, Shea (1989) has independently reiterated our point that humans are *not* neotenic in origin. He outlines a number of reasons for this widespread erroneous inference: (1) confusion of neoteny with progenesis, (2) conflation of growth prolongation in time with shape retardation, (3) failure to move beyond diagnosis using superficial shape similarities to underlying homologous growth porcesses, and (4) failure to appreciate the extent and significance of the many non-paedomorphic features of human evolution. We agree that all of these have been factors in the human neoteny myth, but especially the general confusion of heterochronic terms and diagnosis. Shea (1989, p. 97) concludes: "a hypothesis of general and pervasive human neoteny is clearly no longer viable."

Bogin (1990) has written a somewhat popularized overview on the evolution of human childhood, where he reviews the evidence for delayed maturation and our unique growth phase. He presents data (Fig. 9-2) that reinforces our view of human females as progenetic relative to human males. He expands our own presentation with the depiction of velocity curves (Fig. 9-2) that show that females' smaller size is largely due to their earlier termination of the various growth stages (sequential progenesis) and not from any significant decrease in growth rate per se. Human males only delay the pre-adolescent "spurt," leading to a higher rate when that peak occurs. Therefore, the ultimate cause of males' larger size is the delay itself, which can lead to higher rates ("acceleration") as a byproduct of a prolonged growth phase (see Chapters 3 and 8). Such processes (acceleration via delay in sequential stages) have much to do with the reason heterochrony has been avoided and confused in the past. Yet if we are to understand development in evolution, we must persist in trying to diagnose what is going on, even if nature rarely provides us with conveniently simple growth patterns.

It is symptomatic of the sad state of heterochronic knowledge that Bogin's (1990) article does not even mention the word heterochrony or any of the various processes, such as hypermorphosis. This omission is extremely common in many evolutionary studies and we hope that this book will prompt the realization that such terms are not unneeded impediments, but can improve communication and understanding.

Finally, in reading the manuscript for this volume, one reviewer made the interesting observation that in rare cases, human brain cells do not terminate mitosis. This condition, megalencephaly, is dangerous to the developing juvenile in the extreme case, but we speculate whether it might not be a possible mechanism for "local" brain enlargement. In other words, in this book, we have linked brain-size increase, a prolonged learning stage, longer lifespan, and other hypermorphic traits as a "package deal." One

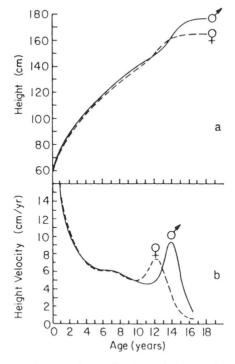

Figure 9-2. Body-size ontogeny in human males and females. a, simple growth curve; b, velocity curve. From Bogin (1990).

rationale for this was that neuron mitosis occurs only at a very early stage in the fetus so that delay of that stage is necessary for a larger brain. It would be interesting to know the genetic and biochemical cause of megalencephaly and whether it is a viable option for brain increase without the attendant need to prolong entire growth stages.

6. CHAPTERS 5, 6, AND 8: SIZE, LIFE HISTORY, AND ONTOGENETIC EVOLUTION

6.1. Life History (r-K-Stress) and Ontogeny

Throughout much of this book we have used the traditional r-K framework as a useful life-history gradient to classify environments to which ontogenies are adapted via selection for timing, rate, shape, and size. However, we have tried to make clear that this gradient is not entirely satisfactory. Here we would like to amend our earlier comments with the addition of concepts that we think considerably improve upon the r-K model.

First is the excellent work of Begon, as discussed in Begon *et al.* (1990). He makes the key distinction that environments may be *size selecting, size neutral,* or both. The first case is selfexplanatory, while size neutrality refers to environments that select for timing

Table 9-1. Environmental Correlates Fostering Larger Body Size, in (Very) Roughly
Descending Order of Decreasing Importance[a]

Correlate	Group	Source
(B) Regularly abundant food	All	See text
(B) Low nutrient food	All	See text
(A) Cool ambient temperature	Warm-blooded	See text
(A) Warm ambient temperature	Cold-blooded	Stevenson, 1985
(B) Prey size	Predators	Peters, 1983
(B) Predation	Marine organisms	Gerritsen, 1982
(A) Seasonality	All	Lindstedt and Boyce, 1985
(B) Sex selection	Many	Woolbright, 1983
(B) Female fecundity	Many	Shine, 1988
(A) Post-arboreality	Vertebrates	Taylor *et al.*, 1972
(A) Water density	Plankton	Malmgren *et al.*, 1981
(A) Nutrient starvation	Benthic forams	Hallock, 1985

[a] "See text" used where references used are too numerous to list. (A) = abiotic selection, (B) = biotic. This is nonexhaustive, intended only as a sampler of the more prominent selection pressures. For references, see McKinney (1990).

of maturation (growth) alone so that size is not important. For example, a "classic" *r*-selecting situation would be size neutral, selecting for early maturation (progenesis) and small size would be "purely" a byproduct. In other situations, such as where competition or predation pressures are major selection agents, size itself may be selected for and timing is less important. However, in many cases, the third alternative is found: both timing and size are under selection. For instance, an environment may not only favor early maturation (because it is unpredictable) but growth may be accelerated as well because a larger size is favored (e.g., for physiological advantages) than truncation (progenesis) alone will yield.

The ontogenetic mechanisms for these ideas are implicit in what we have said earlier: growth trajectories may be "adjusted" far beyond the "pristine" heterochronies of simple offset, onset, or rate changes. In other words, an ancestral ontogeny can be modified to show both late offset and (say) acceleration. Given such "mixed" and "pure" ontogenetic modulations, we are classifying here the three basic ecological milieus (selection on size, timing, or both) that may then select these modulations. Of course, empirically teasing apart which of these three selective regimes has operated is extremely difficult and is just one more problem that confronts life-history research. An example is the interesting case just cited above of the phenotypic change in the snail. Does the plastic response toward larger size and delayed maturation in the presence of a predator occur because large size, late maturation, or both are actively favored? A timely symposium volume (Bruton, 1989) is devoted to many such case studies, examining "the ways epigenesis . . . shapes the life-history styles of plants and animals." A reading of these and other analyses confirms that there are no ready generalizations: ontogeny is often a trade-off of competing pressures, such as the need for a rapid juvenile phase to escape predators combined with the need to attain large size for optimal physiological functioning (scaling) or reproductive advantages (e.g., greater fecundity), among many possible reasons (see Table 9-1). Some interesting generalities no doubt exist, but only with more refined studies to generate a larger data base can we untangle the often complex rate, size, timing, and shape relationships.

Such studies are exemplified by *allometric engineering,* the phrase given by Sinervo and Huey (1990) to their experimental test of allometric constraints in a lizard population. By removing egg mass, they were able to manipulate adult body size in populations of lizards. With such "engineering," they found that intrapopulational differences in running speed were a direct result of body-size differences. However, stamina was not well-correlated with size. This provides excellent documentation of allometric "coupling" and "de-coupling," discussed throughout much of this book. Also see McKinney (1990) for further discussion of the rate, size, timing, and shape problem.

6.1.1. Stress Tolerators

A second amendment to our earlier discussion concerns the work of Grime (1989 and references therein). Grime has generated considerable interest in proposing an expansion of the *r-K* dichotomy with the addition of a third "axis," that of "stress adaptation" (or "stress tolerance"). Stress-tolerant selection favors individuals that can withstand the effects of persistent, long-term stress. There has been some confusion over just what "stress" means. Grime (1989, p. 4) defines stress as "external constraints limiting the rates of resource acquisition and growth or reproduction of organisms." We have tried to illustrate stress in Fig. 9-3 (top) as a deviation from an optimum in one or more (but only one is shown for simplicity) environmental parameter(s), such as ambient temperature, water salinity, and so on. Given all the environmental parameters influencing life (as modeled in Hutchinson's "*n*-dimensional" niche concept) there are as many kinds of stress as there are parameters. In any case, the point, shown in Fig. 9-3 (top), is that as deviations from the optimum increase (stress), species performance is impaired. First affected is reproduction, then growth, then survival.

Most relevant to heterochrony and life history is that stress can therefore affect timing of reproduction (usually by delaying it) and growth (usually slowing it), as shown in the bottom of Fig. 9-3. Ecophenotypic responses are the first ones to appear, but if the stress persists and the species can survive, selection will begin to favor genetic aspects of development that will eventually lead to a stress-tolerant growth program. Specifically, stress-tolerant species tend to be long-lived, show slow growth, and are poor competitors in less stressful environments (Grime, 1989; Calow and Berry, 1989).

Figure 9-4 (left side) attempts to illustrate how stress-tolerant selection fits into the *r-K* dichotomy. Whereas *r-* and *K*-selection represent extreme poles of an axis of environmental stability (disturbance frequency), *K-* and stress-selection represent extreme poles of an axis of environmental stress (or optimality). In other words, any environment on this latter axis is predictable (stable, generally constant), but it may be predictably suboptimal for one or more parameters. Commonly cited examples of extreme stress-selective environments are toxic soils (and the consequent evolution of metal-tolerant plants) from mine tailings and harsh subarctic tundras. Marine examples would be those organisms tolerant of brackish conditions. As with *r-K,* we emphasize that the *K*-stress axis is a gradation, and intermediate conditions and thus selective pressures are to be expected, as shown in Fig. 9-4 (left side).

Figure 9-4 (right side) illustrates some of the possible ontogenetic responses to the three *r-K*-stress end-member environments. Figure 9-4 thus depicts end-member re-

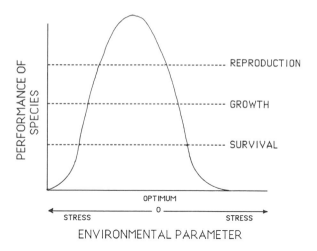

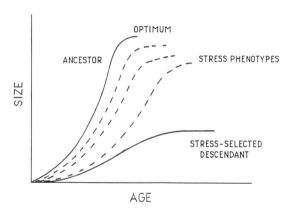

Figure 9-3. Top, effects of stress on reproduction, growth, and survival. Modified from Begon *et al.* (1990). Bottom, effects of stress as manifested in the ontogenetic trajectory of individuals. Phenotypic (or "plastic") effects are distinguished from long-term genetically established selection effects.

sponses to end-member environments; intermediate ontogenetic trajectories are to be expected. As discussed earlier in the book (especially Chapters 6 and 8), the unpredictability of *r*-selective environments will tend to favor early maturation (progenesis). The variation in the *r*-selective curves is because (in addition to the gradient nature of *r-K*) other factors will often be at play so that growth may or may not be accelerated or slowed in addition to the early offset. [At the risk of being tiresome, we reiterate that early growth offset and acceleration of growth rate (before that offset) are independent processes, though they are commonly equated.] *K*-selection will tend to promote larger organisms (via competitive, predatory, or other density-dependent advantages) that tend to mature later, although again, late offset and slower growth are not to be equated. Thus, as implied by the multiple curves in Fig. 9-4 (right), large size can be reached by late offset, rapid growth, both late offset and rapid growth, or some intermediate combination. In this case, larger size is size peramorphosis. Other traits may or may not allometrically "tag-along," depending on growth field covariation and de-coupling abilities and needs.

Finally, stress tolerators tend to be characterized by ontogenies that have slow growth rates ("neotenic"), due to the exigencies of the environment. This often results in size paedomorphs, as with *r*-selection, but the smaller size takes longer to reach. However, there seems to be at least some variation here in that at least some (and perhaps most) stress tolerators show later maturation than shown in the figure. We are not certain, since so little work has been done on stress tolerance, especially on animals and from the ontogenetic view. It does seem clear however that stress tolerators show much less ecophenotypic plasticity compared to r- and K-selected organisms. In any case, we believe that consideration of stress selection adds an important and needed dimension to life-history research, including the role of ontogeny. This is shown in the increasing interest in stress by ecologists. In his recent review, Parsons (1989, p. 42) noted that "an integrated approach to environmental stress and life-history variation is emerging, based on an underlying dependence on metabolic rate." For example, deep-sea barnacles have slower growth than shallow-water barnacles, but can accelerate growth dramatically when food increases. We find such ideas exciting not only from an ecological viewpoint but for what they may also offer evolutionists. The interested reader is referred to Calow and Berry (1989) for a stimulating collection of papers addressing this relationship of evolution, ecology, and stress.

6.1.2. Decoupling Life History Stages: Suspended Aging

A basic theme encountered in a number of places in this book, including human evolution, is that rate or timing changes in ontogeny are often sequential. Thus, human ontogeny largely represents a series of offset delays (hypermorphosis) of the various primate developmental stages. Similarly, large organisms tend to have (on a coarse scale) proportionately longer life-history stages than smaller organisms: longer juvenile stages, longer lifespan, and so on. This pattern led us to assume implicitly that correlation and

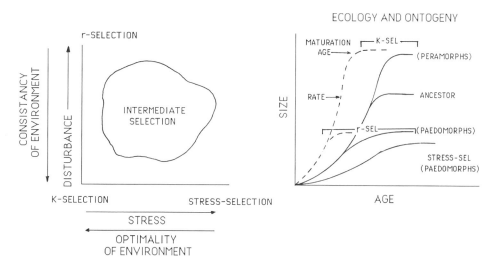

Figure 9-4. Left, relationship of r-K and stress selection, as three end-members of two axes: disturbance frequency (vertical axis) and stress (horizontal axis). Right, ontogenetic curves favored by the three kinds of selection. Just a few of the many possible permutations are shown. Peramorphs, "overdeveloped" for size relative to ancestor; paedomorphs, "underdeveloped."

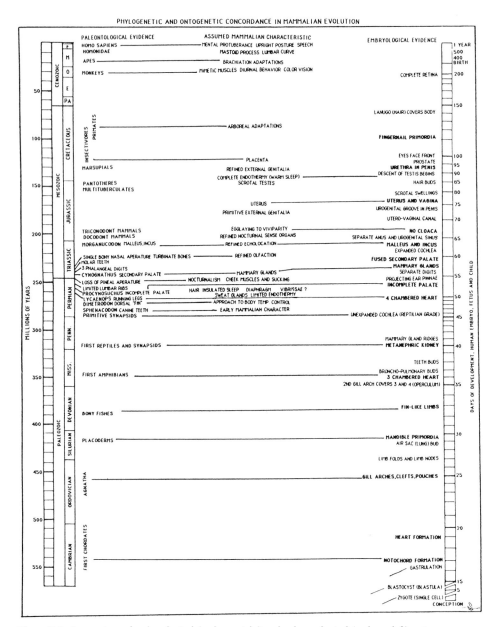

PHYLOGENETIC AND ONTOGENETIC CONCORDANCE IN MAMMALIAN EVOLUTION

Figure 9-5. Comparison of embryological (scale on right) and paleontological (scale on left) trait appearance in mammalian evolution. From Swan (1990).

proportionality among changes in life-history stages was a general pattern. This pattern would be accounted for by the internal "metabolic clock" discussed in Chapter 8. Slowing or prolonging one part of the clock would have a cascade effect on other parts of the organism's life span.

However, Miller and Hadfield (1990) have recently reported a fascinating example that defies this notion. The length of larval life of a nudibranch is determined by a chance encounter with a specific stimulus that causes metamorphosis. A developmental hiatus begins at the onset of metamorphic competence and ends at metamorphosis. Yet the duration of post-larval lifespan is unaffected by the length of this hiatus. In effect, this means that *aging is suspended* during the pre-metamorphic hiatus. This ability to "turn off" the metabolic clock may have major implications for the role of aging and life history in evolution, depending on how common is this ability to suspend aging.

6.2. Resurrecting Recapitulation

While Haeckel's recapitulation theory is obviously a gross oversimplification of evolution, we have argued in this book that much of the overall pattern of evolution can be explained by terminal addition. Evolution is a branching tree of modified ontogenies and the easiest way by far to modify ontogenies (cellular differentiation, multiplication, and assembly) is at or near the end of them, minimizing cascading effects on later contingencies. One particularly important arrow of evolution, that of increasing complexity, seems to owe much of its existence to this process (Chapter 8).

Clearly, recapitulation is not the whole story but our view is that evolutionists have overreacted to Haeckel's oversimplifications. Gould (1977) has also noted this, applying the doubly apt metaphor that they have thrown out the baby with the bathwater on this count. The most recent concurrence with this view is that of Swan (1990) who proposes that we use the phrase ontogeny "concords" with phylogeny. This removes the connotations that burden the term "recapitulation." In addition, it conveys the idea that the relationship is not one of exact repetition by ontogeny of phylogeny, but a less precise one of "agreement." As shown in Fig. 9-5, this agreement seems to be striking, according to Swan's (1990) comparison of mammalian evolution with mammalian embryology. The timing of appearance of various traits in the mammalian embryo seems to recount the appearance in the fossil record remarkably well.

We have a number of reservations about Fig. 9-5 and are skeptical about how well it would withstand intense scrutiny. For instance, the dating of much of the paleontological evidence is considerably less certain than implied by the figure. Similarly, there are mammalian traits, not shown, whose ontogenetic appearance does not precisely concord with the fossil appearance. Nevertheless, as a general heuristic device, we present it here because it captures the rough pattern of ontogenetic and phylogenetic concordance that occurs in some (perhaps many) mammalian traits.

Glossary of Major Terms and Concepts

The following terms and concepts form the core of much of this book. Generally, they appear in boldface type in the text. The chapters in which they are more fully discussed are also given.

Acceleration: Faster rate of developmental events (at any level: cell, organ, individual) in the descendant; produces peramorphic traits when expressed in the adult phenotype. (Chapters 1 and 2)

Allometric heterochrony: Change in a trait as a function of size (as opposed to a function of age in "true" heterochrony); it is useful as a descriptor and can be interpretive considering that body size is often a better metric of "intrinsic" age than chronological age. The same categories apply as in "true" heterochrony, only the adjective "allometric" is prefixed: e.g., "allometric progenesis" is when the descendant terminates growth at a smaller size (as opposed to age) than the ancestor. (Chapter 2)

Allometry: The study of size and shape, usually using biometric data; the change in size and shape observed. Basic kinds of allometric change include the following. **Complex allometry**: occurs when the ratio of the specific growth rates of the traits compared are not constant, a log–log plot comparing the traits will not yield a straight line (i.e., "k" in the allometric formula is inconstant). **Isometry**: occurs when there is no change in shape with size increase; that is, when the traits being compared on a log–log plot yield a straight line with a slope (k in the allometric formula) that is effectively equal to 1. **Negative** and **positive allometry**: occur where there is significant change in shape with size increase; in negative allometry, trait y on a log–log plot is increasing more slowly than trait x (slope, $k < 1$); in positive allometry, the opposite is true (slope, $k > 1$). **Vertical allometry**: the "noise" around the best-fit line estimating size and shape change; such deviations represent important information about "local" effects (e.g., local environmental influence on growth rates, dietary differences). Basic kinds of allometric comparisons include the following. **Cross-sectional**: size and shape change when comparing different individuals of different ages. **Longitudinal**: size and shape change in one individual as it grows through time. **Static**: size and shape change when comparing individuals of roughly the same age or stage (usually adults). All three kinds of comparisons can compare intraspecific or interspecific change. Of the three comparisons, longitudinal is usually preferred, but the most difficult to make. (Chapter 2)

Astogeny: Growth and development of a colony (e.g., of bryozoans, corals), as opposed to the individual; like individuals, colonial development can show heterochronies. (Chapter 6)

Atavism: Appearance of an ancestral structure that had formerly been "suppressed" or unexpressed; in other words, a "throwback," e.g., horses born with lateral "toes" or snakes with rudimentary legs. (Chapter 5)

Bauplan: The basic architecture ("body plan") of a major group (usually phylum), established early in the group's history, usually in the late Precambrian to early Paleozoic metazoan radiation. (Chapters 6 and 8)

Bimaturism: Dimorphic size and shape differences caused by different timing of maturation in males versus females; common in primates, including humans. (Chapters 4 and 7)

Biogenetic law: Haeckel's famous "ontogeny recapitulates phylogeny," although that is not his exact formulation; in a very coarse way, there is some validity to this since evolutionary complexity must usually build on previous contingencies. See **recapitulation**. (Chapters 1 and 8)

Breadth: The number of cells or traits affected by an ontogenetic change; contrast with **depth**. (Chapters 3 and 8)

Canalization: Buffering of developmental "program" against perturbations. (Chapter 4)

Cladogenetic asymmetry: The "nowhere but up" process whereby a group (clade) originates at a character state (e.g., size) that is physically restricted to expansion mainly in one direction (e.g., a mammal can only get so small before encountering prohibitive metabolic problems); the descendant radiation is therefore asymmetric in that character state (e.g., "larger size"), the resulting "trend" is an increase in variance or maximum value of state. (Chapters 6 and 8)

Condensation: The shortening of ontogeny by "telescoping" stages; strict recapitulationists thought it necessary, to keep ontogenies from becoming too lengthy as terminal additions accumulated. (Chapter 1)

Cope's rule: Evolutionary tendency of many fossil lineages to increase in body size, and of many clades to increase in maximum body size of largest species in clade ("increase in variance," see **cladogenetic asymmetry**). (Chapters 6 and 8)

Critical period: A limited time when an outside stimulus is required for development of a behavior. (Chapter 7)

C-value: The mass of DNA in an unreplicated genome, also known as **genome size**; affects cell size and developmental rates. (Chapter 6)

Depth: The degree to which cells or traits are changed, i.e., the amount of acceleration, or early onset (for example), relative to ancestral cells or traits; contrast with **breadth**. (Chapters 3 and 8)

Developmental constraint: Role of ontogenetic "rules" in determining the course of evolution; the degree of "intrinsic" control over evolutionary direction (in contrast to "extrinsic" natural selection). The importance of constraint, relative to selection, is a matter of much current debate, awaiting further study of developmental systems, especially with the aid of quantitative genetics. (Chapters 4–6 and 8)

Developmental conversion: Specific environmental cues activate one of a number of alternative genetic programs controlling development. (Chapter 4)

Developmental gene: A gene active only during development; half of an oversimplified dichotomy with **maintenance gene**. (Chapter 3)

D-gene: A "developmental" gene; this includes both the genes that directly produce enzymes (e.g., **morphogen**) participating in development (structural D-gene), and the genes that regulate them (regulator D-gene); contrast with **R-gene**. (Chapter 3)

Differentiative heterochrony: Change in rate or timing of development before final differentiation of tissue/organ occurs; therefore, "novel" changes can occur, other than simply size/shape change (as occurs in **growth heterochronies**). There are two kinds of differentiative heterochronies. **Novel differentiative heterochronies** result in significant tissue alterations (cell juxtapositions; "disjunctions" in ancestral morphospace) by creating new cell interactions (e.g., inductions by changing time of migration of some cells) and/or preventing old ones; in regulative developmental programs, this can be very "creative" due to cell pluripotency. **Size differentiative heterochronies** result in only size/shape ("allometric") change; even though they act before differentiation, such heterochronies (unlike novel differentiative heterochronies) do not change kinds of cell interactions, only the end result of old ones (e.g., number of cells allocated to organs). All types can act with varying degrees of extent, or **breadth**. (Chapter 3)

Differentiative phase: The second phase of development (after **neofertilization** and before the **growth phase**); the phase when the zygotic genome begins to take over from maternal control, and culminating in organogenesis and tissue differentiation; includes a complex series of mitosis, cytodifferentiation and cell migration. (Chapter 3)

Diffuse coevolution: The tendency for some large groups (e.g., higher taxa such as mammalian predators and ungulates) to affect each other's evolution in a general, loosely integrated way (see Futuyma, 1986a). (Chapter 8)

Direct size selection: Natural selection acting directly on body size; contrast with **indirect size selection**, where selection is on life history events and body size change is a by-product. (Chapter 6)

Dissociated heterochrony: An ontogenetic change in rate or timing of a trait (e.g., organ, or even biochemical trait) that does not occur in some other traits; a change that occurs only in a local **growth field**; contrast with **global heterochrony**, see **heterochronocline (dissociated)**. (Chapter 2)

Dwarfism: Evolutionary size decrease with no change in shape; the smaller descendant is the same in all proportions (isometric); opposite is **giantism**. Both dwarfism and giantism are probably rare, as significant size change requires allometric (disproportional) scaling of body parts. (Chapters 2 and 6)

Ecological and evolutionary cascades: The sometimes "rapid" cascading effects on an ecosystem (and hence its selective regime and hence its evolution) caused by the introduction of some "new" biotic element (Gray, 1988); humans do this artificially: to what extent does development introduce such "new" inputs? (Chapter 8)

Ecophenotypic plasticity: The degree to which a single genotype may show phenotypic variation under differing environmental conditions (contrast to "phenotypic variation"). (Chapter 4)

Evolutionary ratchet: As organisms evolve, especially those that become more complex, there is an accumulation of contingencies: an increasing interdependence among its components, making radical change more difficult, i.e., a "hardening" occurs. Levinton (1988) suggests three kinds of evolutionary ratchets. The **genetic ratchet** refers to genetic contingencies: genes in the genome become more interactive and integrated through time. The **epigenetic ratchet** refers to the contingencies of the ontogenetic construction of the organism: the increasing interdependence of steps in the developmental process. The **selection ratchet** refers to the interdependence of functional traits in the anatomical phenotype, e.g., radius and ulna; since all these ratchets are subject to selection, we prefer the term "functional" ratchet for this last category. (Chapter 8)

Genetic homeostasis: The property of a population to equilibrate its genetic composition and resist sudden change. See **genetic revolution**. (Chapter 5)

Genetic revolution: Rapid change in the genetic structure of a population as it reestablishes homeostasis; the rapid restructuring of the genome. (Chapter 5)

Genome size: The mass of DNA in an unreplicated genome; also known as **C-value**; affects cell size and developmental rates (Chapter 6)

Giantism: Evolutionary increase in size but no change in shape; the giant is proportional (isometric) in all respects to the smaller ancestor; opposite is **dwarfism**. Both dwarfism and giantism are probably rare, as significant size change requires allometric (disproportional) scaling of body parts. (Chapters 2 and 6)

Global heterochrony: An ontogenetic change in rate or timing that affects the "entire" individual, a whole-body change; obviously, this is a relative term in that not every single trait will be affected, at least in the same way. Roughly speaking, large-scale changes (e.g., allometric extrapolations from body size increase) may be termed "global"; contrast with **dissociated heterochrony**. (Chapter 2)

Growth field: A loosely defined term referring to a tissue, organ, or cell mass with growth parameters (rate, timing) distinct from surrounding fields; see Fig. 3-9. (Chapter 3)

Growth heterochrony: Change in rate or timing of development after cells/tissues have undergone final differentiation, i.e., only size/shape ("allometric") changes (in cell mitosis, migration) of trait, organ, individual result; can act either **globally** or locally (**dissociated**). Contrast with **differentiative heterochrony**. (Chapter 3)

Growth phase: The third and final phase of development (after the **differentiative phase**), characterized by cell migration and especially mitosis, after tissues/organs have undergone final differentiation. (Chapter 3)

Heterochronic morphotypes: Morphologies within species, produced by heterochrony. (Chapter 4)

Heterochronocline: An evolutionary sequence wherein ontogenies show regressive or progressive heterochronic changes, either paedomorphosis (**paedomorphocline**) or peramorphosis (**peramorphocline**); often there is a spatial component wherein the sequence is adapting to an environmental gradient. **Anagenetic** heterochrono-clines occur when there is no temporal overlap among the species in the sequence (and thus, no increase in species number, only one successively evolving species). **Cladogenetic** heterochronoclines occur when there is evolutionary branching: a descendant heterochronic-variant species coexists with the parent species. A **stepped** anagenetic or cladogenetic heterochronocline occurs where evolution is rapid followed by a period of stasis. **Mosaic** heterochronoclines occur when some traits are affected by paedomorphic processes while others show the opposite pat-tern, peramorphosis. **Dissociated** heterochronoclines are similar but in this case the different processes all produce the same patterns: all are peramorphic or all are paedomorphic. Cladogenetic heterochronoclines are generally **autocatakinetic**: they are self-generating in a direction because the ancestral species "blocks," through competition, the descendant from moving in its direction. Anagenetic heterochronoclines also may fit this category where intraspecific competition drives it (as opposed to externally directed predation). (Chapter 5)

Heterochrony: Literally, "different time"; more precisely, change in timing or rate of developmental events, relative to the same events in the ancestor. In this book, we explicitly include events at fine (e.g., cellular) levels and in early, even embryonic, stages. Different kinds of heterochronies are recognized: **dissociated, global, growth, differentiative, sequential, organizational** (see separate definitions). (Chapters 1–3)

Hierarchy: Any system of organization based on levels or ranks. Types of hierarchies include the following. **Nested** hierarchy: when higher levels are composed of, and contain, lower levels (e.g., tissues of cells, cells of atoms). **Vertical** hierarchy: "higher" levels are made of "lower" levels, the latter containing "smaller" ele-ments (e.g., tissue, cell, atom). **Horizontal** hierarchy: nested elements occur only on one level (no "smaller" elements), e.g., growth fields all composed of cells, all nested within other growth fields. We may also distinguish hierarchies of **static** versus **dynamic** phenomena, the latter being of more interest to classifying onto-genetic change. (Chapter 3)

Hypermorphosis: Late cessation ("offset") of developmental events (at any level: cell, organ, individual) in the descendant; produces peramorphic traits when expressed in the adult phenotype. Not restricted to late sexual maturation (offset of growth for whole individual) as often stated; can also refer to late cessation of local processes, e.g., growth field. (Chapters 1 and 2)

Indirect size selection: Natural selection for a trait (especially life history) that indirectly alters body size, e.g., selection for early reproduction (global progenesis); contrast with **direct size selection**. (Chapter 6)

Induction: The evocative action of one cell (or tissue) upon another (usually applied to the invoking of differentiation: to produce a third cell/tissue type). (Chapter 3)

K-selection: Selection occurring in relatively constant, stable environments where com-petition and other density-dependent factors are important; traits favored include:

large size, delayed reproduction, repeated reproduction, few offspring, parental care of offspring. (Chapters 6 and 8)

Macroevolution: In this book, simply defined as the origin of higher (supraspecies) taxa, but a number of definitions have been proposed (see Levinton, 1988). (Chapters 5 and 8)

Maintenance gene: A gene active beyond development, into the adult stage, usually in physiological processes; half of an oversimplified dichotomy with **developmental gene**. (Chapter 3)

Mitogen: A type of **morphogen** that specifically stimulates mitosis. (Chapter 3)

Morphogen: The signal which transmits **positional information**, stimulating cells to divide, migrate, or differentiate; often depicted as some kind of biochemical gradient. See also **mitogen**. (Chapter 3)

Morphogenetic tree: A branching model used to illustrate development: tissues (or cells, or organs) share a common history until differentiation into two occurs. (Chapters 3 and 8)

Morphologic covariation: Covarying patterns of growth among different traits so that heterochronic effects are not always localized; often linked to genetic covariance and quantitative genetics. (Chapters 6 and 8)

Mosaic development: i.e., "invariant," "determinate" development wherein determination of cell fate occurs relatively early: later information (cell interactions) has little or no effect on cell fate. Contrast with **regulative development**. (Chapter 3)

Nearshore innovation: The tendency for clades (higher taxa) to appear first in nearshore environments and then spread offshore (see Bottjer and Jablonski, 1988). (Chapter 8)

Neofertilization phase: The earliest phase of development, just after the egg is fertilized; maternal (oocytic) information is very important in governing processes. (Chapter 3)

Neoteny: Slower rate of developmental events (at any level: cell, organ, individual) in the descendant; produces paedomorphic traits when expressed in the adult phenotype. (Chapters 1 and 2)

Neural competition and regression: The death of neurons in the brain that have not made a sufficient number of connections with other neurons. Change in rate or timing of dendrite and axon growth is a major mode of brain "sculpting," i.e., modifying neuron numbers and networks. (Chapter 7)

n-selection: Selection where fitness is determined largely by net reproductive rate, as opposed to competitiveness, predation-resistance, and the usual Darwinian criteria; occurs mainly in small, isolated populations. Contrast with **w-selection**. (Chapter 8)

Ontogenetic niche: The ecological niche occupied by an organism at a given stage in its ontogeny; in many organisms, the ontogenetic niche will change as the individual grows (see Werner and Gilliam, 1984, for review article). (Chapters 5 and 8)

Ontogeny: Growth (usually referring to size increase) and development (usually referring to both size increase and differentiation of traits) of the individual. (All chapters)

Organizational heterochrony: Change in rate or timing of tissue or organ organization, i.e., connectance into an integrated whole. The human brain grows faster than in the ape but our behavior temporarily lags behind because neural components are more slowly connected by neural circuitry. Similar delays or changes can be seen in morphological "stages" or "shapes" that have component parts but have not "differentiated" into a recognizable organ or state. (Chapters 2 and 7)

Paedomorphocline: An evolutionary sequence wherein adult morphologies become progressively more paedomorphic ("juvenile"); often such a **heterochronocline** has a spatial component wherein the sequence is adapting to an environmental gradient. (Chapter 5)

Paedomorphosis: Literally, "child formation"; the retention of subadult ancestral traits in the descendant adult; "underdevelopment," as can be shown on a trait (size) versus age plot (contrast to "peramorphosis"). (Chapters 1 and 2)

Parameterization problem: The extreme difficulty of measuring and interrelating size, shape, and age "parameters" in the Alberch *et al.* (1979) scheme; phrase coined and discussed by Atchley (1987). (Chapter 2)

Pearson's rule: States that correlation (covariation) of body parts increases with their proximity to one another; see also **growth field**. (Chapter 3)

Peramorphocline: An evolutionary sequence wherein morphologies become progressively more peramorphic ("overdeveloped"); often such a **heterochronocline** has a spatial component wherein the sequence is adapting to an environmental gradient. (Chapter 5)

Peramorphosis: Development of traits beyond that of the ancestral adult; "overdevelopment," as can be shown on a trait (size) versus age plot (contrast to "paedomorphosis"). (Chapters 1 and 2)

Phenotypic modulation: Phenotypic variation from environmental alteration of rate or timing of development (heterochrony); however, the genetic program is itself not altered. (Chapter 4)

Phenotypic variation: Variation in a population arising from intrinsic alterations to the developmental program (contrast to "ecophenotypic variation"). Unlike "polymorphism," variation is continuous within the population. (Chapter 4)

Polymely: The frequent appearance of supernumerary ("extra") legs and/or digits in a population. (Chapter 5)

Polymorphism: According to Mayr (1970), the occurrence of several strikingly different discontinuous phenotypes in one interbreeding population. Unlike "phenotypic variation," variation here is discrete. (Chapter 4)

Positional information: A much-abused term, but it generally refers to the information that tells a cell "where it is," thus stimulating it (via gene response) to "behave" accordingly, i.e., when and where to differentiate, migrate, or divide. **Morphogens** are major transmitters of such information. (Chapter 3)

Postdisplacement: Late initiation ("onset") of developmental events (at any level: cell, organ, individual) in the descendant; produces paedomorphosis when expressed in the adult phenotype. (Chapters 1 and 2)

Predisplacement: Early initiation ("onset") of developmental events (at any level: cell, organ, individual) in the descendant; produces peramorphosis when expressed in the adult phenotype. (Chapters 1 and 2)

Progenesis: Early cessation ("offset") of developmental events (at any level: cell, organ, individual) in the descendant; produces paedomorphic traits when expressed in the adult phenotype. Not restricted to early sexual maturation (offset of growth for whole individual) as often stated; can also refer to early cessation of local processes, e.g., growth field. (Chapters 1 and 2)

Recapitulation: The concept that as "higher" organisms (with humans usually at the apex) develop, they pass through the adult stages of "lower" organisms. It was often tied to various metaphysical concepts, such as growth of the soul in humans. The idea is incorrect for a number of reasons. An important one is that evolution does not proceed only by "terminal addition" of traits to ontogenies. That is, mutations in developmental genes can alter earlier phases of ontogenies, not only the very last one. Nevertheless, development is highly conservative and early occurring contingencies are often retained. Further, increases in ontogenetic complexity do result from a general tendency to "add on" more pathways (e.g., cell type and number). Therefore, human embryos (for example) share similarities with the embryos of other groups, even distantly related ones (e.g., in having gill-like structures). While such "higher organisms" do not pass precisely through the adult stages of their ancestors (as extreme recapitulationists held), they often pass through similar stages of their ontogeny. More generally, there is a very coarse pattern that the more closely related two groups are, the more similar their ontogenies, and the later in development the two ontogenies diverge. (Chapters 1 and 8)

Regulative development: Development characterized by late determination of cell fate; cells are "pluripotent" (capable of assuming a number of final states) for some time, until stimulated by positional information. Probably necessary for development of "complex" organisms, as precise "programming" would require too much information, among other reasons discussed in text. Contrast with **mosaic development**. (Chapter 3)

Regulatory gene: A gene that produces substances controlling structural genes, perhaps by binding "upstream" to the structural gene's transcription; half of an oversimplified dichotomy with **structural gene** (see **D-gene**, **R-gene**, **S-gene**). (Chapter 3)

R-gene: A gene that regulates another gene, an **S-gene** or "structural gene," where the enzyme produced by the S-gene is used in "maintenance," not development; in contrast to **D-gene**. (Chapter 3)

r-selection: Selection occurring in an unpredictable environment, where physical, density-independent factors are important; traits selected for include: small size, early maturation, single reproduction, with large numbers of offspring, little parental care. (Chapter 6)

Saltation: A rapid morphological transformation; a "macromutation," "hopeful monster." (Chapters 5 and 8)

Sequential heterochrony: A change that affects a number of sequential ontogenetic events ("stages") in the same way, e.g., sequential hypermorphosis or delay in a number of stages of development; changes that occur in a number of phases in a multiphasic growth curve in the same way. The human brain shows sequential, organizational heterochrony; human body growth shows sequential heterochrony. See also McNamara (1983a) for example with trilobite molt stages. (Chapters 2 and 7)

S-gene: See **R-gene**.

Size: An often highly subjective term; best measured by some multivariate method, such as Principle Component Analysis, where size is defined as the "general vector which best accounts for all observed covariances" (Bookstein *et al.,* 1985). Deviations from this vector may be thought of as "shape" changes, also a highly subjective term. (Chapter 2)

Structural gene: A gene that codes for proteins used in cellular machinery during growth or maintenance; half of an oversimplified dichotomy with **regulatory gene** (see also **D-gene, R-gene, S-gene**). (Chapter 3)

Terminal addition: The alteration of development by the addition of stages to the otherwise unaltered ancestral ontogeny; the "adding on" of new ontogenetic contingencies, or pathways. Contrary to early beliefs, this is not the only way for evolution to occur: mutations of developmental genes can alter even early pathways and still produce a viable organism, as discussed in Chapter 3. See **recapitulation**. (Chapter 1)

Velocity curve: A derivative of a simple trait versus age growth curve: the curve created when plotting trait gain per unit age; especially useful in identifying the processes of acceleration and neoteny, even within growth phases of a multiphasic curve (as most growth curves are). (Chapter 2)

W-selection: Selection for competitive ability, predation-resistance, and other criteria for fitness common in a highly populated, diverse community. Contrast with **n-selection**. (Chapter 8)

References

Abbott, B. D., Adamson, E. D., and Pratt, R. M., 1988, Retinoic acid alters EGF receptor expression during palatogenesis, *Development* **102**:853–867.

Abbott, R. T., 1968, *Indo-Pacific Mollusca,* Academy of Natural Sciences, Philadelphia.

Adamson, E. D., and Meek, J., 1984, The ontogeny of epidermal growth factor during mouse development, *Dev. Biol.* **103**:62–70.

Agassiz, L., 1857, Essay on classification, in: *Contributions to the Natural History of the United States,* Volume 1 (E. Lurie, ed.), Harvard University Press, Cambridge, Mass., 1962.

Aiken, D. E., 1980, Molting and growth, in: *The Biology and Management of Lobsters* (J. S. Cobb and B. F. Phillips, eds.), Academic Press, New York.

Alberch, P., 1980, Ontogenesis and morphological diversification, *Am. Zool.* **20**:653–667.

Alberch, P., 1982, Developmental constraints in evolutionary processes, in: *Evolution and Development* (J. T. Bonner, ed.), Springer-Verlag, Berlin.

Alberch, P., 1983, Morphological variation in the neotropical salamander genus *Bolitoglossa, Evolution* **37**:906–916.

Alberch, P., 1985a, Problems with the interpretation of developmental sequences, *Syst. Zool.* **34**:46–58.

Alberch, P., 1985b, Developmental constraints: Why St. Bernards often have an extra digit and poodles never do, *Am. Nat.* **126**:430–433.

Alberch, P., 1987, Evolution of a developmental process: Irreversibility and redundancy in amphibian metamorphosis, in: *Development as an Evolutionary Process* (R. Raff and E. Raff, eds.), Liss, New York.

Alberch, P., 1989, The logic of monsters: evidence for internal constraint in development and evolution, *Geobios Mem. Spec.* **12**:21–57.

Alberch, P., and Alberch, J., 1981, Heterochronic mechanisms of morphological diversification and evolutionary change in the neotropical salamander, *Bolitoglossa occidentalis, J. Morphol.* **167**:249–264.

Alberch, P., and Gale, E. A., 1985, A developmental analysis of an evolutionary trend: Digital reduction in amphibians, *Evolution* **39**:8–23.

Alberch, P., Gould, S. J., Oster, G. F., and Wake, D. B., 1979, Size and shape in ontogeny and phylogeny, *Paleobiology* **5**:296–317.

Alcock, J., 1983, *Animal Behavior,* 3rd ed., Sinauer, Sunderland, Mass.

Allen, T. F. H., and Starr, T. B., 1982, *Hierarchy: Perspectives for Ecological Complexity,* University of Chicago Press, Chicago.

Ambros, V., 1988, Genetic basis for heterochronic variation, in: *Heterochrony in Evolution: A Multidisciplinary Approach* (M. L. McKinney, ed.), Plenum Press, New York.

Ambros, V., and Horvitz, H., 1984, Heterochronic mutants of the nematode *Caenorhabditis elegans, Science* **226**:409–416.

Anderson, D. T., 1987, Developmental pathways and evolutionary rates, in: *Rates of Evolution* (K. S. W. Campbell and M. F. Day, eds.), Allen & Unwin, London.

Andrews, R. C., 1921, A remarkable case of external hind limbs in a humpback whale, *Am. Mus. Novit.* **9**:1–16.

Anstey, R. L., 1987, Astogeny and phylogeny: Evolutionary heterochrony in Paleozoic bryozoans, *Paleobiology* **13**:20–43.

Aono, H., and Ide, H., 1988, A gradient of responsiveness to the growth-promoting activity of ZPA (zone of polarizing activity) in the chick limb bud, *Dev. Biol.* **128**:136–141.

Archer, M., 1974, The development of premolar and molar crowns of *Antechinus flavipes* (Marsupialia, Dasyuridae) and the significance of cusp ontogeny in mammalian teeth, *J. R. Soc. West. Aust.* **57**:118–125.

Armstrong, E., 1983, Relative brain size and metabolism in mammals, *Science* **220**:1302–1304.

Arthur, W., 1984, *Mechanisms of Morphological Evolution,* Wiley, New York.

Arthur, W., 1987, *The Niche in Competition and Evolution,* Wiley, New York.

Ashburner, M., 1980, Chromosomal action of ecdysone, *Nature* **285**:435–436.

Aslin, R. N., 1981, Experimental influences in sensitive periods in perceptual development: A unified model, in: *Development of Perception* (R. N. Aslin, J. R. Alberts, and M. R. Petersen, eds.), Academic Press, New York.

Atchley, W., 1983, Some genetic aspects of morphometric variation, in: *Numerical Taxonomy* (J. Felsenstein, ed.), Springer-Verlag, Berlin.

Atchley, W., 1984, Ontogeny, timing of development, and genetic variance–covariance structure, *Am. Nat.* **123**:519–540.

Atchley, W. R., 1987, Developmental quantitative genetics and the evolution of ontogenies, *Evolution* **41**:316–330.

Atchley, W. R., 1989, Introduction to the symposium, *Am. Nat.* **134**:437–439.

Atchley, W., Rutledge, J., and Cowley, D., 1981, Genetic components of size and shape. II. Multivariate covariance patterns in the rat and mouse skull, *Evolution* **35**:1037–1055.

Bakker, R., 1977, Cycles of diversity and extinction, in: *Patterns of Evolution* (A. Hallam, ed.), Elsevier, Amsterdam.

Bambach, R., 1985, Classes and adaptive variety: The ecology of diversification in marine faunas, in: *Phanerozoic Diversity Patterns* (J. Valentine, ed.), Princeton University Press, Princeton, N.J.

Bambach, R. K., 1986, Phanerozoic marine communities, in: *Patterns and Processes in the History of Life* (D. M. Raup and D. Jablonski, eds.), Springer-Verlag, Berlin.

Bard, J. B. L., 1977, A unity underlying the different zebra striping patterns, *J. Zool.* **183**:527–539.

Barrington, E. J. W., 1968, Metamorphosis in lower chordates, in: *Metamorphosis* (W. Etkins and L. I. Gilbert, eds.), North-Holland, Amsterdam.

Barton, N. H., Jones, J. S., and Mallet, J., 1988, No barriers to speciation, *Nature* **336**:13–14.

Bateson, P., 1988, The active role of behavior in evolution, in: *Evolutionary Processes and Metaphors* (M. Ho and S. Fox, eds.), Wiley, New York.

Batten, R. L., 1975, The Scissurellidae—Are they neotenously derived fissurellids? (Archaeogastropoda), *Am. Mus. Novit.* **2567**:1–37.

Beadle, S. C., 1989, Heterochrony and eccentricity in sand dollars, *Paleobiology* **15**:205–222.

Beckmann, M. P., Betshiltz, C., Heldin, C.-H., Westermark, B., Di Marco, E., Di Fiore, P. P., Robbins, K. C., and Aaronson, S. A., 1988, Comparison of biological properties and transforming potential of human PDGF-A and PDGF-B chains, *Science* **241**:1346–1347.

Bedo, G., Santisteban, P., and Aranda, O., 1989, Retinoic acid regulates growth hormone gene expression, *Nature* **339**:231–234.

Beecher, C. E., 1893, Some correlations of ontogeny and phylogeny in the Brachiopoda, *Am. Nat.* **27**:599–604.

Beecher, C. E., 1897, Outline of a natural classification of the trilobites, *Am. J. Sci.* Ser. 4, **3**:89–106, 181–207.

Begon, M., Harper, J., and Townsend, C., 1986, *Ecology,* Sinauer, Sunderland, Mass.

Begon, M., Harper, J., and Townsend, C., 1990, *Ecology,* Blackwell, Boston, Mass.

Bell, M. A., 1988, Stickleback fishes: Bridging the gap between population biology and paleobiology, *Trends Ecol. Evol.* **3**:320–325.

Bemis, W. E., 1984, Paedomorphosis and the evolution of the Dipnoi, *Paleobiology* **10**:293–307.

Bennett, M. D., 1972, Nuclear DNA content and minimum generation time in herbaceous plants, *Proc. R. Soc. London Ser. B* **181**:109–135.

Benton, M. J., 1984, Tooth form, growth, and function in Triassic rhynchosaurs (Reptilia, Diapsida), *Palaeontology* **27**:737–776.

Benton, M. J., 1988a, Bringing up baby, *Nature* **334**:566.

Benton, M. J., 1988b, The nature of adaptive radiation, *Trends Ecol. Evol.* **3**:127–128.

Berge, C., Didier, M., Chaline, J., and Dommergues, J.-L., 1986, La bipèdie des hominides: comparisons des itinéraires ontogénétiques des dimensions femoropelviennes des pongides, australopithèques et hommes, in: *Ontogenèse et Evolution* (J. Chaline and B. Laurin, eds.), C.N.R.S., Paris.

Berneys, E. A., 1986, Diet-induced head allometry among foliage-chewing insects and its importance for graminovores, *Science* **231**:495–497.

Berrill, N. J., 1955, *The Origin of Vertebrates*, Oxford University Press (Clarendon), Oxford.

Berrill, N. J., 1971, *Developmental Biology*, McGraw–Hill, New York.

Berven, K. A., 1982a, The genetic basis of altidudinal variation in the wood frog, *Rana sylvatica*. I. An experimental analysis of life history traits, *Evolution* **36**:962–983.

Berven, K. A., 1982b, The genetic basis of altitudinal variation in the wood frog, *Rana sylvatica*. II. An experimental analysis of larval development, *Oecologia* **52**:360–369.

Berven, E. K., 1987, The heritable basis of variation in larval developmental patterns within populations of the wood frog (*Rana sylvatica*), *Evolution* **41**:1088–1097.

Berven, K. A., Gill, D. E., and Smith-Gill, S. J., 1979, Countergradient selection in the green frog, *Rana clamitans*, *Evolution* **33**:609–623.

Berzin, A. A., 1972, *The Sperm Whale*, Keter Press, Jerusalem.

Beynon, A. D., and Dean, M. C., 1988, Distinct dental development patterns in early fossil hominids, *Nature* **335**:509–514.

Beynon, A. D., and Wood, B. A., 1987, Pattern and rates of enamel growth in molar teeth of early hominids, *Nature* **326**:493–496.

Bier, K., 1954, Uber den Einfluss der Konigen auf die Arbeiterinnen-Fertilitat im Ameisenstaat, *Insectes Soc.* **1**:7–19.

Bier, K., 1956, Arbeiterinnfertilitat und Aufzucht von Geschlechstieren als Regulationsleistung, *Insectes Soc.* **3**:177–184.

Blackstone, N. W., 1987a, Size and time, *Syst. Zool.* **36**:211–215.

Blackstone, N. W., 1987b, Allometry and relative growth: Patterns and process in evolutionary studies, *Syst. Zool.* **36**:76–78.

Blackstone, N., and Yund, P., 1989, Morphological variation in a colonial marine hydroid: A comparison of size-based and age-based heterochrony, *Paleobiology* **15**:1–10.

Bleiweiss, R., 1987, Development and evolution of avian racket plumes: Fine structure and serial homology of the wire, *J. Morphol.* **194**:23–39.

Blount, R. F., 1950, The effects of heteroplastic hypophysed grafts upon the axolotl, *J. Exp. Zool.* **113**:717–739.

Blundell, T. L., and Humbel, R. E., 1980, Hormone families: Pancreatic hormones and homologous growth factors, *Nature* **287**:781–787.

Bock, W. J., 1970, Microevolutionary processes as a fundamental concept in macroevolutionary models, *Evolution* **24**:704–722.

Bock, W. J., and Von Wahlert, G., 1965, Adaptation and the form–function complex, *Evolution* **19**:269–299.

Bogin, B., 1990, The evolution of human childhood, *Bioscience* **40**:16–25.

Bolk, L., 1926, *Das Problem der Menschwerdung*, Fischer, Jena.

Bolk, L., 1929, Origin of racial characteristics in man, *Am. J. Phys. Anthropol.* **13**:1–28.

Boltovsky, E., 1988, Size change in the phylogeny of Foraminifera, *Lethaia* **21**:375–382.

Bonner, J. T., 1968, Size change in development and evolution, in: *Paleobiological Aspects of Growth and Development—A Symposium* (D. B. Macurda, ed.), Paleontological Society Memoir 2.

Bonner, J. T. (ed.), 1982, *Evolution and Development*, Springer-Verlag, Berlin.

Bonner, J. T., 1988, *The Evolution of Complexity*, Princeton University Press, Princeton, N.J.

Bonner, J. T., and Horn, H. S., 1982, Selection for size, shape and developmental timing, in: *Evolution and Development* (J. T. Bonner, ed.), Springer-Verlag, Berlin.

Bookstein, F. L., Chernoff, B., Elder, R. L., Humphries, J. M., Smith, G. R., and Strauss, R. E., 1985, Morphometrics in evolutionary biology, *Acad. Nat. Sci. Philos. Spec. Publ.* **15**.

Borrelli, E., Heyman, R. A., Arias, C., Sawchenko, P. E., and Evans, R. M., 1989, Transgenic mice with inducible dwarfism, *Nature* **339**:538–541.

Bottjer, D. J., and Jablonski, D., 1988, Paleoenvironmental patterns in the evolution of post-Palaeozoic benthic marine invertebrates, *Palaios* **3**:540–560.

Bounhiol, J.-J., 1936, Métamorphose après ablation des corpora allata chez le ver à soie (*Bombyx mori*, L.), *C. R. Acad. Sci.* **203**:388–389.

Bounhiol, J.-J., 1937, Métamorphose prématurée par ablation des corpora allata chez le jeune ver à soie, *C. R. Acad. Sci.* **205**:175–177.

Bower, T. G. R., 1974, *Development in Infancy,* Freeman, San Francisco.

Box, H. O., 1984, *Primate Behaviour and Social Ecology,* Chapman & Hall, London.

Boy, J. A., 1971, Zur Problematik der Branchiosaurier (Amphibia, Karbon-Perm), *Palaeontol. Z.* **45**:107–119.

Boy, J. A., 1972, Die Branchiosaurier (Amphibia) des saarpfälzischen Rotliegenden (Perm, SW-Deutschland), *Abh. Hess. Landesamtes Bodenforsch.* **65**:1–137.

Boyce, M. S., 1984, Restitution of *r*- and *k*-selection as a model of density-dependent natural selection, *Annu. Rev. Ecol. Syst.* **15**:427–449.

Brace, C. L., Rosenberg, K. R., and Hunt, K. D., 1987, Gradual change in human tooth size in the Late Pleistocene and post-Pleistocene, *Evolution* **41**:705–720.

Bradshaw, A. D., 1965, Evolutionary significance of phenotypic plasticity in plants, *Adv. Genet.* **13**:115–155.

Brasier, M. D., 1982, Architecture and evolution of the foraminiferid test—A theoretical approach, in: *Aspects of Micropalaeontology* (P. T. Bauner and A. R. Lords, eds.), Allen & Unwin, London.

Braude, P., Bolton, V., and Moore, S., 1988, Human gene expression first occurs between the four- and eight-cell stages of preimplantation development, *Nature* **332**:459–462.

Brian, M. V., 1965, Studies of caste differentiation in *Myrmica rubra* L. 6. Factors influencing the course of female development in the early third instar, *Insectes Soc.* **10**:91–102.

Briggs, J. C., 1953, The behaviour and reproduction of salmonid fishes in a small coastal stream, *Calif. Fish Bull.* **94**:1–62.

Britten, R. J., and Davidson, E. H., 1969, Gene regulation for higher cells: A theory, *Science* **165**:349–357.

Britten, R. J., and Davidson, E. H., 1971, Repetitive and non-repetitive DNA sequences and a speculation on the origins of evolutionary novelty, *Q. Rev. Biol.* **46**:111–138.

Brocchi, G., 1814, *Conchiliologia Fossile Subapennina,* Milan.

Bromage, T. G., 1985, Taung facial remodelling: A growth and development study, in: *Hominid Evolution: Past, Present and Future,* Liss, New York.

Bromage, T. G., 1987, The biological and chronological maturation of early hominids, *J. Hum. Evol.* **16**:257–272.

Bromage, T. G., and Dean, M. C., 1985, Re-evaluation of the age at death of immature fossil ape-men, *Nature* **317**:525–527.

Brower, J. C., 1976, *Promelocrinus* from the Wenlock at Dudley, *Palaeontology* **19**:651–680.

Brower, J. C., 1988, Ontogeny and phylogeny in primitive calceocrinid crinoids, *J. Paleontol.* **62**:917–934.

Brower, J. C., 1990, Ontogeny and phylogeny of the dorsal cup in calceocrinid crinoids, *J. Paleont.* **64**:300–318.

Brown, D. S., 1981, The English Upper Jurassic Plesiosauridea (Reptilia) and a review of the phylogeny and classification of the Plesiosauria, *Bull. Br. Mus. Nat. Hist. Geol.* **35**:253–347.

Brown, R. E., 1985, Hormone and paternal behaviour in vertebrates, *Am. Zool.* **25**:895–910.

Brown, V., and Davies, R. G., 1972, Allometric growth in two species of *Ectobius* (Dictyoptera: Blattidae), *J. Zool.* **166**:97–132.

Brues, A. M., 1977, *People and Races,* Macmillan Co., New York.

Bruns, T. D., Fogel, R., White, T. J., and Palmer, J. D., 1989, Accelerated evolution of a false-truffle from a mushroom ancestor, *Nature* **339**:140–142.

Bruton, M. (ed.), 1983, *Alternative Life History Styles of Animals,* Kluwer, Norwell, Mass.

Bryant, P. J., and Simpson, P., 1984, Intrinsic and extrinsic control of growth in developing organs, *Q. Rev. Biol.* **59**:387–415.

Brylski, P., and Hall, B. K., 1988, Ontogeny of a macroevolutionary phenotype: The external cheek pouches of geomyoid rodents, *Evolution* **42**:391–395.

Burghardt, G., and Bekoff, M. (eds.), 1978, *The Development of Behavior,* Garland, New York.

Burns, J., Birkbeck, J. A., and Roberts, D. F., 1975, Early fetal brain growth, *Hum. Biol.* **47**:511–522.

Buss, L. W., 1987, *The Evolution of Individuality,* Princeton University Press, Princeton, N.J.

Buss, L. W., 1988, Diversification and germ-line determination, *Paleobiology* **14**:313–321.

Butler, C. G., 1959, Queen substance, *Bee World* **40**:269–275.

Butler, C. G., 1967, Insect pheromones, *Biol. Rev.* **42**:42–87.

Byrne, R., and Whiten, A. (eds.), 1988, *Machiavellian Intelligence: Social Expertise and the Evolution of Intellect in Monkeys, Apes, and Humans,* Oxford University Press (Clarendon), Oxford.

Cadee, G. C., 1988, The use of size-frequency distribution in paleoecology, *Lethaia* **21**:289–290.

Calder, W., 1984, *Size, Function and Life History,* Harvard University Press, Cambridge, Mass.

Callomon, J. H., 1963, Sexual dimorphism in Jurassic ammonites, *Trans. Leicester Lit. Philos. Soc.* **57**:21–56.

Callomon, J. H., 1969, Dimorphism in Jurassic ammonites, in: *Sexual Dimorphism in Fossil Metazoa and Taxonomic Implications* (G. E. G. Westermann, ed.), International Union of Geological Science, Stuttgart.

Callomon, J. H., 1981, Dimorphism in ammonoids, in: *The Ammonoidea* (M. R. House and J. R. Senior, eds.), Systematic Association Special Volume 18, Academic Press, New York.

Calow, P., 1976, *Biological Machines: A Cybernetic Approach to Life,* Arnold, London.

Calow, P., and Berry, R. (eds.), 1989, *Evolution, Ecology, and Environmental Stress,* Academic Press, London.

Campbell, K. S. W., and Day, M. F. (eds.), 1987, *Rates of Evolution,* Allen & Unwin, London.

Campbell, K. S. W., and Marshall, C. R., 1987, Rates of evolution among Palaeozoic echinoderms, in: *Rates of Evolution* (K. S. W. Campbell and M. F. Day, eds.), Allen & Unwin, London.

Carpenter, G., and Cohen, S., 1979, Epidermal growth, *Annu. Rev. Biochem.* **48**:193–216.

Cassin, C., and Capuron, A., 1979, Buccal organogenesis in *Pleurodeles waltlii* Micah (Urodele, Amphibia), study by intrablastocoelic transplantation and *in vitro* culture, *J. Biol. Buccale* **7**:61–76.

Cavalier-Smith, T., 1978, Nuclear volume control by nucleoskeletal DNA, selection for cell volume and cell growth rate, and the solution of the DNA C-value paradox, *J. Cell Sci.* **34**:247–278.

Cavalier-Smith, T. (ed.), 1985, *The Evolution of Genome Size,* Wiley, New York.

Cavener, D. R., 1989, Evolution of developmental regulation, *Am. Nat.* **134**:459–473.

Chalfie, M., Horvitz, H., and Sulston, J., 1981, Mutations that lead to reiterations in the cell lineages of *C. elegans, Cell* **24**:59–69.

Chaline, J., Marchand, D., and Berge, C., 1986, L'évolution de l'homme: un model gradualiste ou ponctualiste? *Bull. Soc. R. Belge Anthropol. Prehist.* **97**:77–97.

Charlesworth, B., 1980, *Evolution in Age-structured Populations,* Cambridge University Press, Cambridge.

Charlesworth, B., Lande, R., and Slatkin, M., 1982, A neo-Darwinian commentary on macroevolution, *Evolution* **36**:474–498.

Chatterton, B. D. E., 1971, Taxonomy and ontogeny of Siluro-Devonian trilobites from near Yass, New South Wales, *Palaeontographica* (A) **137**:1–108.

Chatterton, B. D. E., and Speyer, S. E., 1989, Larval ecology, life history strategies, and patterns of extinction and survivorship among Ordovician trilobites, *Paleobiology* **15**:118–132.

Cherry, L. M., Case, S. M., and Wilson, A. C., 1978, Frog perspective on the morphological difference between humans and chimpanzees, *Science* **200**:209–211.

Cheverud, J., 1982, Relationships among ontogenetic, static, and evolutionary allometry, *Am. J. Phys. Anthropol.* **59**:139–149.

Cheverud, J. M., 1984, Quantitative genetics and developmental constraints on evolution by selection, *J. Theor. Biol.* **110**:155–171.

Cheverud, J. M., 1988, A comparison of genetic and phenotypic correlations, *Evolution* **42**:958–968.

Cisne, J. L., Molenock, J., and Rabe, B. D., 1980, Evolution in a cline: The trilobite *Triarthrus* along an Ordovician depth gradient, *Lethaia* **13**:47–59.

Cisne, J. L., Chandlee, G. O., Rabe, B. D., and Cohen, J. A., 1982, Clinal variation, episodic evolution, and possible parapatric speciation: the trilobite *Flexicalymene senaria* along an Ordovician depth gradient, *Lethaia* **15**:325–341.

Clark, A. M., 1976, Tropical epizoic echinoderms and their distribution, *Micronesica* **12**:111–117.

Clark, H. L., 1946, *The Echinoderm Fauna of Australia, Its Composition and Its Origin,* Carnegie Institution of Washington Publication 566, Washington, D.C.

Clarke, P. G. H., 1985, Neuronal death in the development of the vertebrate nervous system, *Trends Neurosci.* **8**:346–349.

Clarkson, E. N. K., 1971, On the early schizochroal eyes of *Ormathops* (Trilobita, Zelinskellinae), *Mem. Bur. Rech. Geol. Min.* **73**:51–63.

Clarkson, E. N. K., 1975, The evolution of the eye in trilobites, *Fossils Strata* **4**:7–31.

Clutton-Brock, T. H., and Harvey, P. H., 1979, Comparison and adaptation, *Proc. R. Soc. London Ser. B* **205**:547–565.

Clutton-Brock, T. H., Guiness, F. E., and Albon, S. D., 1982, *Red Deer, Behavior and Ecology of Two Sexes,* University of Chicago Press, Chicago.

Coates, A. G., and Oliver, W. A., 1973, Colonality in zoantharian corals, in: *Animal Colonies: Development and Structure through Time* (R. S. Boardman, A. H. Cheetham, and W. A. Oliver, eds.), Dowden, Hutchinson & Ross, Stroudsburg, Pa.

Cock, A. G., 1966, Genetical aspects of metrical growth and form in animals, *Q. Rev. Biol.* **41**:131–190.

Cohen, J., and Massey, B. D., 1983, Larvae and the origins of major phyla, *Biol. J. Linn. Soc.* **19**:321–328.

Coleman, S., Silberstein, G. B., and Daniel, C. W., 1988, Ductal morphogenesis in the mouse mammary gland: Evidence supporting a role for epidermal growth factor, *Dev. Biol.* **127**:304–315.

Conroy, G. C., and Vannier, M. W., 1987, Dental development of the Taung skull from computerized tomography, *Nature,* **329**:625–627.

Conway Morris, S., and Crompton, D. W. T., 1982, The origins and evolution of the Acanthocephala, *Biol. Rev.* **57**:85–115.

Conway Morris, S., and Fritz, W. H., 1984, *Lapworthella filigrana* n. sp. (incertae sedis) from the Lower Cambrian of the Cassiar Mountains, northern British Columbia, Canada, with comments on possible levels of competition in the early Cambrian, *Palaeont. Z.* **58**:197–209.

Conway Morris, S., and Harper, E., 1988, Genome size in conodonts (Chordata): Inferred variations during 270 million years, *Science* **241**:1230–1232.

Cooke, J., 1988, The early embryo and the formation of body pattern, *Am. Sci.* **76**:35–41.

Cope, E. D., 1874, On the homologies and origin of the types of molar teeth of Mammalia Educabilia, *J. Acad. Nat. Sci. Philad.* **8**:71–89.

Cowie, J., and McNamara, K. J., 1978, *Olenellus* (Trilobita) from the Lower Cambrian strata of north-west Scotland, *Palaeontology* **21**:615–634.

Creighton, G. K., and Strauss, R. E., 1986, Comparative patterns of growth and development in cricetine rodents and the evolution of ontogeny, *Evolution* **40**:94–106.

Crews, D., Diamond, M. A., Whittier, J., and Mason, R., 1985, Small male body size in garter snakes depends on testes, *Am. J. Physiol.* **249**:R62–R66.

Crick, F. H. C., 1970, Diffusion in embryogenesis, *Nature* **225**:420–422.

Crowl, T., and Covich, A., 1990, Predator-induced life history shifts in a freshwater snail, *Science* **247**:949–951.

Cumber, R. A., 1949, The biology of bumble-bees with special reference to the production of the worker caste, *Trans. R. Entomol. Soc. London* **100**:1–45.

Damsky, C. H., Knudsen, K. A., and Buck, C. A., 1984, Integral membrane glycoproteins in cell–cell and cell–substratum adhesion, in: *The Biology of Glycoproteins* (R. J. Ivatt, ed.), Plenum Press, New York.

Darragh, T. A., 1985, Molluscan biogeography and biostratigraphy of the Tertiary of southeastern Australia, *Alcheringa* **9**:83–116.

Darwin, C., 1845, *Journal of Researches into the natural history and geology of the countries visited during the voyage of H. M. S. Beagle round the world,* 2nd ed., Freeman, London.

Darwin, C., 1871, *The Descent of Man, and Selection in relation to Sex,* Murray, London.

Davenport, R., 1979, *An Outline of Animal Development,* Addison–Wesley, Reading, Mass.

David, B., 1989, Jeu en mosaïque des hétérochronies: variation et diversité chez les Pourtalesiidae (echinides abyssaux), *Geobios Mem. Spec.* **12**:115–131.

Davies, A. M., 1987, Molecular and cellular aspects of patterning sensory neurone connections in the vertebrate nervous system, *Development* **101**:185–208.

Davies, A. M., 1989, Intrinsic differences in the growth rate of early nerve fibres related to target distance, *Nature* **337**:553–555.

Davis, J. C., 1986, *Statistics and Data Analysis,* Wiley, New York.

Davis, S. J., 1981, The effects of temperature change and domestication on the body size of Late Pleistocene to Holocene mammals of Israel, *Paleobiology* **7**:101–114.

Dawkins, R., 1986, *The Blind Watchmaker,* Longman, Harlow.

Dean, M. C., and Wood, B. A., 1981, Developing pongid dentition and its use for ageing individual crania in comparative cross-sectional growth studies, *Folia Primatol.* **36**:111–127.

DeAngelis, D. L., and Waterhouse, J. C., 1987, Equilibrium and nonequilibrium concepts in ecological models, *Ecol. Monogr.* **57**:1–21.

DeAngelis, D., Post, M., and Travis, C., 1986, *Positive Feedback in Natural Systems,* Springer-Verlag, Berlin.

de Beer, G. R., 1930, *Embryology and Evolution,* Oxford University Press (Clarendon), Oxford.

de Beer, G., 1956, The evolution of ratites, *Bull. Br. Mus. Nat. Hist. Zool.* **4**:59–70.

de Beer, G., 1958, *Embryos and Ancestors,* Oxford University Press (Clarendon), Oxford.

DeChiara, T., Efstratiadis, A., and Roberson, E., 1990, A growth deficiency phenotype in heterozygous mice carrying an insulin-like growth factor II gene disrupted by targeting, *Science* **345**:78–80.

De Ghett, V. J., 1972, The behavioral and morphological development of the Mongolian gerbil., Ph.D. dissertation, Bowling Green State University.

Deleurance, E., 1946, Une regulation a base sensorielle peripherique. L'inhibition de la ponte des ouvrieres par la presence de la fondatrice des Polistes (Hymenopte–Vespidae), C. R. Acad. Sci. **223**:871–872.

Demment, M. W., and van Soest, P. J., 1985, A nutritional explanation for body-size patterns of ruminant and non-ruminant herbivores, Am. Nat. **125**:641–672.

de Queiroz, K., 1985, The ontogenetic method for determining character polarity and its relevance to phylogenetic systematics, Syst. Zool. **34**:280–299.

Deroux, G., 1960, Formation regulierede males murs, de taille et d'organisation larvaire chez en Eulamellibranche commensal (*Montacuta phascolionis* Dautz.), C. R. Acad. Sci. **250**:2264–2266.

Devreotes, P., 1989, *Dictyostelium discoideum*: A model system for cell–cell interactions in development, Science **245**:1054–1058.

Dial, K. P., and Marzluff, J. M., 1988, Are the smallest organisms the most diverse? Ecology **69**:1620–1624.

Diamond, J., 1984, "Normal" extinctions of isolated populations, in: *Extinctions* (M. H. Nitecki, ed.), University of Chicago Press, Chicago.

Dienske, H., 1986, A comparative approach to the question of why human infants develop so slowly, in: *Primate Ontogeny, Cognition and Social Behaviour* (J. G. Else and P. C. Lee, eds.), Cambridge University Press, Cambridge.

Doane, W. W., 1973, Role of hormones in insect development, in: *Developmental Systems: Insects,* Volume 2 (S. J. Counce and C. H. Waddington, eds.), Academic Press, New York.

Dobzhansky, T., 1942, Biological adaptation, Sci. Mon. **55**:391–402.

Dobzhansky, T., 1956, What is an adaptive trait? Am. Nat. **90**:337–347.

Dobzhansky, T., 1968, Adaptedness and fitness, in: *Population Biology and Evolution* (R. C. Lewontin, ed.), Syracuse University Press, Syracuse, N.Y.

Dobzhansky, T., 1969, *The Biology of Ultimate Concern,* Rapp & Whiting, London.

Dobzhansky, T., 1977, *Evolution,* Freeman, San Francisco.

Doherty, P., Mann, D. A., and Walsh, F. S., 1987, Cell–cell interactions modulate the responsiveness of PC12 cells to nerve growth factor, Development **101**:605–615.

Dommergues, J.-L., 1988, Can ribs and septa provide an alternative standard for age in ammonite ontogenetic studies? Lethaia **21**:243–256.

Dommergues, J.-L., and Meister, C., 1989, Trajectoires ontogénétiques et hétérochronies complexes chex des ammonites (Harpoceratinae) du Jurassique inférieur (Domerian), Geobios Mem. Spec. **12**:157–166.

Donovan, D. T., 1973, The influence of theoretical ideas on ammonite classification from Hyatt to Trueman, Univ. Kansas Paleontol. Contrib. **62**:1–16.

Doolittle, W. F., and Sapienza, C., 1980, Selfish genes, the phenotype paradigm and genome evolution, Nature **284**:601–603.

Driever, W., Thoma, G., and Nüsslein-Volhard, C., 1989, Determination of spatial domains of zygotic gene expression in the *Drosophila* embryo by the affinity of binding sites for the bicoid morphogen, Nature **340**:363–367.

Ducibella, T., 1974, The occurrence of biochemical metamorphic events without anatomical metamorphosis in the axolotl, Dev. Biol. **38**:175–186.

Duellman, W. E., and Trueb, L., 1986, *Biology of the Amphibians,* McGraw–Hill, New York.

Dunbar, R. I. M., 1982, Adaptation, fitness and the evolutionary tautology, in: *Current Problems in Sociobiology* (King's College Sociobiology Group, eds.), Cambridge University Press, Cambridge.

Ebenman, B., and Persson, L., 1988, Dynamics of size-structured populations: An overview, in: *Size-structured Populations* (B. Ebenman and L. Persson, eds.), Springer-Verlag, Berlin.

Ede, D. A., 1978, *An Introduction to Developmental Biology,* Wiley, New York.

Edelman, G., 1986, Evolution and morphogenesis: The regulator hypothesis, in: *Genetics, Development and Evolution* (J. P. Gustafson, G. L. Stebbins, and F. J. Ayala, eds.), Plenum Press, New York.

Edgecombe, G. D., and Chatterton, B. D. E., 1987, Heterochrony in the Silurian radiation of encrinurine trilobites, Lethaia **20**:337–351.

Edgecombe, G. D., Speyer, S. E., and Chatterton, B. D. E., 1988, Protaspid larvae and phylogenetics of encrinurid trilobites, J. Paleontol. **62**:779–799.

Edley, M. T., and Law, R., 1988, Evolution of life histories and yields in experimental populations of *Daphnia magna,* Biol. J. Linn. Soc. **34**:309–326.

Edmonds, H. W., and Sawin, P. B., 1936, Variation of the branches of the aortic arch in rabbits, *Am. Nat.* **70:**65–66.

Edmunds, G. F., and Edmunds, C. H., 1980, Predation, climate, and emergence and mating of mayflies, in: *Advances in Ephemeroptera* (J. F. Flannagan and K. E. Marshal, eds.), Plenum Press, New York.

Edwards, D., 1979, A late Silurian flora from the Lower Old Red sandstone of south-west Dyfed, *Palaeontology* **22:**23–52.

Edwards, D., Feehan, J., and Smith, D. G., 1983, A late Wenlock flora from Co. Tipperary, Ireland, *Bot. J. Linn. Soc.* **86:**19–36.

Eigenmann, J. E., Patterson, J., Zapf, J., and Froesch, E. R., 1984, Insulin-like growth factor I in the dog: A study in different dog breeds and in dogs with growth hormone elevation, *Acta Endocrinol.* **105:**294–301.

Ekstig, B., 1985, The regular condensation of developmental stages as a mechanism of the evolution of man and culture, *Evol. Theory* **7:**195–204.

Eldredge, N., 1977, Trilobites and evolutionary patterns, in: *Patterns of Evolution, as illustrated by the Fossil Record* (A. Hallam, ed.), Elsevier, Amsterdam.

Eldredge, N., and Cracraft, J., 1980, *Phylogenetic Patterns and the Evolutionary Process,* Columbia University Press, New York.

Eldredge, N., and Gould, S. J., 1972, Punctuated equilibria: An alternative method to phyletic gradualism, in: *Models in Paleobiology* (T. J. M. Schopf, ed.), Freeman, Cooper, San Francisco.

Emerson, S. B., 1986, Heterochrony and frogs: The relationship of a life history trait to morphological form, *Am. Nat.* **127:**167–183.

Emerson, S. B., 1988, Testing for historical patterns of change: A case study with frog pectoral girdles, *Paleobiology* **14:**174–186.

Emerson, S. B., Travis, J., and Blouin, M., 1988, Evaluating a hypothesis about heterochrony, *Evolution* **42:**68–78.

Emlet, R. B., McEdward, L. R., and Strathmann, R. R., 1986, Echinoderm larval ecology viewed from the egg, in: *Echinoderm Studies* (J. Lawrence and M. Jangoux, eds.), A. A. Balkema, Rotterdam.

Endler, J. A., 1986, *Natural Selection in the Wild,* Princeton University Press, Princeton, N.J.

Epstein, H. T., 1979, Correlated brain and intelligence development in humans, in: *Development and Evolution of Brain Size* (M. E. Hahn, C. Jensen, and B. C. Dudeck, eds.), Academic Press, New York.

Erwin, D. H., 1989, The End-Permian Mass extinction, *Trends in Ecology and Evolution* **4:**225–229.

Erwin, D. H., Valentine, J. W., and Sepkoski, J. J., 1987, A comparative study of diversification events: The early Paleozoic versus the Mesozoic, *Evolution* **41:**1177–1186.

Escobedo, J. A., and Williams, L. T., 1988. A PDGF receptor domain essential for mitogenesis but not for many other responses to PDGF, *Nature* **335:**85–87.

Etkin, W., 1968, Hormonal control of amphibian metamorphosis, in: *Metamorphosis, A Problem in Developmental Biology* (W. Etkin and L. I. Gilbert, eds.), Appleton–Century–Crofts, New York.

Ettensohn, C., 1990, Cell interactions in the sea urchin embryo, *Science* **248:**1115–1118.

Fallon, J. F., and Cameron, J., 1977, Interdigital cell death during limb development of the turtle with an interpretation of evolutionary significance, *J. Embryol. Exp. Morphol.* **40:**285–289.

Feder, J. L., Chilcore, C. A., and Bosh, G. L., 1988, Genetic differentiation between sympatric host races of the apple maggot fly and *Rhagoletis pomonella, Nature* **336:**61–64.

Fell, H. B., and Pawson, D. L., 1966, General biology of echinoderms, in: *Physiology of Echinodermata* (R. A. Boolootian, ed.), Interscience, New York.

Ferber, I., and Lawrence, J. M., 1976, Distribution, substratum preference and burrowing behaviour of *Lovenia elongata* (Gray) (Echinoidea: Spatangoida) in the Gulf of Elat ('Aqaba), Red Sea, *J. Exp. Mar. Biol. Ecol.* **22:**207–225.

Fernandez Castro, N., and Lucas, A., 1987, Variability of the frequency of male neoteny in *Ostrea puelchana* (Mollusca: Bivalvia), *Mar. Biol.* **96:**359–365.

Fink, W. L., 1982, The conceptual relationship between ontogeny and phylogeny, *Paleobiology* **8:**254–264.

Fink, W. L., 1988, Phylogenetic analysis and the detection of ontogenetic patterns, in: *Heterochrony in Evolution: A Multidisciplinary Approach* (M. L. McKinney, ed.), Plenum Press, New York.

Finlay, B. L., Wikler, K. C., and Sengelaub, D. R., 1987, Regressive events in brain development and scenarios for vertebrate brain evolution, *Brain Behav. Evol.* **30:**102–117.

Finney, S. C., 1986, Heterochrony, punctuated equilibrium, and graptolite zonal boundaries, in: *Palaeoecology and Biostratigraphy of Graptolites* (C. P. Hughes and R. B. Rickards, eds.), Geological Society Special Publication No. 20.

Forbes, T., and Lopez, G., 1989, Determination of critical periods in ontogenetic trajectories, *Funct. Ecol.* **3**:625–632.

Ford, C., and Beach, F., 1952, *Patterns of Sexual Behaviour*, Eyre & Spottiswoode, London.

Ford, E. B., 1940, Polymorphism and taxonomy, in: *The New Systematics* (J. Huxley, ed.), Oxford University Press (Clarendon), Oxford.

Ford, E. B., and Huxley, J. S., 1929, Genetic rate factors in *Gammarus*, *Arch. Entwicklungsmech. Org.* **117**:67–79.

Fortey, R. A., and Chatterton, B. D. E., 1988, Classification of the trilobite suborder Asaphina, *Palaeontology* **31**:165–222.

Foster, A. B., 1985, Contrasting evolutionary patterns in two coral reefs and their possible relationship to life history traits, *Geol. Soc. Am. Abstr. Prog.* **17**(7):585.

Foster, A. B., Johnson, K. G., and Schultz, L. L., 1988, Allometric shape change and heterochrony in the free living coral *Trachyphyllia bilobata* (Duncan), *Coral Reefs* **7**:37–44.

Foster, D. W., and Kaesler, R. L., 1988, Shape analysis from the Ostracoda, in: *Heterochrony in Evolution: A Multidisciplinary Approach* (M. L. McKinney, ed.), Plenum Press, New York.

Foster, S. A., 1986, On the adaptive value of large seeds for tropical forest trees: a review and synthesis, *Bot. Rev.* **52**:260–299.

Fox, R. F., 1988, *Energy and the Evolution of Life*, Freeman, New York.

Franklin, I. R., 1987, Population biology and evolutionary change, in: *Rates of Evolution* (K. S. W. Campbell and M. F. Day, eds.), Allen & Unwin, London.

Frazzetta, T., 1975, *Complex Adaptations in Evolving Populations*, Sinauer, Sunderland, Mass.

Freeman, G. L., 1982, What does the comparative study of development tell us about evolution? in: *Evolution and Development* (J. T. Bonner, ed.), Springer-Verlag, Berlin.

French, V., 1983, Development and evolution of the insect segment, in: *Development and Evolution* (B. C. Goodwin, N. J. Holder, and C. C. Wylie, eds.), Cambridge University Press, Cambridge.

Frest, T. J., Strimple, H. L., and McGinnis, M. R., 1979, Two new crinoids from the Ordovician of Virginia and Oklahoma, with notes on pinnulation in the Disparida, *J. Paleontol.* **53**:399–415.

Friedmann, H., 1929, *The Cowbirds*, Thomas, Springfield, Ill.

Frost, D. O., 1984, Axonal growth and target selection during development: Retinal projections to the ventrobasal complex and other 'nonvisual' structures in neonatal Syrian hamsters, *J. Comp. Neurol.* **230**:576–592.

Frost, D. O., So, K.-F., and Schneider, G. E., 1979, Postnatal development of retinal projections in Syrian hamsters: A study using autoradiographic and anterograde degeneration techniques, *Neuroscience* **4**:1649–1677.

Frost, G. T., 1987, How did big brains evolve? The role of neonatal body size, *Hum. Evol.* **2**:193–203.

Futuyma, D. J., 1986a, Evolution and coevolution in communities, in: *Patterns and Processes in the History of Life* (D. M. Raup and D. Jablonski, eds.), Springer-Verlag, Berlin.

Futuyma, D., 1986b, *Evolutionary Biology*, Sinauer, Sunderland, Mass.

Futuyma, D. J., 1988, *Sturm und Drang* and the evolutionary synthesis, *Evolution* **42**:217–226.

Garstang, W., 1922, The theory of recapitulation: A critical restatement of the biogenetic law, *J. Linn. Soc. Zool.* **35**:81–101.

Garstang, W., 1928, The morphology of the Tunicata, and its bearing on the phylogeny of the Chordata, *Q. J. Microsc. Sci.* **75**:51–187.

Garstang, W., 1929, The origin and evolution of larval forms, *Br. Assoc. Adv. Sci. Rep.* **1928**:77–98.

Garstang, W., 1946, The morphology and relations of the Siphonophora, *Q. J. Microsc. Sci.* **87**:103–193.

Garstang, W., 1951, *Larval Forms and Other Zoological Verses*, Blackwell, Oxford.

Gavan, J., 1953, Growth and development of the chimpanzee: A longitudinal and comparative study, *Hum. Biol.* **25**:93–143.

Geary, D., 1988, Heterochrony in gastropods: A paleontological view, in: *Heterochrony in Evolution: A Multidisciplinary Approach* (M. L. McKinney, ed.), Plenum Press, New York.

Gehlsen, K. R., Dillner, L., Engvall, E., and Ruoslahti, E., 1988, The human laminin receptor is a member of the integrin family of cell adhesion receptors, *Science* **241**:1228–1229.

Geist, V., 1971, *Mountain Sheep*, University of Chicago Press, Chicago.

Geist, V., 1986, The paradox of the great Irish stags, *Nat. Hist.* **95**:54–65.

Giard, A., 1887, La castration parasitaire et son influence sur les caractères extérieurs du sexe male chez les crustacés décapodes, *Bull. Sci. Dep. du Nord* **18**:1–28.

Gibbs, H. L., and Grant, P. R., 1987, Oscillating selection on Darwin's finches, *Nature* **327**:511–513.

Gibson, K. R., 1986, Cognition, brain size and the extraction of embedded food resources, in: *Primate Ontogeny, Cognition and Social Behaviour* (J. G. Else and P. C. Lee, eds.), Cambridge University Press, Cambridge.

Gilbert, L. I., 1974, Endocrine action during insect growth, *Recent Prog. Horm. Res.* **30**:347–390.

Gilbert, L. I., and King, D. S., 1973, Physiology of growth and development: Endocrine aspects, in: *The Physiology of Insecta,* 2nd ed. (M. Rockstein, ed.), Academic Press, New York.

Gilbert, S. F., 1985, *Developmental Biology,* Sinauer, Sunderland, Mass.

Gingerich, P., 1985, Species in the fossil record: Concepts, trends and transitions, *Paleobiology* **11**:27–41.

Gittleman, J. L., 1986, Carnivore life history patterns: Allometric, ecological and phylogenetic associations, *Am. Nat.* **217**:744–771.

Gliddon, C. J., and Gouyon, P.-H., 1989, The units of selection, *Trends Ecol. Evol.* **4**:204–208.

Goetinck, P. F., and Carlone, D. L., 1988, Altered proteoglycan synthesis disrupts feather pattern formation in chick embryonic skin, *Dev. Biol.* **127**:179–186.

Goss, R., 1972, *Regulation of Organ and Tissue Growth,* Academic Press, New York.

Gould, S. J., 1966, Allometry and size in ontogeny and phylogeny, *Biol. Rev.* **41**:587–680.

Gould, S. J., 1969, An evolutionary microcosm: Pleistocene and Recent history of the land snail P. (*Poecilozonites*) in Bermuda, *Bull. Mus. Comp. Zool.* **138**:407–532.

Gould, S. J., 1974, The evolutionary significance of 'bizarre' structures: Antler size and skull size in the 'Irish Elk', *Megaloceras gigantans, Evolution* **28**:191–220.

Gould, S. J., 1975, Allometry in primates, with emphasis on scaling and the evolution of the brain, *Contrib. Primatol.* **5**:244–292.

Gould, S. J., 1977, *Ontogeny and Phylogeny,* Harvard University Press, Cambridge, Mass.

Gould, S. J., 1979, A biological homage to Mickey Mouse, *Nat. Hist.* **88**(5):30–36.

Gould, S. J., 1980, The promise of paleobiology as a nomothetic, evolutionary discipline, *Paleobiology* **6**:96–118.

Gould, S. J., 1981, *The Mismeasure of Man,* Norton, New York.

Gould, S. J., 1982, Change in developmental timing as a mechanism of macroevolution, in: *Evolution and Development* (J. T. Bonner, ed.), Springer-Verlag, Berlin.

Gould, S. J., 1985, The paradox of the first tier: An agenda for paleobiology, *Paleobiology* **11**:2–12.

Gould, S. J., 1988a, The uses of heterochrony, in: *Heterochrony in Evolution: A Multidisciplinary Approach* (M. L. McKinney, ed.), Plenum Press, New York.

Gould, S. J., 1988b, Trends as change in variance: A new slant on progress and directionality in evolution, *J. Paleontol.* **62**:319–329.

Gould, S. J., 1989, A developmental constraint in *Cerion,* with comments on the definition and interpretation of constraint in evolution, *Evolution* **43**:516–539.

Gould, S. J., and Lewontin, R. C., 1979, The spandrels of San Marco and the Panglossian paradigm: A critique of the adaptionist programme, *Proc. R. Soc. London Ser. B* **205**:581–598.

Gould, S. J., and Vrba, E. S., 1982, Exaptation—a missing term in the science of form, *Paleobiology* **8**:4–15.

Gould, S. J., Gilinsky, N. L., and German, R. Z., 1987, Asymmetry of lineages and the direction of evolutionary time, *Science* **236**:1437–1441.

Goux, J. M., and Roubaud, P., 1978, Un concept central de la biologie: l'Adaptation, *La Pensee* **202**:49–65.

Grant, V., 1963, *The Origin of Adaptations,* Columbia University Press, New York.

Grant, V., 1985, *The Evolutionary Process,* Columbia University Press, New York.

Gray, R., 1988, Metaphors and methods: Behavioral ecology, panbiogeography, and the evolving synthesis, in: *Evolutionary Processes and Metaphors* (M. Ho and S. Fox, eds.), Wiley, New York.

Gregory, W. K., 1947, The monotremes and the palimpsest theory, *Bull. Am. Mus. Nat. Hist.* **88**:1–52.

Grime, J. P., 1989, The stress debate: symptom of impending synthesis?, in: *Evolution, Ecology, and Environmental Stress* (P. Calow and R. Berry, eds.), Academic Press, London.

Guerrant, E. O., 1982, Neotenic evolution of *Delphinium nudicaule* (Ranunculaceae): A hummingbird-pollinated larkspur, *Evolution* **36**:699–712.

Guerrant, E. O., 1988, Heterochrony in plants: The intersection of evolution, ecology and ontogeny, in: *Heterochrony in Evolution: A Multidisciplinary Approach* (M. L. McKinney, ed.), Plenum Press, New York.

Gurney, M. E., and Konishi, M., 1980, Hormone-induced sexual differentiation of brain and behavior in zebra finches, *Science* **208**:1380–1383.

Haeckel, E., 1866, *Generelle Morphologie der Organismen: Allgemeine Grundzüge der organischen Formen-Wis-*

senschaft, mechanisch begründet durch die von Charles Darwin reformirte Descendenz-Theorie, Reimer, Berlin.

Hafner, J. C., and Hafner, M. S., 1988, Heterochrony in rodents, in: *Heterochrony in Evolution: A Multidisciplinary Approach* (M. L. McKinney, ed.), Plenum Press, New York.

Haldane, J. B. S., 1932, *The Causes of Evolution,* Longmans, London.

Haldane, J. B. S., 1949, Suggestions as to quantitative measurements of rates of evolution, *Evolution* **3**:51–56.

Hall, B. K., 1975, Evolutionary consequences of skeletal differentiation, *Am. Zool.* **15**:329–350.

Hall, B. K., 1983, Epigenetic control in development and evolution, in: *Development and Evolution* (B. C. Goodwin, N. J. Holder, and C. C. Wylie, eds.), Cambridge University Press, Cambridge.

Hall, B. K., 1984a, Developmental processes underlying heterochrony as an evolutionary mechanism, *Can. J. Zool.* **62**:1–7.

Hall, B. K., 1984b, Developmental mechanisms underlying the formation of atavisms, *Biol. Rev.* **59**:89–124.

Hall, B. K., 1988, The embryonic development of bone, *Am. Sci.* **76**:174–181.

Hallam, A., 1989, Heterochrony as an alternative to species selection in the generation of phyletic trends, *Geobios Mem. Spec.* **12**:193–198.

Hallam, A., in press, Biotic and abiotic factors in the evolution of early Mesozoic marine molluscs, in: *The Causes of Evolution* (R. Ross and W. Allmon, eds.), University of Chicago Press, Chicago.

Hallas, S. E. A., 1988, The ontogeny of behaviour in *Portia fimbriata, P. labiata* and *P. schultzi,* web-building jumping spiders (Araneae: Salticidae), *J. Zool.* **215**:231–238.

Halle, L. J., 1977, *Out of Chaos,* Houghton Mifflin, Boston.

Hamburger, V., 1980, Embryology and the modern synthesis in evolutionary theory, in: *The Evolutionary Synthesis: Perspectives on the Unification of Biology* (E. Mayr and W. Provine, eds.), Harvard University Press, Cambridge, Mass.

Hanken, J., 1985, Morphological novelty in the limb skeleton accompanies miniaturization in salamanders, *Science* **229**:871–873.

Hanken, J., 1989, Development and evolution in amphibians, *Am. Sci.* **77**:336–343.

Hanken, J., and Hall, B. K., 1988, Skull development during anuran metamorphosis: 1. Early development of the first three bones to form—the exoccipital, the parasphenoid, and the frontoparietal, *J. Morphol.* **195**:247–256.

Haq, S. M., 1965, Development of the copepod *Euterpina acutifrons* with special reference to dimorphism in the male, *Proc. Zool. Soc. London* **144**:175–201.

Haq, S. M., 1972, Breeding of *Euterpina acutifrons,* a harpacticoid copepod, with special reference to dimorphic males, *Mar. Biol.* **15**:221–235.

Harris, R. N., 1987, Density-dependent paedomorphosis in the salamander *Notophthalmus viridescens dorsalis, Ecology* **68**:705–712.

Harrison, G., Tanner, J., Pilbeam, D., and Baker, P., 1988, *Human Biology,* Oxford University Press, Oxford.

Haselbacher, G. K., Schwab, M. E., Pasi, A., and Humbel, R. E., 1985, Insulin-like growth factor II (IGF-II) in human brain: Regional distribution of IGF-II and higher molecular mass forms, *Proc. Natl. Acad. Sci. USA* **82**:2153–2157.

Hata, R. I., and Slavkin, H. C., 1978, *De novo* induction of a gene product during heterologous epithelial–mesenchymal interactions *in vitro, Proc. Natl. Acad. Sci. USA* **75**:2790–2794.

Hayami, I., 1978, Notes on the rates and patterns of size change in evolution, *Paleobiology* **4**:252–260.

Heintz, A., and Garutt, V. E., 1965, Determination of the absolute age of the fossil remains of mammoth and wooly rhinoceros from the permafrost in Siberia by the help of radiocarbon (C$_{14}$), *Nor. Geol. Tidsskr.* **45**:73–79.

Heron, A. C., 1972, Population ecology of a colonizing species, *Oecologia* **10**:269–293.

Hersh, A. H., 1934, Evolutionary relative growth in the Titanotheres, *Am. Nat.* **68**:537–561.

Hewitt, R. A., and Westermann, G. E. G., 1987, Function of complexly fluted septa in ammonoid shells. II. Septal evolution and conclusions, *Neves Jahrb. Geol. Palaeontol. Abh.* **174**:135–169.

Hickey, D. R., 1987, Shell shape plasticity in Late Pennsylvanian myalinids (Bivalvia), *J. Paleontol.* **61**:290–311.

Hiernaux, J., 1977, Long-term biological effects of human migration from the African savana to the equatorial forest: A case study of human adaptation to a hot and wet climate, in: *Population Structure and Human Variation* (G. A. Harrison, ed.), Cambridge University Press, Cambridge.

Hinchliffe, J. R., and Johnson, D. R., 1980, *The Development of the Vertebrate Limb,* Oxford University Press (Clarendon), Oxford.

Hinegardner, R., 1976, Evolution of genome size, in: *Molecular Evolution* (F. J. Ayala, ed.), Sinauer, Sunderland, Mass.

Hintz, R. L., 1985, Control mechanisms of prenatal bone growth, in: *Normal and Abnormal Bone Growth* (A. Dixon and B. G. Sarnat, eds.), Liss, New York.

Hodgkin, J., 1986, Notes on developmental biology and genetics of the nematode, *C. elegans,* in: *Cellular and Molecular Aspects of Developmental Biology* (M. Fougereau and R. Stora, eds.), North-Holland, Amsterdam.

Hoffman, A., 1981, *Biological controls of the punctuated versus gradual mode of species evolution,* International Symposium on Concepts and Methods in Paleontology, Barcelona.

Hoffman, A., 1982a, Punctuated-equilibrium evolutionary model and paleoecology, *Soc. Geol. Pologne Ann.* **48**:327–331.

Hoffman, A., 1982b, Punctuated versus gradual mode of evolution—A reconsideration, *Evol. Biol.* **15**:411–436.

Holder, N., 1983, The vertebrate limb: Patterns and constraints in development and evolution, in: *Development and Evolution* (B. C. Goodwin, N. Holder, and C. C. Wylie, eds.), Cambridge University Press, Cambridge.

Holloway, R. L., 1979, Brain size, allometry, and reorganization: Towards a synthesis, in: *Development and Evolution of Brain Size* (M. E. Hahn, C. Jensen, and B. C. Dudek, eds.), Academic Press, New York.

Horner, H. A., and Macgregor, H. C., 1983, C-value and cell-volume: Their significance in the evolution and development of amphibians, *J. Cell Sci.* **63**:135–146.

Howe, R. W., 1967, Temperature effects on embryonic development in insects, *Annu. Rev. Entomol.* **12**:15–42.

Hughes, R. N., and Hughes, P. I., 1981, Morphological and behavioural aspects of feeding in the Cassidae (Tonnacea, Mesogastropoda), *Malacologia* **20**:385–402.

Hughes, T. P., 1984, Population dynamics based on individual size rather than age, *Am. Nat.* **123**:778–795.

Humphreys, W. F., and Lützen, J., 1972, Studies on parasitic gastropods from echinoderms. I. On the structure and biology of the parasitic gastropod *Megadenus cantharelloides* n. sp., with comparisons on *Paramegadenus* n.g., *Det. Kong. Danske Vid. Sels. Bial. Skr.* **19**:1–26.

Hutchinson, G. E., 1959, Homage to Santa Rosalia *or* why are there so many kinds of animals? *Am. Nat.* **93**:145–159.

Huxley, J. S., 1932, *Problems of Relative Growth,* Methuen, London.

Huxley, J. S., 1963, *Evolution, The Modern Synthesis,* Prentice–Hall, Englewood Cliffs, N.J.

Huxley, J. S., and Tessier, G., 1936, Terminology of relative growth, *Nature* **137**:780.

Hyatt, A., 1889, Genesis of the Arietidae, *Bull. Mus. Comp. Zool.* **16**:1–238.

Hyatt, A., 1893, Phylogeny of an acquired characteristic, *Proc. Am. Philos. Soc.* **32**:349–647.

Imbrie, J., 1956, Biometrical methods in the study of invertebrate fossils, *Bull. Am. Mus. Nat. Hist.* **108**:219–252.

Imms, A. D., 1957, *A General Textbook of Entomology,* 9th ed. (revised by O. W. Richards and R. G. Davies), Methuen & Co., London.

Irwin, R. E., 1988, The evolutionary importance of behavioural development: The ontogeny and phylogeny of bird song, *Anim. Behav.* **36**:814–824.

Ishikawa, A., and Namikawa, T., 1987, Postnatal growth and development in laboratory strains of large and small musk shrews (*Suncusmurinus*), *J. Mammal.* **68**:766–774.

Ismail, G., 1988, Conservation of the giant *Rafflesia* in Sabah, Malaysia, *Trends Ecol. Evol.* **3**:316–317.

Ivanov, A. N., 1975, Late ontogeny of ammonites and its peculiarities among micro-, macro- and megaconchs [In Russian], *Sb. Nauchn. Tr. Gos. Ped. Inst.* **142**:5–57.

Jaanusson, V., 1973, Morphological discontinuities in the evolution of graptolite colonies, in: *Animal Colonies: Their Development and Function through Time* (R. S. Boardman, A. H. Cheerham, and W. J. Oliver, eds.), Dowden, Hutchinson & Ross, Stroudsburg, Pa.

Jablonski, D., 1982, Evolutionary rates and modes in Late Cretaceous gastropods: Role of larvae ecology, *Third North American Paleontological Convention, 1982, Proceedings* **1**:257–262.

Jablonski, D., and Bottjer, D. J., in press, Onshore–offshore trends in marine invertebrate evolution, in: *The Causes of Evolution* (R. M. Ross and W. D. Allmon, eds.), University of Chicago Press, Chicago.

Jablonski, D., Gould, S. J., and Raup, D., 1986, The nature of the fossil record: A biological perspective, in: *Patterns and Processes in the History of Life* (D. M. Raup and D. Jablonski, eds.), Springer-Verlag, Berlin.

Jackson, J. B. C., 1979, Morphological strategies of sessile organisms, in: *Biology and Systematics of Colonial Organisms* (G. Larwood and B. R. Rosen, eds.), Systematics Association Special Volume 11.

Jackson, R. T., 1890, Phylogeny of the Pelecypoda: The Aviculidae and their allies, *Mem. Boston Soc. Nat. Hist.* **4**(8):277–400.

Jackson, R. T., 1912, Phylogeny of the Echini, with a review of Palaeozoic species, *Mem. Boston Soc. Nat. Hist.* **7**:1–443.

Jacob, F., 1977, Evolution and tinkering, *Science* **196**:1161–1166.

Jacobson, A. G., 1966, Inductive processes in embryonic development, *Science* **152**:25–34.

James, H. F., and Olsen, S. L., 1983, Flightless birds, *Nat. Hist.* **92**:30–40.

Jamieson, I. G., 1989, Behavioral heterochrony and the evolution of birds' helping at the nest: An unselected consequence of communal breeding? *Am. Nat.* **133**:394–406.

Jarvik, E., 1980, *Basic Structure and Evolution of Vertebrates,* Vol. 1, Academic Press, London.

Jeffries, R. P. S., 1986, *The Ancestry of the Vertebrates,* British Museum (Natural History), London.

Jell, P. A., 1978, Trilobite respiration and genal caeca, *Alcheringa* **2**:251–260.

Jenner, C. E., and McCrary, A. B., 1967, Sexual dimorphism in erycinacean bivalves, *Rep. Am. Malacol. Union* **35**:43.

Jerison, H. J., 1973, *The Evolution of the Brain and Intelligence,* Academic Press, New York.

Johnson, A. L. A., 1981, Detection of ecophenotypic variation in fossils and its application to a Jurassic scallop, *Lethaia* **14**:277–285.

Jolicoeur, P., 1989, A simplified model for bivariate complex allometry, *J. Theoret. Biol.* **140**:41–49.

Jolicoeur, P., and Pirlot, P., 1988, Asymptotic growth and complex allometry of the brain and body in the white rat, *Growth Dev. Aging* **52**:3–9.

Jolly, A., 1964, Prosimians' manipulation of simple object problems, *Anim. Behav.* **12**:560–570.

Jones, B., 1987, Biostatistics in paleontology, *Geosci. Can.* **15**:3–22.

Jones, D. S., 1988, Sclerochronology and the size versus age problem, in: *Heterochrony in Evolution: A Multidisciplinary Approach* (M. L. McKinney, ed.), Plenum Press, New York.

Jones, G. P., 1987, Competitive interactions among adults and juveniles in a coral reef fish, *Ecology* **68**:1534–1547.

Kardami, E., Spector, D., and Strohman, R. C., 1988, Heparin inhibits skeletal muscle growth *in vitro, Dev. Biol.* **126**:19–28.

Katz, M. J., 1980, Allometry formula: A cellular model, *Growth* **44**:89–96.

Katz, M. J., 1987, Is evolution random? in: *Development as an Evolutionary Process* (R. A. Raff and E. Raff, eds.), Liss, New York.

Kauffman, S. A., 1983, Developmental constraints: Internal factors in evolution, in: *Development and Evolution* (B. C. Goodwin, N. Holder, and C. C. Wylie, eds.), Cambridge University Press, Cambridge.

Kauffman, S. A., 1990, *Origins of Order: Self-organization and Selection in Evolution,* Oxford University Press, Oxford.

Kear, J., 1962, Food selection in finches, *Proc. Zool. Soc. London* **138**:163–204.

Kember, N. F., 1978, Cell kinetics and the control of growth in long bones, *Cell Tissue Kinet.* **11**:477–485.

Kemp, R. B., and Hinchliffe, J. R. (eds.), 1984, *Matrices and Cell Differentiation,* Liss, New York.

Kennedy, W. J., and Wright, C. W., 1985, Evolutionary patterns in Late Cretaceous ammonites, *Spec. Pap. Palaeontol.* **33**:131–143.

Kerfoot, W. C., and Sih, A. (eds.), 1987, *Predation—Direct and Indirect Impacts on Aquatic Communities,* University Press of New England, Hanover, N.H.

Kieny, M., and Pautou, M. P., 1976, Experimental analysis of excedentary regulation in xenoplastic quail–chick limb bud recombinants, *Wilhelm Roux Arch. Entwicklungsmech. Org.* **179**:327–338.

King, A. P., and West, M. J., 1977, Species identification in the North American cowbird: Appropriate responses to abnormal song, *Science* **195**:1002–1004.

King, M. C., and Wilson, A. C., 1975, Evolution at two levels in humans and chimpanzees, *Science* **188**:107–116.

Kitchell, J. A., Clark, D. L., and Gombos, A. M., 1986, Biological selectivity of extinction: A link between background and mass extinction, *Palaios* **1**:504–511.

Kluge, A. G., and Strauss, R. E., 1985, Ontogeny and systematics, *Annu. Rev. Ecol. Syst.* **16**:247–268.

Koch, P. L., 1986, Clinal variation in mammals: Implications for the study of chronoclines, *Paleobiology* **12**:269–281.

Koestler, A., 1969, Beyond atomism and holism—The concept of the holon, in: *Beyond Reductionism* (A. Koestler and J. R. Smythies, eds.), Hutchinson, London.

Kollar, E. J., and Fisher, C., 1980, Tooth induction in chick epithelium: Expression of quiescent genes for enamel synthesis, *Science* **207**:993–995.

Kollman, J., 1885, Das Ueberwintern von europäischen Frosch- und Tritonlarven und die Umwandlung des mexikanischen Axolotl, *Verh. Naturforsch. Ges. Basel* **7**:387–398.

Kollman, J., 1905, Neue Gedanken uber das alter Problem von der Abstammung des Menschen, *Corresp. Bl. Dtsch. Ges. Anthropol. Ethnol. Urges* **36**:9–20.

Konishi, M., 1965, The role of auditory feedback on the control of vocalization in the white-crowned sparrow, *Z. Tierpsychol.* **22**:77–78.

Konishi, M., and Akutagawa, E., 1985, Neuronal growth atrophy and death in a sexually dimorphic song nucleus in the zebra finch brain, *Nature* **315**:145–147.

Koops, W. J., 1986, Multiphase growth curve analysis, *Growth* **50**:169–177.

Korsching, S., and Thoenen, H., 1988, Developmental changes of nerve growth factor levels in sympathetic ganglia and their target organs, *Dev. Biol.* **126**:40–46.

Kowalski, C. J., and Guire, K. E., 1974, Longitudinal data analysis, *Growth* **38**:131–169.

Krauss, F., 1988, An empirical evaluation of the use of the ontogeny polarization criterion in phylogenetic inference, *Syst. Zool.* **37**:106–141.

Krimbas, C. B., 1984, On adaptation, neo-Darwinian tautology, and population fitness, *Evol. Biol.* **17**:1–57.

Kurtén, B., 1959, On the bears of the Holsteinian Interglacial, *Stockholm Contrib. Geol.* **2**:73–102.

Kurtén, B., 1968, *Pleistocene Mammals of Europe,* Weidenfield & Nicholson, London.

LaBarbera, M., 1986, The evolution and ecology of body size, in: *Patterns and Processes in the History of Life* (D. M. Raup and D. Jablonski, eds.), Springer-Verlag, Berlin.

LaBarbera, M., 1989, Analyzing body size as a factor in ecology and evolution, *Annu. Rev. Ecol. and Syst.* **20**:97–117.

Laird, A. K., 1965, Dynamics of relative growth, *Growth* **29**:249–263.

Laitman, J. T., and Heimbuch, R. C., 1982, The basicranium of Plio-Pleistocene hominids as an indicator of their upper respiratory systems, *Am. J. Phys. Anthropol.* **59**:323–343.

Lalli, C. M., and Conover, R. J., 1973, Reproduction and development of *Paedoclione doliiformis,* and a comparison with *Clione limacina* (Opisthobranchia: Gymnosomata), *Mar. Biol.* **19**:13–22.

Lamendella, J. T., 1976, Relations between ontogeny and phylogeny of language: A neo-recapitulationist view, *Ann. N. Y. Acad. Sci.* **280**:396–412.

Lampitt, R. S., 1990, Directly measured rapid growth of a deep-sea barnacle, *Nature* **345**:805–807.

Lande, R., 1978, Evolutionary mechanisms of limb loss in tetrapods, *Evolution* **32**:73–92.

Lande, R., 1979, Quantitative genetic analysis of multivariate evolution, applied to brain:body allometry, *Evolution* **33**:402–416.

Lande, R., 1980, Genetic variation and phenotypic evolution during allopatric speciation, *Am. Nat.* **116**:463–478.

Landman, N. H., 1988, Heterochrony in ammonites, in: *Heterochrony in Evolution: A Multidisciplinary Approach* (M. L. McKinney, ed.), Plenum Press, New York.

Lane, N. G., and Sevastopulo, G. D., 1982, Microcrinoids from the Middle Pennsylvanian of Indiana, *J. Paleontol.* **56**:103–115.

Lane, N. G., Sevastopulo, G. D., and Strimple, H. L., 1985, *Amphipsalidocrinus:* A monocyclic camerate microcrinoid, *J. Paleontol.* **59**:79–84.

Larson, A., 1980, Paedomorphosis in relation to rates of morphological and molecular evolution in the salamander *Aneides flavipunctatus* (Amphibia, Plethodontidae), *Evolution* **34**:1–17.

Lauder, J. M., 1978, Effects of early hypo- and hyper thyroidism on development of rat cerebellar cortex. IV. The parallel fibers, *Brain Res.* **142**:25–39.

Lavandier, P., 1988, Semivoltinisme dans des populations de haute montagne de *Baetis alpinus* Pictet (Ephemeroptera), *Bull. Soc. Hist. Nat. Toulouse* **124**:61–64.

Lawton, J. H., and Warren, P. H., 1988, Static and dynamic explanations for patterns in food webs, *Trends Ecol. Evol.* **3**:242–245.

Lawton, M. F., and Lawton, R. O., 1986, Heterochrony, deferred breeding and avian sociality, in: *Current Ornithology,* Volume 3 (R. F. Johnston, ed.), Plenum Press, New York.

Leamy, L., 1988, Genetic and maternal influences on brain and body size in randombred house mice, *Evolution* **42**:42–53.

Lee, D. E., 1980, Cenozoic and Recent rhynchonellide brachiopods of New Zealand: Systematics and variation in the genus *Tegulorhynchia, J. R. Soc. N.Z.* **10**:223–245.

Lerner, I. M., 1954, *Genetic Homeostasis,* Oliver & Boyd, Edinburgh.

Leutenegger, W., and Larson, S., 1985, Sexual development of the postcranial skeleton of New World monkeys, *Folio. Primatol.* **44**:82–95.

Levinton, J. S., 1983, Stasis in progress: The empirical basis of macroevolution, *Annu. Rev. Ecol. Syst.* **14**:103–137.

Levinton, J., 1988, *Genetics, Paleontology and Macroevolution,* Cambridge University Press, Cambridge.

Levinton, J. S., and Simon, C. M., 1980, A critique of the punctuated equilibria model and implications for the detection of speciation in the fossil record, *Syst. Zool.* **29**:130–142.

Lewin, R., 1987, The earliest "humans" were more like apes, *Science* **236**:1061–1063.

Lewin, R., 1988a, Life history patterns emerge in primate study, *Science* **242**:1636–1637.

Lewin, R., 1988b, Living in the fast track makes for small brains, *Science* **242**:513–514.

Lewis, P. D., Patel, A. J., Johnson, A. L., and Balazs, R., 1976, Effect of thyroid deficiency in the postnatal rat brain: A quantitative histological study, *Brain Res.* **104**:49–62.

Lewontin, R. C., 1957, The adaptations of populations to varying environments, *Cold Spring Harbor Symp. Quant. Biol.* **22**:395–408.

Lewontin, R. C., 1977, Adattamento, in: *Enciclopedia del Novecento,* Volume 1, Instituto dell'Enciclopedia Italiana.

Lewontin, R. C., 1978, Adaptations, *Sci. Am.* **239**(Sept.):156–169.

Lidgard, S., 1986, Ontogeny in animal colonies: A persistent trend in the bryozoan fossil record, *Science* **232**:230–232.

Lieberman, P., 1984, *The Biology and Evolution of Language,* Harvard University Press, Cambridge, Mass.

Lindberg, D. R., 1988, Heterochrony in gastropods, in: *Heterochrony in Evolution: A Multidisciplinary Approach* (M. L. McKinney, ed.), Plenum Press, New York.

Lobb, R., Sasse, J., Sullivan, R., Shing, Y., D'Amore, P., Jacobs, J., and Klagsburn, M., 1986, Purification and characterization of heparin-binding endothelial growth factors, *J. Biol. Chem.* **261**:1924–1928.

Loch, A. R., and McLaren, I. A., 1970, The effect of varying and constant temperature on the size of a marine copepod, *Limnol. Oceanogr.* **15**:638–640.

Loher, W., 1961, The chemical acceleration of the maturation process and its hormonal control in the male of the desert locust, *Proc. R. Soc. London Ser. B* **153**:380–397.

Lomolino, M., 1985, Body size of mammals on islands, *Am. Nat.* **125**:310–316.

Long, J. A., 1990, Heterochrony and the origin of tetrapods, *Lethaia* **23**:157–166.

Lord, E. M., 1984, Cleistogamy: A comparative study of intraspecific floral variation, in: *Contemporary Problems in Plant Anatomy* (R. A. White and W. C. Dickson, eds.), Academic Press, New York.

Lorenz, K. Z., 1965, *Evolution and Modification of Behavior,* University of Chicago Press, Chicago.

Lorenz, K. Z., 1970, *Studies on Animal and Human Behavior,* Harvard University Press, Cambridge, Mass.

Ludvigsen, R., and Chatterton, B. D. E., 1980, The ontogeny of *Failleana* and the origin of the Bumastinae (Trilobita), *Geol. Mag.* **117**:471–478.

Lützen, J., 1968, Unisexuality in the parasitic family Entoconchidae (Gastropoda: Prosobranchia), *Malacologia* **7**:7–15.

Lützen, J., 1972, Studies on parasitic gastropods from echinoderms. II. *Stilifer* Broderip, *Dtsch. Kong. Danske Videns. Sels. Biol. Skr.* **19**:1–18.

Lützen, J., 1976, On a new genus and two new species of Prosobranchia (Mollusca), parasitic on the tropical sea urchin *Echinometra mathaei, Isr. J. Zool.* **25**:38–51.

MacArthur, R. H., and Wilson, E. O., 1967, *The Theory of Island Biogeography,* Princeton University Press, Princeton, N.J.

MacFadden, B. J., 1986, Fossil horses from "Eohippus" (*Hyracotherium*) to *Equus*: Scaling, Cope's law, and the evolution of body size, *Paleobiology* **12**:355–369.

McGowan, C., 1984, Evolutionary relationships of ratites and carinates: Evidence from ontogeny of the tarsus, *Nature* **307**:733–735.

McGowan, C., 1986, A putative ancestor for the sword-like ichthyosaur *Eurhinosaurus, Nature* **322**:454–456.

Mackie, E. J., Tucker, R. P., Halfter, W., Chiquet-Ehrisman, R., and Epperlein, H. H., 1988, The distribution of tenascin coincides with pathways of neural crest cell migration, *Development* **102**:237–250.

McKinney, M. L., 1984, Allometry and heterochrony in an Eocene echinoid lineage: Morphological change as a byproduct of size selection, *Paleobiology* **10**:407–419.

McKinney, M. L., 1986, Ecological causation of heterochrony: A test and implications for evolutionary theory, *Paleobiology* **12**:282–289.

McKinney, M. L., 1987, Taxonomic selectivity and continuous variation in mass and background extinctions, *Nature* **325**:143–145.

McKinney, M. L., 1988a, Classifying heterochrony: Allometry, size and time, in: *Heterochrony in Evolution: A Multidisciplinary Approach* (M. L. McKinney, ed.), Plenum Press, New York.

McKinney, M. L., 1988b, Heterochrony in evolution: An overview, in: *Heterochrony in Evolution: A Multidisciplinary Approach* (M. L. McKinney, ed.), Plenum Press, New York.

McKinney, M. L., 1990, Trends in body-size evolution, in: *Evolutionary Trends* (K. J. McNamara, ed.), Belhaven Press, London.

McKinney, M. L., and Oyen, C. W., 1989, Causation and nonrandomness in biological and geological time series: Temperature as a proximal control of extinction and diversity, *Palaios* **4**:3–15.

McKinney, M. L., and Schoch, R. M., 1985, Titanothere allometry, heterochrony, and biomechanics: Revising an evolutionary classic, *Evolution* **39**:1352–1363.

McKinney, M. L., McNamara, K. J., and Zachos, L. G., 1990, Heterochronic hierarchies: Application and theory in evolution, *Hist. Biol.* **3**:269–287.

McLean, J. H., 1984, A case for the derivation of the Fissurellidae from the Bellerophontacea, *Malacologia* **25**:635–655.

Maclean, N., and Hall, B. K., 1987, *Cell Commitment and Differentiation,* Cambridge University Press, Cambridge.

McMahon, T. A., and Bonner, J. T., 1983, *On Size and Life,* Freeman, San Francisco.

McNamara, K. J., 1978, Paedomorphosis in Scottish olenellid trilobites (early Cambrian), *Palaeontology* **21**:635–655.

McNamara, K. J., 1981, Paedomorphosis in Middle Cambrian xystridurine trilobites from northern Australia, *Alcheringa* **5**:209–224.

McNamara, K. J., 1982a, Taxonomy and evolution of living species of *Breynia* (Echinoidea: Spatangoida) from Australia, *Rec. West. Aust. Mus.* **10**:167–197.

McNamara, K. J., 1982b, Heterochrony and phylogenetic trends, *Paleobiology* **8**:130–142.

McNamara, K. J., 1983a, Progenesis in trilobites, in: *Trilobites and Other Early Arthropoda: Papers in Honour of Professor H. B. Whittington, FRS* (D. E. G. Briggs and P. D. Lane, eds.), Special Papers in Palaeontology 30.

McNamara, K. J., 1983b, The earliest *Tegulorhynchia* (Brachiopoda: Rhynchonellida) and its evolutionary significance, *J. Paleontol.* **57**:461–473.

McNamara, K. J., 1985a, Taxonomy and evolution of the Cainozoic spatangoid echinoid *Protenaster*, *Palaeontology* **28**:311–330.

McNamara, K. J., 1985b, A new micromorph ammonite genus from the Albian of South Australia, *Spec. Publ. S. Aust. Dep. Mines Energy* **5**:263–268.

McNamara, K. J., 1986a, A guide to the nomenclature of heterochrony, *J. Paleontol.* **60**:4–13.

McNamara, K. J., 1986b, The role of heterochrony in the evolution of Cambrian trilobites, *Biol. Rev.* **61**:121–156.

McNamara, K. J., 1987a, Plate translocation in spatangoid echinoids: Its morphological, functional and phylogenetic significance, *Paleobiology* **13**:312–325.

McNamara, K. J., 1987b, Taxonomy, evolution, and functional morphology of southern Australian Tertiary hemiasterid echinoids, *Palaeontology* **30**:319–352.

McNamara, K. J., 1988a, Heterochrony and the evolution of echinoids, in: *Echinoderm Phylogeny and Evolutionary Biology* (C. R. C. Paul and A. B. Smith, eds.), Oxford University Press, Oxford.

McNamara, K. J., 1988b, The abundance of heterochrony in the fossil record, in: *Heterochrony in Evolution: A Multidisciplinary Approach* (M. L. McKinney, ed.), Plenum Press, New York.

McNamara, K. J., 1989, The role of heterochrony in the evolution of spatangoid echinoids, *Geobios Mem. Spec.* **12**:283–295.

McNamara, K. J., 1990 (ed.), *Evolutionary Trends,* Belhaven Press, London.

McNamara, K. J., and Philip, G. M., 1980a, Australian Tertiary schizasterid echinoids, *Alcheringa* **4**:47–65.

McNamara, K. J., and Philip, G. M., 1980b, Living Australian schizasterid echinoids, *Proc. Linn. Soc. N. S. W.* **104**:127–146.

McNamara, K. J., and Philip, G. M., 1984, A revision of the spatangoid echinoid *Pericosmus* for the Tertiary of Australia. *Rec. West. Aust. Mus.* **11**:319–356.

McPherson, B. A., Smith, D. C., and Berlocher, S. H., 1988, Genetic differences between host races of *Rhagoletis pomonella*, *Nature* **336**:64–66.

Maddison, W. P., Donoghue, M. J., and Maddison, D. R., 1984, Outgroup analysis and parsimony, *Syst. Zool.* **33:**83–103.

Maderson, P. F. A., 1975, Embryonic tissue interactions as the basis for morphological change, *Am. Zool.* **15:**315–328.

Maderson, P. F. A., Alberch, P., Goodwin, B. C., Gould, S. J., Hoffman, A., Murray, J. D., Raup, D. M., de Rieglès, A., Wagner, G. P., and Wake, D. B. 1982, The role of development in macroevolutionary change, in: *Evolution and Development* (J. T. Bonner, ed.), Springer-Verlag, Berlin.

Madge, P. E., 1956, The ecology of *Oncopera fasciculata* (Walker). II. The influence of temperature and moisture on speeds of development and survival rate of the eggs, *Aust. J. Zool.* **4:**327–345.

Maglio, V. J., 1972, Evolution of mastication in the Elephantidae, *Evolution* **26:**638–658.

Majima, R., 1985, Intraspecific variation in three species of *Glossaulax* (Gastropoda, Naticidae) from the Late Cenozoic strata in central and southwest Japan, *Trans. Proc. Palaeontol. Soc. Jpn.* (N.S.) **138:**111–137.

Makowski, H., 1963, Problems of sexual dimorphism in ammonites, *Palaeontol. Pol.* **12:**1–92.

Makowski, H., 1971, Some remarks on the ontogenetic development and sexual dimorphism in the Ammonoidea, *Acta Geol. Pol.* **21:**321–340.

Malmgren, B. A., Berggren, W. A., and Lohmann, G. P., 1983, Evidence for punctuated gradualism in the Late Neogene *Globorotalia tumida* lineage of planktonic foraminifera, *Paleobiology* **9:**377–389.

Mancini, E. A., 1978, Origin of micromorph faunas in the geologic record, *J. Paleontol.* **52:**311–322.

Mandelbrot, B., 1977, *The Fractal Geometry of Nature,* Freeman, San Francisco.

Manley-Buser, K. A., 1986, A heterochronic study of the human foot, *Am. J. Phys. Anthropol.* **69:**235.

Manly, B. F. J., 1985, *The Statistics of Natural Selection,* Chapman & Hall, London.

Mann, A. E., 1975, *Some Paleodemographic Aspects of the South African Australopithecine,* University of Pennsylvania Publications in Anthropology No. 1, Philadelphia.

Mann, M. D., Glickman, S. E., and Towe, A. L., 1988, Brain/body relations among myomorph rodents, *Brain Behav. Evol.* **31:**111–124.

Manton, S. M., and Anderson, D. T., 1979, Polyphyly and the evolution of arthropods, in: *The Origin of Major Invertebrate Groups* (M. R. House, ed.), Academic Press, New York.

Marcotte, B. M., 1983, The imperatives of copepod diversity: Perception, cognition, competition and predation, in: *Crustacean Phylogeny* (F. R. Schram, ed.), Balkema, Rotterdam.

Mares, M., 1983, Desert rodent adaptation and community structure, *Great Basin Nat. Mem.* **7:**30–43.

Margulis, L., and Sagan, D., 1986, *Origins of Sex,* Yale University Press, New Haven, Conn.

Martin, P. S., and Klein, R. G. (eds.), 1984, *Quaternary Extinctions: A Prehistoric Revolution,* University of Arizona Press, Tucson.

Martin, R. A., 1986, Energy, ecology, and cotton rat evolution, *Paleobiology* **12:**370–382.

Martin, R. D., 1983, *Human Brain Evolution in an Ecological Context. 52nd James Arthur Lecture on the Evolution of the Human Brain,* American Museum of Natural History, New York.

Marx, J. L., 1989, The cell cycle coming under control, *Science* **245:**252–255.

Masoud, I. M., Moses, A. C., and Shapiro, F. D., 1985, Skeletal growth in the normal rabbit: A longitudinal study of serum somatomedin-C and skeletal development, in: *Normal and Abnormal Growth* (A. D. Dixon and B. G. Sarnat, eds.), Liss, New York.

Masterton, R., 1976, *Evolution of Brain and Behaviour in Vertebrates,* Wiley, New York.

Matsuda, R., 1979, Abnormal metamorphosis and arthropod evolution, in: *Arthropod Phylogeny* (A. P. Gupta, ed.), Van Nostrand–Reinhold, Princeton, N.J.

Matsuda, R., 1987, *Animal Evolution in Changing Environments,* Wiley, New York.

Mattfeldt, T., and Mall, G., 1987, Statistical methods for growth allometric studies, *Growth* **51:**86–102.

Matyja, B. A., 1986, Developmental polymorphism in Oxfordian ammonites, *Acta Geol. Pol.* **36:**37–68.

May, R. M., 1978, The dynamics and diversity of insect faunas, in: *Diversity in Insect Faunas* (L. A. Mound and N. Waloff, eds.), Blackwell, Oxford.

Maynard Smith, J., 1972, *On Evolution,* Edinburgh University Press, Edinburgh.

Maynard Smith, J., Burian, J., Kauffman, S., Alberch, P., Campbell, J., Goodwin, B., Lande, R., Raup, D., and Wolpert, L., 1985, Developmental constraints and evolution, *Q. Rev. Biol.* **60:**265–287.

Mayr, E., 1963, *Animal Species and Evolution,* Harvard University Press, Cambridge, Mass.

Mayr, E., 1970, *Populations, Species, and Evolution,* Harvard University Press, Cambridge, Mass.

Mayr, E., 1982, Adaptation and selection, *Biol. Zentralbl.* **101:**161–174.

Mayr, E., 1983, How to carry out the adaptionist program? *Am. Nat.* **121:**324–334.

Meckel, J. F., 1821, *System der vergleichenden Anatomie,* 7 volumes, Rengerschen Buchhandlung, Halle.

Medawar, P. B., 1951, Problems of adaptation, *New Biol.* **11**:10–26.

Medawar, P. B., 1960, *The Future of Man,* Basic Books, New York.

Mercola, M., and Stiles, C. D., 1988, Growth factor superfamilies and mammalian embryogenesis, *Development* **102**:451–460.

Mercola, M., Melton, D. A., and Stiles, C. D., 1988, Platelet-derived growth factor A chain is maternally encoded in *Xenopus* embryos, *Science* **241**:1223–1225.

Merimee, T. J., and Rimoin, D. L., 1986, Growth hormone and insulin-like growth factors in the Western pygmy, in: *African Pygmies* (L. L. Cavalli-Sforza, ed.), Academic Press, New York.

Merimee, T. J., Zapf, J., and Froesch, E. R., 1982, Insulin-like growth factors (IGFs) in pygmies and subjects with the pygmy trait: Characterization of the metabolic actions of IGFI and IGFII in man, *J. Clin. Endocrinol. Metab.* **55**:1081–1088.

Meyer, A., 1987, Phenotypic plasticity and heterochrony in *Cichlasoma managuense* (Pisces, Cichlidae) and their implications for speciation in cichlid fishes, *Evolution* **41**:1357–1369.

Michaelson, J., 1987, Cell selection in development, *Biol. Rev.* **62**:115–139.

Mickevich, M. F., 1981, Quantitative phyletic biogeography, in: *Advances in Cladistics* (V. A. Funk and D. R. Brooks, eds.), Proceedings of the First Meeting of the Willi Hennig Society, New York Botanical Gardens, Bronx.

Mickevich, M. F., 1982, Transformation series analysis, *Syst. Zool.* **31**:461–478.

Millendorf, S. A., 1979, The functional morphology and life habits of the Devonian blastoid *Eleutherocrinus cassedayi* Shumard and Yandell, *J. Paleontol.* **53**:553–561.

Miller, S. E., and Hadfield, M. G., 1990, Developmental arrest during larval life and lifespan extension in a marine mollusc, *Science* **248**:356–358.

Milner, A., 1982, Late extinctions of amphibians, *Nature* **338**:117.

Milner, A. R., 1989, Small temnospondyl amphibians from the Middle Pennsylvanian of Illinois, *Palaeontology* **25**:125–141.

Minelli, A., and Bortoletto, S., 1988, Myriapod metamerism and arthropod segmentation, *Biol. J. Linn. Soc.* **33**:323–343.

Mishler, B. D., 1986, Ontogeny and phylogeny in *Tortula* (Musci: Pottiaceae), *Syst. Bot.* **11**:189–208.

Mitchell, P. J., and Tjian, R., 1989, Transcriptional regulation in mammalian cells by sequence-specific DNA binding proteins, *Science* **245**:371–378.

Miyazaki, J. M., and Mickevich, M. F., 1982, Evolution of *Chesapecten* (Mollusca: Bivalvia, Miocene–Pliocene) and the Biogenetic Law, *Evol. Biol.* **15**:369–409.

Moeur, J. E., and Istock, C. A., 1980, Ecology and evolution of the pitcher-plant mosquito. IV. Larval influence over adult reproductive performance and longevity, *J. Anim. Ecol.* **49**:775–792.

Montagu, M. F. A., 1981, *Growing Young,* McGraw–Hill, New York.

Moore, J. A., 1942, The role of temperature in speciation of frogs, *Biol. Symp.* **6**:189–213.

Morel, L., Van der Meer, R. K., and Lavine, B. K., 1988, Ontogeny of nestmate recognition cues in the red carpenter ant (*Camponotus floridanus*), *Behav. Ecol. Sociobiol.* **22**:175–183.

Morescalchi, A., 1979, New developments in vertebrate cytotaxonomy. I. Cytotaxonomy of the amphibians, *Genetica* **50**:179–193.

Morescalchi, A., and Serra, V., 1974, DNA renaturation kinetics in some paedogenetic urodeles, *Experientia* **30**:487–489.

Morgan, J. A., and Tyler, A., 1938, The relation between the entrance point of the spermatozoan and bilaterality of the egg in *Chaetopterus, Biol. Bull.* **74**:401–402.

Morriconi, E., and Calvo, J., 1983, *Diferentes modalidades reproductivas en Ostrea puelchana de dos localidades cercanas (Golfo San Matias, Rio Negro, Argentina),* VIII Simposio Latinoamericano sobre Oceanografia Biologica, Montevideo, Uraguay: Asociacion Latinoamericano de Investigacion en Ciencas del Mar.

Mortensen, T., 1933, Papers from Dr. Th. Mortensen's Pacific Expedition 1914–16—Biological observations on ophiurids, *Vidensk. Medd. Dan. Naturh. Forenist.* **93**:171–194.

Mortensen, T., 1936, Echinoidea and Ophiuroidea, *Discovery Rep.* **12**:199–348.

Morton, B., 1972, Some aspects of the functional morphology and biology of *Pseudopythina subsinuata* (Bivalvia: Leptonacea) commensal on stomatopod crustaceans, *J. Zool.* **166**:79–96.

Morton, B., 1981, The biology and functional morphology of *Chlamydoconcha orcutti* with a discussion on the taxonomic status of the Chlamydoconchacea (Mollusca: Bivalvia), *J. Zool.* **195**:81–121.

Moss, M. L., and Moss-Salentijn, L., 1983, Vertebrate cartilages, in: *Cartilage,* Volume 1 (B. K. Hall, ed.), Academic Press, New York.

Moss, M. L., Moss-Salentijn, L., Vilmann, H., and Newell-Morriss, L., 1982, Neuro-skeletal topology of the primate basicranium: Its implications for the "fetalization hypothesis," *Gegenbaurs Morphol. Jahrb. Leipzig* **128**:58–67.

Moss-Salentijn, L., 1974, Studies on long bone growth. 1. Determination of differential elongation in paired growth plates, *Acta Anat.* **90**:145–160.

Murray, J., 1981, Prepattern formation mechanism for animal coat patterns, *J. Theor. Biol.* **88**:161–199.

Needham, J., 1959, *A History of Embryology,* Cambridge University Press, Cambridge.

Nelson, G. J., 1978, Ontogeny, phylogeny, paleontology, and the biogenetic law, *Syst. Zool.* **27**:324–345.

Nesis, K. N., 1977, Population structure in the squid *Sthenoteuthis oyalaniensis* (Lesson, 1930) (Ommastrephidae) in the western tropical Pacific [in Russian], *Proc. Inst. Oceanol. Acad. Sci. USSR* **10**:15–29.

Nesis, K. N., and Nigmatullin, C. M., 1977, The distribution and biology of the genus *Ornithoteuthis* Okada, 1927, and *Hyaloteuthis* Gray, 1849 (Cephalopoda, Oegopsida) [in Russian], *Biul. Mosk. Ob. Isp. Prir. Otd. Biol.* **84**:50–63.

Newell, N. D., 1949, Phyletic size increase, an important trend illustrated by fossil invertebrates, *Evolution* **3**:103–124.

Newman, R. A., 1988, Adaptive plasticity in development of *Scaphiopus couchii* tadpoles in desert ponds, *Evolution* **42**:774–783.

Nieto-Sampedro, M., 1988, Astrocyte mitogen inhibitor related to epidermal growth factor receptor, *Science* **240**:1784–1786.

Nilsson, A., Isgaard, J., Lindhal, A., Dahlstrom, A., Skottner, A., and Isaksson, O. G. P., 1986, Regulation by growth hormone of number of chondrocytes containing IGF-I in rat growth plate, *Science* **233**:571–574.

Noble, J. P. A., and Lee, D.-J., 1990, Ontogenies and astogenies and their significance in some favositid heliolitid corals, *J. Paleontol.* **64**:515–523.

Noble, M., Murray, K., Stroobant, P., Waterfield, M. D., and Riddle, P., 1988, Platelet-derived growth factor promotes division and motility and inhibits premature differentiation of the oligodendrocyte/type-2 astrocyte progenitor cell, *Nature* **333**:560–562.

Norris, M. J., 1954, Sexual maturation in the desert locust (*Schistocerca gregaria* (Forsk.)), *Anti-Locust Bulletin No. 18.*

Norris, M. J., and Pener, M. P., 1965, An inhibitory effect of allatectomized males and females on the sexual maturation of young adult males of *Schistocerca gregaria* (Forsk.) (Orthoptera: Acrididae), *Nature* **208**:1122.

Nottebohm, F., 1978, The "critical period" for song learning, in: *Critical Periods* (J. P. Scott, ed.), Dowden, Hutchinson & Ross, Stroudsburg, Pa.

Ockelman, K. W., 1964, *Turtonia minuta* (Fabricus), a neotenous veneracean bivalve, *Ophelia* **1**:121–146.

Oeldorf, E., Nishioka, M., and Bachmann, K., 1978, Nuclear DNA amounts and developmental rate in holarctic anura, *Z. Zool. Syst. Evolutionsforsch.* **16**:216–224.

O Foighil, D., 1985, Form, function, and origin of temporary dwarf males in *Pseudopythina rugifera* (Carpenter, 1864) (Bivalvia: Galeommatacea), *Veliger* **27**:245–252.

O'Grady, R. T., 1985, Ontogenetic sequences and the phylogenetics of parasitic flatworm life cycles, *Cladistics* **1**:159–170.

Ohno, S., 1980, So much "junk" DNA in our genome, *Brookhaven Symp. Biol.* **23**:366–370.

Oken, L., 1847, *Elements of Physiophilosophy* (tr. Alfred Tulk), Ray Society, London.

Olempska, E., 1989, Gradual evolutionary transformations of ontogeny in an Ordovician ostracod lineage, *Lethaia* **22**:159–168.

Olmo, E., 1983, Nucleotype and cell size in vertebrates: A review, *Basic Appl. Histochem.* **27**:227–256.

O'Neill, R. V., DeAngelis, D. L., Waide, J. B., and Allen, T. F. H., 1986, *A Hierarchical Concept of Ecosystems,* Princeton University Press, Princeton, N.J.

Ordway, E., 1965, Caste differentiation in *Augochlorella* (Hymenoptera, Halictidae), *Insectes Soc.* **12**:291–308.

Orgel, L. E., and Crick, F. H. C., 1980, Selfish DNA: The ultimate parasite, *Nature* **284**:604–607.

Oster, G., and Alberch, P., 1982, Evolution and bifurcation of developmental programs, *Evolution* **36**:444–459.

Oster, G. F., Shubin, N., Murray, J. D., and Alberch, P., 1988, Evolution and morphogenetic rules: The shape of the vertebrate limb in ontogeny and phylogeny, *Evolution* **42**:862–884.

Pachut, J. F., 1989, Heritability and intraspecific heterochrony in Ordovician bryozoans from environments differing in diversity, *J. Paleontol.* **63**:182–194.

Pachut, J. F., and Anstey, R. L., 1979, A developmental explanation of stability–diversity–variation hypotheses: Morphogenetic regulation in Ordovician bryozoan colonies, *Paleobiology* **5**:168–187.

Pagel, M. D., and Harvey, P. H., 1988, How mammals produce large-brained offspring, *Evolution* **42**:948–957.

Pagel, M. D., and Harvey, P. H., 1989, Taxonomic differences in the scaling of brain on body weight among mammals, *Science* **244**:1589–1591.

Paigen, K., 1980, Temporal genes and other developmental regulators in mammals, in: *The Molecular Genetics of Development* (T. Leighton and W. Loomis, eds.), Academic Press, New York.

Paigen, K., 1989, Experimental approaches to the study of regulatory evolution, *Am. Nat.* **134**:440–458.

Pain, J., 1961, Sur la pheromone des reines d'abeilles et ses effects physiologiques, *Ann. Abeille* **4**:73–152.

Palmiter, R. D., Brinster, R. L., Hammer, R. E., Trumbauer, M. E., Rosenfeld, M. G., Birnberg, N. C., and Evans, R. M., 1982, Dramatic growth of mice that develop from eggs microinjected with metallothionein-growth hormone fusion genes, *Nature* **300**:611–615.

Pandolfi, J. M., 1988, Heterochrony in colonial animals, in: *Heterochrony in Evolution: A Multidisciplinary Approach* (M. L. McKinney, ed.), Plenum Press, New York.

Parker, S. T., 1976, A comparative longitudinal study of the sensorimotor development in a macaque, a gorilla, and a human infant from a Piagetian perspective, Paper presented at the Animal Behaviour Society Conference, Boulder.

Parker, S. T., 1977a, Piaget's sensorimotor series in an infant macaque: The organisation of non-stereotyped behaviour in the evolution of intelligence, Ph.D. thesis, University of California, Berkeley.

Parker, S. T., 1977b, Piaget's sensorimotor period series in an infant macaque: A model for comparing unstereotyped behavior and intelligence in human and non-human primates., in: *Primate Biosocial Development* (S. Chevalier-Skolnikoff and F. E. Poirer, eds.), Garland Press, New York.

Parker, S. T., and Gibson, K. R., 1979, A developmental model for the evolution of language and intelligence in early hominids, *J. Hum. Evol.* **2**:367–408.

Parks, A. L., Parr, B. A., Chin, J.-E., Leaf, D. S., and Raff, R. A., 1988, Molecular analysis of heterochronic changes in the evolution of direct developing sea urchins, *J. Evol. Biol.* **1**:27–44.

Parsons, P. A., 1989, Environmental stresses and conservation of natural populations, *Annu. Rev. Ecol. Syst.* **20**:29–49.

Pashley, D. P., 1988, Quantitative genetics, development, and physiological adaptation in host strains of fall armyworm, *Evolution* **42**:93–102.

Passingham, R. E., 1975, The brain and intelligence, *Brain Behav. Evol.* **11**:1–15.

Paul, C. R. C., 1985, The adequacy of the fossil record reconsidered, *Spec. Pap. Palaeontol.* **33**:7–15.

Pavlov, A. P., 1901, Le Crétace inférieur de la Russie et sa faune, *Nouv. Mem. Soc. Imp. Nat. Moscow* **16**:87.

Perris, R., von Boxberg, Y., and Löfberg, J., 1988, Local embryonic matrices determine region-specific phenotypes in neural crest cells, *Science* **241**:86–89.

Peters, R. H., 1983, *The Ecological Implications of Body Size,* Cambridge University Press, Cambridge.

Peters, W. L., and Peters, J. G., 1988, The secret swarm, *Nat. Hist.* **97**(5):8–12.

Pflugfelder, O., 1937, Bau, Entwicklung und Funktion der Corpora allata und cardiaca von *Dixippus morosus* Br., *Z. Wiss. Zool.* **149**:477–512.

Piaget, J., 1952, *The Origins of Intelligence in Children,* Norton, New York.

Piaget, J., 1970, *Genetic Epistemology,* Columbia University Press, New York.

Piantadosi, S., 1987, Generalizing growth functions assuming parameter heterogeneity, *Growth* **51**:50–63.

Pilbeam, D. R., and Gould, S. J., 1974, Size and scaling in human evolution, *Science* **186**:892–901.

Pimm, S. L., Jones, H. L., and Diamond, J., 1988, On the risk of extinction, *Am. Nat.* **132**:757–785.

Plotnick, R. E., 1989, Application of bootstrap methods to reduced major axis line fitting, *Syst. Zool.* **38**:144–153.

Poethig, S., 1988, A non-cell-autonomous mutation regulating juvenility in maize, *Nature* **336**:82–83.

Pojeta, J. 1971. Review of Ordovician pelecypods. *U.S. Geol. Surv. Prof. Pap.* **695**:1–46.

Portmann, A., 1941, Die Tragzeiten der Primaten und die Dauer der Schwangerschaft beim Menschen: ein Problem der vergleichenden Biologie, *Rev. Suisse Zool.* **48**:511–518.

Primack, R. B., 1987, Relationships among flowers, fruits and seeds, *Annu. Rev. Ecol. Syst.* **18**:409–430.

Prothero, D. R., and Sereno, P. C., 1982, Allometry and paleoecology of medial Miocene dwarf rhinoceroses from the Texas Gulf Coastal Plain, *Paleobiology* **8**:16–30.

Prothero, J., and Jurgens, K., 1988, Scaling of maximal lifespan in mammals, in: *Evolution of Longevity in Animals* (A. Woodhead and K. Thompson, eds.), Plenum Press, New York.

Prout, T., and McChesney, F., 1985, Competition among immatures affects their adult fertility: Population dynamics, *Am. Nat.* **126**:521–558.

Purves, D., 1988, *Body and Brain*, Harvard University Press, Cambridge, Mass.

Rachootin, S. P., and Thompson, K. S., 1981, Epigenetics, paleontology and evolution, in: *Evolution Today* (G. G. E. Scudder and J. L. Revel, eds.), Proceedings of the Second International Congress of Systematics and Evolutionary Biology.

Raff, M. C., Lillien, L. E., Richardson, W. D., Burne, J. F., and Noble, M. D., 1988, Platelet-derived growth factor from astrocytes drives the clock that times oligodendrocyte development in culture, *Nature* **333**:562–565.

Raff, R. A., 1987, Constraint, flexibility, and phylogenetic history in the evolution of direct development in sea urchins, *Dev. Biol.* **119**:6–19.

Raff, R. A., 1989, Review of a theory of the evolution of development, *Nature* **337**:518.

Raff, R. A., and Kaufman, T. C., 1983, *Embryos, Genes, and Evolution,* Macmillan Co., New York.

Raff, R. A., Anstrom, J. A., En Chin, J., Field, K. G., Ghiselin, M. T., Lane, D. J., Olsen, G. J., Pace, N. R., Parks, A. L., and Raff, E. C., 1987, Molecular and developmental correlates of macroevolution, in: *Development as an Evolutionary Process* (R. Raff and E. Raff, eds.), Liss, New York.

Rakic, P., 1974, Neurons in rhesus monkey visual cortex: Systematic relation between time of origin and eventual disposition, *Science* **183**:425–427.

Rakic, P., and Riley, K. P., 1983, Overproduction and elimination of retinal axons in the fetal rhesus monkey, *Science* **219**:1441–1444.

Rall, L. B., Scott, J., Bell, G. I., Crawford, R. A., Penschow, J. D., Niall, M. D., and Coghlan, J. P., 1985, Mouse prepro-epidermal growth factor synthesis by the kidney and other tissues, *Nature* **313**:228–230.

Ramsköld, L., 1988, Heterochrony in Silurian phacopid trilobites as suggested by the ontogeny of *Acernaspis, Lethaia* **21**:307–318.

Raup, D. M., and Gould, S. J., 1974, Stochastic simulation and evolution of morphology—Towards a nomothetic paleontology, *Syst. Zool.* **23**:305–322.

Rayner, A. D. M., 1988, Life in a collective: Lessons from the fungi, *New Sci.* **120**(1639):49–53.

Rayner, A. D. M., Boddy, L., and Dowson, C. G., 1987, Genetic interactions and developmental versatility during establishment of decomposer basidiomycetes in wood and tree litter, in: *Ecology of Microbial Communities* (M. Fletcher, T. R. G. Gray, and J. G. Jones, eds.), Cambridge University Press, Cambridge.

Reilly, S. M., and Lauder, G. V., 1988, Atavisms and the homology of hyobranchial elements in lower vertebrates, *J. Morphol.* **195**:237–245.

Reiss, J. O., 1989, The meaning of developmental time: Attempt at a metric for comparative embryology, *Am. Nat.* **134**:170–189.

Rendel, J. M., 1968, Canalisation and gene control, Logos Press, London.

Rendel, J. M., 1979, Canalisation and selection, in: *Quantitative Genetic Variation* (J. N. Thompson and J. M. Thoday, eds.), Academic Press, New York.

Riedl, R., 1978, *Order in Living Organisms* (transl. by R. P. S. Jefferies), Wiley, New York.

Rieppel, O., 1985, Ontogeny and the hierarchy of types, *Cladistics* **1**:234–246.

Riska, B., 1986, Some models for development, growth, and morphometric correlation, *Evolution* **40**:1303–1311.

Riska, B., 1989, Composite traits, selection response, and evolution, *Evolution* **43**:1172–1191.

Riska, B., and Atchley, W. R., 1985, Genetics of growth predict patterns of brain size evolution, *Science* **229**:668–671.

Roberts, A. B., Anzano, M. A., Wakefield, L. M., Roche, N. S., Stern, D. F., and Sporn, M. B., 1985, Type B transforming growth factor: A bifunctional regulator of cell growth, *Proc. Natl. Acad. Sci. USA* **82**:119–123.

Robertson, R., 1985, Archaeogastropod biology and the systematics of the genus *Tricolia* (Trachacea: Tricoliidae) in the Indo-West-Pacific, *Monogr. Mar. Mollusca* **31**:1–103.

Roisin, Y., 1988, Morphology, development and evolutionary significance of the working stages in the caste system of *Prorhinotermes* (Insecta, Isoptera), *Zoomorphology* **107**:339–347.

Rose, K. D., and Bown, T. M., 1984, Gradual phyletic evolution at the generic level in early Eocene omomyid primates, *Nature* **309**:250–252.

Ross, C., 1988, The intrinsic rate of natural increase and reproductive effort in primates, *J. Zool.* **214**:199–211.

Ross, R., and Allmon, W. (eds.), 1990, *The Causes of Evolution,* University of Chicago Press, Chicago.

Rothwell, G. W., 1987, The role of development in plant phylogeny: A paleobotanical perspective, *Rev. Palaeobot. Palynol.* **50**:97–114.

Roughgarden, J., 1979, *Theory of Population Genetics and Evolutionary Ecology: An Introduction,* Macmillan Co., New York.

Rozengurt, E., 1986, Early signals in the mitogenic response, *Science* **234**:161–166.

Runnegar, B., and Bentley, C., 1983, Anatomy, ecology, and affinities of the Australian early Cambrian bivalve *Pojetaia runnegari* Jell, *J. Paleontol.* **57**:73–92.

Ruse, M., 1971, Functional statements in biology, *Philos. Sci.* **38**:87–95.

Ruse, M., 1972, Discussion: Biological adaptation, *Philos. Sci.* **39**:525–528.

Russell, E. S., 1916, *Form and Function: A Contribution to the History of Animal Morphology,* John Murray, London.

Ruvkun, G., and Giusto, J., 1989, The *Chaenorhabdites elegans* heterochronic gene lin-14 encodes a nuclear protein that forms a temporal developmental switch, *Nature* **338**:313–319.

Sadler, P. M., 1981, Sediment accumulation rates and the completeness of stratigraphic sections, *J. Geol.* **89**:569–584.

Sagan, C., 1980, *Cosmos,* Random House, New York.

Sage, R. D., and Selander, R. K., 1975, Trophic radiation through polymorphism in cichlid fishes, *Proc. Natl. Acad. Sci. USA* **72**:4669–4673.

Sakagami, S. F., 1954, Occurrence of an aggressive behaviour on queenless hives, with considerations on the social organisation of honeybee, *Insectes Soc.* **1**:331–343.

Sakagami, S. F., Beig, D., Zucchi, R., and Akahira, Y., 1963, Occurrence of ovary-developed workers in queenright colonies of stingless bees, *Rev. bras. Biol.* **23**:115–129.

Salthe, S. N., 1985, *Evolving Hierarchical Systems: Their Structure and Representation,* Columbia University Press, New York.

Sander, K., 1982, Oogenesis and embryonic pattern formation; known and missing links, in: *Ontogeny and Phylogeny* (B. Goodwin and R. Whittle, eds.), Cambridge University Press, Cambridge.

Sauer, J. R., and Slade, N. A., 1987, Uinta ground squirrel demography: Is body mass a better categorical variable than age? *Ecology* **68**:642–650.

Sawin, P. B., and Edmonds, H. W., 1949, Morphogenetic studies of the rabbit. VII. Aortic arch variations in relation to regionally specific growth differences, *Anat. Rec.* **105**:377–396.

Scarpini, E., Ross, A. H., Rosen, J. L., Brown, M. J., Rostami, A., Koprowski, H., and Lisak, R. P., 1988, Expression of nerve growth factor receptor during human peripheral nerve development, *Dev. Biol.* **125**:301–310.

Schindel, D. E., 1980, Microstratigraphic sampling and the limits of paleontologic resolution, *Paleobiology* **6**:408–426.

Schindel, D. E., 1982, Resolution analysis: A new approach to the gaps in the fossil record, *Paleobiology* **8**:340–353.

Schindewolf, O., 1929, Ontogenie und phylogenie, *Palaeontol. Z.* **10**:54–67.

Schmalhausen, I. I., 1949, *Factors of Evolution: The Theory of Stabilizing Selection,* McGraw–Hill (Blakiston), New York.

Schmidt, H., 1909, *Das biogenetische Grundgesetz Erust Haeckels und seine Gegner,* Neuer Frankfurter Verlag, Frankfurt.

Schoch, R., 1986, *Phylogeny Reconstruction in Paleontology,* Van Nostrand, Princeton, N.J.

Schoener, T. W., 1967, The ecological significance of sexual dimorphism in size in the lizard *Anolis conspersus, Science* **155**:474–477.

Schoener, T. W., 1983, Field experiments on interspecific competition, *Am. Nat.* **122**:240–285.

Schopf, T. J. M., Raup, D. M., Gould, S. J., and Simberloff, D. S., 1975, Genomic versus morphologic rates of evolution: Influence of morphologic complexity, *Paleobiology* **1**:63–70.

Schultz, A. H., 1969, *The Life of Primates,* Universe Books, New York.

Schwaner, T. D., 1985, Population structure of black tiger snakes, *Notechis ater niger,* on offshore islands of South Australia, in: *Biology of Australasian Frogs and Reptiles* (G. Grigg, R. Shine, and H. Ehmann, eds.), Royal Zoological Society of New South Wales.

Schwartz, G. G., and Rosenblum, L. A., 1981, Allometry of primate hair density and the evolution of human hairlessness, *Am. J. Phys. Anthropol.* **55**:7–12.

Scott, G. H., 1982, Tempo and stratigraphic record of speciation in *Globorotalia puncticulata, J. Foraminiferal Res.* **12**:1–12.

Scott, J. P., 1978, Critical periods for the development of social behaviour in dogs, in: *Critical Periods* (J. P. Scott, ed.), Dowden, Hutchinson & Ross, Stroudsburg, Pa.

Seed, J., Olwin, B. B., and Hauschka, S. D., 1988, Fibroblast growth factor levels in the whole embryo and limb bud during chick development, *Dev. Biol.* **128**:50–57.

Selander, R. K., 1966, Sexual dimorphism and differential niche utilization in birds, *Condor* **68**:113–151.

Semlitsch, R. D., and Gibbons, J. W., 1985, Phenotypic variation in metamorphosis and paedomorphosis in the salamander *Ambystoma talpoideum, Ecology* **66**:1123–1130.

Semlitsch, R. D., Scott, D. E., and Pechmann, J. H. K., 1988, Time and size at metamorphosis related to adult fitness in *Ambystoma talpoideum, Ecology* **69**:184–192.

Sepkoski, J. J., 1988, Alpha, beta, or gamma: Where does all the diversity go? *Paleobiology* **14**:221–234.

Service, E. R., 1971, *Primitive Social Organization: An Evolutionary Perspective,* Random House, New York.

Sessions, S. K., and Larson, A., 1987, Developmental correlates of genome size in plethodontid salamanders and their implications for genome evolution, *Evolution* **41**:1239–1251.

Shapovalov, L., and Taft, A. C., 1954, The life histories of the steelhead rainbow trout (*Salmo gairdneri gairdneri*) and silver salmon (*Oncorhynchus kisutch*) with special reference to Waddell Creek, California, and recommendations regarding their management, *Calif. Fish Bull.* **98**:1–375.

Shaw, A. B., 1964, *Time in Stratigraphy,* McGraw–Hill, New York.

Shea, B. T., 1983, Allometry and heterochrony in the African apes, *Am. J. Phys. Anthropol.* **62**:275–289.

Shea, B. T., 1985a, Bivariate and multivariate growth allometry: Statistical and biological considerations, *J. Zool.* **206**:367–390.

Shea, B. T., 1985b, Ontogenetic allometry and scaling: A discussion based on the growth and form of the skull in African apes, in: *Size and Scaling in Primate Biology* (W. L. Jungers, ed.), Plenum Press, New York.

Shea, B. T., 1988, Heterochrony in primates, in: *Heterochrony in Evolution: A Multidisciplinary Approach* (M. L. McKinney, ed.), Plenum Press, New York.

Shea, B. T., 1989, Heterochrony in human evolution: the case for neoteny reconsidered, *Yearbook of Phys. Anthro.* **32**:69–101.

Shea, B. T., 1990, Neoteny, in: *The Cambridge Encyclopedia of Human Evolution* (J. S. Jones, R. D. Martin, and D. Pilbeam, eds.), Cambridge University Press, Cambridge.

Shea, B. T., in press, The developmental control of skeletal growth allometries: Evidence from giant transgenic (MT-rGH) and dwarf mutant (dw/dw) mice, *Evolution.*

Shea, B. T., and Pagezy, H., 1988, Allometric analyses of body form in Central African pygmies, *Am. J. Phys. Anthropol.* **75**:269–270.

Shea, B. T., Hammer, R. E., Brinster, R. L., and Ravosa, M. J., in press, Relative growth of the skull and postcranium in giant transgenic mice, *Genet. Res.*

Shine, R., 1978, Growth rates and sexual maturation in six species of Australian elapid snakes, *Herpetologica* **34**:73–79.

Shine, R., 1986, Sexual differences in morphology and niche utilization in an aquatic snake, *Acrochordus arafurae, Oecologia* **69**:260–267.

Shine, R., and Crews, D., 1988, Why male garter snakes have small heads: The evolution and endocrine control of sexual dimorphism, *Evolution* **42**:1105–1110.

Shubin, N., and Alberch, P., 1986, A morphogenetic approach to the origin and basic organization of the tetrapod limb, *Evol. Biol.* **20**:319–387.

Sih, A., 1987, Predators and prey lifestyles: An evolutionary and ecological overview, in: *Predation–Direct and Indirect Impacts on Aquatic Communities* (W. C. Kerfoot and A. Sih, eds.), University Press of New England, Hanover, N.H.

Silvertown, J., 1989, The paradox of seed size and adaptation, *Tr. Ecol. Evol.* **4**:24–26.

Simms, M. J., 1986, Contrasting life styles in Lower Jurassic crinoids: A comparison of benthic and pseudopelagic Isocrinida, *Paleontology* **29**:475–493.

Simms, M. J., 1988, The role of heterochrony in the evolution of post-Palaeozoic crinoids, in: *Echinoderm Biology* (R. D. Burke, P. V. Mladenov, P. Lambert, and R. L. Parsley, eds.), Balkema, Rotterdam.

Simon, H. A., 1962, The architecture of complexity, *Proc. Am. Philos. Soc.* **106**:467–482.

Simpson, G. G., 1944, *Tempo and Mode in Evolution,* Columbia University Press, New York.

Simpson, G. G., 1953, *The Major Features of Evolution,* Columbia University Press, New York.

Sinervo, B., and Huey, R., 1990, Allometric engineering: An experimental test of the causes of interpopulational differences in performances, *Science* **248**:1106–1109.

Singh-Pruthi, H., 1924, Studies on insect metamorphosis. I. Prothetely in mealworms (*Tenebrio mollitor*) and other insects: Effects of different temperatures, *Biol. Rev.* **1**:139–147.

Sinnott, E. W., 1921, The relation between body size and organ size in plants, *Am. Nat.* **55**:385–403.

Sinnott, M., 1963, *The Problem of Organic Form,* Yale University Press, New Haven, Conn.

Slatkin, M., 1987, Quantitative genetics of heterochrony, *Evolution* **41**:799–811.

Smith, A. B., 1981, Implications of lantern morphology for the phylogeny of post-Palaeozoic echinoids, *Palaeontology* **24**:779–801.

Smith, D. C., 1988, Heritable divergence of *Rhagoletis pomonella* host races by seasonal asynchrony, *Nature* **336**:66–67.

Smith, J. M., 1983, Evolution and development, in: *Development and Evolution* (B. C. Goodwin, N. Holder, and C. C. Wylie, eds.), Cambridge University Press, Cambridge.

Smith, J. P., 1914, *Acceleration of Development in Fossil Cephalopoda,* Leland Stanford Junior University Publications, University Series, Stanford.

Smith, L. J., 1985, Embryonic axis orientation in the mouse and its correlation with blastocyst relationships to the uterus, *J. Embryol. Exp. Morphol.* **89**:15–35.

Smith, M. M., and Heemstra, P. C., 1986, *Smiths' Sea Fishes,* Macmillan Co., New York.

Smith-Gill, S. J., 1983, Developmental plasticity: Developmental conversion versus phenotypic modulation, *Am. Zool.* **23**:47–55.

Smith-Gill, S. J., and Berven, K. A., 1979, Predicting amphibian metamorphosis, *Am. Nat.* **113**:563–585.

Snyder, J., and Bretsky, P. W., 1971, Life habits of diminutive bivalve molluscs in the Maquoketa Formation (Upper Ordovician), *Am. J. Sci.* **271**:227–251.

Soldán, T., 1979, Internal anatomy of *Dolania americana* (Ephemeroptera: Behningiidae), *Ann. Entomol. Soc. Am.* **72**:636–641.

Sondaar, P. Y., 1977, Insularity and its effect on mammal evolution, in: *Major Patterns in Vertebrate Evolution* (M. K. Hecht, P. C. Goody, and B. M. Hecht, eds.), Plenum Press, New York.

Sondaar, P. Y., 1986, The island sweepstakes, *Nat. Hist.* **95**(9):50–57.

Spinozzi, G., and Natale, F., 1986, The interaction between prehension and locomotion in macaque, gorilla and child cognitive development, in: *Primate Ontogeny, Cognition and Social Behaviour* (J. G. Else and P. C. Lee, eds.), Cambridge University Press, Cambridge.

Sporn, M. B., and Roberts, A. B., 1986, Peptide growth factors and inflammation, tissue repair, and cancer. *J. Clin. Inves.* **78**:329–332.

Sprinkle, J., and Bell, B. M., 1978, Paedomorphosis in edrioasteroid echinoderms, *Paleobiology* **4**:82–88.

Sprules, W. G., 1974, The adaptive significance of paedogenesis in North American species of *Ambystoma* (Amphibia: Caudata): An hypothesis, *Can. J. Zool.* **52**:393–400.

Staal, G. B., 1968, Experimental evidence on the role of hormone in the larval development of insects, *Proc. Int. Congr. Entomol., 13th, Moscow* p. 442.

Stanley, S. M., 1972, Functional morphology and evolution of byssally attached bivalve mollusks, *J. Paleontol.* **46**:165–212.

Stanley, S. M., 1973, An explanation for Cope's rule, *Evolution* **27**:1–26.

Stanley, S. M., 1975, A theory of evolution above the species level, *Proc. Natl. Acad. Sci. USA* **72**:646–650.

Stanley, S. M., 1977, Trends, rates, and patterns of evolution in the Bivalvia, in: *Patterns of Evolution, As Illustrated by the Fossil Record* (A. Hallam, ed.), Elsevier, Amsterdam.

Stanley, S. M., 1979, *Macroevolution, Pattern and Process,* Freeman, San Francisco.

Stanley, S. M., 1990, The general correlation between rate of speciation and rate of extinction, in: *The Causes of Evolution* (R. Ross and W. Allmon, eds.), University of Chicago Press, Chicago.

Stanley, S. M., and Yang, X., 1987, Approximate evolutionary stasis for bivalve morphology over millions of years: A multivariate, multilineage study, *Paleobiology* **13**:113–139.

Stauffer, D., 1985, *Introduction to Percolation Theory,* Taylor & Francis, London.

Stearns, S. C., 1976, Life history tactics: a review of the ideas, *Q. Rev. Biol.* **51**:3–47.

Stearns, S. C., 1986, Natural selection and fitness, adaptation and constraint, in: *Patterns and Processes in the History of Life* (D. M. Raup and D. Jablonski, eds.), Springer-Verlag, Berlin.

Stehouwer, D. J., 1988, Metamorphosis of behavior in the bullfrog (*Rana catesbeiana*), *Dev. Psychobiol.* **21**:383–395.

Stern, J. T., 1970, The meaning of "adaptation" and its relation to the phenomenon of natural selection, *Evol. Biol.* **4**:39–66.

Stevens, P. F., 1980, Evolutionary polarity of character states, *Annu. Rev. Ecol. Syst.* **11**:333–358.

Storr, G., 1989, A new *Pseudonaja* (Serpentes: Elapidae) from Western Australia, *Rec. West. Aust. Mus.* **14**:421–423.

Stubblefield, C. J., 1959, Evolution of trilobites, *Q. J. Geol. Soc. London* **115**:145–162.

Surlyk, F., 1972, Morphological adaptations and population structures of the Danish chalk brachiopods (Maastrichtian, Upper Cretaceous), *Biol. Skr. K. Dan. Vidensk. Selsk.* **19**:1–57.

Susman, R., 1984, *The Pygmy Chimpanzee*, Plenum Press, New York.

Swan, A. R. H., 1988, Heterochronic trends in Namurian ammonoid evolution, *Palaeontology* **31**:1033–1051.

Swan, L. W., 1990, The concordance of ontogeny with phylogeny, *Bioscience* **40**:376–384.

Sweeney, B. W., and Vannote, R. L., 1982, Population synchrony in mayflies: A predator satiation hypothesis, *Evolution* **36**:810–821.

Szarski, H., 1983, Cell size and the concept of wasteful and frugal evolutionary strategies, *J. Theor. Biol.* **105**:201–209.

Takeichi, M., 1988, The cadherins: Cell–cell adhesion molecules controlling animal morphogenesis, *Development* **102**:639–655.

Takhtajan, A., 1972, Patterns of ontogenetic alterations in the evolution of higher plants, *Phytomorphology* **22**:164–171.

Tanner, J. M., 1978, *Foetus into Man: Physical Growth from Conception to Maturity*, Harvard University Press, Cambridge, Mass.

Tarburton, M. K., and Minot, E. O., 1987, A novel strategy of incubation in birds, *Anim. Behav.* **35**:1898–1899.

Teixidó, J., Gilmore, R., Lee, D. C., and Massagué, J., 1987, Integral membrane glycoprotein properties of the prohormone pro-transforming growth factor-a, *Nature* **326**:883–885.

Thatcher, R. W., Walker, R. A., and Giudice, S., 1987, Human cerebral hemispheres develop at different rates and ages, *Science* **236**:1110–1113.

Thiery, J.-P., Brackenbury, R., Rutishauser, U., and Edelman, G. M., 1977, Adhesion among neural cells of the chick embryo. II. Purification and characterization of a cell adhesion molecule from neural retina, *J. Biol. Chem.* **252**:6841.

Thom, R., 1975, *Structural Stability and Morphogenesis*, Benjamin, New York.

Thomson, K. S., 1972, Estimation of cell size and DNA content in fossil fishes and amphibians, *J. Exp. Zool.* **205**:315–320.

Thomson, K. S., 1988, *Morphogenetics and Evolution*, Oxford University Press, Oxford.

Thorngren, K.-G., and Hansson, L. I., 1973, Cell kinetics and morphology of the growth plate in the normal and hypophysectomized rat, *Calcif. Tissue Res.* **13**:113–129.

Thorngren, K.-G., and Hansson, L. I., 1981, Cell production of different growth plates in the rabbit, *Acta Anat.* **110**:121–125.

Threlkeld, S., 1990, Book review of: *Size-structured Populations: Ecology and Evolution*, *Bioscience* **40**:318–319.

Thulborn, R. A., 1985, Birds as neotenous dinosaurs, *Rec. N. Z. Geol. Surv.* **9**:90–92.

Tickle, C., Summerbell, D., and Wolpert, L., 1975, Positional signalling and specification of digits in chick limb morphogenesis, *Nature* **254**:199–200.

Tissot, B. N., 1988, Multivariate analysis, in: *Heterochrony in Evolution: A Multidisciplinary Approach* (M. L. McKinney, ed.), Plenum Press, New York.

Tomlinson, P. B., 1970, Monocotyledons—Towards an understanding of their morphology and anatomy, in: *Advances in Botanical Research*, Volume 3 (R. D. Preston, ed.), Cambridge University Press, Cambridge.

Tomlinson, P. B., 1982, Chance and design in the construction of plants, in: *Axioms and Principles of Plant Construction* (R. Sattler, ed.), Nijhoff, The Hague.

Tomlinson, P. B., 1987, Architecture of tropical plants, *Annu. Rev. Ecol. Syst.* **18**:1–21.

Tompkins, R., 1978, Genic control of axolotl metamorphosis, *Am. Zool.* **18**:313–319.

Tower, W. L., 1901, An abnormal clypeasteroid echinoid, *Zool. Anz.* **24**:188–191.

Travis, J., 1988, Differential fertility as a major mode of selection, *Trends Ecol. Evol.* **3**:227–230.

Tschanz, K., 1988, Allometry and heterochrony in the growth of the neck of Triassic prolacertiform reptiles, *Palaeontology* **31**:997–1011.

Tuomi, J., and Vuorisalo, T., 1989, Hierarchical selection in modular organisms, *Trends Ecol. Evol.* **4**:209–212.

Turing, A., 1952, The chemical basis for morphogenesis, *Phil. Trans. R. Soc. London Ser. B* **237**:37–72.

Turner, R. D., and Yakovlev, Y., 1983, Dwarf males in the Teredinidae (Bivalvia, Pholadacea), *Science* **219**:1077–1078.

Turvey, S. P., 1981, Aspects of the biology of the freshwater crayfish *Euastacus spinifer* (Heller) (Decapoda: Parastacidae), M.Sc. thesis (unpubl.), University of Sydney.

Twitty, V. C., and Schwind, J. L., 1931, The growth of eyes and limbs transplanted heteroplastically between two species of *Amblystoma, J. Exp. Zool.* **59**:61–86.

Valentine, J. W., 1969, Niche diversity and niche size pattern in marine fossils, *J. Paleontol.* **43**:905–915.

Valentine, J. W., 1980, Determinants of diversity in higher taxonomic categories, *Paleobiology* **6**:444–450.

Valentine, J. W. (ed.), 1985, *Phanerozoic Diversity Patterns,* Princeton University Press, Princeton, N.J.

Valentine, J. W., 1986, Fossil record of the origin of Baupläne and its implications, in: *Patterns and Processes in the History of Life* (D. M. Raup and D. Jablonski, eds.), Springer-Verlag, Berlin.

Valentine, J. W., and Erwin, D. H., 1987, Interpreting great developmental experiments: The fossil record, in: *Development as an Evolutionary Process* (R. A. Raff and E. C. Raff, eds.), Liss, New York.

van de Koppel, J. M. H., and Hewlett, B. S., 1986, Growth of Aka pygmies and Bagandus of the Central African Republic, in: *African Pygmies* (L. L. Cavalli-Sforza, eds.), Academic Press, New York.

van Valen, L., 1974, A natural model for the origin of some higher taxa, *J. Herpetol.* **8**:109–121.

van Valen, L., 1975, Group selection, sex, and fossils, *Evolution* **29**:87–94.

Vermeij, G. J., 1973, Morphological patterns in high intertidal gastropods: Adaptative strategies and their limitations, *Mar. Biol.* **20**:319–346.

Vermeij, G. J., 1974, Adaptation, variability, and evolution, *Syst. Zool.* **22**:466–477.

Vermeij, G. J., 1978, *Biogeography and Adaptation,* Harvard University Press, Cambridge, Mass.

Vermeij, G. J., 1982, Unsuccessful predation and evolution, *Am. Nat.* **120**:701–720.

Vermeij, G. J., 1987, *Evolution and Escalation—An Ecological History of Life,* Princeton University Press, Princeton, N.J.

Voitkevitch, A. A., 1958, Spatial relations between true limbs and accessory limbs developing under natural conditions, *Dok. Akad. Nauk SSSR* **128**:28–31.

von Bertalanffy, L., 1968, *General Systems Theory,* Braziller, New York.

Vrba, E. S., 1980, Evolution, species and fossils: How does life evolve? *S. Afr. J. Sci.* **76**:61–84.

Vrba, E. S., and Eldredge, N., 1984, Individual, hierarchies and processes: Towards a more complete evolutionary theory, *Paleobiology* **10**:146–171.

Vrba, E. S., and Gould, S. J., 1986, The hierarchical expansion of sorting and selection: Sorting and selection cannot be equated, *Paleobiology* **12**:217–228.

Waddington, C. H., 1940, *Organizers and Genes,* Cambridge University Press, Cambridge.

Waddington, C. H., 1957, *The Strategy of the Genes,* Allen & Unwin, London.

Waddington, C. H., 1960, Evolutionary adaptation, in: *Evolution after Darwin* (S. Tax, ed.), Volume 1, University of Chicago Press, Chicago.

Waddington, C. H., 1962, *New Patterns in Genetics and Development,* Columbia University Press, New York.

Waddington, C. H., 1975, *Evolution of an Evolutionist,* Cornell University Press, Ithaca, N.Y.

Wade, G. N., 1976, Sex hormones, regulatory behaviors, and body weight, *Adv. Stud. Behav.* **6**:201–279.

Wake, D. B., 1963, Comparative osteology of the plethodontid salamander genus *Aneides, J. Morphol.* **113**:77–118.

Wake, D. B., 1966, Comparative osteology and evolution of the lungless salamanders, family Plethodontidae, *Mem. S. Calif. Acad. Sci.* **4**:1–111.

Wake, D., 1989, Phylogenetic implications of ontogenetic data, *Geobios Mem. Spec.* **12**:369–378.

Wake, D. B., and Larson, A., 1987, Multidimensional analysis of an evolving lineage, *Science* **238**:42–47.

Waller, D. M., 1982, Factors influencing seed weight in *Impatiens capensis* (Balsaminaceae), *Am. J. Bot.* **69**:1470–1475.

Ward, L. W., and Blackwelder, B. W., 1975, *Chesapecten,* a new genus of Pectinidae (Mollusca: Bivalvia) from the Miocene and Pliocene of eastern North America, *Bull. U.S. Geol. Surv.* No. 861.

Waren, A., 1983, A generic revision of the family Eulimidae (Gastropoda, Prosobranchia), *J. Moll. Stud. Suppl.* **13**:1–96.

Warren, A. A., and Hutchinson, M. N., 1988, A new capitosaurid amphibian from the early Triassic of Queensland, and the ontogeny of the capitosaur skull, *Palaeontology* **31**:857–876.

Wassersug, R. J., and Duellman, W. E., 1984, Oral structures and their development in egg-brooding hylid frog embryos and larvae: Evolutionary and ecological implications, *J. Morphol.* **182**:1–37.

Wassersug, R., and Hoff, K., 1982, Developmental changes in the orientation of the anuran jaw suspension: A preliminary exploration into the evolution of anuran metamorphosis, in: *Evolutionary Biology*, Volume 15 (M. K. Hecht, B. Wallace, and G. T. Prance, eds.), Plenum Press, New York.

Waters, J. A., Horowitz, A. S., and Macurda, D. B., 1985, Ontogeny and phylogeny of the Carboniferous blastoid *Pentremites*, *J. Paleontol.* **59**:701–712.

Wayne, R. K., 1986, Cranial morphology of domestic and wild canids: The influence of development on morphological change, *Evolution* **40**:243–261.

Weaver, C., 1963, An amazing parasitic mollusk, *Hawaiian Shell News* **11**(16):1.

Weitz, C., 1979, Introduction to Physical Anthropology and Archaeology, Prentice-Hall, Englewood Cliffs, New Jersey.

Welker, W., 1976, Brain evolution in mammals, in: *Evolution of Brain and Behavior in Vertebrates* (R. Masterton ed.), Wiley, New York.

Wells, M. J., and Wells, J., 1977, Cephalopoda: Octopoda, in: *Reproduction of Marine Invertebrates* (A. C. Giese and J. S. Pearse, eds.), Academic Press, New York.

Werker, J. F., 1989, Becoming a native listener, *Am. Sci.* **77**:54–59.

Werner, E. E., and Gilliam, J. F., 1984, The ontogenetic niche and species interactions in size-structured populations, *Annu. Rev. Ecol. Syst.* **15**:393–425.

Werner, E. E., and Hall, D. J., 1988, Ontogenetic habitat shifts in bluegill: The foraging rate–predation risk trade-off, *Ecology* **69**:1352–1366.

West-Eberhard, M. J., 1989, Phenotypic plasticity and the origins of diversity, *Annu. Rev. Ecol. Syst.* **20**:249–278.

Wheeler, D. E., and Nijhout, H. F., 1981, Soldier determination in ants: New role for juvenile hormone, *Science* **213**:361–363.

White, J. F., and Gould, S. J., 1965, Interpretation of the coefficient in the allometric equation, *Am. Nat.* **99**:5–18.

Whitney, R. R., 1961, The Bairdiella, *Bairdiella icistius* (Jordan and Gilbert), in: *The Ecology of the Salton Sea, California, in Relation to the Sportfishery* (B. W. Walker, ed.), Fish. Bull. St. Calif. Dep. Fish Game **113**:105–151.

Whittington, H. B., 1956, Silicified Middle Ordovician trilobites: The Odontopleuridae, *Bull. Mus. Comp. Zool.* **144**:155–288.

Whittington, H. B., 1981, Paedomorphosis and cryptogenesis in trilobites, *Geol. Mag.* **118**:591–602.

Whittow, G. C., 1976, Energy metabolism, in: *Avian Physiology*, 3rd ed. (P. C. Sturkie, ed.), Springer-Verlag, Berlin.

Whyte, L. L., 1965, *Internal Factors in Evolution*, Braziller, New York.

Wigglesworth, V. B., 1933, The physiology of the cuticle and of ecdysis in *Rhodnius prolixus*, with special reference to the function of the oenocytes and of the dermal glands, *Q. J. Microsc. Sci.* **76**:269–318.

Wigglesworth, V. B., 1936, The function of the corpus allatum in the growth and reproduction of *Rhodnius prolixus*, *Q. J. Microsc. Sci.* **79**:91–121.

Wigglesworth, V. B., 1954, *The Physiology of Insect Metamorphosis*, Cambridge University Press, Cambridge.

Wigglesworth, V. B., 1961, Some observations on the juvenile hormone effects of farnesol in *Rhodnius prolixus*, Stal., *J. Insect Physiol.* **7**:73–78.

Wigglesworth, V. B., 1965, *The Principles of Insect Physiology*, 6th ed., Methuen, London.

Wikler, K. C., Kirn, J., Windrem, M. S., and Finlay, B. L., 1986, Control of cell number in the developing visual system. III. Partial tectal ablation, *Dev. Brain Res.* **28**:23–32.

Wilbur, H. M., and Collins, J. P., 1983, Ecological aspects of amphibian metamorphosis, *Science* **182**:1305–1314.

Wild, R., 1973, Die Triasfauna der Tessiner Kalkalpen. XXIII. *Tanystropheus longobardicus* (Bassani), Neue Ergebnisse, *Schweiz. Palaeontol. Abh.* **95**:1–160.

Williams, C. M., and Kafatos, F. C., 1971, Theoretical aspects of the action of juvenile hormones, *Bull. Soc. Entomol. Suisse* **44**:151–162.

Williamson, P. G., 1981, Palaeontological documentation of speciation in Cenozoic molluscs from Turkana Basin, *Nature* **293**:437–443.

Williamson, P. G., 1987, Selection or constraint?: A proposal on the mechanism for stasis, in: *Rates of Evolution* (K. S. W. Campbell and M. F. Day, eds.), Allen & Unwin, London.

Willis, J. H., 1974, Morphogenetic action of insect hormones, *Annu. Rev. Entomol.* **19**:97–116.

Wilson, J. G., 1988, Resource partitioning and predation as a limit to size in *Nucula turgida* (Leckenby & Marshall), *Funct. Ecol.* **2**:63–66.

Wolpert, L., 1969, Positional information and the spatial pattern of cellular differentiation, *J. Theor. Biol.* **25**:1–47.

Wolpert, L., 1978, Pattern formation in biological development, *Sci. Am.* **239**(4):124–137.

Wolpert, L., 1982, Pattern formation and change, in: *Evolution and Development* (J. T. Bonner, ed.), Springer-Verlag, Berlin.

Wolpert, L., Lewis, J., and Summerhill, D., 1975, Morphogenesis of the vertebrate limb, in: *Cell Patterning: Ciba Foundation Symposium No. 29*, London.

Wood, 1986, see Shea, B. T., 1986, Ontogenetic approaches to sexual dimorphism in anthropoids, *Hum. Evol.* **1**:97–110.

Wootton, J. T., 1987, The effects of body mass, phylogeny, habitat, and trophic level on mammalian age at first reproduction, *Evolution* **41**:732–749.

Wright, S., 1932, The role of mutation, inbreeding, crossbreeding and selection in evolution, *Proc. Sixth Int. Congr. Genet.* pp. 356–366.

Wright, S., 1982, Character change, speciation, and the higher taxa, *Evolution* **36**:427–443.

Wulff, R. D., 1985, Effect of seed size on heteroblastic development in seedlings of *Desmodium paniculatum*, *Am. J. Bot.* **72**:1684–1686.

Wyatt, G. R., 1972, Insect hormones, in: *Biochemical Actions of Hormones*, Volume 2 (G. Litwack, ed.), Academic Press, New York.

Xiang, X., Benson, K., and Chada, K., 1990, Mini-mouse: disruption of the pygmy locust in a transgenic insertional mutant, *Science* **247**:967–969.

Yakovlev, P. I., 1967, Development of the nuclei of the dorsal thalamus and the cerebral cortex: Morphogenetic, and tectogenetic correlation, in: *Modern Neurology* (D. Locke, ed.), Little, Brown, Boston.

Yip, J. W., and Klein, K., 1984, Innervation of brachial sympathetic ganglion cells in normal and wing-extirpated chicks, *Neurosci. Abstr.* **10**:769.

Zakharov, Y. D., 1978, *Lower Triassic Ammonoids of East USSR* [in Russian], Nauka, Moscow.

Zar, J., 1968, Calculation and miscalculation of the allometric equation as a model in biological data, *Bioscience* **18**:1118–1120.

Zelditch, M. L., 1988, Ontogenetic variation in patterns of phenotypic integration in the laboratory rat, *Evolution* **42**:28–41.

Zihlman, A. L., Cronin, J. E., Cramer, D. L., and Sarich, V. M., 1978, Pygmy chimpanzees as a possible prototype for the common ancestor of humans, chimpanzees and gorillas, *Nature* **275**:744–746.

Zuckerkandl, E., 1976, Programs of gene action and progressive evolution, in: *Molecular Anthropology* (M. Goodman, R. Tashian, and J. Tashian, eds.), Plenum Press, New York.

Zuev, G. V., Nigmatullin, C. M., and Nikolsky, V. N., 1979, Growth and life span of *Sthenoteuthis pteropus* in the east-central Atlantic [in Russian], *Zool. Zh.* **58**:1632–1641.

Zwilling, E., 1955, Teratogenesis, in: *Analysis of Development* (B. H. Willier, P. A. Wiess, and V. Hamburger, eds.), Saunders, Philadelphia.

Author Index

Italic indicates a figure or table.

Subject Index

Italic indicates glossary reference. Parenthesis indicates figure/table reference

DATE DUE

DEMCO, INC. 38-2971